Mathematics and Visualization

Series Editors
Gerald Farin
Hans-Christian Hege
David Hoffman
Christopher R. Johnson
Konrad Polthier

Springer
Berlin
Heidelberg
New York
Hong Kong
London
Milan
Paris
Tokyo

Guido Brunnett
Bernd Hamann
Heinrich Müller
Lars Linsen
(Editors)

Geometric Modeling for Scientific Visualization

 Springer

Guido Brunnett
Department of Computer Science
Technical University of Chemnitz
09107 Chemnitz, Germany
e-mail:
brunnett@informatik.tu-chemnitz.de

Heinrich Müller
Informatik VII (Computer Graphics)
University of Dortmund
44221 Dortmund, Germany
e-mail:
mueller@ls7.cs.uni-dortmund.de

Bernd Hamann
Department of Computer Science
University of California, Davis
One Shields Avenue
Davis, CA 95616-8562, USA
e-mail:
hamann@cs.ucdavis.edu

Lars Linsen
Center for Image Processing
and Integrated Computing
University of California, Davis
One Shields Avenue
Davis, CA 95616-8562, USA
e-mail:
llinsen@ucdavis.edu

Cataloging-in-Publication Data applied for

A catalog record for this book is available from the Library of Congress.

Bibliographic information published by Die Deutsche Bibliothek
Die Deutsche Bibliothek lists this publication in the Deutsche Nationalbibliografie;
detailed bibliographic data is available in the Internet at http://dnb.ddb.de

Mathematics Subject Classification (2000): 68-02

ISSN 1612-3786

ISBN 3-540-40116-4 Springer-Verlag Berlin Heidelberg New York

This work is subject to copyright. All rights are reserved, whether the whole or part of the material is concerned, specifically the rights of translation, reprinting, reuse of illustrations, recitation, broadcasting, reproduction on microfilm or in any other way, and storage in data banks. Duplication of this publication or parts thereof is permitted only under the provisions of the German Copyright Law of September 9, 1965, in its current version, and permission for use must always be obtained from Springer-Verlag. Violations are liable for prosecution under the German Copyright Law.

Springer-Verlag Berlin Heidelberg New York
Springer-Verlag is a part of Springer Science+Business Media

springeronline.com

© Springer-Verlag Berlin Heidelberg 2004
Printed in Germany

The use of general descriptive names, registered names, trademarks, etc. in this publication does not imply, even in the absence of a specific statement, that such names are exempt from the relevant protective laws and regulations and therefore free for general use.

Typeset in TEX by the authors
Cover design: *design & production* GmbH, Heidelberg

Printed on acid-free paper   46/3142db - 5 4 3 2 1 0 -

Preface

Geometric Modeling and Scientific Visualization are both established disciplines, each with their own series of workshops, conferences, and journals. But clearly both disciplines overlap; this observation led to the idea of composing a book on Geometric Modeling for Scientific Visualization.

Experts in both fields from all over the world have been invited to participate in the book. We received 39 submissions of high-quality research and survey papers, from which we could only allow the 27 strongest to be published in this book. All papers underwent a strict refereeing process. The topics covered in this collection include

- Surface Reconstruction and Interpolation
- Surface Interrogation and Modeling
- Wavelets and Compression on Surfaces
- Topology, Distance Fields and Solid Modeling
- Multiresolution Data Representation
- Biomedical and Physical Applications

The editors would like to thank all the submitting authors for their contribution as well as all the referees for their efforts and their helpful comments. Moreover, we thank the Springer-Verlag for making the book possible and, in particular, Ute McCrory for good cooperation.

Chemnitz, Germany   *Guido Brunnett*
Davis, California   *Bernd Hamann*
Dortmund, Germany   *Heinrich Müller*
Davis, California   *Lars Linsen*
May, 2003

Table of Contents

Part I Surface Reconstruction and Interpolation

Reconstruction from Unorganized Point Sets Using Gamma Shapes
Marietta E. Cameron, Kenneth R. Sloan, Ying Sun 3

Isometric Embedding for a Discrete Metric
Ingrid Hotz, Hans Hagen 19

Mesh-Independent Surface Interpolation
David Levin 37

Empirical Analysis of Surface Interpolation by Spatial Environment Graphs
Robert Mencl, Heinrich Müller 51

Part II Surface Interrogation and Modeling

Smooth Polylines on Polygon Meshes
Georges-Pierre Bonneau, Stefanie Hahmann 69

Progressive Precision Surface Design
Mark A. Duchaineau, Kenneth I. Joy 85

Access to Surface Properties up to Order Two for Visualization Algorithms
Helwig Hauser, Thomas Theußl, Eduard Gröller 107

Modeling Rough Surfaces
Yootai Kim, Raghu Machiraju, David Thompson 123

A Feature Based Method for Rigid Registration of Anatomical Surfaces
Georgios Stylianou 139

Part III Wavelets and Compression on Surfaces

Lifting Biorthogonal B-spline Wavelets
Martin Bertram 153

Tree-based Data Structures for Triangle Mesh Connectivity Encoding
Ioannis Ivrissimtzis, Christian Rössl, Hans-Peter Seidel 171

Compression of Normal Meshes
Andrei Khodakovsky, Igor Guskov 189

New Results in Signal Processing and Compression of Polygon Meshes
Gabriel Taubin 207

Part IV Topology, Distance Fields, and Solid Modeling

Adaptively Represented Complete Distance Fields of Polygonal Models
Jian Huang, Roger Crawfis 225

Fully Dynamic Constrained Delaunay Triangulations
Marcelo Kallmann, Hanspeter Bieri, Daniel Thalmann 241

EVM: A Complete Solid Model for Surface Rendering
Jorge Rodríguez, Dolors Ayala, Antonio Aguilera 259

Topology Simplification of Symmetric, Second-Order 2D Tensor Fields
Xavier Tricoche, Gerik Scheuermann 275

Automating Transfer Function Design Based on Topology Analysis
Gunther H. Weber, Gerik Scheuermann 293

Part V Multiresolution Data Representation

Simplicial-based Multiresolution Volume Datasets Management: An Overview
Rita Borgo, Paolo Cignoni, Roberto Scopigno 309

Selective Refinement on Nested Tetrahedral Meshes
Leila De Floriani, Michael Lee 329

Divisive Parallel Clustering for Multiresolution Analysis
Bjoern Heckel, Bernd Hamann 345

Hierarchical Large-scale Volume Representation with $\sqrt[3]{2}$ Subdivision and Trivariate B-spline Wavelets
Lars Linsen, Jevan T. Gray, Valerio Pascucci, Mark A. Duchaineau, Bernd Hamann, Kenneth I. Joy 359

Multiresolution Surface and Volume Representations
Oliver G. Staadt 379

Part VI Biomedical and Physical Applications

Geometric Methods for Vessel Visualization and Quantification - A Survey
Katja Bühler, Petr Felkel, Alexandra La Cruz 399

An Application for Dealing with Missing Data in Medical Images, with Application to Left Ventricle SPECT Data
Oscar Civit Flores, Isabel Navazo, Àlvar Vinacua 421

Constraint-Based Astronometric Modeling Tools
Andrew J. Hanson, Chi-Wing Fu, Priscilla C. Frisch 437

Geometric Modelling for Virtual Colon Unfolding
Anna Vilanova, Eduard Gröller 453

Appendix: Color Plates 469

Part I

Surface Reconstruction and Interpolation

Reconstruction from Unorganized Point Sets Using Gamma Shapes

Marietta E. Cameron[1], Kenneth R. Sloan[2], and Ying Sun[3]

[1] Birmingham-Southern College Birmingham, Alabama *mcameron@bsc.edu*
[2] University of Alabama at Birmingham, Birmingham, Alabama
  sloan@cis.uab.edu
[3] University of Alabama at Birmingham, Birmingham, Alabama
  suny@cis.uab.edu

1 Introduction

The problem of reconstructing a three-dimensional surface from data sampled within a volume is an important issue in several diverse fields. Biologists, physicists, clinicians, weather forecasters and industrial quality control personnel share a need for this basic visualization tool.

There are three basic approaches to the problem, primarily motivated by three different classes of data: gridded data, contour data, and unorganized data. Gridded data is very structured, typically consisting of a scalar value at every point on a three dimensional grid, densely and isotropically filling the volume. Contour data consists of a collection of points organized along curves arising from the intersection of a surface and several planes of section. Unorganized data is completely unstructured; usually these points lie on the surface of the object, but sometimes they are simply scattered measurements distributed throughout the volume.

When dense gridded data are available, there are relatively well understood methods for extracting surfaces [15]. The density and the simple structure of gridded data simplify the problem. Typically, there is a single scale (the sampling density) which determines the smallest surface feature which can be represented by the data.

A time-honored method of gathering data about an object is to slice the object, creating a set of 2D sections. The intersections of the object's surface with these multiple slicing planes are a set of contour curves. The reconstruction of a surface from contour data usually leads to a single surface, rather than a family of surfaces (i.e., there is no user controlled variable). This style of data gathering has been strongly motivated by the technology available, especially in biomedical research, and has been well studied [6] [7] [12] [14] [16] [18]. There are a few important problems still to be solved, but the practical significance of these methods is on the wane.

While there are many details still to be investigated, and both gridded-based and contour-based methods are still applicable in a wide range of applications, we believe that the balance has now tipped towards applications and problems based on completely arbitrary scattered data. There are a variety of data sources which produce original data in this form; perhaps the example to keep in mind is that of a range-sensing device.

In this paper, we present a novel method of reconstructing 3D shapes from unorganized sample points. The method produces a new family of shapes, which we call *Gamma Shapes*. Gamma Shapes is an extension to Alpha Shapes [10] [11] [17]. The motivation is similar to that for *Weighted Alpha Shapes* [9], but the mechanism and the results are different. The key idea is to select a scale factor which is locally modulated by a scalar field. Efficient methods for determining this scalar field are part of the idea. One method which appears promising uses intermediate results from the work on *Crust* [3] [4]. Like the volume data case, and different than the contour data case, the Gamma Shape method produces a family of shapes which can be controlled by a user-selected parameter. Experiments indicate that a good choice for the scalar field does a very good job of homing in on the "correct" surface, but it is convenient to have the extra design handle which works in exactly the same manner as the interactive handle for Alpha Shapes.

Key features of this Gamma Shapes are:
1. The shape generation is based completely on a unstructured point cloud
2. The automatic selection of a local scale factors - γ
3. A user-controlled selection of a global scale factor - α

2 Related Work

2.1 Delaunay Triangulations

Given a collection of points S in 2D or 3D. The Delaunay Triangulation is a unique triangulation that ignores the co-incidental arrangements of the points and that satisfies several criteria for "nice" triangles (tetrahedra). The criteria which is most relevant here is: every triangle (tetrahedron) has an associated circumscribed circle (sphere) which is empty. In other words, there are no other points from the point cloud inside the circumscribing circle (sphere). The Delaunay Triangulation carves up space into mutually exclusive pieces, in a way which guarantees "nice" pieces. Each triangle (tetrahedron) covers a compact, local region of space. The vertices of each simplex are "neighbors". Note that the distance between neighbors gives some idea of the local sampling density - a local scale.

2.2 Voronoi Diagram

The Voronoi diagram of a collection of points S is a partition of space into convex cells. For each point in S, there is a cell which is the set of points that are closer to that point than to any other points in S.

Thus, the Delaunay Triangulation produces neighborhoods of points which are connected to form a simplex(point, edge, triangle or tetraheron), and the Voronoi Diagram produces neighborhoods of space which are associated with single points. These descriptions are dual to each other. As it turns out, the Delaunay Triangulation and the Voronoi Diagram are also duals of each other. See Figure 1. Compute one, and the other is easy to derive.

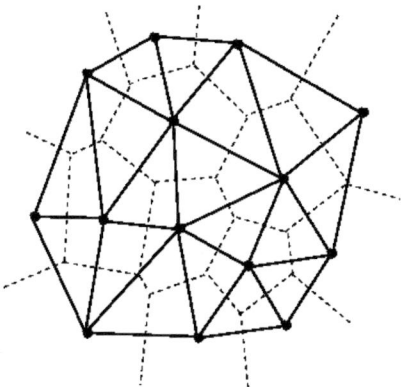

Fig. 1. Delaunay Triangulations(solid lines) and Voronoi Diagram(dotted lines)

We will start by using only the information in the Delaunay Triangulation, and only near the end require the extra geometric information provided by the Voronoi Diagram. In particular, we will use the Voronoi vertices as our primary measure of local scale.

2.3 Alpha Shapes

Review Alpha Shapes can be viewed as a generalization of the concept of Convex Hull. Consider a cutting tool which carves away space, but which cannot carve away any of the points in the given point cloud. If the cutting tool is a plane, then the space left untouched is just the convex hull of the point cloud. If, on the other hand, the cutting tool is a ball, it can sometimes fit in between points and carve out regions of space which were inaccessible to the planar tool which generates the convex hull. ¿From this point of view, the original planar cutting tool is simply a ball with infinite radius (i.e., a halfspace).

The Alpha Shape construction associates a family of polyhedral shapes with an unorganized set of points parameterized by α, the radius of the cutting ball. As we have seen, when α is infinite, the Alpha Shape is precisely the convex hull. At the other extreme, when α is zero, the Alpha Shape is just the original set of points since the infinitely small cutting ball destroys everything else. Our problem is to choose a value of α which *exposes* the surface, without poking holes in the surface or *detaching* elements of the object. On the other hand, we do want to separate distinct objects in the scene (or, pieces of a complicated non-convex object).

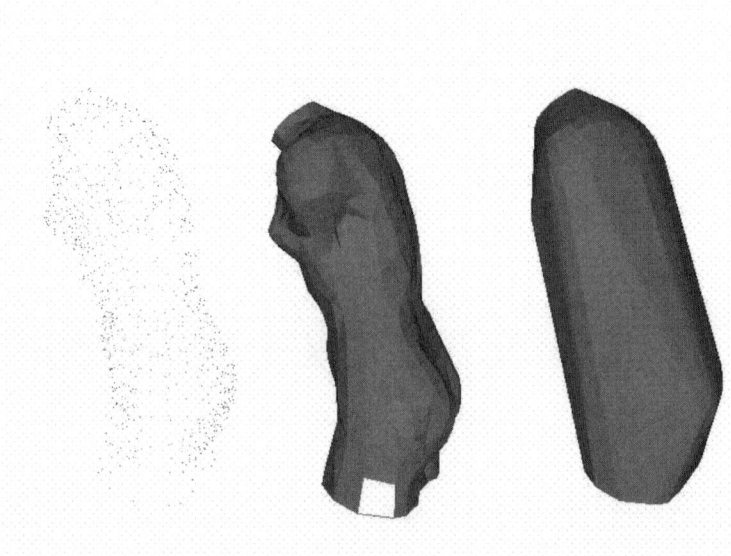

Fig. 2. Three Alpha Shapes. On the left is the original point cloud, corresponding to $\alpha = 0$. In the middle, a subjectively chosen "best" value for α; this "best" value produces a generally good surface, but there are still holes is some spots and missing detail in others. On the right, $\alpha = \infty$ yields the convex hull.

A Limitation of Alpha Shapes Given a empty ball with radius α, a simplex (point, edge, triangle, or tetrahedron) of the point set belongs to the individual shape associated with α if the simplex's vertices lie within in the shell of the ball. As α changes, the status of each simplex changes. The beauty of the Alpha Shape idea is that the entire family of shapes along with the complete history of the status of each and every simplex can be represented by an annotated version of the Delaunay Triangulation. For each simplex in the Delaunay Triangulation, it is sufficient to note a few intervals (of values for α) which determine the status of that simplex. Once this annotated Delaunay Triangulation is produced, the Alpha Shape corresponding to any particular α can be produced quickly and easily. Thus, after some pre-processing, there is an effective procedure to produce any member of the Alpha Shape family associated with a particular point set. It is feasible to allow an interactive user to move a slider controlling α and watch the resulting Alpha Shapes morph from convex hull to disconnected point set, and back. The question is: which member of this family represents the intended surface? Harder questions are:

is there a single value for α which works everywhere on the object? everywhere in a scene involving several distinct objects?

Note that α is a direct measure of sampling density. Points which are far apart as compared with α are not neighbors since the cutting ball will carve away the intervening space. Points which are close together as compared with α form a barrier to the cutting ball and end up defining a patch of surface (e.g., a triangle which is exposed, but not destroyed). The original Alpha Shape idea contains the implicit assumption that the point set is uniformly dense everywhere. If this assumption does not hold, then it may be impossible to select one global value for α which will generate the right surface everywhere. For each choice of α, there will be both areas where the details of the densely sampled parts will remain hidden or undesired holes and disconnected parts will appear in the sparsely sampled areas. See Figures 2 and 3.

"Weighted Alpha Shapes" addressed these issues by assigning interactively specified weights to each point. Higher weights are assigned in the sparse regions and lower weights are assigned in densely sampled regions, and the weights are *added* as an offset to the base value of α. This has the right qualitative feel, and we follow in this general direction, but uses a slightly different formulation which appears to work a bit better. And, we demonstrate that there is at least one natural way to automatically determine how to modulate α. Both of these points distinguish "Gamma Shapes" from "Weighted Alpha Shapes", although the connection is obviously very strong.

3 General Position

In our experiments, the input points are assumed to be in general position. The following conditions must hold:

1. No four neighboring points lie in a common plane. This condition means that there are no degenerate simplices (flat tetrahedra, triangles with all three vertices colinear) that may be formed.
2. No five neighboring points lie on a common sphere. This condition insures that there are no ties; therefore, the Delaunay triangulation of P is unique.
3. For any fixed α, the smallest sphere through any 2, 3, or 4 points of P has a radius different from α. This condition insures that each alpha shape in the resulting family of shapes is unique.

Input data rarely come in general position. However, there are methods for handling this inconvenience. We will assume that the general position conditions are met.

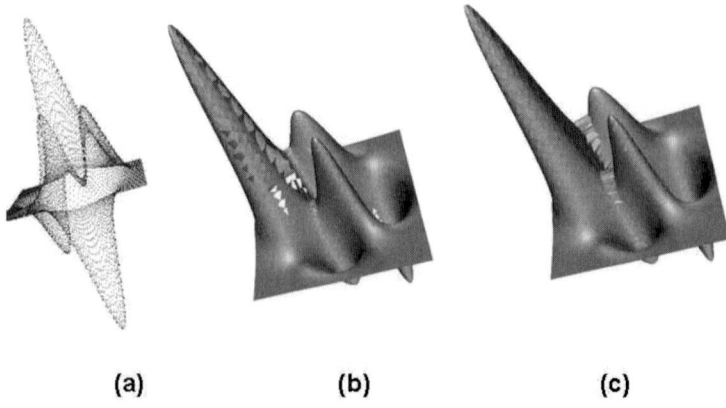

Fig. 3. Alpha Shapes Limitations. (a)A point cloud generated using the peaks function from MatLab. The 10,000 points are uniform in two coordinates not in third coordinate.(b) Surface for $\alpha = 0.169$. Note this shape does not contain "bridges" between the peaks. However, fulfilling the no bridge criteria introduces holes. (c) Surface for $\alpha = 0.313$. This shape does not contain holes but "bridges" now exist between the peaks. There is not a value of alpha that produces a shape without holes *and* without "bridges"

4 Gamma: A Local Scale Factor

As we have seen, an α-shape of a point set is formed by removing pieces of an enclosing polytope via a spherical eraser with fixed radius α. The result is called the α-hull. Simplifying the α-hull to straight edges and triangular faces yields the α-shape. The value of α is a global scale factor, reflecting the sampling density (or, equivalently, the size of the smallest interesting feature). perhaps controlled by an interactive viewer. The point set produces a family of shapes, parameterized by the value of α. After some initial pre-processing, it is possible to interactively select α to explore the entire family.

Given a particular value of α, the corresponding α-shape can be quickly computed from an annotated version of the Delauney Triangulation. Each edge, triangle, and tetrahedron in the Delauney Triangulation has an associated interval (a, b). The edge, triangle, or tetrahedron is $\alpha - exposed$ if, and only if, the radius of the cutting ball, α is inside this interval.

This all works very well, in a uniform world. But, the world is rarely uniform. Sampling densities vary from place to place, for a variety of reasons.

Gamma Shapes are an attempt to deal with this non-uniformity. The idea is not new, but the details of our method are different. Essentially, we want to locally modulate the global scale factor, α, with a local scale factor, which we will call γ.

Suppose that we define a scalar field $\gamma(x, y, z)$ over the volume in which the data points (and our desired surface) are contained. The intuition is that γ is a local scale factor. As the cutting ball moves through the volume, imagine that it changes size (the alternate view is that the cutting ball remains constant, but space is warped.) For the moment, assume that $\gamma(x, y, z)$ is somehow already defined. We will consider methods for determining $\gamma(x, y, z)$ below.

Consider what happens as α (the user-selected radius of a spherical cutting tool) is modulated by γ (the local scale factor). The new condition for an edge, triangle, or tetrahedron to be $\gamma - exposed$ is that $\alpha\gamma$ lies in the interval (a, b). An equivalent condition is that α lies in the interval $(\frac{a}{\gamma}, \frac{b}{\gamma})$

Note that this is almost (but not quite) the mechanism used in "Weighted Alpha Shapes" [9] and "Anisotropic Density-Scaled Alpha Shapes" [19] . In " Weighted Alpha Shapes" the local scale information is added to α. The "Density-Scaled Alpha Shape" uses a local scale factor based on point density. However, the density method introduces another interactive parameter that is to be used in addition to the alpha parameter. With "Gamma Shapes", the local scale information is multiplied by α without introducing an additional user-controlled parameter. *Gamma Shapes*'s advantage is not that the method produces superior surfaces to the surfaces generated by"Weighted Alpha Shapes" and the "Anisotropic Density-Scaled Alpha Shapes." The method's advantages are that it offers some insight on the automatic generation of the gamma scalar field, it can provide a reasonable"educated guess" on which value of α produces the ideal shape, and it does not offer any additional parameters.

5 Determining Gamma

Alpha Shapes work well, and α is a useful design handle which allows for navigation through a convenient family of shapes. $\gamma(x, y, z)$ is well motivated as a local scale factor, or a local warping of space. The two parameters interact nicely. The annotated Delaunay Triangulation continues to serve as an effective way to represent and produce on demand the entire family of shapes determined by the original point cloud and $\gamma(x, y, z)$.

But, where does $\gamma(x, y, z)$ come from? So far, we have simply assumed that it exists, and has the properties which we need. We have also glossed over a minor detail.

First, the detail. When a simplex (e.g., a triangle) is classified by the unmodified Alpha Shape method, there is a single value, α, which is compared with the interval (a, b). In the Gamma Shape case, we have a function, $\gamma(x, y, z)$ which may vary over the extent of the simplex (space may appear to be warped differently at one corner than it is at another corner). This is a problem for edges, triangles, and tetrahedra - but not for points. One way to solve this problem is by evaluating γ at the points and specifying that higher dimension objects inherit the MINIMUM γ of its components. Thus, an edge has associated with it a γ which is the MINIMUM of the $\gamma(x, y, z)$ at the two endpoints - and so on up through triangles and tetrahedra.

But, this still depends on a pre-existing $\gamma(x, y, z)$. Where does it come from? How is it computed? These are questions which we propose to investigate further; at the moment there are two interesting candidates which we have been using in preliminary experiments.

5.1 Poles

One recent line or work on reconstructing surfaces from unorganized point clouds is based on the idea of a "crust" (a collection of triangles, all of which are good candidates to be *surface triangles*.) We will make use of the "Voronoi pole computation" which is a central part of many "crust" algorithms [1] [2].

Consider the Voronoi Diagram for a point cloud which arises from a *properly sampled* surface. The "properly sampled" condition is important. We assume (probably with considerable loss of generality) that the surface has been "properly sampled".

Edges of the Voronoi Diagram separate neighboring points (note that these neighboring points are connected together, and participate together in one or more simplices in the Delaunay Triangulation).

Vertices of the Voronoi Diagram emerge as intersections of these edges. The Voronoi region associated with a "site" is defined by the surrounding Voronoi Vertices. The shape of this Voronoi region is a powerful indication of the distribution of neighboring sample points. When the local sampling is uniform and isotropic, the Voronoi region is nearly circular (spherical). When the local sampling is anisotropic, the orientation of the region and the relative distances to the several Voronoi Vertices (as measured from the central site) give strong clues about the local behavior of the surface.

For a sample point s, the *poles* are two extreme Voronoi Vertices on either side of the presumed surface. The poles for a point s, p^+ and p^-, are computed as follows:

1. If s is not on the convex hull, p^+ is the farthest Voronoi vertex of V_s from s. Let n^+ be the vector sp^+.
2. If s lies on the convex hull, there is no p^+. Let n^+ be the average normals of the triangles incident to s.
3. p^- is the Voronoi vertex of V_s with negative projection on n^+ that is farthest from s.

When we consider a *properly sampled* point on the surface of an object, we see that the Voronoi region associated with that point is elongated in a direction more or less perpendicular to the surface. This fact can be used as a clue to the surface normal - but we will not need this at the moment. Considering the global Voronoi Diagram, we note the observation in [1] that the entire collection of poles approximates the medial axis of the surface. When we use the $\alpha-ball$ to carve out our object, we want the ball to fit into the region between the surface and the medial axis.

As a crude approximation, in order to guide our intuition, consider the likely behavior of the Voronoi region, and the poles, for a sample point on the desired surface:

1. p^+ (if it exists) is *outside* the surface (or perhaps far into the interior of the object). At least one of the sample points whose Voronoi region shares this Voronoi vertex comes from a different object, or a distant, separate part of the same object.
2. p^- (which always exists) is *inside* (and near) the surface. The sample points whose Voronoi regions share this Voronoi vertex are all true "neighbors".
3. the distance from the selected sample point to it's associated negative pole, p^-, is a reasonable (over)estimate of the local sampling density. It is slightly *larger* than half the prevailing distance between two true neighbors.

Thus, the information contained in the poles gives us a strong indication of the local scale. A cutting ball with a radius comparable to "distance to the nearest pole" will not penetrate the local piece of surface (it is just barely too big) but it will fit into the space immediately adjacent to the local piece of surface.

This is our current method of assigning γ. Figure 5 show three members from the family of gamma shapes. Figure 4 shows demonstrates how this method compares to ?Alpha Shapes? on the peaks function.

We continue to investigate improvements to this pole generating the scalar field $\gamma(x,y,z)$. We contend that there is an improvement that will produce the ideal reconstruction at $\alpha = 1.0$. As shown in Figure 6, The current implementation generates artifacts between the peaks at $\alpha = 1.0$. The pole method approximates the medial axis. Given a sample point s, it is possible the closet point p, on the medial axis to s is not one of the generated poles.

5.2 Fitting an Ellipsoid to The Voronoi Region

Given a convex set of points in 3D, we can fit an ellipsoid to the region defined by those points. Good methods are available in this research area [13] [5].

We fit the ellipsoids to the Voronoi Cells using the moments of the Voronoi vertices. Consider the covariance matrix of the Voronoi points :

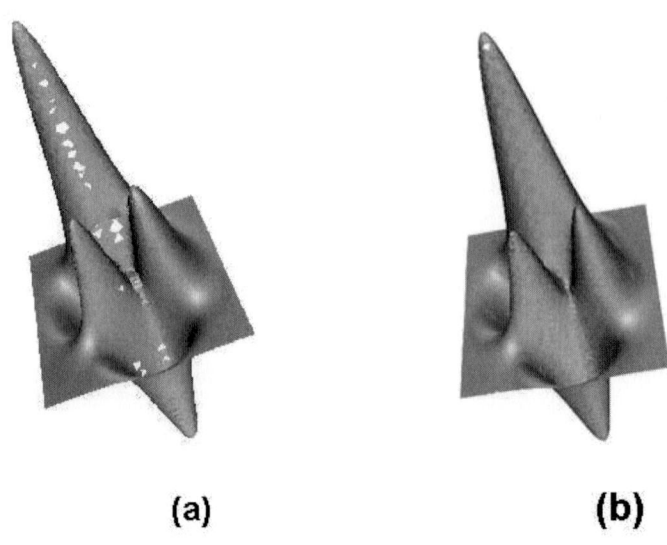

Fig. 4. (a) The ?ideal? reconstruction from Alpha Shapes $\alpha = 0.207$ for Peaks Function. (b) The ?ideal? reconstruction from Gamma Shapes $\alpha = 0.899$

$$C = \sum_{i=1}^{k} \begin{pmatrix} X_i^2 & X_i Y_c & X_i Z_i \\ Y_i X_i & Y_i^2 & Y_i Z_i \\ Z_i X_i & Z_i Y_i & Z_i^2 \end{pmatrix}$$

where

$$X_i = x_i - \bar{x}$$
$$Y_i = y_i - \bar{y}$$
$$Z_i = z_i - \bar{z}$$
$$\bar{x} = \frac{1}{k} \sum_{i=1}^{k} x_i$$
$$\bar{y} = \frac{1}{k} \sum_{i=1}^{k} y_i$$

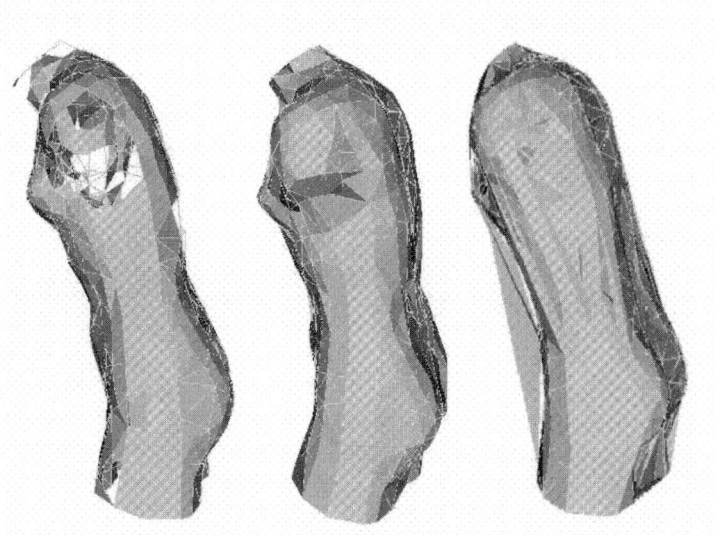

Fig. 5. Three Gamma Shapes. For all three, γ was automatically selected using Voronoi Pole information. The three choices come from selecting $\alpha = \frac{1}{2}, 1.0, 11.0$.

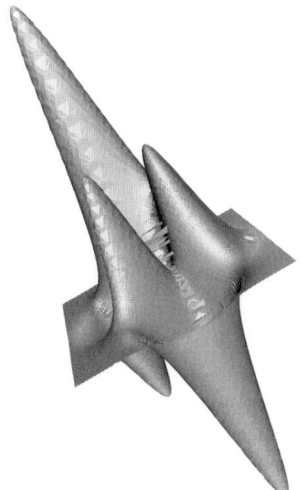

Fig. 6. Gamma Shape for proposed "ideal" value $\alpha = 1.0$. Note the bridging triangles between the front peaks

$$\bar{z} = \frac{1}{k}\sum_{i=1}^{k} z_i$$

Because C is a real symmetric matrix, it is orthogonal, and thus if it is non-zero, it is guaranteed to have orthogonal eigenvectors. We compute C, find the eigenvectors and eigenvalues of C, and set the three axes of the ellipsoid to the three eigenvectors. This yields the directions and the ratios of the lengths of the three axes of an ellipsoid. For a perfect ellipsoid(with an infinite number of points), the ratio of the three eigenvalues is exactly:

$$\lambda_1 : \lambda_2 : \lambda_3 = A^2 : B^2 : C^2$$

This can be seen by computing the continuous moment matrix :

$$M_c = \begin{pmatrix} \mathbb{N} x^2 d_x d_y d_z & \mathbb{N} xy d_x d_y d_z & \mathbb{N} xz d_x d_y d_z \\ \mathbb{N} yx d_x d_y d_z & \mathbb{N} y^2 d_x d_y d_z & \mathbb{N} yz d_x d_y d_z \\ \mathbb{N} zx d_x d_y d_z & \mathbb{N} zy d_x d_y d_z & \mathbb{N} z^2 d_x d_y d_z \end{pmatrix}$$

Now, we have an ellipsoid with center at the centroid of the Voronoi Region, and the direction and relative lengths of the axes determined by the eigenvectors. For any Voronoi vertex V_i, there is a unique scale value S such that the ellipsoid passes through that point.

$$S_i = \sqrt{\frac{(x_i - \bar{x})^2}{1} + \frac{(y_i - \bar{y})^2}{\frac{(\lambda_2^i)^2}{(\lambda_1^i)^2}} + \frac{(z_i - \bar{z})^2}{\frac{(\lambda_3^i)^2}{(\lambda_1^i)^2}}}$$

We need a scale factor to fully determine the lengths of the ellipsoid's axes. Each voronoi point V_i will generate a different scale value S_i. Get the average of

$$S = \frac{1}{k}\sum_{i=1}^{k} S_i$$

and set it to the scale value of the ellipsoid.

There are still a few details to be worked out.

When the data point is on the convex hull of the set of the data points, the Voronoi region is not closed. (i.e., the positive pole p^+ does not exist). One idea is to insert a constructed p^+ to close the Voronoi region. For example, we might use:

$$p^+ = P + (P - p^-)$$

Even when the data point is not *on* the convex hull, it may be that p^+ is very, very far away. This may distort our notion of the local neighborhood.

In these cases, it may make sense to replace p^+ by moving it closer to the centroid of the Voronoi region.

If we examine the three axes of the ellipsoid, and classify each one as "long", "medium", and "small" (LMS), we can identify a few ellipsoid types:

LSS - long and skinny, L is the surface normal, the surface is sampled isotropically.

LMS - the long axis gives us the pole - the medium axis may tell us something about the surface.

MMS - a "thick disc". This case requires more exploration on concerning when it occurs. How can we use it? For starters, the "pole" extracted from such a Voronoi region appears to be useful for giving a SCALE(What Gamma Shapes does) but NOT useful for providing a DIRECTION (what CoCone does [8]).

MMM - a sphere. Here we have a data point with neighbors equidistant in every direction. Orienting the surface through this point appears to be hopeless. We need more advanced methods.. What are they??

This method is clearly still in the early stages of development. However, it is interesting to look at the ellipsoids generated from our running example.

6 Future Work

There are three major items on our agenda for future work:

1. Develop a testbed and implement a full-scale working version of Gamma Shapes. The current implementation is cobbled together by combining bits and pieces of other programs, along with a shell script to combine the necessary information. As a result, some computations are currently unnecessarily duplicated, and there is no respectable user interface.
2. Experiment with both artificial and real data. While we have two candidates for $\gamma(x, y, z)$ which appear to work reasonably well, the Gamma Shape idea does not critically depend on the choice of method for determining $\gamma(x, y, z)$. In fact, a strength of the idea is that most of the scheme is independent of this estimate. The choice of $\gamma(x, y, z)$ may well be domain-dependent. We plan to use the testbed developed above to experiment with a wide variety of data sources. And, of course, we will pay some attention to the computational resources required for a real-world implementation.
3. Analyze the properties of Gamma Shapes. Virtually everything that is known now about Gamma Shapes follows directly and simply from previous work on Alpha Shapes. A major part of the proposed research will involve mathematical analysis of the properties of Gamma Shapes (and how these properties depend on the choice of $\gamma(x, y, z)$.

    We have presented a new scheme for constructing surface meshes from unorganized point clouds. The Gamma Shape method consists of two

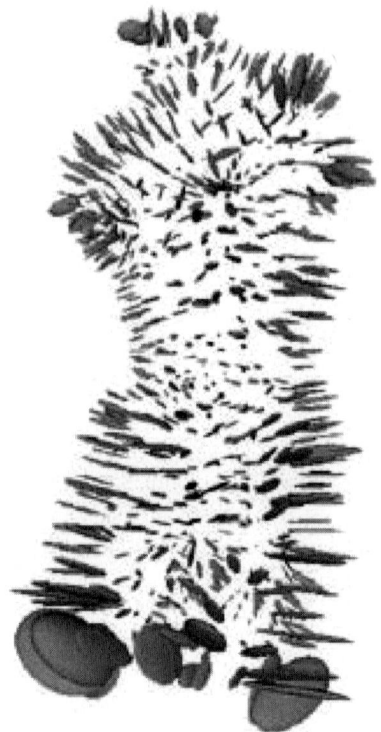

Fig. 7. Ellipsoids.

parts: a variation on Alpha Shapes which pays attention to a local scale factor, γ, and methods for deriving good values for γ. While we have only preliminary experimental experience with Gamma Shapes, the results thus far are encouraging.

References

1. Nina Amenta, Marshall Bern, and Manolis Kamvysselis. A new voronoi-based surface reconstruction algorithm. *Proceedings of the 25th Annual Conference on Computer Graphics*, pages 415–421, Jul 1998.
2. Nina Amenta, S. Choi, and N Leekha. A simple algorithm for homeomorphic surface reconstruction. *Proceedings of the Sixteenth Annual Symposium on Computational Geometry*, pages 213–222, Jun 2000.
3. Nina Amenta, Sunghee Choi, and Ravi Kolluri. The power crust. *Proceedings of 6th ACM Symposium on Solid Modeling*, pages 249–260, 2001.
4. Nina Amenta, Sunghee Choi, and Ravi Kolluri. The power crust,unionsof balls, and the medial axis transform. *Computational Geometry: Theory andApplications*, pages 127–153, 2001.

5. Li. B. The moment calculation of polyhedra. *Pattern Recognition*, 26:1229–1233, Aug 1993.
6. Jean-Daniel Boissonnat. Shape reconstruction from planar cross sections. *Computer Vision, Graphics, and Image Processing*, 44(1):1–29, 1988.
7. H. N. Christiansen and T. W. Sederberg. Conversion of complex contour line definitions into polygonal element mosaics. *Computer Graphics*, 12(2):187–192, Aug 1978. Proceedings of SIGGRAPH '78.
8. T. K. Dey, J. Giesen, S. Goswami, J. Hudson, R. Wenger, and W.Zhao. Undersampling and oversampling in sample based shape modeling. *Proc. IEEE Visualization 2001*, pages 83–90, 2001.
9. Herbert Edelsbrunner. Weighted alpha shapes. Technical Report UIUCDCS-R-92-1760, Univerisity of Illinois at Urbana-Champaign, May 1992.
10. Herbert Edelsbrunner and David G. Kirkpatrick. On the shape of a set of points in the plane. *IEEE Transactions on Information Theory*, 29:551–559, Jul 1983.
11. Herbert Edelsbrunner and Ernst P. Mucke. Three-dimensional alpha shapes. *ACM Transactions on Graphics*, 13:43–72, Jan 1994.
12. H. Fuchs, Z. M. Kedem, and S. P. Uselton. Optimal surface reconstruction from planar contours. *Communications of the ACM*, 20(10):693–702, Oct 1977.
13. X. Y. Jiang and H Bunke. Simple and fast computation of moments. *Pattern Recognition*, 24:801–806, Aug 1991.
14. E. Keppel. Approximating complex surfaces by triangulation of contour lines. *IBM Journal of Research and Development*, 19:2–11, Jan 1975.
15. W. E. Lorensen and H. E. Cline. Marching cubes: A high resolution 3d surface reconstruction algorithm. *Computer Graphics*, 21(4):163–169, July 1987. Proceedings of SIGGRAPH '87.
16. David Meyers, Shelley Skinner, and Kenneth Sloan. Surfaces from contours. *ACM Transactions on Graphics*, 11(3):228–258, Jul 1992.
17. Ernst Peter Mucke. *Shapes and Implementations in Three-Dimensional Geometry*. PhD thesis, University of Illinois at Urbana-Champaign, 1993.
18. Kenneth R. Sloan, Jr. and James Painter. From contours to surfaces: Testbed and initial results. In *Proceedings: CHI+GI '87*, pages 115–120, Toronto, Canada, Apr 1987.
19. M. Teichmann and M. Capps. Surface reconstruction with anisotropic density-scaled alpha-shapes. *In IEEE Visualization '98 Proceedings*, pages 67–72, 1998.

Isometric Embedding for a Discrete Metric

Ingrid Hotz and Hans Hagen

University of Kaiserslautern
Department of Computer Science
Germany
{*hotz, hagen*}@*informatik.uni-kl.de*

Summary. Symmetrical second order tensor fields are of great importance in computational physics. Applications are widely spread over many areas of physics and engineering, ranging from fluid flow over mechanical properties of solids to numerical relativity. Sometimes it is interesting to interpret them in a geometrical way, as distortions of a flat metric. The visualization of an isometric embedding of the corresponding surface can help to get an intuitive access to such metrics. The embedding is given by nonlinear partial differential equations for the surface coordinates which are hard to solve. In general the interesting tensors are results of numerical simulations or measurements, that means we have discrete data. In this paper we describe an algorithmic method to construct a surface that corresponds to these discrete values. The fundamental idea is to use Euclidean distances to build up the surface ring by ring. The needed distances are based upon a curvature estimation of the surface and the metric.

1 Introduction

In general one has no intuitive understanding of the abstract metric itself. In the two dimensional case or when considering two dimensional slices of some higher dimensional space, this problem can be solved by visualizing a two dimensional surface embedded in the three dimensional Euclidean space, with the same intrinsic geometry as the original surface. This isometric embedding is given by a system of partial nonlinear differential equations for the surface coordinates. There are no satisfactory standard algorithms to solve these differential equations for arbitrary surfaces. The numerical methods depend on the category of the differential equations: elliptic, parabolic or hyperbolic which is determined by the sign of the Gaussian curvature. Thus only surfaces with a fixed sign of the Gaussian curvature can be treated. An alternative method, the direct construction of the three dimensional points of a wire frame leads to a system of nonlinear algebraic equations. Having a sufficiently good initial guess these systems can be solved using the Newton-Raphson method [8]. For surfaces with a spherical topology Nollert and Herold used this approach [7]. In general it is not easy to obtain good start values. Poor start values can lead to nearly arbitrary results, without any quality control. Another algorithm for surfaces with spherical topology was given by Bondarescu, Alcubierre and Seidel [1]. Their algorithm appears to work well for their special task of computing the embedding of apparent horizons of black holes with the very restricting property of being star-like. In this paper we follow a constructive approach, computing the surface ring by ring around a point of interest. The focus is thereby not a high accuracy, because the results are heavily dependent on the chosen interpolation. Thus we look for an easy fast algorithm which results in a surface that allows an intuitive interpretation, as demanded by our application for tensor field visualization [4]. Another advantage of the point by point algorithm is the automatic break off if the embedding does not exist any longer. The algorithm can be divided into two parts: The computation of the distances of neighboring surface points and the surface construction. In this paper we focus on the first part the computation of the distances.

2 Mathematical Basics

We start with a section on basics form differential geometry where we introduce the notations which we will use later. For more details we refer to books on Differential Geometry [2].

A regular parametric surface $S \subset I\!R^3$ is a two dimensional subset of $I\!R^3$. In general it is given by a parametrization $X : I\!R^2 \supset U \to S \subset I\!R^3$ with $U \ni p = (u,v) \mapsto X(u,v) = (x,(u,v),y(u,v),z(u,v)) =: P$ (see figure 1), where the two partial derivatives of X at $p = (u,v)$, denoted by $\boldsymbol{X}_u := \partial X/\partial u$

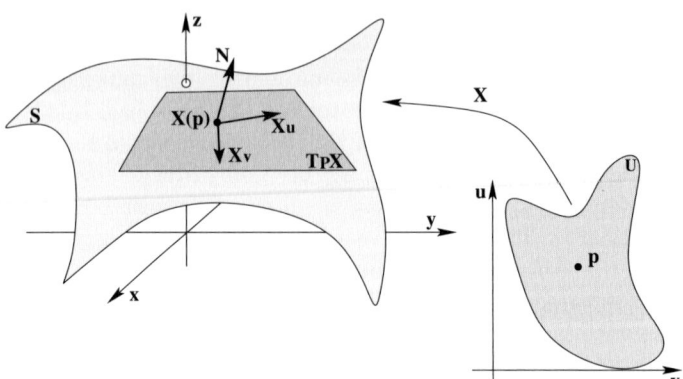

Fig. 1. Parametrized surface

and $\boldsymbol{X}_v := \partial X/\partial v$ are linearly independent for all $p \in U$. The unit normal vector $\boldsymbol{N}(p)$ is given by $\boldsymbol{N}(p) := \frac{\boldsymbol{X}_u \wedge \boldsymbol{X}_v}{||\boldsymbol{X}_u \wedge \boldsymbol{X}_v||}$, where $.\wedge.$ denotes the cross product.

The geometry of the surface is especially characterized by the metric and the curvature. A natural metric g on the surface is induced by the Euclidean inner product of $\mathbb{R}^3$. The so defined bilinear form on T_pS is the first fundamental form I_p of the surface at p. Its matrix representation I_p with respect to the basis $\{\boldsymbol{X}_u, \boldsymbol{X}_v\}$ of T_pS is given by:

$$(g_{ij}) = \begin{pmatrix} <\boldsymbol{X}_u, \boldsymbol{X}_u> & <\boldsymbol{X}_u, \boldsymbol{X}_v> \\ <\boldsymbol{X}_v, \boldsymbol{X}_u> & <\boldsymbol{X}_v, \boldsymbol{X}_v> \end{pmatrix} \equiv \begin{pmatrix} E & F \\ F & G \end{pmatrix} \tag{1}$$

It allows measurements to be made on the surface (e.g. length of curves, angles of tangent vectors, areas of regions) without referring back to the embedding space $\mathbb{R}^3$. The second characteristic property of the surface is the curvature. It measures the rate of change of the unit normal vector field $\boldsymbol{N}$ in a neighborhood of a point $P \in S$. This behavior is expressed by the second fundamental form II_p defined on T_pS by $II_p(\boldsymbol{w}_1, \boldsymbol{w}_2) = <L(\boldsymbol{w}_1), \boldsymbol{w}_2>$ $\forall \boldsymbol{w}_1, \boldsymbol{w}_2 \in T_pS$, where $L := -d\boldsymbol{N}$. Its matrix representation in the basis $\{\boldsymbol{X}_u, \boldsymbol{X}_v\}$ of T_pS is denoted by:

$$(h_{ij}) = -\begin{pmatrix} <\boldsymbol{N}_u, \boldsymbol{X}_u> & <\boldsymbol{N}_u, \boldsymbol{X}_v> \\ <\boldsymbol{N}_v, \boldsymbol{X}_u> & <\boldsymbol{N}_v, \boldsymbol{X}_v> \end{pmatrix} =: \begin{pmatrix} e & f \\ f & g \end{pmatrix} \tag{2}$$

This matrix is symmetric and real. Therefore it has two real eigenvalues k_1, k_2, the principle curvatures, with corresponding orthogonal eigenvectors, the principal directions.

At every point one can define the normal section in a direction $\boldsymbol{w}$ as the intersection curve of the surface S with the plane given by the surface normal $\boldsymbol{N}(P)$ and $\boldsymbol{w}$. Its curvature is called the normal curvature k_n in direction $\boldsymbol{w}$. The extreme values of k_n correspond to the principle curvatures of the surface at P. The product of the principle curvatures $K = k_1 k_2$ is the Gaussian curvature. The Gaussian curvature can be used to classify the surface points: Points with $K > 0$ are called elliptic points, points with $K < 0$ hyperbolic points, points with $K = 0$ parabolic points. If both principle curvatures are zero they are flat points. Except the Gaussian curvature, the entities of the curvature depend on the outer space.

3 The Problem

We now switch the view from an outer perspective to an inner point of view. Then the only entities we know are the metric and values depending on the metric g and its derivations. Starting with these informations we look for an embedded surface S with the intrinsic geometry. This means there exists a parametrization $X : \mathbb{R}^2 \supset U \to S \subset \mathbb{R}^3$ with $(u,v) \mapsto X(u,v) = (x,(u,v), y(u,v), z(u,v))$ such that differential arc length corresponds to the given metric

$$ds^2 = g_{11} \, du^2 + 2 \, g_{12} \, du \, dv + g_{22} \, dv^2 \ . \tag{3}$$

This condition is equivalent to the following system of partial differential equations (PDE's):

$$g_{11} = x_u^2 + y_u^2 + z_u^2$$
$$g_{12} = x_u x_v + y_u y_v + z_u z_v \qquad (4)$$
$$g_{22} = x_v^2 + y_v^2 + z_v^2$$

According to the theorem of Janet-Cartan-Burstin [6] a local embedding in $I\!R^3$ always exists when the functions g_{ij} are real and analytic. But in general there will be no global solution. This means we will need several patches to cover the whole domain.

The metrics, we are interested in, are not given as analytical functions, but as discrete values from simulations or measurements. There are two fundamentally different approaches to obtain the embedded surface:

- Interpolation of the discrete data and solution of the PDE's:
  This approach leads to the following problems: Not only the shape of the resulting surface, but also the existence of the embedding depends on the chosen interpolation. Even for linear interpolations these PDE's are not easy to solve.
- Direct computation of a wire frame of surface points:
  The discretization of the differential equations leads to a system of algebraic equations for the discrete points. Like the differential equations the algebraic equations are not linear.

Because our data are not analytical the second approach is more natural. The discretization of equation (3) leads to distance conditions for neighboring points X_i and X_j:

$$|X_i - X_j| = d_{ij} \qquad (5)$$

Because these equations only involve absolute values of distances, the point coordinates are not determined uniquely. An example of two different surfaces with the same distances between neighboring points is shown in figure 2. To

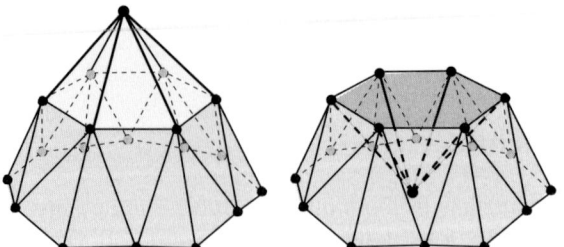

Fig. 2. Two surfaces with the same distances between neighboring points

get rid of this ambiguity use the Gaussian curvature which is an entity of the intrinsic geometry and therefore available for us.

Definition 1.
A discrete surface is an isometric embedding of a discrete metric, if the surface points fulfill the following properties:

- *The Gaussian curvature of the surface in the vertices corresponds to the Gaussian curvature determined by the metric.*
- *The distances of neighboring points correspond to the metric.*

The values for the distances and the Gaussian curvature are no fixed numbers but allow some variance according to the uncertainty in the distance estimation.

Summarizing the situation:
Given is a two dimensional domain $U \in I\!R^2$ and the metric as 2×2 matrices on a finite set $D = \{p_1, ..., p_n\} \subset U$ according to some basis e_u and e_v of $I\!R^2$.

$$g : D \to I\!R^{2,2} \qquad p_i \mapsto \begin{pmatrix} g^i_{11} & g^i_{12} \\ g^i_{12} & g^i_{22} \end{pmatrix} \qquad (6)$$

We look for a local embedding in the neighborhood of a point p of interest in terms of the above definition. The focus is not a high accuracy but an easy and fast algorithm that results in a surface, that allows an intuitive understanding of the metric. We demand that the algorithm stops automatically if the embedding does not exist any longer.

4 The Idea

The solution of the system of algebraic equations (5) with numerical standard methods as Newton-Raphson is a global approach. Starting with some initial grid all surface points are determined at the same time. Going this way no local embeddings are possible. Therefore we follow a constructive approach where the surface is build up ring by ring around a point of interest as long as the embedding exists. The computation of new points is based on distances to other points that are already computed. The estimation of the distances of the points is based upon the geodesic distances in combination with a curvature estimation. The construction of the surface is an iterative process where alternating new points are computed and the curvature approximation is improved.

- A first estimation of the curvature in the new points is computed. This estimation is based on the curvature of neighboring points and the local Gaussian curvature which we obtain from the metric.
- All relevant distances for the computation of the new points are computed based on the first curvature estimation.
- The error tolerance for each distance is determined, based on an estimation of the uncertainty in the distance computation.

- A first approximation for the new points is calculated using these distances.
- The curvature estimation is improved based on the first approximation for the surface points.
- Distances and error tolerance are computed again, with the improved curvature values.
- Final computation of the new points.

To get a better result the last two steps can be repeated. The algorithm can be divided in two parts. The destination of the distances of the surface points and the surface construction.

5 The Surface Construction

The surface points are computed ring wise around a point of interest. The choice of the basic structure of the parameter space is based on the following criteria: The construction of the surface ought to be as symmetrical as possible and represent the ring structure. For N points we have $3N$ degrees of freedom. Each distance condition delivers one equation, which involves two points. Therefore at least $6N$ equations are necessary. Thus 6 distances are needed for each point. These requirements suggest a uniform triangulation with equilateral triangles as shown in figure 3. The length of the edges is the characteristic value of the triangulation and is a measure for the resolution.

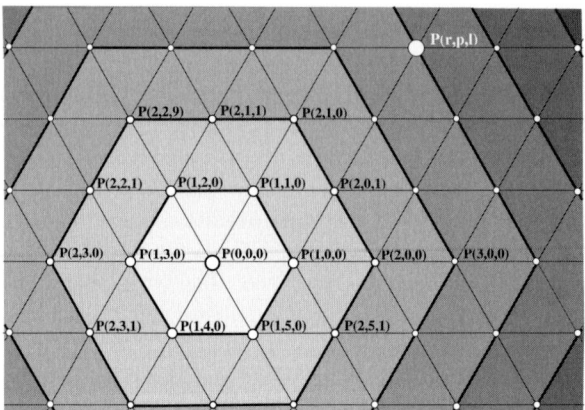

Fig. 3. Triangulation of the parameter space

The computation of a new point is based on distances to three direct neighbors that are yet computed. This computation corresponds to the construction of a tetrahedron which determines the position of the new point up to a sign with respect to the basic triangle. Distances to neighbors of second order are used to decide which of the two possible points is chosen. If an exact solution does not exist an optimum point is computed. The new point is accepted if the error of the distances does not exceed a given limit. The error tolerances are given by a maximum percentage error. The tolerance value is taken from the scope in the distance estimation, that depends on the variation of the metric in the involved points. Because in general the distances to further points are more erroneous, the tolerance value depends also on the order of the connection. In the final embedding algorithm the error tolerance is varied from a greater value in the first iteration step to a smaller value for the last iteration.

If we have exact Euclidean distance this method leads to an accurate reconstruction of a given surface. For further details of the surface construction we refer to [5].

6 Computation of the Distances

For the surface construction we need Euclidean distances between neighboring points $P_i = X(p_i)$ and $P_j = X(p_j)$ on the surface.

$$|P_i - P_j|^2 = (x_i - x_j)^2 + (y_i - y_j)^2 + (z_i - z_j)^2 \tag{7}$$

This distance is a value of the outer geometry and not direct available for us, but can be approximated using the metric and a curvature estimation. The first idea is to integrate the differential arc length along a straight line $p(t) = p_i + t(p_j - p_i)$ with $t \in [0, 1]$ in the parameter space from p_i to p_j:

$$|P_i - P_j|^2 \stackrel{?}{=} \mathbb{N}_{p_i}^{p_j} \, ds = s \tag{8}$$

For a linear interpolation of the metric between the two end points this computation is easily done. The differential arc length is given by:

$$ds^2 = g_{uu} \, du^2 + g_{vv} \, dv^2 + 2 g_{uv} \, du \, dv \tag{9}$$

For a linear interpolation of the components of the metric $g(t) = g_i + t(g_j - g_i)$ between the two points we get:

$$s = \begin{cases} \dfrac{2}{3} \dfrac{(\Delta p \cdot g_j \cdot \Delta p)^{3/2} - (\Delta p \cdot g_i \cdot \Delta p)^{3/2}}{\Delta p \cdot (g_j - g_i) \cdot \Delta p} & \text{for } g_j \neq g_i \\ \sqrt{\Delta p \cdot g \cdot \Delta p} & \text{for } g_j = g_i = g \end{cases} \tag{10}$$

with $\Delta p = p_j - p_i$. In general the image of a straight line will not be a straight line on the surface anymore (see figure 4). This means that beside the interpolation error we make two essential mistakes:

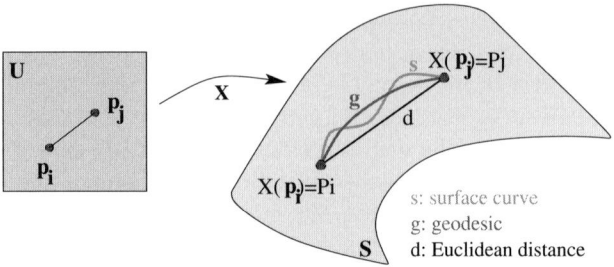

Fig. 4. Mapping of a straight line from parameter space to the surface

"Detours on the surface": (corresponds to the geodesic curvature k_g) This problem can be solved by computing the geodesic distance. The run of the geodesic connection between two points is given unique by the metric and is therefore computable with our knowledge. But this computation has to be done numerically even for linear interpolations and is very expensive (boundary value problem).

"Detours with the surface": (corresponds to the normal curvature k_n of the surface.) This error is the major problem, because we do not have any knowledge about the curvature of the surface except the Gaussian curvature, which is an entity of the intrinsic geometry.

To get a term for the correction of the arclength we expand the surface curve around its center $P_c = (0, 0, 0)$ using the local Frenet basis in this point.

$$\begin{aligned} x(s) &\simeq l - \frac{k^2 l^3}{6} + O(l^4) \\ y(s) &\simeq \frac{k l^3}{2} + \frac{k' l^3}{6} + O(l^4) \\ z(s) &\simeq -\frac{k\tau l^3}{6} + O(l^4) \end{aligned} \qquad (11)$$

Where l is the arclength, k the curvature and τ the torsion in P_c. With $P_i = P(-\frac{s}{2})$ and $P_j = P(\frac{s}{2})$ we get:

$$d_{ij} = |P_i - P_j| \simeq s \cdot (1 - \frac{k^2 s^2}{24}) + O(s^5) \qquad (12)$$

As curvature in P_c we uses the mean value for the curvature in the points P_i and P_j.

7 Curvature Estimation

According to the theorem of Meusnier the curvature k is composed of the normal curvature k_n and the geodesic curvature k_g:

$$k^2 = k_g^2 + k_n^2 \qquad (13)$$

7.1 The Geodesic Curvature k_g

Since the computation of the geodesics as boundary value problem is a high numerical effort, we decided to restrict ourselves to the computation of the geodesic curvature at the two points P_i and P_j respectively.

The geodesic curvature k_g of a curve $\alpha(t) = X(t) = X(u(t), v(t))$ is defined as the algebraic value of the covariant derivation of the normalized tangent-vector-field. $\alpha'(t) = u'(t)X_u(t) + v'(t)X_v(t)$ with $|\alpha'| = 1$. The sign depends on the orientation of the surface and is not important for our purpose.

$$\frac{D\alpha'}{dt} = \left(u'' + \Gamma^1_{11}(u')^2 + 2\Gamma^1_{12}u'v' + \Gamma^1_{22}(v')^2\right) \cdot X_u \quad (14)$$
$$+ \left(v'' + \Gamma^2_{11}(u')^2 + 2\Gamma^2_{12}u'v' + \Gamma^2_{22}(v')^2\right) \cdot X_v$$

We have for the geodesic curvature:

$$|k_g(t)|^2 = \left|\frac{D\alpha'}{dt}\right|^2 = E \cdot \left(u'' + \Gamma^1_{11}(u')^2 + 2\Gamma^1_{12}u'v' + \Gamma^1_{22}(v')^2\right)^2$$
$$+ 2F \cdot \left(u'' + \Gamma^1_{11}(u')^2 + 2\Gamma^1_{12}u'v' + \Gamma^1_{22}(v')^2\right)$$
$$\cdot \left(v'' + \Gamma^2_{11}(u')^2 + 2\Gamma^2_{12}u'v' + \Gamma^2_{22}(v')^2\right) \quad (15)$$
$$+ G \cdot \left(v'' + \Gamma^2_{11}(u')^2 + 2\Gamma^2_{12}u'v' + \Gamma^2_{22}(v')^2\right)^2$$

With

$$\frac{u'}{v'} = \frac{\Delta p_u}{\Delta p_v}$$

that means, it exists a $c(t) \in \mathbb{R}$ with

$$u' = \frac{1}{\sqrt{c}} \cdot \Delta p_u \qquad v' = \frac{1}{\sqrt{c}} \cdot \Delta p_v \quad (16)$$

The value of $c(t)$ is determined by the requirement, that α is parameterized by arc length

$$c = E(\Delta p_u)^2 + 2F\Delta p_u \Delta p_v + G(\Delta p_v)^2 \quad (17)$$

For the second derivative of u and v we have

$$u'' = -\frac{\Delta p_u \cdot c'}{2\,c^{3/2}} = -\frac{\Delta p_u \cdot (c_u \cdot \Delta p_u + c_v \cdot \Delta p_v)}{2\,c^2}$$
$$v'' = -\frac{\Delta p_v \cdot c'}{2\,c^{3/2}} = -\frac{\Delta p_v \cdot (c_u \cdot \Delta p_u + c_v \cdot \Delta p_v)}{2\,c^2}$$

and finally for the geodesic curvature (summation over repeated indices):

$$k_g(t)^2 = \frac{g_{ij}}{c^2} \cdot \left(\Gamma^i_{kl}\Delta p_l - \frac{c_k}{2\,c}\Delta p_i\right) \cdot \Delta p_k$$
$$\cdot \left(\Gamma^j_{mn}\Delta p_n - \frac{c_m}{2\,c}\Delta p_j\right) \cdot \Delta p_m \quad (18)$$

7.2 The Gaussian Curvature

The Gaussian curvature is given by:

$$K = \frac{-1}{2EG}\left\{G_{uu} + E_{vv} - \frac{G_u^2 + E_v G_v}{2G} - \frac{E_v^2 + E_u G_u}{2E}\right\} \quad (19)$$

For a linear interpolation this leads in the interior of the triangles to:

$$K = \frac{-1}{2EG}\left\{-\frac{G_u^2 + E_v G_v}{2G} - \frac{E_v^2 + E_u G_u}{2E}\right\} \quad (20)$$

On the edges we use the differences coefficients:

$$\begin{aligned} E_u(u_0, v_0) &= \frac{E(u_0 + \Delta u) - E(u_0 - \Delta u)}{2\Delta u} \\ E_{uu}(u_0, v_0) &= \frac{E_u(u_0 + \Delta u) - E_u(u_0 - \Delta u)}{2\Delta u} \\ &= \frac{E(u_0 + 2\Delta u) - 2E(u_0 - \Delta u) + E(u_0 - 2\Delta u)}{4\Delta^2 u} \end{aligned} \quad (21)$$

Similar for the other partial derivatives.

7.3 The Normal Curvature k_n

The curvature of the surface is represented by the second fundamental form defined in equation (2). It is not an entity of the intrinsic geometry and therefore not directly given by the metric. The Mainardi-Codazzi equations and the Gauss equation give a connection of the curvature and the metric. These equations are two non linear partial differential equations and one non linear algebraic equation. Because the category of the partial differential equations is connected with the sign of the Gaussian curvature, these equations can only be solved for surfaces with a constant sign of the Gaussian curvature.

We follow another way to get an estimation for the curvature of the surface in a surface point. As a first approximation we use the curvature of neighboring points, where the curvature is already computed. Then we adapt it to the local Gaussian curvature K. Let κ_1 and κ_2 be the principal curvatures taken from the neighboring points. Then the principal curvatures at the actual point k_1 and k_2 are determined by the conditions:

$$k_1 k_2 = K \quad \text{and} \quad k_1 \cdot \kappa_2 = k_2 \cdot \kappa_1 \quad (22)$$

Because we are only interested in the absolute value of the curvature it is not important whether we choose k_1 positive or negative. The principal directions are adapted from the neighboring points. We use this curvature approximation to compute the normal curvature, which we need for the distance correction in section 6, with Euler's formula:

$$k_n = k_1 \cos^2 \varphi + k_2 \sin^2 \varphi \quad (23)$$

When all six neighbors of a point are computed, the principal curvatures and their corresponding directions are estimated based on the resulting mesh.

8 Estimation of the Curvature of a Discrete Surface

The question of estimating the surface curvature appears in many computer graphics applications. The definitions of curvatures of a surface involve the second derivatives of the surface parametrization. For a discrete approximation of a smooth surface there is no unique generalization of these definitions. There exist various approaches to this topic. Hamann [3] proposed a least-squares paraboloid fitting of the adjacent vertices to determine the principle curvatures and their associated directions. Taubin [9] constructs a quadratic form at each vertex of the polyhedral surfaces and then computes the eigenvalues and eigenvectors of the resulting form. We use a different approach which is adapted to our requirements. There are many possibilities how a surface can be fitted to the mesh with different resulting curvatures. Therefore it is not the question which solution is the most accurate but which delivers a somehow natural result. Because we have to compute the curvature estimation very often one requirement is a fast and easy method. Also, we do not deal with arbitrary triangulations and we can use the structure of the mesh.

For our estimation only the direct neighbors will be considered. Here we consider the case of six neighbors. The method works also for another number of neighbors greater than two. Our approach consists of three steps. First we select an appropriate tangent plane. Then we compute normal curvatures in the directions of six adjacent points of the mesh. Then the Euler formula (23) is fitted to these values to determine the principal directions and principal curvatures.

8.1 Choice of the Surface Normal in a Vertex

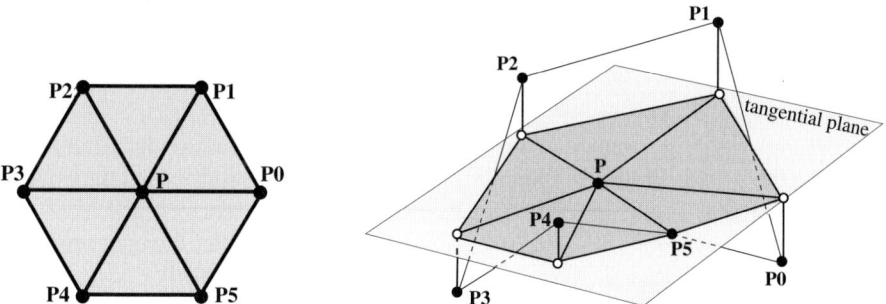

Fig. 5. Left: Topology in the parameter space, right: on the tangential plane

To determine the normal in a vertex P we use the six direct neighbors. Their topology shown in figure (5). For the tangent plane we demand that the projection of the sexangle $(P_0, ..., P_5)$ on the tangential plane is star like respective P. For our purpose there will always be a plane with this property, if the chosen triangulation is fine enough. There are two principle different approaches:

I Normal as weighted sum of the triangle normals:

$$N = \frac{1}{|...|} \sum_{i=0}^{5} w_i \cdot n_i \qquad (24)$$

The w_i are weights and n_i triangle normals:

$$n_i = n_{\triangle(P,P_i,P_{i+1})} = \frac{1}{|...|} (P_i - P) \times (P_{i+1} - P)$$

II Normal as weighted sum of the edges:

$$N = \frac{1}{|...|} \sum_{i=0}^{5} w_i \cdot s_i \qquad \text{with } s_i = \frac{P_i - P}{|P_i - P|} \qquad (25)$$

These approaches define a unique normal as long as the weighted sum is not zero. Each approach has its own advantages, the first is better for flat configurations the second for peaked. Thus, we prefer the weighted average of the adjacent triangle normals. Possible criteria for weighting could be for example: equal weighting, distances of the points, apex angle of the triangles or area of the triangles. Our normal vector should be defined locally and thus be independent on the area of the adjacent triangles and the distances of the neighboring points. Since the points can be non-uniform spread around the middle point, equal weighting leads to a non-uniform consideration of the different directions. This imbalance can be avoided by using the apex angle of the triangles as weights. A similar approach was chosen by Thuermer and Wuetherich [10].

$$w_i = \frac{\alpha_i}{\sum_j \alpha_j} \qquad (26)$$

where α_i is the apex angle of the corresponding triangle.

To evaluate this approach we computed the "discrete normals" of analytical surfaces and compared them with the analytical normals based on various triangulations. In all cases the results were very satisfying.

8.2 Computation of the Normal Curvatures for a Given Normal

In the second step we estimate the normal curvature of the surface in the directions of the points $P_0, .., P_5$. The normal curvature is defined as the curvature of the intersection curve of the surface and a plane spanned by the surface normal and the corresponding direction. Since we demanded that the projection of the neighbor points on the tangent plane are star-like, the six adjacent triangles can be represented as a height function over the tangent plane. Therefore it is useful to express the surface using cylinder coordinates $(r, \varphi, z(r, \varphi))$. The middle point assigns the origin and the normal the z-direction. For a discrete surface the normal section curve is represented by

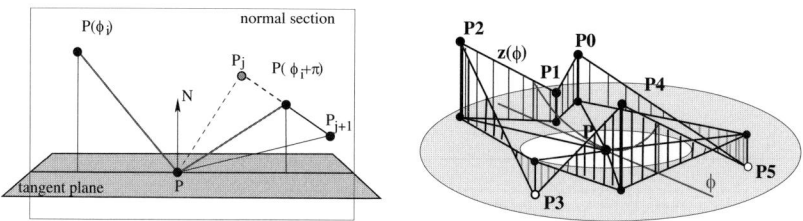

Fig. 6. Normal section of the surface in direction of P_i

three points $P = (0, 0, 0)$, $P(\varphi_i) = (r(\varphi_i), \varphi, z(\varphi_i))$ and $P(\varphi_i + \pi) := (r(\varphi_i + \pi), \varphi_i + \pi, z(\varphi_i + \pi))$ (see figure 6). To estimate the curvature in P we choose two quadratic functions: one for $\varphi = \varphi_i$ and one for $\varphi = \varphi_i + \pi$. Than we average the curvatures at the middle point.

$$\begin{aligned} z_1(x) &= \tfrac{z(\varphi_i)}{r(\varphi_i)^2} x^2 & \text{for } \varphi = \varphi_i \\ z_2(x) &= \tfrac{z(\varphi_i + \pi)}{r(\varphi_i + \pi)^2} x^2 & \text{for } \varphi = \varphi_i + \pi \end{aligned} \tag{27}$$

Thus, the curvature at the zero point is:

$$k_n(\varphi_i) = \left\{ \frac{z(\varphi + \pi)}{r^2(\varphi + \pi)} + \frac{z(\varphi)}{r^2(\varphi)} \right\} \tag{28}$$

8.3 Determination of the Principle Directions and Curvatures

Finally we have six values for the normal curvature depending on the angle φ. Through these values we fit a function according to the Euler-equation.

$$k_n(\varphi) = k_1 \sin^2(\varphi - \varphi_0) + k_2 \cos^2(\varphi - \varphi_0) \tag{29}$$

To determine the constant φ_0 we extend the normal curvature around the extremal values.

$$\begin{aligned} k_n(\varphi) &= k_{II} + (k_I - k_{II})(\varphi - \varphi_0)^2 & \text{near } \varphi_0 \\ k_n(\varphi) &= k_I - (k_I - k_{II})(\varphi - \varphi_0)^2 & \text{near } \varphi_0 + \pi/2 \end{aligned} \tag{30} \tag{31}$$

Thus, we get two quadratic functions of φ for the normal curvature in the neighborhood of the extremal value:

$$k_n(\varphi) =: \alpha \varphi^2 + \beta \varphi + \gamma \tag{32}$$

Its coefficients α, β and γ depend on the three constants φ_0, k_1 and k_2. They are uniquely determined by three values for $k_n(\varphi)$. For the first equation (30) we us the maximum value of k_n and the two values left and right of it. For the second equation (31) we take the values around the minimum. This approximation is only used for the determination of φ_0 as average of the two solutions. We obtain the constants k_1 and k_2 by minimizing the quadratic error Δ for a fixed φ_0.

$$\Delta := \sum_{i=0}^{5} (k_1 \sin^2(\varphi_i - \varphi_0) + k_2 \cos^2(\varphi_i - \varphi_0) - k_i)^2 \tag{33}$$

The error is a polynomial of second order in k_1 and k_2, thus there exists just one extremal value. It is a minimum because the limit of Δ for k to ∞ is ∞. If we equate the two partial derivatives with zero, we obtain a system of linear equations for the values k_1 and k_2 resulting in:

$$k_I = \frac{\sum_{i<j}^{5}(k_i \cos^2(\varphi_j-\varphi_0)-k_j \cos^2(\varphi_i-\varphi_0))(\sin^2(\varphi_i-\varphi_0)-\sin^2(\varphi_j-\varphi_0))}{\sum_{i<j}^{5}(\sin^2(\varphi_i-\varphi_0)-\sin^2(\varphi_j-\varphi_0))}$$

$$k_{II} = \frac{\sum_{i<j}^{5}(k_j \sin^2(\varphi_i-\varphi_0)-k_i \sin^2(\varphi_j-\varphi_0))(\sin^2(\varphi_i-\varphi_0)-\sin^2(\varphi_j-\varphi_0))}{\sum_{i<j}^{5}(\sin^2(\varphi_i-\varphi_0)-\sin^2(\varphi_j-\varphi_0))}$$

$$\tag{34}$$

9 Results

9.1 Curvature Estimation

We evaluated this curvature estimation for different surfaces, also with points not uniformly spread around the midpoint. It cannot be expected that we get exactly the analytical values for the curvature. But the comparison with the analytical values allows us to evaluate the results. The worst result we got for isolated flat points, which in general will not be recognized.

A first example is the paraboloid. The result can be seen in figure (7). Principal directions correspond exactly to the analytical value. The smaller principal curvature (according to the amount) matches nearly the analytical value. The error for the greater value of the estimated curvature has its origin at the missing values in the corresponding direction. A second example with negative Gaussian curvature is the hyperbolic paraboloid (see figure (7)). Again the principle directions fit very good to the analytical directions, but the absolute values of the curvature are smaller. We also got similar results for the monkey saddle and Enneper's minimal surface.

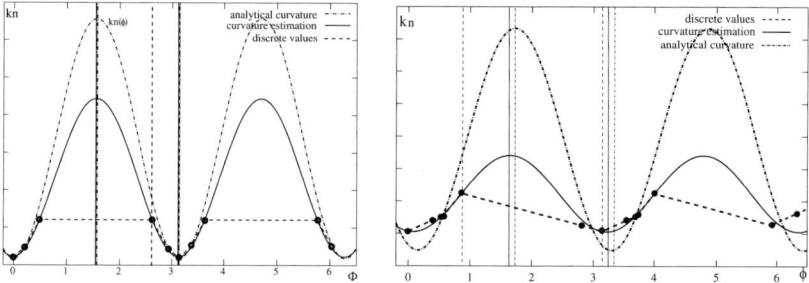

Fig. 7. Euler function: left for a paraboloid, right for a hyperbolic paraboloid with not uniformly spread points

9.2 Surface Embedding

To evaluate the algorithm it makes sense to start with well known surfaces. For the metric of these surfaces we know for sure, that the embedding exists. We also know how the result should look like. Using accurate distances from such surfaces the algorithm provides an exact reconstruction. The result for the paraboloid can be seen in figure (8), for the torus in figure (9). In general

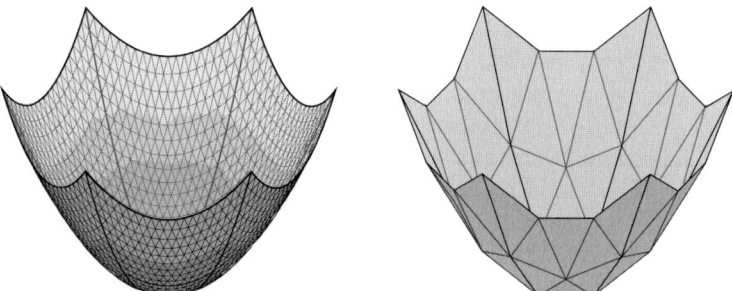

Fig. 8. Paraboloid left with a resolution of 20 rings, right with a resolution of 3 rings

the distance estimation does not deliver accurate values. Therefore the next step is the investigation of the influence of this deviation on the existence and the shape of the result. We imposed an statistical error Gaussian distribution around zero on the distances. The variance of the distribution is given as parameter. The effect of the distortion is demonstrated for the example of the paraboloid with different error variances. Because the differences in the shape of similar surfaces are difficult to see, we show the projection of the paraboloid in the xy-plane (see figure 10). We start with a small error and let it grow slowly. It can be observed that at the beginning the surface does not change much. But from a certain threshold the shape of the surface changes

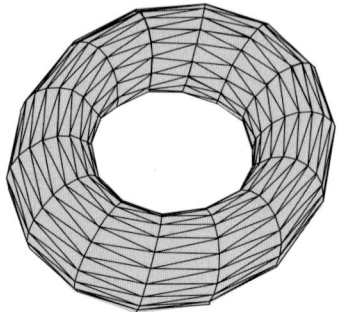

Fig. 9. Torus with a resolution of 14 rings

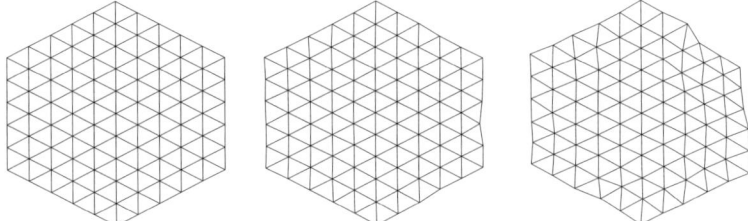

Fig. 10. Error variance varying from 0.1% (left) to 1%(right)

abruptly and the surface will not be smooth anymore. The threshold depends on the resolution and the curvature of the surface.

For more details on the investigation of the influence of the error in the distance estimation on the final surface we refer to a paper were we deal with this topic [5].

In contrast to the statistical perturbation for real data the errors from neighboring distances are correlated. Thus the threshold value up to which the construction is stable is higher than in the statistical case. As our first test surface we used the plane. Because all z-values of the plane are equal to zero, errors can be detected easily. Even for a high number of rings the plane was rebuild exactly. The second example is again the paraboloid. The following tabular lists the percentage error of the distances of neighboring points on the resulting surface in comparison with the analytical surface. The resolution of the surface is given by the number of rings.

Solution - number of rings	average error at %	number of computed distances
5	0.140	216
10	0.118	1026
20	0.231	4446
30	0.234	10266

For other surfaces we got similar results.

10 Conclusion

Our constructive algorithm allows a fast computation of the embedding of an abstract metric of surfaces with arbitrary Gaussian curvature. The algorithm solves the problem of the ambiguity of other methods that are based on the computation of distances of surface points. Because the surface is constructed point by point our algorithm can also be used if the embedding exists only locally. We do not need an initial guess of the surface. But therefore it is not possible to construct a surface with a given topology. Until now the data must be given on an regular grid, but a generalization to a more flexible structure is no principal problem.

The main problem concerning the embedding of a surface with an arbitrary metric is the sensitivity of the global existence of the embedding on small variations of the metric. Often many patches are needed to cover the complete domain. In general these patches cannot be connected without cracks. Therefore the aim is to find a balance between allowed errors and a minimum number of patches. For that purpose it could be suggestive to allow a greater initial variance of the distance. The result could be used as start surface of an optimization process which spreads the error evenly over the complete surface.

References

1. Bondarescu, M., Alcubierre, M., Seidel, E.(2001): Isometric embeddings of black hole horizons in three-dimensional flat space. arXiv:gr-qc/0109093
2. DoCarmo, M.P.(1976): Differential geometry of curves and surfaces. Prentice Hall
3. Hamann, B. (1993): Curvature approximation for triangulated surfaces. In Farin, G. et al., editor. Geometric Modelling. pages 139-153,
4. Hotz, I. (2001): Visualizing second order symmetric tensor fields by metric surfaces. IEEE Visualization Conference. Work in Progress San Diego 2001
5. Hotz, I. (2002): Isometric embedding by surface reconstruction from distances. IEEE Visualization Conference, Boston 2002, pages 251-257
6. Janet, M. (1931): Sur la possibilite de plonger un espace riemannian donne das un espace euclidean, In Ann. Soc. Math. Pol., 5, pages 74-85
7. Nollert, H.-P., Heinz Herold, H. (1996): Visualization in curved spacetimes - Visualization of Surfaces via Embedding. In Hehl, F.W., Puntigam, R.A., Ruder, H. (Eds.). Relativity and Scientific Computing. Springer Berlin Heidelberg, pages 330-351
8. Press, W.H., Teukolsky, S.H., Vetterling, W.T., Flannery, B.P.(1992): Numerical recipes in C. 2nd edition. Cambridge University Press
9. Taubin G. (1995): Estimating the tensor of curvature of a surface from a polyhedral approximation. In Proc. 5th Intl. Conf. on Computer Vision (ICCV 1995), pages 902-907
10. Thuermer, G., Wuetherich, Ch. (1998): Computing vertex normals from polygonal facets. Journal of Graphics Tools. 3(1) pages 43-46

Mesh-Independent Surface Interpolation

David Levin

School of Mathematical Sciences, Sackler Faculty of Exact Sciences, Tel Aviv University, Tel Aviv 69978, Israel *levin@post.tau.ac.il*

Summary. Smooth interpolation of unstructured surface data is usually achieved by joining local patches, where each patch is an approximation (usually parametric) defined on a local reference domain. A basic mesh-independent projection strategy for general surface interpolation is proposed here. The projection is based upon the 'Moving-Least-Squares' (MLS) approach, and the resulting surface is C^∞ smooth. The projection involves a first stage of defining a local reference domain and a second stage of constructing an MLS approximation with respect to the reference domain. The approach is presented for the general problem of approximating a $(d-1)$-dimensional manifold in $\mathbb{R}^d$, $d \geq 2$. The approach is applicable for interpolating or smoothing curve and surface data, as demonstrated here by some graphical examples.

1 Introduction

The problem of interpolating or approximating a function on $\mathbb{R}^d$ using scattered data values has many nice and well established solutions [4], [10]. Some of the methods use meshing strategies, but the preferred methods are those which are mesh-free, such as polynomial approximation, approximations by shifts of radial basis functions [3], and moving least-squares approximations [8].

The situation is quite different in the problem of surface approximation in $\mathbb{R}^d$, given scattered points on a surface. In general, it is not possible to find a natural **global** reference domain, or a parametric domain, which may be used as a base for launching one of the standard global approximation tools. The common practice in this case is to use a collection of **local** reference domains, related to some meshing of the data (e.g. triangulation), and the approximating surface is obtained as the collection of patches defined over those reference domains. The patches are usually defined as piecewise polynomial or piecewise rational parametric patches, smoothly joined together, and, in most cases, the resulting surface depends upon the specific mesh defining the patches. This long established approach works very well in numerous applications, and the Statue of Liberty is a fine example of a surface generated by patches [2].

The main goal of this work is to develop a mesh-independent method for smooth surface interpolation (or approximation) from unstructured scattered data. Such data sets are common in reverse engineering processes, where the surface of a sculptured object is measured by a laser-scanner or by a coordinate measurement machine. A mesh-independent surface approximation may be valuable, for example, serving as a reference surface for comparing different patching approximations of the surface.

To achieve the above goal we present in this work a different paradigm for surface approximation, namely, a projection procedure. This is an approach which seems more complex on the one hand, but, as shown later, may also be considered as simpler and more natural, on the other hand. It is based upon the basic notions of surfaces in differential geometry, namely a local reference system and a local mapping function for each point of the surface. The main tool used here for realizing this approach is based on the moving least-squares idea, which seems appropriate since it uses local approximations.

Let us first recall the definition of the moving least-squares approximation for the case of function approximation [8]. Let $\{x_i\}_{i \in I}$ be a set of distinct data points in $\mathbb{R}^d$, and let $\{f(x_i)\}_{i \in I}$ be some data values at these points. The moving least-squares approximation of degree m at a point $x \in \mathbb{R}^d$ is the value $\tilde{p}(x)$ where $\tilde{p} \in \Pi_m^d$ is minimizing, among all $p \in \Pi_m^d$, the weighted least-squares error

$$\sum_{i \in I} (p(x_i) - f(x_i))^2 \theta(\|x - x_i\|) \,. \tag{1.1}$$

Throughout the paper θ is a non-negative weight function, $\|\cdot\|$ is the Euclidean distance in $\mathbb{R}^d$ and Π_m^d is the space of polynomials of total degree m in $\mathbb{R}^d$. The approximation is made local if $\theta(s)$ is rapidly decreasing as $s \to \infty$, or is of finite support, and interpolation is achieved if $\lim_{s \to 0} \theta(s) = \infty$.

The above function approximation method is adopted here for defining a surface approximation strategy which we name MLS (the initials MLS may stand for Moving Least-Squares, or for Moving Local System, or for Meshless Surface). To prepare the presentation of the MLS approach we start in Section 2 with a basic MLS procedure for data smoothing, which is interesting and powerful by itself. Already here it is made clear that the MLS approach is applicable in any dimension in the most natural way. In section 3 we present a modified strategy, resulting is an MLS projection procedure. This procedure is the basis for the smoothing and the interpolation methods defined in Section 4. It is argued that the resulting approximating surfaces are mesh-independent, localized and infinitely smooth. We stress that the purpose of this work is to present the projection idea for the construction of surface approximants. The MLS approach should be viewed as one of the possible tools for implementing this idea. Some theoretical problems concerning the projection idea and the MLS approach remain open at this stage.

2 Smoothing Noisy Surface Data

Let S be a $(d-1)$-dimensional manifold in $\mathbb{R}^d$, let $\{r_i\}_{i \in I}$ be points on S or situated near S, e.g., points obtained from some measurements of S. An interpolating approximation is a manifold passing through $\{r_i\}_{i \in I}$, while a smoothing approximation is a manifold passing 'near' the data points $\{r_i\}_{i \in I}$. Considering the smoothing problem, rather than looking for a smoothing manifold, let us try to approximate the projection of the data points r_j, $j \in I$, and of points near the data set, onto S. The projection procedure suggested here involves two steps: Given a point r near S, we first find a local approximation to S by a hyperplane in $\mathbb{R}^d$. Then we 'project' the point r on a local polynomial approximation of S, defined over that hyperplane. Formally, the process is as follows:

The Basic MLS Procedure:

Step 1 – The local approximating hyperplane. Find a hyperplane $H_r = \{x \mid \langle a, x \rangle - D = 0, \ x \in \mathbb{R}^d\}$, $a \in \mathbb{R}^d$, $\|a\| = 1$, such that the following quantity is minimized,

$$\sum_{i \in I} (\langle a, r_i \rangle - D)^2 \theta(\|r_i - r\|) , \qquad (2.1)$$

where $\langle \cdot, \cdot \rangle$ is the standard inner product in $\mathbb{R}^d$. Since the weights $\{\theta(\|r_i - r\|)\}$ decrease as the distance $\|r_i - r\|$ increases, the resulting hyperplane H_r approximates a tangent hyperplane to S near the point r. In general, there may be several local minima of (2.1). We choose the one which is the closest to r, namely, such that $|\langle a, r \rangle - D|$ is the smallest.

Step 2 – The approximated projection $\tilde{P}_m$. Let $\{x_i\}_{i \in I}$ be the orthogonal projections of the points $\{r_i\}_{i \in I}$ onto H_r, represented in a specific orthonormal coordinate system on H_r, and let $f_i = \langle r_i, a \rangle - D$, $i \in I$, be the heights of the points $\{r_i\}_{i \in I}$ over H_r. Also, let q be the orthogonal projection of r onto H_r, and let us choose the origin of the orthonormal coordinate system on H_r to be at q. We define a local approximation of degree m to S by a polynomial $\tilde{p} \in \Pi_m^{d-1}$ minimizing, among all $p \in \Pi_m^{d-1}$, the weighted least-squares error

$$\sum_{i \in I} (p(x_i) - f_i)^2 \theta(\|r_i - r\|) . \qquad (2.2)$$

The value $\tilde{p}(0)$ approximates the height of S over H_r at the origin, hence the point $\tilde{r} = q + \tilde{p}(0)a$ is defined to be the approximation of the projection of r on S. The result may be denoted in an operator notation as $\tilde{r} = \tilde{P}_m(r)$.

Remark 1. The operator $\tilde{P}_m$ defined by the basic MLS procedure is not a projection.

It is essential to note that the distances defining the weights in (2.2) are defined by the distances from the data points $\{r_i\}$, rather than the distances from the points x_i as in (1.1). Also, the distances are taken from the point r which is not on H_r. The last issue plays a central role later on. It actually implies that the above projection step is not really a projection, namely, in general, the projection of $\tilde{r}$ is not going to stay $\tilde{r}$.

Example 1. (Curve smoothing) The applications in mind are of course for surfaces in $\mathbb{R}^3$, but the procedure is best understood when applied to the approximation of curves in $\mathbb{R}^2$. In the upper part of Figure 1 we display the data points $\{r_i\}$, drawn with the polygonal line connecting them, and the local approximating line H_r for some data point $r = r_j$. In the lower part the points are rotated so that the y- coordinate is in the direction of the local normal to H_r, and the local approximating quadratic and cubic polynomial approximations are drawn. In all our examples, the weight function used in the computation of the MLS approximations is $\theta(s) = e^{-s^2/h^2}$ where h is an average separation distance between the data points.

Fig. 1. Upper part: the noisy data and a local approximating line. Lower part: the data points rotated to the local normal, and the local 2nd and 3rd degree polynomial approximations.

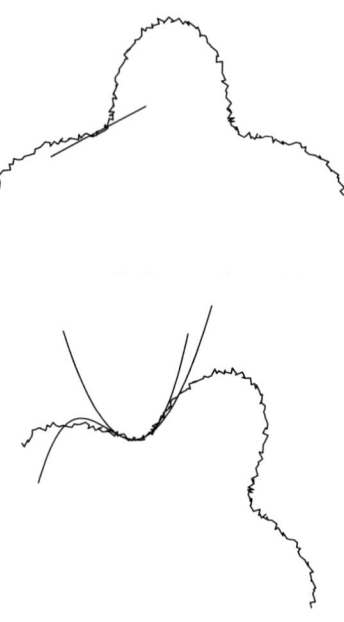

3 The MLS Projection

We recall that the main objective of this work is to develop a method for smooth surface interpolation from unstructured scattered data. As shown later this is achieved by using a proper projection procedure. The basic MLS procedure for approximated projection described above is very effective for thinning data sets. I.e., given a cloud of points representing noisy curve or surface data, the cloud is made thinner by applying the basic MLS procedure to each of the data points. This has already been implemented in [6]. Yet, as explained in Remark 1, the basic MLS procedure does not actually define a projection operator. Another important argument against the basic MLS procedure, related to Remark 1, is the following:

Remark 2. The basic MLS procedure in $\mathbb{R}^d$ is not a mapping onto a $d-1$ dimensional manifold.

Instead of proving this statement in general, let us try to explain it and to demonstrate it for the case $d = 3$. Let $\{r_i\}$ be data points near a surface S in $\mathbb{R}^3$, and consider the application of the basic MLS procedure for the approximate projection of a point r onto S, $\tilde{r} = \tilde{P}_m(r)$. Let $B(s, \delta)$ denote the closed ball of radius δ centered at s, and consider the image of $B(r, \delta)$ under $\tilde{P}_m$:

$$\tilde{P}_m(B(s,\delta)) = \{\tilde{P}_m(r) \mid r \in B(s,\delta)\} .$$

The statement in Remark 2 claims that $\tilde{P}_m(B(s,\delta))$ is not a two dimensional manifold. In general it contains interior points. To visualize this consider S to be the x-y plane in $\mathbb{R}^3$ and the points $\{r_i\}$ to be $\{r_i\} = \{(i_1 h, i_2 h, (-1)^{i_1+i_2} h)\}_{i=(i_1,i_2)\in\mathbb{Z}^2}$ with $h \ll \delta$. Also, let $m = 0$ and $s = (0,0,0)$. As we project a point $r \in B(s,\delta)$, the weights used for determining the local plane fit are decaying with the distance from r. Hence, the best fitted plane is going to be above the x-y plane if r is above the x-y plane, and vice versa. The same applies to the next step of a local approximation by a constant (m=0). As a result, the image set $\tilde{P}_0(B(s,\delta))$ of the ball $B(s,\delta)$ is going to be lentil-shaped.

Remarks 1 and 2 set the goals in defining a modified MLS procedure for approximated projection in $\mathbb{R}^d$ below; First, is should be a projection, second, it should project the points onto a $d-1$ dimensional manifold.

The MLS Projection Procedure:

Given a data set of points $\{r_i\}_{i\in I}$ on a $d-1$ hypersurface S in $\mathbb{R}^d$, or near S, and given a point r near S, the projection of r with respect to $\{r_i\}_{i\in I}$ is defined as follows:

Step 1 – The local approximating hyperplane. Find a hyperplane $H = \{x \mid \langle a, x \rangle - D = 0, \ x \in \mathbb{R}^d\}$, $a \in \mathbb{R}^d$, $\|a\| = 1$, and a point q on H, i.e., $\langle a, q \rangle = D$, such that $(r - q) \parallel a$, i.e.,

$$q = r + ta, \ t \in \mathbb{R}, \tag{3.1}$$

and such that the following quantity is locally minimized,

$$\sum_{i \in I} (\langle a, r_i \rangle - D)^2 \theta(\|r_i - q\|). \tag{3.2}$$

Reformulating this, we look for a direction $a \in \mathbb{R}^d$, $\|a\| = 1$, and a finite distance $t \in \mathbb{R}$ such that

$$\sum_{i \in I} \langle a, r_i - r - ta \rangle^2 \theta(\|r_i - r - ta\|), \tag{3.3}$$

is locally minimized. Usually, there may exist more than one pair $\{a, t\}$ locally minimizing the above quantity. The pair $\{a, t\}$ is then chosen to be the one with the minimal $|t|$.

For later use we introduce the notation $q = Q(r)$ and $a = A(r)$.

Step 2 – The MLS projection P_m. Let $\{x_i\}_{i \in I}$ be the orthogonal projections of the points $\{r_i\}_{i \in I}$ onto H, represented in an orthonormal coordinate system on H so defined that r is projected to the origin. Also, let $f_i = \langle r_i, a \rangle - D$, $i \in I$, be the heights of the points $\{r_i\}_{i \in I}$ over H. Find a polynomial $\tilde{p} \in \Pi_m^{d-1}$ minimizing, among all $p \in \Pi_m^{d-1}$, the weighted least-squares error

$$\sum_{i \in I} (p(x_i) - f_i)^2 \theta(\|r_i - q\|). \tag{3.4}$$

The projection of r is defined as

$$P_m(r) \equiv q + \tilde{p}(0)a. \tag{3.5}$$

An important observation, related to the projection property, follows directly from (3.3):

Proposition 1. *If the pair $\{a, t\}$ minimizes (3.3) for some $r = r^* \in \mathbb{R}^d$, then the pair $\{a, t - s\}$ minimizes (3.3) for $r = r^* + sa$.*

Proposition 2. *Q is a projection operator. Furthermore, if there exists a subset $U \subset \mathbb{R}^d$ such that for any $r \in U$ the minimization problem (3.3) has a unique global solution, and $P_m : U \to U$, then $P_m|_U$ is also a projection operator.*

Proof. Since $q = r + ta$ then, by Proposition 1, $A(q) = A(r)$ and $Q(q) = Q(r) = q$. Also, note that $P_m(r) = q + \tilde{p}(0)a = r + ta + \tilde{p}(0)a = r + sa$ with

$s = t + \tilde{p}(0)$. Hence, by Proposition 1, and the uniqueness assumption, it follows that
$$A(P_m(r)) = A(r) ,$$
and
$$Q(P_m(r)) = P_m(r) + (t - (t + \tilde{p}(0)))a = q + \tilde{p}(0)a - \tilde{p}(0)a = q = Q(r) .$$
Now, since only $a = A(r)$ and $q = Q(r)$ are used to define $P_m(r)$, the result follows. ∎

The subset U in Proposition 2 is going to be a neighborhood of the hypersurface S to be approximated.

Let us now repeat the previous examples, using the MLS projection.

Example 2. (Curve smoothing by the MLS projection) In Figure 2 we depict a noisy data set (the same one used in Example 1), a line segment L near the data set, and its MLS projection $P_2(L)$. The data points are drawn with the polygonal line connecting them, and the projected curve is drawn as the polygonal line connecting the projections of 30 equidistant points on L. To visualize the projection, one of the points on the line segment is connected with its projection.

Fig. 2. The noisy data, a line segment, and its MLS projection

Let S be a smooth hypersurface in $\mathbb{R}^d$, and let $\{r_i\}_{i \in I}$ be points on S. We say that the data set $R = \{r_i\}_{i \in I}$ has a mesh size h if h is the minimal

value for which $R \cap B(s, h/2)\ n(e^-) \neq \emptyset$ for any $s \in S$, where $B(s, \delta) \equiv \{x \mid \|x - s\| \leq \delta\}$. A point $P_m(r)$ defined by the MLS projection is a generic point on the approximating hypersurface defined by the data set $\{r_i\}_{i \in I}$. Based upon Proposition 2, $P_m(r) = P_m(Q(r)) = P_m(Q(q)) = P_m(q)$. For a data set $R = \{r_i\}_{i \in I}$ of a mesh size h on a smooth hypersurface S, we introduce the following definition:

Definition. (The MLS approximating hypersurface) *For a given data set R of mesh size h let*

$$\tilde{Q} = \{r \mid Q(r) = r\ ,\ R \cap B(r, h)\ n(e^-) \neq \emptyset\}\ . \tag{3.7}$$

Then the MLS approximating hypersurface induced by the data set $\{r_i\}_{i \in I}$ is defined as

$$\tilde{S} = \{P_m(r) \mid r \in \tilde{Q}\}\ . \tag{3.8}$$

The above formal definition seems quite strange, and it raises many questions, such as:

- Is $\tilde{S}$ really a $d - 1$ dimensional hypersurface?
- How smooth is this hypersurface?
- What is the approximation order?
- How can one reconstruct $\tilde{S}$?

Before we approach some of these questions, let us try to gain some intuition for the above definition. We note that in case the underlying hypersurface S is a hyperplane, then $\tilde{Q} = S$ and $\tilde{S} = S$. Also, for any $r \in \mathbb{R}^d$, $Q(r)$ is just the orthogonal projection of r onto S and $P_m(r) = Q(r)$. If the hypersurface S is smooth, and the mesh size h tends to zero, while the weight function θ is of finite support of size $O(h)$, then it can be shown that $\tilde{Q} \to S$ linearly in h.

Let us consider the system of equations defining the points in $\tilde{Q}$. These are derived by considering the minimization problem (3.3), and with the extra condition that the minimum is attained for $t = 0$. The necessary conditions for a minimum provide $d + 2$ equations for the $2d + 1$ unknowns which are the components of a and of r and the Lagrange multiplier for the side condition $\|a\| = 1$. A naive counting implies that the solution is a $d - 1$ parameter family, and thus $\tilde{Q}$ is expected to be a $d - 1$ dimensional manifold.

The issues of smoothness and convergence rate are not analyzed here. Yet, based upon the results in [7] we expect the following:

Conjecture (Smoothness and approximation order). Let us assume S is C^{m+1} and that the mesh size h tends to zero, while the weight function θ is infinitely smooth and is of finite support of size $O(h)$. Then $\tilde{S}$ is a C^∞ surface and $\tilde{S} \to S$ at a rate $O(h^{m+1})$, where m is the degree of the polynomials used in the MLS procedure.

4 The MLS Interpolation Scheme and Mesh Independence in $\mathbb{R}^3$

In this section we aim at the two goals set in the title of this paper, namely, interpolation and mesh-independence. As remarked in the Introduction, in the case of function approximation, interpolation is obtained if the weight function is chosen so that $\lim_{s \to 0} \theta(s) = \infty$. Doing just this is not enough for achieving interpolation in the case of surface data. In addition, Step 2 of the MLS projection method should be revised as follows:

Let $\{r_i\}_{i \in I}$, $\{x_i\}_{i \in I}$, $\{f_i\}_{i \in I}$, $q = Q(r)$ be as defined in the MLS projection procedure, and define $q_i = Q(r_i)$, $i \in I$.

In the revised MLS projection procedure the local polynomial approximation $\tilde{p} \in \Pi_m^{d-1}$ is obtained by minimizing the weighted least-squares defined with distances measured from the points $\{q_i\}_{i \in I}$, namely,

$$\sum_{i \in I} (p(x_i) - f_i)^2 \theta(\|q_i - q\|) \ . \tag{4.1}$$

The projection of r is thus defined as

$$P_m^*(r) \equiv q + \tilde{p}(0)a \ . \tag{4.2}$$

Assume we are given a data set $R = \{r_i\}_{i \in I}$, of mesh size h, of points on a hypersurface S in $\mathbb{R}^d$, and let $\tilde{Q}$ and $\tilde{S}$ be defined by (3.7) and (3.8). If $r \in \tilde{Q}$ then, following Proposition 1, $Q(r + tA(r)) = r$ for a small enough $|t|$. This observation induces the following definition:

Definition. (Equivalence sets) *The equivalence set $E(r)$ of a point $r \in \tilde{Q}$ is the connected set containing r of all the points of the form $r + tA(r)$ for which Q is uniquely defined and $Q(r + tA(r)) = r$. Consequently, the equivalence set $E(\tilde{Q})$ is defined as $E(\tilde{Q}) = \cup_{r \in \tilde{Q}} E(r)$.*

Proposition 3. (MLS interpolation) *Let $R = \{r_i\}_{i \in I}$ be a data set on a $d-1$ hypersurface S in $\mathbb{R}^d$ and let the MLS projection be defined as above, by (4.2), using (4.1) with a decreasing weight function θ satisfying $\lim_{s \to 0} \theta(s) = \infty$. Then,*

$$P_m^*(r) = r_j \quad \forall \, r \in E(Q(r_j)), \quad j \in I \ . \tag{4.3}$$

Proof. The proof follows directly from the fact that $q = Q(r) = Q(r_j) = q_j$ for any $r \in E(Q(r_j))$. Together with the condition $\lim_{s \to 0} \theta(s) = \infty$, this enforces $\tilde{p}$ to satisfy $\tilde{p}(0) = \tilde{p}(x_j) = f_j$, and thus $P_m^*(r) = q + \tilde{p}(0)a = q_j + f_j a = r_j$. ∎

The other goal set in the title is to achieve a method which is mesh-independent. Here we restrict the discussion to surfaces in $\mathbb{R}^3$. Assume we are given a data set $R = \{r_i\}_{i \in I}$, of mesh size h, of points on a surface S in $\mathbb{R}^3$, or near S. In practice the MLS approximating surface is to be defined by using some underlying parametrization domain, but the resulting surface is going to be independent of the choice of the parametrization domain.

Consider a proper triangulation $\mathcal{T} = \{T_k\}_{k \in K}$ of the data set R. I.e., each T_k, $k \in K$, is a triangle in $\mathbb{R}^3$ with vertices in $\{r_i\}_{i \in I}$, and the vertices and edges in $\mathcal{T}$ define a planar graph. We denote the boundary of $\mathcal{T}$ by $\partial \mathcal{T}$, i.e., the set of all edges in $\mathcal{T}$ which belong to one triangle in $\mathcal{T}$ only.

The linear surface defined by a triangulation $\mathcal{T}$ is an approximation of S, also denoted by $\mathcal{T}$, and it may serve as a base for defining the MLS approximating surface.

Definition. (The MLS approximation $\tilde{S}_\mathcal{T}$) Let $\mathcal{T}$ be a triangulation of $R = \{r_i\}_{i \in I}$ and assume $\mathcal{T} \subset E(\tilde{Q})$. The MLS approximation based upon $\mathcal{T}$ is defined as

$$\tilde{S}_\mathcal{T} = \{P_m(x) \mid x \in \mathcal{T}\} \equiv P_m(\mathcal{T})$$

i.e., the collection of the MLS projections of all the points on $\mathcal{T}$, with respect to the data set $\{r_i\}_{i \in I}$.

Practically, we have $\mathcal{T} \subset E(\tilde{Q})$ if there are enough data points where the underlying surface S has high curvature or is nearly self intersecting, and if the edges in $\mathcal{T}$ are short enough. In particular, we assume that $R \not\subset \partial \mathcal{T}$.

Proposition 4. (Mesh independence) *Assume $\tilde{S}$ is a two-dimensional submanifold. Let $\mathcal{T}_1$ and $\mathcal{T}_2$ be two simply connected triangulations of $R = \{r_i\}_{i \in I}$ sharing the same boundary, $\partial \mathcal{T}_1 = \partial \mathcal{T}_2$, such that $\mathcal{T}_j \subset E(\tilde{Q})$, and $\forall r \in \partial \mathcal{T}_j$ $E(Q(r)) \cap \mathcal{T}_j = r$, $j = 1, 2$. Then $P_m(\mathcal{T}_1) = P_m(\mathcal{T}_2)$.*

Proof. Clearly,
$$\tilde{S}_{\mathcal{T}_j} = P_m(\mathcal{T}_j) \subset \tilde{S} \ , \ j = 1, 2 \ ,$$
and
$$P_m(\partial \mathcal{T}_1) = P_m(\partial \mathcal{T}_2) \ .$$

Also we note that the MLS projection is a continuous operator over a triangulation $\mathcal{T}_j$, $j = 1, 2$. That is, $\|P_m(r_1) - P_m(r_2)\| < \varepsilon$ if $\|r_1 - r_2\| < \delta(\varepsilon)$ and $r_1, r_2 \in \mathcal{T}_j$. Hence, $P_m(\partial \mathcal{T}_j)$ is a simple closed curve in $\mathbb{R}^3$. The condition $E(Q(r)) \cap \mathcal{T}_j = r$ $\forall r \in \partial \mathcal{T}_j$ implies that each point in $P_m(\partial \mathcal{T}_j)$ has a unique source in $\mathcal{T}$. In particular, $P_m(\partial \mathcal{T}_j) \cap P_m(int(\mathcal{T}_j)) = \emptyset$ where $int(\mathcal{T}) \equiv \mathcal{T} \setminus \partial \mathcal{T}$. It also implies that $P_m(\partial \mathcal{T}_j)$ is a simple closed curve on $\tilde{S}$. Given a continuous map on a simply connected domain, which is a one-to-one mapping on the boundary, we know from differential topology [5], that its image is also a simply connected domain. It thus follows that $P_m(\mathcal{T}_j)$, $j = 1, 2$, are both simply connected domains, sharing the same boundary. Since they also share some interior points, and are both subsets of $\tilde{S}$ which is assumed to be a submanifold, they must be identical. ∎

Remark 3. The conditions of Proposition 4 seem quite difficult to verify beforehand, but they can be checked while computing the projection.

Example 3. (Surface approximation by the MLS projection) The application of the MLS projection for surface approximation is presented in Figure 3. The data points are depicted by small circles, and in the upper figure we also see a rectangular plane segment M near the data points. This rectangular domain is used as a local parametric base domain for the projection procedure. In the lower figure we see the mesh of points which is the P_2 projection of a rectangular mesh on M.

Some computational hints. Step 1 of the MLS projection procedure involves minimizing the quantity in (3.3), subject to $\|a\| = 1$, and with the additional constraint that $|t|$ is small. This is a non-linear optimization problem, and we have used an iterative method to solve it. To handle the constraints we represent the unknown parameters in terms of unconstrained parameters u, v and w. The constant h defined the maximal value we allow for $|t|$.

$$a = \big(\cos(u)\cos(v), \sin(u)\cos(v), \sin(v)\big) \ , \quad t = h \cdot \sin(w) \ . \qquad (4.4)$$

5 Discussion and Conclusions

We have presented a projection approach to surface approximation from unstructured point cloud data. The method is motivated by the MLS approach, where both a local coordinate system and a local polynomial approximation are computed by MLS for the projection of each point. The method has already proved to be useful in practical computer graphics applications for the visualization and simplification of point-sampled surfaces [1] [9]. Both [1] and [9] present strategies for making the ideas presented here more practical, and they also present very nice figures of surface data smoothing by the MLS projection.

From a theoretical point of view, there is an open question regarding the assumption in Proposition 2 about the existence of a subset U which satisfies certain implicit properties. It is not clear under what circumstances this assumption holds, or how to check it, and this calls for further investigation. Yet, the main contribution of the paper is in presenting the idea of a projection operator for surface reconstruction. The challenge set here is to find other projection operators which may be more robust or easier to compute.

Acknowledgement. I thank Ed Nadler for helpful discussions on data smoothing, and I am grateful to my son Adi Levin for his valuable assistance with the computer applications.

References

1. Alexa, M., Behr, J., Cohen-Or, D., Fleishman, S., Levin, D., Silva, T. (2001): Point set surfaces. IEEE Visualization 2001

Fig. 3. Upper part - the data points, and a plane segment L near it. Lower part - the projection $P_2(L)$.

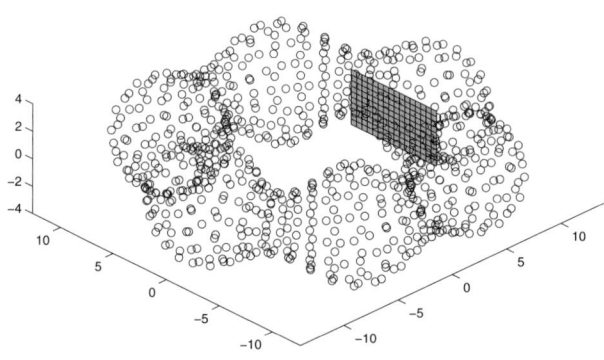

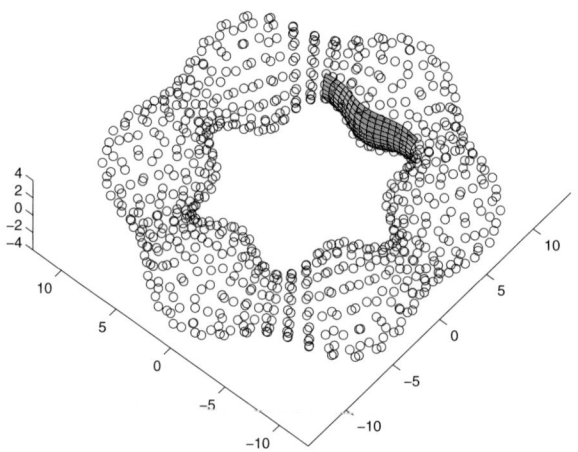

2. Boehm, W., Farin, G., Kahmann, J. (1984): A survey of curve and surface methods in CAGD. Computer Aided Geometric Design, **1**, 10–60
3. Dyn, N. (1987): Interpolation and approximation by radial and related functions. In: Chui, C.K., Schumaker, L.L., Ward, J.D. (eds) Approximation Theory VI. Academic Press, pp. 211–234
4. Franke, R. (1982): Scattered data interpolation: Tests of some methods. Math. Comp., **38**, No. 157, 181–200
5. Hirsch, M.W. (1976): Differential Topology. Springer-Verlag
6. Lee, I.-Q. Curve reconstruction from unorganized points. Computer Aided Geometric Design, to appear

7. Levin, D. (1998): The approximation power of moving least-squares. Math. Comp., **67**, 1517–1531
8. McLain, D.H. (1976): Two dimensional interpolation from random data. The Computer Journal, **19**, 178–181
9. Pauly, M., Gross, M., Kobbelt, L. (2002): Efficient simplification of point-sampled surfaces. IEEE Visualization 2002
10. Powell, M.J.D. A review of methods for multivariate interpolation at scattered data points. Report DAMTP 1996/NA11, University of Cambridge

Empirical Analysis of Surface Interpolation by Spatial Environment Graphs

Robert Mencl and Heinrich Müller

Informatik VII, University of Dortmund, D-44221 Dortmund, Germany

Summary. An empirical analysis of a new algorithm for reconstruction of surfaces from three-dimensional point clouds is presented. The particular features of the algorithm are the reconstruction of open surfaces with boundaries from data sets of variable density, and the treatment of sharp edges, that is, locations of infinite curvature. The empirical data in particular confirm a formal analysis which has been performed for compact surfaces of limited curvature without boundary.

1 Introduction

Let us given a finite set $P = \{\mathbf{p}_1, \ldots, \mathbf{p}_n\}$ of points in 3D-space. The problem of triangular mesh interpolation of scattered data points is to find a non-self-penetrating triangular mesh M which has the points of P as vertices. A triangular mesh consists of vertices, edges, and triangles. At most two triangles are incident at every edge. We assume that the given point set P consists of points of an intersection-free surface S in 3D-space. S needs not necessarily be closed or connected, and thus the resulting mesh has to be neither. The goal of interpolation is to find a mesh M with vertex set P which approximates S reasonably.

There is a wide range of applications for which surface construction from scattered point data is important, for instance scanning of 3D-shapes. With respect to visualization, surfaces open the application of the wide-spread surface-oriented visualization and rendering techniques. For example, surfaces may be used for visualizing additional information e.g. coded in data textures or real textures mapped on the surface.

Many algorithms of different types have been developed in the past. For a survey we refer to [8, 10]. An important reason is that the possibilities of interpolation depend on P. In practice, P often is not a proper sampling of the surface, and the intention of the developers is to let their algorithms find an intuitive solution also in this case. Recently, the question what a proper sampling is, and the development of algorithms which find the desired interpolation under this condition, have found increasing interest. In particular the work of Amenta et al. [2–4] has achieved considerable advance with respect to this question.

In [8, 11] we have presented an approach based on earlier ideas of the authors [7,9]. In a first step, an initial surface description graph (SDG) is calculated which approximates the desired surface, and which defines some sort of skeleton of the desired mesh (Figure 12). So-called clustered β-environment graphs are used for this purpose. In a second step, the SDG is completed to a triangular mesh. The mesh is obtained by incrementally inserting edges and triangles at the current boundary of the mesh, and taking care that the mesh is always feasible.

In [11], we have outlined arguments which show the existence of sample sets for surfaces of bounded curvature, on which the algorithm yields a reconstruction according to an intuitive definition. As an example it has been shown that grid-based sampling of surfaces is a favorable strategy. Furthermore, beyond what we could argue theoretically, the algorithm additionally shows a good behavior at sharp edges, ridges, and boundaries.

The purpose of this paper is to present results of empirical investigations which complement the theoretical analysis. The theoretical analysis has led to assertions which have not all been rigorously proven, but for which arguments based on proofs of weaker properties have been given that they hold with high likelihood, that is with a neglectable number of exceptions. The data of the empirical investigations confirm those arguments, and thus the good behavior of the algorithm. For the requirements of practice this approach appears sufficient because it can be expected that proofs of precise quantitative assertions are of considerable length and complexity. In our opinion this approach is also an acceptable alternative to the common practice of formal treatment of restricted problem versions which are tractable but often not really relevant for applications.

The empirical investigations are based on two types of data sets. The first type are data from "real" surfaces with particular features, like sharp edges, holes, and strongly non-convex regions. They serve as typical examples for real data sets which can be successfully processed by the algorithm. The second type are data sets representing geometric surfaces which typically occur as part of more complex surfaces, like planar regions or regions of a torus with different curvature behavior, and thus are in some sense representative. The data points have been acquired by two strategies, sampling on an approximately uniform mesh, and uniform random sampling. In this paper, we confirm empirically the theoretical assertions expressing the successful behavior of the algorithm, for these sampling strategies. The approach verifies that the algorithm works well in practice on sufficiently densely located data points sampled by those strategies.

In order to make the paper independent from the companion paper [11], we start with a description of the algorithm on a level of detail which should be sufficient in order to understand its principle and its analysis. Section 2 is devoted to the first step, and section 3 to the second step of the algorithm. In section 4, the main arguments of the theoretical analysis are recapitulated, and data of the empirical analysis are presented.

2 Step 1: Calculation of a Surface Description Graph

Let be $\beta \in \mathbb{R}$, $\beta \geq 0$, $e = \overline{\mathbf{pq}}$ be a line segment, and $\mathbf{p}'$, $\mathbf{q}'$ be points on the line induced by e, located symmetrically with respect to the center $\mathbf{m} := (\mathbf{p}+\mathbf{q})/2$ of e, that is $\mathbf{p}' = \beta \mathbf{p} + (1-\beta)\mathbf{m}$, $\mathbf{q}' = \beta \mathbf{q} + (1-\beta)\mathbf{m}$ (Figure 1). Then the β-*environment* of e is defined as the intersection of the ball with center $\mathbf{p}'$ and radius $||\mathbf{q}-\mathbf{p}'||$ with the ball with center $\mathbf{q}'$ and equal radius $||\mathbf{p}-\mathbf{q}'||$. Figure 1 shows examples for $\beta = 0$, 0.5, and 1 represented by $\mathbf{p}' = \mathbf{p}_0, \mathbf{p}_{0.5}, \mathbf{p}_1$ and $\mathbf{q}' = \mathbf{q}_0, \mathbf{q}_{0.5}, \mathbf{q}_1$. The β-*environment graph* (β-EG) of a point set P has P as vertex set, and those line-segments between pairs of vertices as edges which do not have any point of P in the interior of their β-environment.

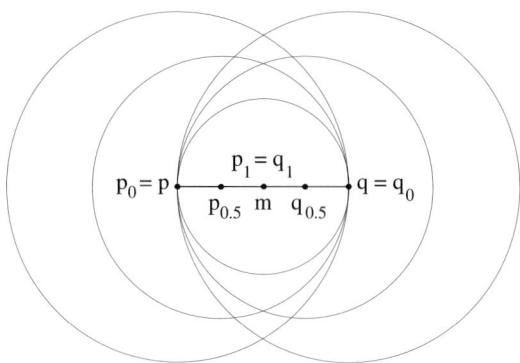

Fig. 1. Construction of the β-environment of a line segment between two points $\mathbf{p}$ and $\mathbf{q}$, depicted in two dimensions at three examples ($\beta = 0, 0.5, 1$). The β-environments are the regions of intersection of the pair of balls with centers $\{\mathbf{p}_0, \mathbf{q}_0\}$, $\{\mathbf{p}_{0.5}, \mathbf{q}_{0.5}\}$, and $\{\mathbf{p}_1, \mathbf{q}_1\}$, respectively.

Similar classes of graphs like the β-EGs have been defined by other authors [1, 5, 13–15]. For a discussion we refer to [8, 11].

The β-EGs are subgraphs of the Delaunay graph of their vertex set. Furthermore, β-EGs are connected if $0 \leq \beta \leq 1$ [8, 11].

Figure 5 shows β-EGs for three of our sample data sets, the corresponding point sets are presented in Fig. 4. We can notice that also in the 3D-case the resulting graph yields a reasonable, almost planar approximation. The number of "crossing" edges is very small, and decreases in β. The examples show that the environment graphs are basically well-suited for interpolation. However, they show two difficulties which may sometimes occur. The first difficulty is that there may still exist *crossing edges*, in particular for β close to 0. The second problem is the occurrence of long "*bridge edges*" which can be noticed in figure 5 in particular for the "pharao".

In order to reduce the unlikely case of intersecting edges further, we use the following concept of χ-*intersecting line segments* generalizing the notion of intersection of line segments from the plane to line segments in space. This concept uses the so-called *dihedral angle* between adjacent triangles t_1 and t_2, which is the angle between the normals n_1 and n_2 of t_1 and t_2 showing to the concave side of the surface defined by t_1 and t_2.

For the definition of χ-*intersection* of two line segments we consider the four triangles having one of the given line segments as an edge, and one of the vertices of the other edge as third vertex. In the plane the dihedral angle between two adjacent triangles is either $0°$ or $180°$. In space we admit angles in intervals $[0°, \chi]$ and $[180° - \chi, 180°]$, defined by a bound $\chi > 0°$. The precise definition is given in [8, 11]. Experimental experience shows that $60° \leq \chi \leq 90°$ is a suitable choice in the case of curved surfaces. The examples of this paper have been constructed with $\chi = 90°$.

In order to remedy the problem of bridge edges, we replace β-EGs with *clustered β-environment graphs*. Clustered β-EGs are subgraphs of β-EGs. They are defined by a greedy clustering scheme which processes the β-EG edges at every vertex **p** according to increasing length. In this way, edges of exceptional local length are recognized and not taken over into the clustered β-EG. The details we refer to [8, 11].

3 Step 2: Triangulation of the Surface Description Graph

The algorithm of triangulation consists of two main sub-steps, *generation of a partial embedding* and *incremental triangulation*.

The goal of sub-step 1 is the generation of very local embeddings of the SDG obtained from step 1 of the reconstruction algorithm. Each local embedding concerns a vertex and its incident edges in the SDG. A local embedding of this type is achieved by defining a sorted arrangement of the incident edges of the vertex. The local surface into which the vertex and its edges are embedded is the "umbrella" of triangles obtained by closing every sector defined by two consecutive edges of the arrangement by a chord. The arrangement is chosen so that the sum of dihedral angles in the umbrella is maximum (Fig. 2).

Sub-step 2 processes the sectors obtained in sub-step 1 according to a suitable priority. Roughly, sectors are preferred which have large dihedral angles with their neighboring sectors in the umbrella described above. Among those the sectors with small opening angle are considered first. The initial sectors are inserted into a priority queue. Later-on, the queue is updated whenever an edge e and an induced triangle are inserted into the mesh M, that is sectors disappearing by edge insertion are removed and new sectors are inserted. The new sectors are obtained by calculating new arrangements at the vertices of e, under consideration of e.

Fig. 2. Examples of optimal dihedral arrangements for five and six edges, and the corresponding umbrellas.

Sectors are removed from the priority queue and processed until the queue is empty. Processing of a sector w is influenced by four parameters, the *line segment candidate region* $C_c(s)$, the *triangle candidate region* $C_c(t)$, the *boundary control angle* γ'_c, and the *line segment intersection control angle* χ_c.

$C_c(s)$ and $C_c(t)$ are sets of points "close to" a line segment s or "lying over" a triangle t, respectively. More details are specified later. γ'_c typically is an angle close to 180° which prevents insertion of edges in order to achieve boundaries. χ_c controls the sensitivity of the intersection test of edges.

For every sector w with edges $e_1 := \overline{\mathbf{pq_1}}$ and $e_2 := \overline{\mathbf{pq_2}}$, the following conditions are successively evaluated, and the action assigned to the first one which holds is executed:

1. *Condition 1:* The angle of w at $\mathbf{p}$ exceeds γ'_c (Figure 3, left).
   Action: None, because the sector belongs to a boundary.
2. *Condition 2:* Points $\mathbf{q}$ over the sector triangle $t(w) := \Delta(\mathbf{p}, \mathbf{q_1}, \mathbf{q_2})$ exist, that is $C_c(t(w)) \neq \emptyset$ (Figure 3, middle left).
   Action: Insert an edge $\overline{\mathbf{pq}}$ into the SDG where $\mathbf{q}$ is one of the points over $t(w)$ closest to $\mathbf{p}$.
3. *Condition 3:* Points close to the edge $e_3 := \overline{\mathbf{q_1 q_2}}$ exist, that is $C_c(e_3) \neq \emptyset$ (Figure 3, middle right).
   Action: Choose the edge e to be inserted as the diagonal inserted obtained min-max-triangulation of the quadrilateral $\mathbf{p}, \mathbf{q_1}, \mathbf{q}, \mathbf{q_2}$ where $\mathbf{q} \in C_c(e_3)$ is a suitable point. Insert e into the SDG. If $e = e_3$, then insert $t(w)$ into the mesh M.

If none of the conditions holds then e_3 is inserted into the SDG, and $t(w)$ into M (Figure 3, right).

The operation of insertion triggers two further actions. The first action is to check for whether the edge e to be inserted χ-intersects any edge already in the SDG. If a χ-intersection occurs then e is not inserted. The second action is a re-calculation of the umbrella at the two vertices incident to e.

This algorithm is based on a large amount of experimental experience which in particular concerns the treatment of sharp edges and ridges of unbounded curvature at which, by suitable choice of the parameters, quite satisfiable results can be achieved [8].

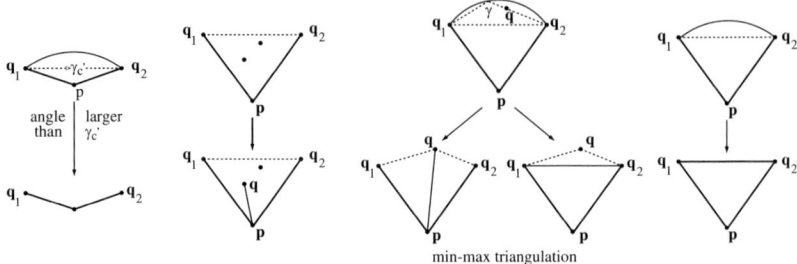

Fig. 3. Depiction of sectors which satisfy condition 1, 2, 3 (from left to right) of the triangulation algorithm, and the corresponding actions. The fourth picture shows the case if none of the conditions hold. The upper row depicts the configuration of the condition, the lower row the result of the action performed.

4 Analysis

Let S be a surface, P a set of sample points on S. A triangular mesh M with vertex set P is called a *reconstruction* of S if the nearest-neighbor image of M on S is non-self-intersecting. The nearest-neighbor image of M is obtained by assigning to every point on an edge or a triangle of M the closest points on S.

A surface type of particular interest with respect to this definition of reconstruction are *safe-fringe* (*SF*) surfaces. A surface S has a safe fringe of thickness $\delta > 0$, if a ball of radius δ can roll on both sides of the surface collision-free and with exactly one contact point everywhere.

For compact SF-surfaces S without boundary it can be shown that a mesh M has a non-self-intersecting nearest neighbor image if its edges are sufficiently short, if the maximum angle of its triangles is sufficiently far from 180°, and if the dihedral angles between adjacent triangles of M are sufficiently close to 180° [8]. Our algorithm has the property that sample sets on SF-surfaces exist so that the calculated mesh has these properties with high likelihood. An outline of arguments, confirmed by empirical data, is given in the following.

For the clustered β-EGs constructed in step 1, sample sets are of interest which yield sufficiently short β-EG-edges, and which do not intersect each other in the nearest-neighbor image.

For a compact SF-surface S without boundary, the existence of a sample set P_0 can be proved so that the lengths of all edges of the β-EG of P_0, except the non-blockable edges, are less than a given arbitrary length bound [8]. *Non-blockable edges* $e = \overline{pq}$ are those which are approximately perpendicular to S at **p** and **q**. This fact implies that, if the length bound is sufficiently smaller than the minimum length of all non-blockable line segments, the clustered β-EG has edges shorter than the given length bound.

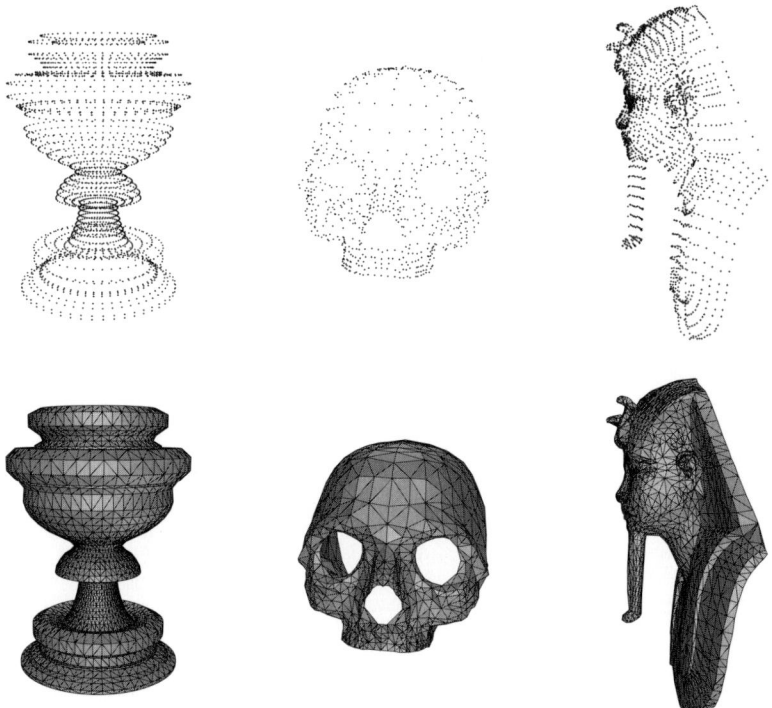

Fig. 4. The first row shows the data of a cup (2650 points), a skull (698 points), and a pharaoh (2286 points). The second row shows the interpolating meshes obtained by the algorithm.

As subgraphs of the Delaunay triangulation, β-EGs are planar graphs for sets of vertices in the plane in general position. If the vertices are on a smooth curved surface, and if the edge lengths are sufficiently small, the vertices of a β-EG are approximately locally flat. Thus it is very likely that nearest-neighbor image of a β-EG is locally planar. More precise arguments are given in [8].

We now come to the analysis of step 2. Its central concept is the γ-environment. The γ-environment of a line segment $\overline{\mathbf{p_1p_2}}$, $0 < \gamma < 180°$, consists of all those points $\mathbf{q}$ in space for which the angle at $\mathbf{q}$, in the triangle $\Delta(\mathbf{p}_1, \mathbf{p}_2, \mathbf{q})$, is less than γ. In the following, the γ-environment is used as line segment candidate region $C_c(s)$ (Figure 3, middle right).

A first observation is that, if γ is close to $180°$, it can be argued that the likelihood is low that the algorithm constructs edges whose γ-environment contains a given sample point [8,11]. Furthermore, it can be shown for compact SF-surfaces S without boundary that, for a given constant $l_1 > 0$, any

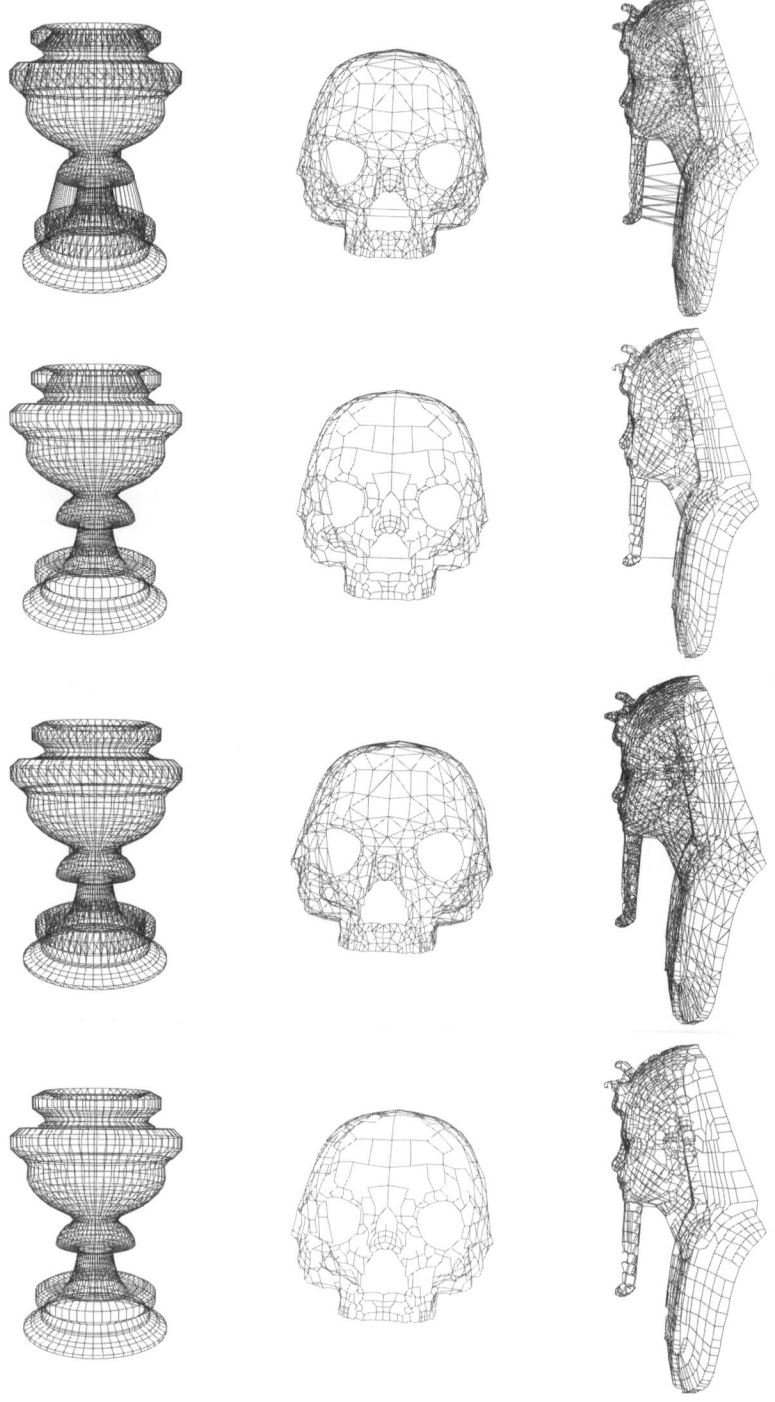

Fig. 5. The first row shows the β-EGs for $\beta = 0$, the second row those for $\beta = 1$. The number of bridge edges did reduce somewhat. The third and fourth row present the clustered β-EGs for $\beta = 0$ and $\beta = 1$, respectively.

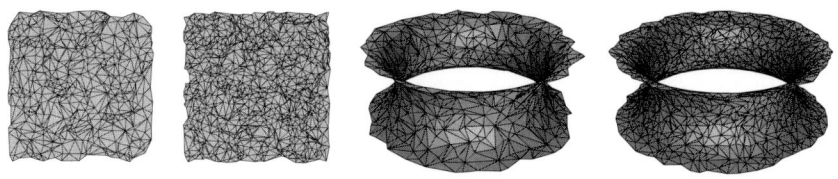

Fig. 6. The reconstructed surfaces for sample sets used for the empirical analysis, with about 500 (first and third picture) and 1000 (second and fourth picture) sample points.

given finite point set – in our case P_0 from step 1 – can be extended to a finite sample set P_1 of S for which each line segment, with vertices in P_1, longer than l_1, but not too long, has a γ-environment containing a sample point. Both together implies a high likelihood that the algorithm applied to P_1 inserts just edges which are short, and do not induce triangles with large angles since the γ-environments of the edges can be expected empty.

In the plane, the triangulation algorithm avoids intersecting triangles and edges by avoiding insertion of triangles over which points are located, and of edges which χ-intersect already existing edges. For sufficiently short edges, the mesh can be considered as approximately flat, so that the likelihood that this behavior changes on curved surfaces is low.

In summary the algorithm of step 2 should yield meshes which are favorable in the sense outlined in the third paragraph of this section.

This analysis gives arguments for the existence of sample sets which are favorable for the algorithm. However, it raises the question how special the sample sets are. One approach to get an answer is to consider the behavior of the algorithm on random sample sets. We followed this approach for a set of representative surfaces: a flat square, a hemisphere and the inner and outer part of a torus. In the following we present results obtained for the flat square and the inner part of the torus. For the data of the other experiments which are quite similar to the inner-torus case, we refer to [8]. Furthermore, we performed a detailed empirical analysis on eight real data sets of which three are shown in Fig. 4 and Fig. 5.

The square is randomly sampled with the desired number of points by a random function delivering x- and y-coordinates scaled to the side length of the square. The sample sets of the inner region of the torus are generated by arranging approximately equal-spaced points on circles in parallel to the "equator" of the torus and covering the inner region. Then the points are jittered by modifying their positions by random offsets along their circle and perpendicular to it. Figure 6 shows the interpolations obtained for data sets with about 500 and 1000 sample points. These data sets will be used later-on. Additionally performed measurements with more points did not show principle differences.

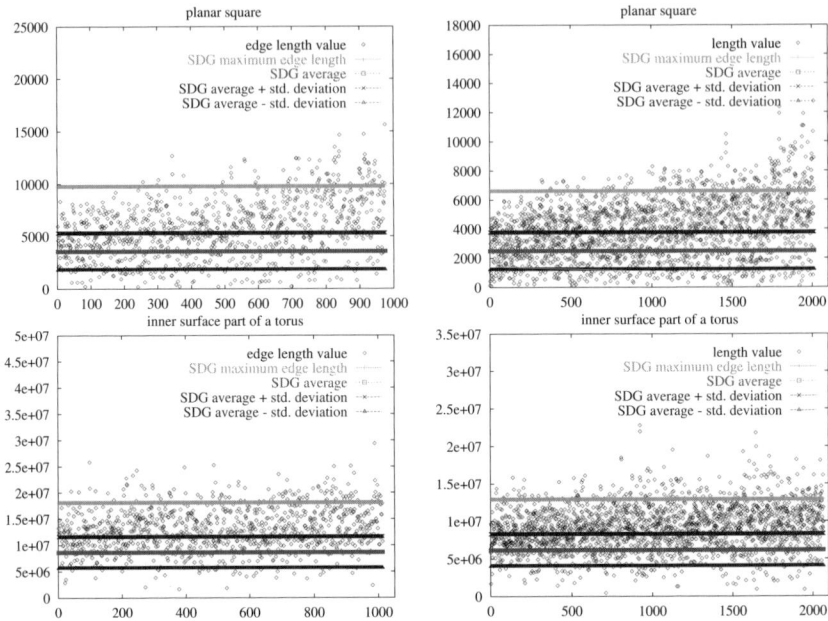

Fig. 7. The lengths of the inserted edges, taken for a sampled square (first row) and the sampled inner region of a torus (second row). For comparison, the average edge length, the standard deviation, and the maximum edge length of the initial SDG are depicted, too. Data sets with about 500 (left column) and about 1000 points (right column) have been used. A color version of the pictures can be found on the color pages.

Figure 4 shows that elimination of bridge edges works well. In the two top rows, 0-EGs and the 1-EGs are shown. The next two rows present the clustered version of the those graphs. The numbers of edges which have been removed are 226, 16, and 55 for the 0-EG, and 50, 6, and 11 for the 1-EG.

For triangulation it is important that the inserted edges are short. Figure 7 shows the lengths of the edges sequentially inserted by the algorithm for the square and the torus region. Here and in the following, $\beta = 1$, $\gamma = \gamma_c = 135°$, and $\chi = 75°$ for the χ-intersection test has been used. Furthermore the average length of the edges of the SDG which is the input of triangulation, the region of standard deviation and the maximum length of an edge in the SDG are depicted in the figure. We can notice that most inserted edges are less than twice as long as those of the SDG. Thus the inserted edges remain short, as desired.

A second important point for triangulation is that the angles of the triangles should not become extreme. Figure 8 shows the size of the sector angles. In the flat case, the sectors are processed according to increasing angles because the dihedral angle is trivially 180°. The reason of the scattering effect

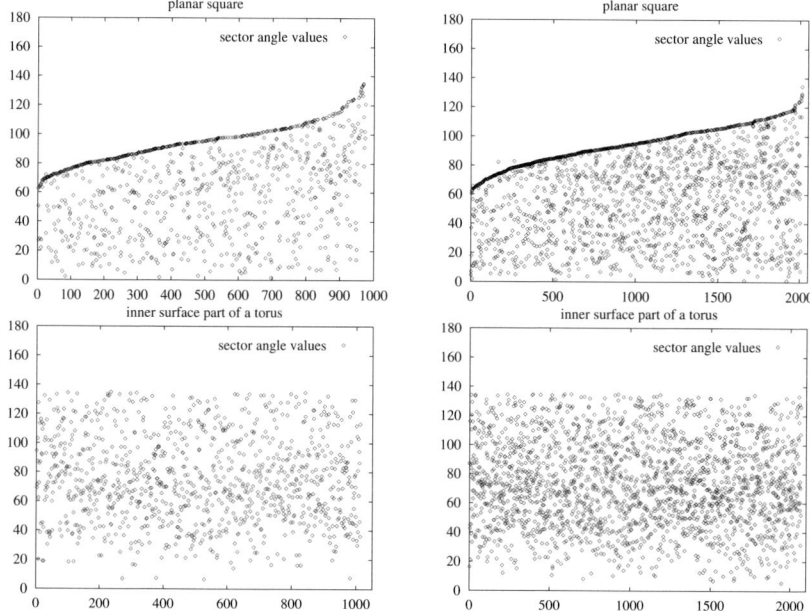

Fig. 8. The size of the angle of every processed sector, measured for a square (first row) and the inner region of a torus (second row), and about 500 (left) and 1000 (right) data points.

is that new sectors which may have smaller angles are generated if an edge is inserted into G. An interesting effect is the tendency that the angle size nevertheless increases during execution of the algorithm. The plot shows that the angles of the processed sectors are indeed bounded by 135°.

In the plots of the planar square, a sorting effect can be noticed. The reason is that in this case the sector angles, and not the dihedral angles to neighboring sectors, define the sorting criterion. Another observation is that very small angles do not occur frequently.

Figure 9 displays the number of occurrences of inserted γ-edges and non-γ-edges, respectively. Almost all of the inserted edges are γ-edges. The number of non-γ-edges is between 0 and 10. This also corresponds to the observation of the theoretical analysis outlined at the beginning of this section. Moreover, the dihedral angles of adjacent triangles of these non-γ-edges always fit smoothly into the surrounding structure. Most of the dihedral angles between adjacent triangles at these edges exceeded 170° in the experiments. The smallest angle ever found has been about 156°. Obviously, the edge selection process chooses the configurations well, so that a good surface mesh is generated.

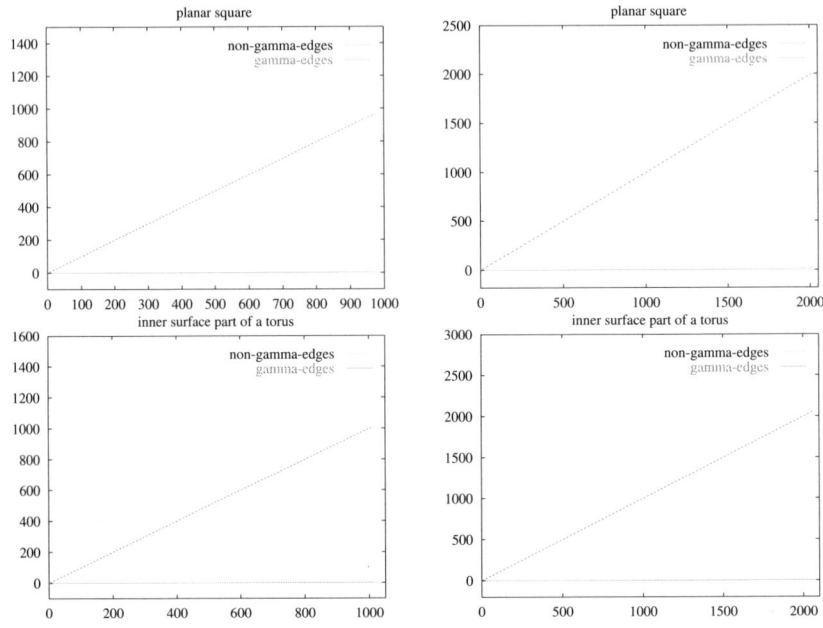

Fig. 9. The number of occurrences of inserted γ-edges and non-γ-edges, respectively, measured for a square (first row) and the inner region of a torus (second row), and about 500 (left) and 1000 (right) data points.

In the triangulation step, the algorithm checks three conditions in order to find out how to treat a given sector. Figure 10 presents the number of occurrences of condition 2 ("over trian"), of condition 3 ("near to edge"), and of the case that none of the conditions hold ("normal"). The "normal" case in particular covers sectors which can be completed to a triangle, which are already completed to a triangle but the triangle has not yet been inserted, and which cause troubles because of intersection. The overwhelming majority of sectors belongs to the "normal" case, and in particular to the first two versions just mentioned. The case of condition 2 does practically not occur, the curve is about constant equal to 0. The curves of condition 3 increase slightly over-proportionally.

Figure 11 depicts the statistics of the result of the action of min-max-triangulation caused if condition 3 holds. The majority of the edges is of type $\overline{pq}$ (Figure 3, middle right, right case of the min-max triangulation results), what could be expected.

The statistics show that the algorithm works well for random sampling. It can be expected to work even better if sometimes occurring longer edges are avoided. As shown in [8, 11], sampling by regular spatial meshes is a possibility in this direction. In this case the interpolation constructed by the

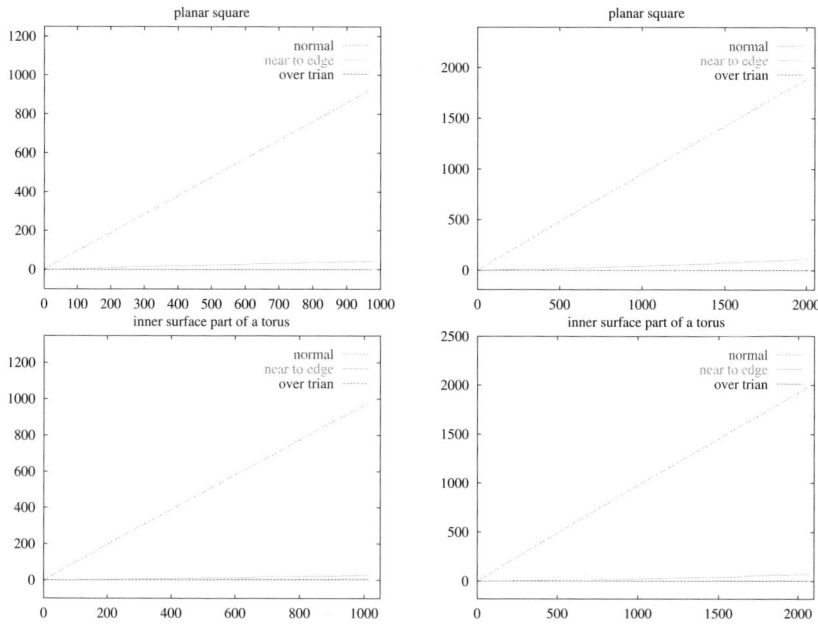

Fig. 10. The number of occurrences of condition 2 ("over trian"), of condition 3 ("near to edge"), and of the case that none of the conditions hold ("normal"). The data are measured for a square (first row) and the inner region of a torus (second row), and about 500 (left) and 1000 (right) data points. A color version of the pictures can be found on the color pages.

algorithm corresponds to a marching cubes interpolation [6]. Furthermore, approximately regular sampling patterns like the vertices of an approximately regular quadrilateral meshing of a surface, like the one used for the cup in Fig. 4, are favorable.

5 Conclusion

We have presented an empirical analysis of a surface reconstruction algorithm which confirms the "heuristic" theoretical analysis presented in [11]. The analysis has been restricted to smooth surfaces without boundary, represented by so-called SF-surfaces. The algorithm, however, also works well for surfaces with sharp edges and ridges. One reason is that the β-EG can be forced to consider sharp edges and ridges if the sampling density is increased at those items. But even in cases of sample sets without particular adaptation, good results may be achieved, due to the heuristics of the triangulation phase. Figure 12 shows a reconstruction of the surface of a cube by the algorithm. Figures 4 and 6 illustrates the ability of the algorithm to reconstruct surfaces

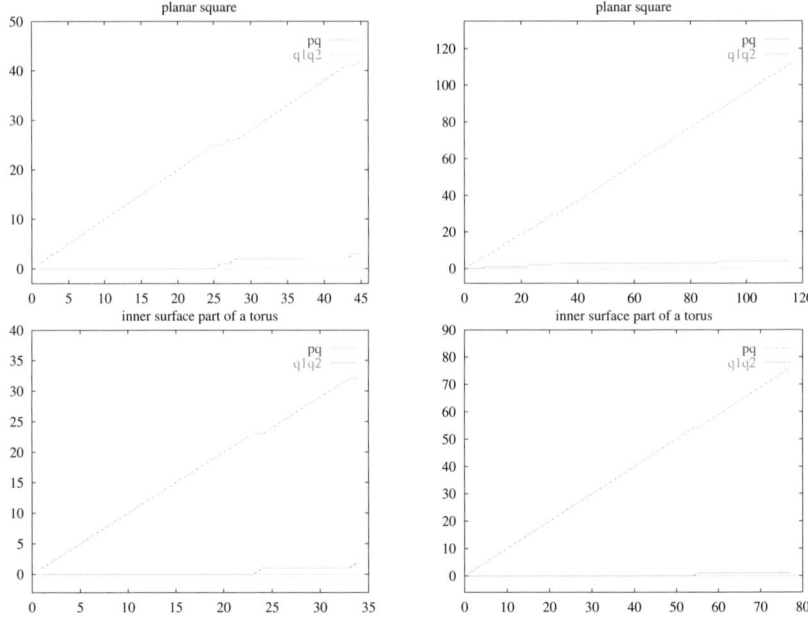

Fig. 11. The number of occurrences of edge types $e_3 = \overline{q_1 q_2}$ and $e = \overline{pq}$ (cf. Fig. 3) over time for the min-max triangulation performed under condition 3 of the algorithm of triangulation. The data are measured for a square (first row) and the sampled inner region of a torus (second row), and about 500 (left) and 1000 (right) data points.

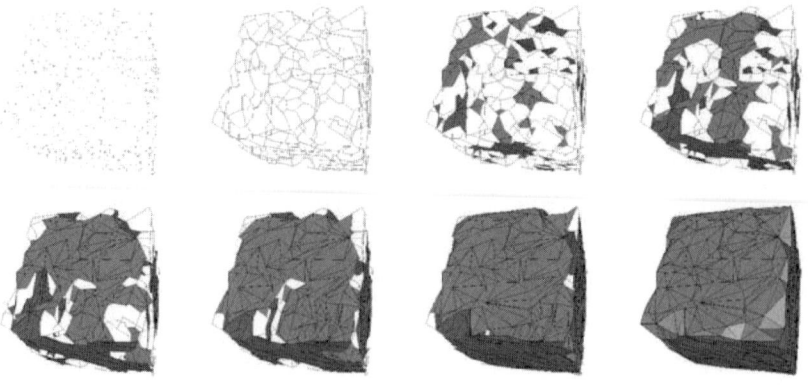

Fig. 12. A set of points sampled from a cube, the 1-EG of the point set, and snapshots from the triangulation phase. The example shows that the algorithm can also reconstruct sharp edges.

with boundary. The theoretical analysis of the treatment of surfaces with sharp edges and boundaries remains a problem for further research.

References

1. U. Adamy, J. Giesen and M. John, *New Techniques for Topologically Correct Surface Reconstruction*, Proceedings of IEEE Visualization 2000, IEEE Computer Society Press, 2000.
2. N. Amenta, M. Bern and M. Kamvysselis, *A new Voronoi-based surface reconstruction algorithm*, Proceedings of SIGGRAPH'98, 1998, 415–421.
3. N. Amenta, S. Choi, T.K. Dey and N. Leekha, *A simple algorithm for homeomorphic surface reconstruction*, In Proc. 16th ACM Sympos. Comput. Geom., 2000.
4. N. Amenta, S. Choi and R. K. Kolluri, *The Power Crust, Unions of Balls, and the Medial Axis Transform*, Computational Geometry: Theory and Applications, 2001, 19(2-3), 127–153.
5. D.G. Kirkpatrick, J.D. Radke, *A framework for computational morphology*, in: G.T. Toussaint (ed.), Computational Geometry, Elsevier Science Publisher B.V., North-Holland, 1985, 217–248.
6. W.E. Lorensen, H.E. Cline, *Marching Cubes: A high resolution 3D-surface construction algorithm*, Computer Graphics, 1987, 21 (4), 163–169.
7. R. Mencl, *A Graph–Based Approach to Surface Reconstruction*, Computer Graphics Forum, 1995, 14(3), 445–456.
8. R. Mencl, *Reconstruction of Surfaces from Unorganized Three-Dimensional Point Clouds*, PhD thesis, Informatik VII, University of Dortmund, Germany, 2001.
9. R. Mencl and H. Müller, *Graph–Based Surface Reconstruction Using Structures in Scattered Point Sets*, Proceedings of CGI '98 (Computer Graphics International), Hannover, Germany, 1998, 298–311.
10. R. Mencl, H. Müller, *Interpolation and Approximation of Surfaces from Three-Dimensional Scattered Data Points*, Proceedings of Scientific Visualization – Dagstuhl '97, IEEE Computer Society Press, 2000.
11. R. Mencl, H. Müller, *Surface interpolation by spatial environment graphs*, In: Data Visualization: The State of the Art, Kluwer Academic Publishers, 2002.
12. F.P. Preparata and M.I. Shamos, *Computational Geometry: An Introduction*, Springer Verlag, 1985.
13. J.D. Radke, *On the shape of a set of points*, in: Computational Morphology, G.T. Toussaint (ed.), Elsevier Science Publisher B.V., North Holland, 1988, 105–136.
14. S.V. Rao, *Some studies on beta-skeletons*, PhD thesis, Dept. of Computer Science & Engineering, Indian Institute of Technology, India, 1998.
15. R.C. Veltkamp, *Closed Object Boundaries from Scattered Points*, Lecture Notes in Computer Science 885, Springer Verlag, 1994.

Part II

Surface Interrogation and Modeling

Smooth Polylines on Polygon Meshes

Georges-Pierre Bonneau[1] and Stefanie Hahmann[2]

[1] iMAGIS-GRAVIR, INRIA Rhône-Alpes,
655 Avenue de l'Europe, F-38330 Montbonnot (France)
Georges-Pierre.Bonneau@imag.fr
[2] Laboratoire LMC-IMAG,
BP. 53, F-38041 Grenoble (France)
Stefanie.Hahmann@imag.fr

Summary. Curves on surfaces can be very useful to visualize surface features at low graphical and memory cost. Curves on surfaces are also used for surface segmentation with possible applications to visualization, reconstruction and parameterization of complex surfaces. In this paper a simple and efficient algorithm for building smooth polylines on triangulated 2D-manifold polygonal meshes is introduced. The algorithm combines geometrical optimization with topological modifications in order to iteratively smooth an initial crude polyline. One key feature of this algorithm is that it relies solely on the geometry of the surface and the polyline. Another key feature is that during the smoothing the polylines always stay on the surface. Different smoothing criteria are proposed.

1 Introduction

Curves on surfaces have applications in many different areas. In visualization, their ability to reveal global surface features or local details at a very low graphical and memory cost has been used to help visualizing large data sets [10]. Curves on surfaces are also important in surface segmentation. The applications of surface segmentation span a large field of research from visualization, to reconstruction and parameterization. Irregular polygonal meshes can be reparameterized on a coarse mesh by first constructing a set of smooth polylines on a fine mesh that yield a segmentation of the mesh into three sided patches. The reconstruction of parametric surfaces from dense polygonal data can be based on the interactive design of a mesh of smooth polylines drawing the boundaries of the patches.

In this paper we present an algorithm for building smooth polylines (PL) on 2D manifold triangular meshes, for definition see section 2. An initial PL is first computed from a set of user specified mesh triangles. This is a crude (non smooth) approximation of the desired PL. Then an iterative execution of geometrical optimization steps and topological steps enables to smooth this initial PL on the surface. During this process the PL always lies exactly on the mesh. A key feature of the algorithm is that it relies solely on the surface geometry and the PL geometry. It is independent of any surface parameterization or any external prescribed smooth curve. The algorithm has the following features:

- It is a purely geometry driven algorithm. This means that the smoothing criteria as well as the error measurements only depend on the intrinsic geometric properties of the surface or the PL.
- Between two user-specified or automatically computed points on the polygonal surface a smooth PL can be computed by using different smoothing filters.
    · We first introduce a "linear" filter which yields a local discrete geodesic. The visual effect is a global straightening of the initial PL.
    · Then several extensions are possible, we choose to implement a "cubic" smoothing filter. Here the global shape of the initial PL is preserved while small details are smoothed out.
- In any case smooth PL are obtained by iteratively moving the vertices of the PL on the mesh edges (geometrical modification) and by passing through mesh vertices if necessary (topological modification).
- The algorithm is easy to implement. However, the topological steps need to be well understood.

Related Works

There are many papers specifically dedicated to compute curves related to geodesics [11], [13], [7], [2], [12]. Geodesics or close-to-geodesics are not the issue of the present paper, rather we are computing smooth curves that need not to be geodesics.

Krishnamurthy and Levoy [9] build smooth PLs on a polygonal mesh using an external smooth 3D curve as a guide . The use of an external guiding curve may fail in high curvature surface regions ([9] figure 7). Also the PL does not exactly lie on the mesh, only the vertices of the PLs are constrained to lie on the mesh. In [8] Krishnamurthy modifies this method in order to avoid the use of an external guiding curve. But still the PLs don't lie exactly on the mesh, and the optimization relies on a set of user given external forces applied to a polynomial model of the PL, whereas our algorithm relies solely on the surface geometry and the PL geometry.

Eck et al. [5] construct a set of smooth PLs on a fine mesh, that segment the mesh into three sided patches. First a crude version of the PL is computed. The PLs are smoothed in the following manner: a special parameterization of the surface around one PL is used to map it onto a planar domain, then the PL is replaced by the inverse image by this map of a straight line in the parameter domain. Praun, Sweldens and Schröder [14] use the same method to straighten the PL, but with a different parameterization. While Eck et al. use a parameterization based on harmonic maps, Praun et al. use the parameterization introduced by Floater [6]. By contrast our algorithm does not depend on any parameterization of the triangle mesh, and is much simpler to implement.

The paper is organized as follows. In section 2, the notations for a PL on a 2D manifold triangulation are given. Section 3 outlines the algorithm. Section 4 is dedicated to the computation of the initial PL. Section 5 describes the geometrical optimization that is used by the algorithm. Section 6 explains the topological steps that are required by the algorithm. Results are given in section 7. Finally section 8 concludes the paper and gives possible future works.

2 Notations

The triangles in the mesh are denoted T_i, and the edges E_i. A *triangle strip* is defined by $n+1$ adjacent triangles $T_1, \ldots, T_{n+1}$ in the mesh. Let $E_1, \ldots, E_n$ denote the common edges between adjacent triangles in the triangle strip. A polyline is defined as the combination of a strip and a set of n scalars $\alpha_1, \ldots, \alpha_n$ that store the barycentric coordinate of its vertices $V_1, \ldots, V_n$ on the edges $E_1, \ldots, E_n$, with respect to the orientation of the triangles $T_1, \ldots, T_n$. A triangle strip can be efficiently stored in memory with one pointer to the first triangle, and n bits to encode the next edge in the strip. Figure 1 illustrates these notations.

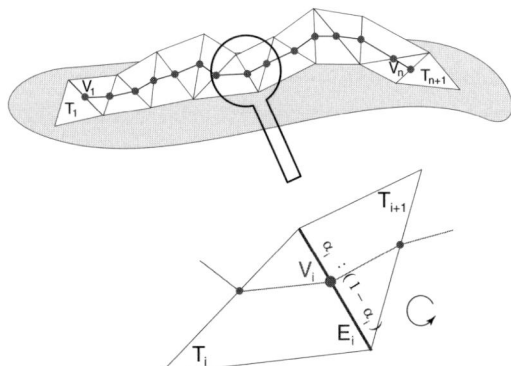

Fig. 1. *Notations. Triangle strip, PL, barycentric coordinates of PL vertices.*

3 Overview of the Algorithm

This section gives an overview of the PL design and smoothing process. This process is summarized in fig. 2 and illustrated in fig. 4.

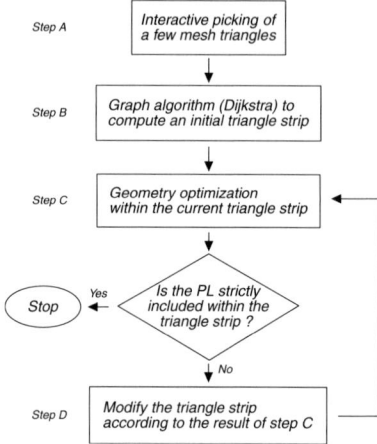

Fig. 2. *Overview of the algorithm.*

<u>Step A:</u> The user specifies a few triangles in the mesh, not necessarily adjacent. The order of the picked triangles is relevant: the initial triangle strip will interpolate these triangles in the given order. Note that these initial triangles can be interpolated by the final PL, if desired.

<u>Step B:</u> A Dijkstra algorithm ([4], [1]) is performed, that outputs a triangle strip, i.e. a sequence of adjacent triangles joining the triangles specified in step A. The output of step B is a triangle strip that gives a crude initial PL. Each vertex of this initial PL is chosen as the mid point of the edges between consecutive triangles in the strip (see fig. 4).

Steps C and D are looped until an exit criterion is fulfilled.

<u>Step C:</u> This is the geometrical part of the algorithm. The PL is optimized within the current triangle strip. The vertices of the PL are constraint to stay on the same edge during step C. Different criteria can be used for this geometrical optimization. The result of step C is a PL that can either be strictly included in the current triangle strip, or that can touch the boundary of this strip. If it is strictly included in the triangle strip, the algorithm stops and outputs the final PL.

<u>Step D:</u> This is a topological part of the algorithm. The output of step C is a PL in which some of the vertices have merged with the mesh vertices. This happens when the geometrical optimization performed in step C has pushed some vertices of the PL towards one of the two end points of the edge where these PL vertices lie on. This situation is illustrated in fig. 3. Around such vertices, a sequence of triangles of the current triangle strip is replaced by a

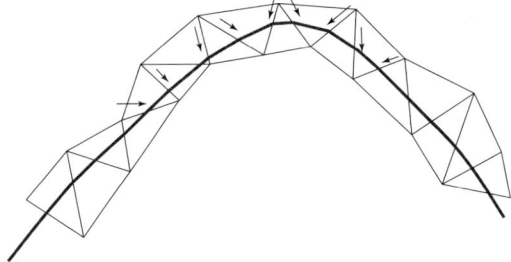

Fig. 3. *The geometrical optimization step C may push some of the PL vertices towards the mesh vertices. In such cases, the triangle strip has to be modified, which is the purpose of step D of the algorithm, see fig. 2.*

new sequence of triangles. The resulting new sequence of triangles is still a triangle strip and it still includes the PL.

Remark: It can happen that the polyline has one or several vertices merging with the mesh vertices. In this case, the exit criterion (see step C) is never fullfilled, since the optimal polyline is not strictly included in the triangle strip. As a consequence, there could be a cycle in steps C, D. In order to avoid such a behaviour, a general method would be to identify such cycles, and to stop the iteration then. In the current implementation the users fixes only a maximal number of iterations. In all examples we have tried, a maximal number of 100 iterations was fully sufficient.

4 Computing an Initial Triangle Strip.

The first step of the algorithm consists of computing the initial crude triangle strip. One could imagine different methods to define an initial triangle strip. One could for example draw a planar curve on the graphic window where the triangle mesh is displayed, and use this curve in order to pick triangles from the mesh. This could be implemented very efficiently using graphic hardware. Another possibility would be to draw a space curve, near the surface mesh, and to select triangles on the surface according to some projection mapping. In either case, the sequence of picked triangles would not be guaranteed to form a triangle strip. In regions of high surface curvature for example, the projection of the planar or the space curve could lead to sequences of triangle strips with "breaks" between them, see fig. 5. Thus it is necessary to define an algorithm that can compute a triangle strip between any two given triangles in the mesh.

We first build the graph in which the nodes are labelled with the mesh triangles, and edges connect nodes that correspond to adjacent triangles. This is an unoriented graph, the maximum degree of nodes is 3 since the triangles

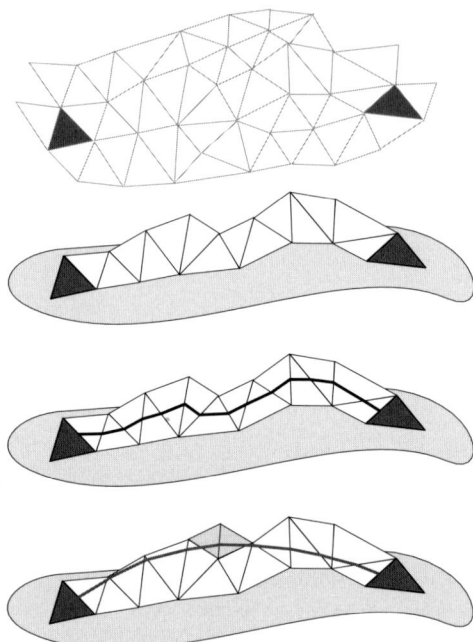

Fig. 4. *Overview of the algorithm. (a) the user selects triangles in the mesh. (b) a initial triangle strip is computed. (c) an initial PL is build from the edge mid points of the triangle strip. (d) geometrical optimization and topological steps yield a smooth PL.*

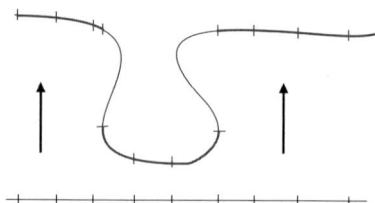

Fig. 5. *Using an external curve in order to pick triangles on a polygonal mesh may fail in regions of high surface curvature. It may produce triangle strips with breaks.*

have up to 3 neighbours. Fig. 6 shows this graph for a simple triangle mesh. In our current implementation, we assign weights to the edges of the graph. These weights are chosen equal to the distance between the barycenter of two adjacent triangles. Since the edges on this graph connect adjacent triangles, a path in this graph corresponds to a triangle strip on the mesh. Then a Dijkstra algorithm is applied to the graph in order to compute the shortest path between the nodes which correspond to the given pair of triangles. This path gives the desired triangle strip.

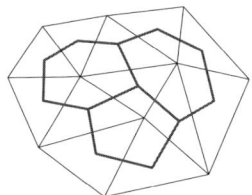

Fig. 6. *The triangulation and the dual graph is used for the computation of the initial triangle strip. All edges of this dual graph are weighted by the distance between the barycenter of the triangles.*

As explained in the overview, we apply this procedure to each pair of triangles (T_i, T_{i+1}) in a sequence of triangles $T_1, \ldots, T_n$ chosen by the user. Fig. 7 shows different triangle strips interpolating the first and last triangle. The top one only interpolates these two triangles. The middle one interpolates also one triangle below the ear, and the bottom one interpolates one triangle between the ear and the horn.

5 Geometry Optimization of the Polyline

The kernel of the algorithm consists of the iterative application of the geometrical optimization of the PL inside its current triangle strip followed by the topological modification of the strip. This section is dedicated to the geometrical optimization step, which is itself also an iterative process. It successively selects vertices of the PL according to some error metric, and it slides the position of this selected PL vertex along its mesh edge, in order to minimize the error. During this process all vertices are constrained to stay on their edge. This is why the triangle strip of the PL is fixed. This process is efficiently implemented using a heap data structure [3]. After the modification of one vertex, the new error in this vertex always vanishes, and the errors at the neighbouring vertices have to be updated. The process stops when the maximum error falls below a user-given threshold value. Note that, in case the user wants to interpolate some vertices, we can simply omit to add these vertices in the heap. Hence these vertices remain fixed in the PL.

Error Criteria
In [9] the error criterion depends on an external space curve, or on some external forces applied to the PL. In [5], [14] the polygon is smoothed using some parameterization of the surface, and modifying the PL in the parameter domain. By contrast our error criterion doesn't depend on any parameterization, it relies solely on local geometrical properties.

We have to compute one error per vertex, and we have to find the new position of the vertex along its edge that minimizes the error. In or-

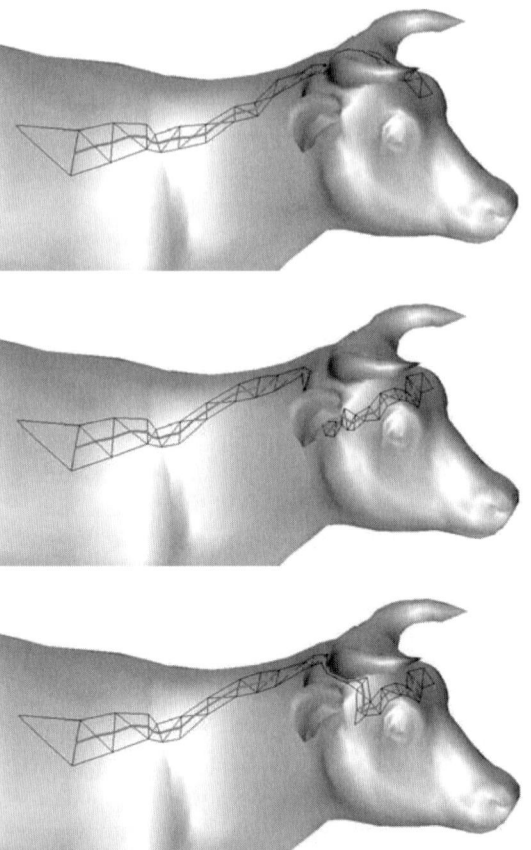

Fig. 7. *Different triangle strips interpolating the same first and last triangles. The top one only interpolates these two triangles. The middle one interpolates in addition one triangle below the ear, and the bottom one interpolates one triangle between the ear and the horn.*

der to do this the following method is applied. Let V_i denote the vertex that has to be optimized, and E_i the mesh edge on which V_i lies. First, a continuously defined curve C is computed from the neighbouring vertices $V_{i-k}, V_{i-k+1}, \ldots, V_i, V_{i+1}, \ldots, V_{i+k}$. Then V_i is replaced by the point on the edge E_i that is closest to the curve C. Note that the curve C doesn't need to be on the triangle mesh. The error in vertex V_i is chosen as the absolute value of the difference between the barycentric coordinate of V_i and the barycentric coordinate of the optimized point. In the current implementation, two different choices of C are offered to the user.

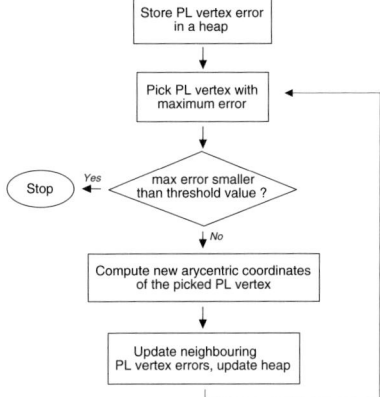

Fig. 8. *Overview of the geometrical optimization step.*

Linear Smoothing Filter

The first choice for C is simply the straight line between the vertices V_{i-1} and V_{i+1}. We will refer to this as the linear choice. This is illustrated in fig. 9. The effect of the linear choice for C is intuitive: it tends to straighten the PL. And in fact, this choice leads to PL that is a straight line if the mesh is planar, and that approximates geodesic paths for non planar meshes. Examples of results are given in section 7.

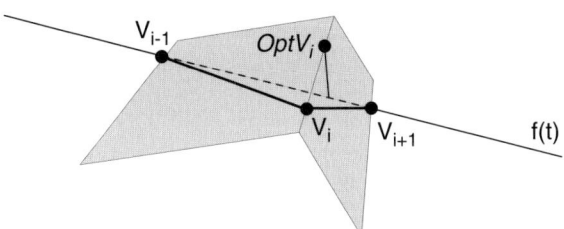

Fig. 9. *The optimal position of V_i with the linear filter is the closest point on the edge to the line V_{i-1}, V_{i+1}.*

Cubic Smoothing Filter

Straightened curves are not always desirable. Instead of straightening the PL, it could be useful to smooth it, while keeping its global shape. This is the case for example if the PL is closed. The linear choice for C would shrink the curve until it collapses to its centre of gravity. For all these reasons, a second choice for the curve C has been introduced: It is the unique cubic curve that interpolates $V_{i-2}, V_{i-1}, V_{i+1}, V_{i+2}$ (with uniform parameterization), see fig.

10. Since in this case C depends on a larger filter of points around V_i, it tends to smooth out the fine details, but doesn't change the overall shape of the PL. In fact if all points of the PL lie on a cubic curve, then the PL is not modified at all. In other words, this choice of C ensures that the scheme has a cubic precision.

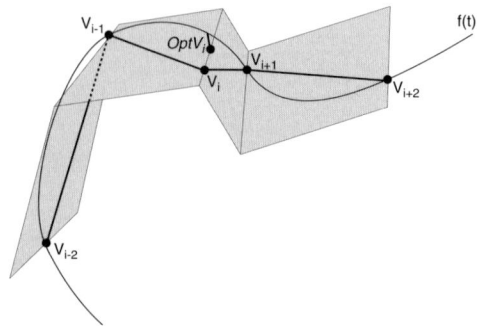

Fig. 10. *The optimal position of V_i with the cubic filter is the closest point on the edge to the unique cubic polynomial interpolating $V_{i-2}, V_{i-1}, V_{i+1}, V_{i+2}$.*

6 Topological Steps

The output of the geometrical optimization step is a PL in which some of the vertices have merged with the mesh vertices. This situation is illustrated in fig. 3. The limit case is shown in fig. 11: the PL has smooth parts strictly included in the current triangle strip, and non-smooth parts that touch the boundary of the strip.

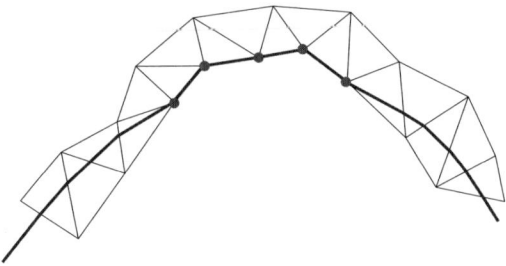

Fig. 11. *The results of the geometrical step is a PL which has smooth parts strictly included in the triangle strip, and non smooth parts along the boundary of the strip.*

The triangle strip has then to be modified, which is a topological step. This topological modification has been implemented in the following manner.

First we select the mesh vertices which have merged with the PL vertices. These vertices are highlighted in red in fig. 12. Then we find the edges of the triangle strip that are adjacent to these vertices. These edges are highlighted in red in fig. 12.

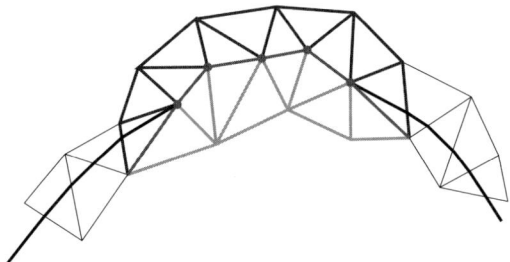

Fig. 12. *Topological step. The red points show the PL vertices that have merged with the mesh vertices. The red edges show the triangle strip edges that are adjacent to these vertices. The green triangles are included in the new triangle strip. They replace all but the first and the last blue triangles. (see colorplate)*

Let $T_1, \ldots, T_k$ denote the whole triangle strip, and $T_i, \ldots, T_{i+n}$ denote the sequence of triangles adjacent to the selected (red) edges, that are part of the strip. $T_i, \ldots, T_{i+n}$ are highlighted in blue in fig. 12. Let $\widetilde{T}_1, \ldots, \widetilde{T}_m$ be the sequence of triangles adjacent to the selected edges, but not part of the strip. The new triangle strip is now $T_1, \ldots, T_i, \widetilde{T}_1, \ldots, \widetilde{T}_m, T_{i+n}, \ldots, T_k$, Note that the new triangle strip still includes the PL. The geometry of the PL is not modified, but its topology is changed: the vertices that have merged with the mesh vertices are now encoded relative to the edges of the new triangle strip. In particular, if the barycentric coordinate of the PL vertices was zero (resp. one) relative to the old triangle strip, it is equal to one (resp. zero) with the new triangle strip.

Fig. 17-right shows an example of topological steps. The PL after geometrical optimization is in black, the blue PL shows the portion of the current triangle strip that has to be removed and the red PL shows the new portion of the triangle strip.

Loops in Triangle Strip
An issue in this topological modification step is the possibility of creating loops in the triangle strip. This happens whenever the triangle strip has a high curvature, see fig. 13. This is a classical problem in Computer Aided Geometric Design, that is related to the computation of offset curves. To avoid this problem in the algorithm, a further modification of the triangle

strip is conducted, that consists in searching for possible loops and deleting them.

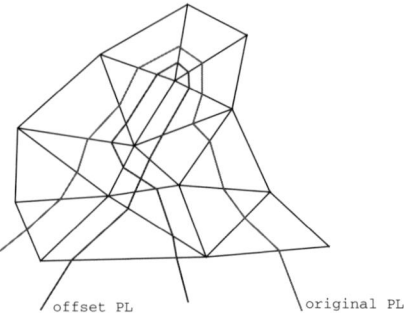

Fig. 13. *The topological modification can lead to loops in the triangle strip. These loops are deleted in an additional step.*

7 Results

Most of the time spent for building smooth PL is due to the user's choice of the input triangles that the PL will interpolate. Once this is done, building the initial triangle strip takes theoretically $O(n \log m)$ where n is the number of mesh vertices and m the number of mesh edges. Practically it took less then 0.5s on a 1 GHz PIII even for the ball joint mesh that has 137062 vertices. The time spent for the iterative optimization process generally depends on the shape of the initial PL. Practically, all optimizations took less than 5s for all the examples of this paper.

Most of the meshes shown in this paper are courtesy Cyberware and can be downloaded at http://www.cyberware.com. Fig. 14 and fig. 15 show a typical example where discrete geodesics can not be used, since the PL is closed. The mesh (a ball joint) has 137062 vertices and 274120 faces. The initial crude PL (left in fig. 14 and 15) interpolates 30 user specified triangles. The images on the right in fig. 14 and 15 show the result of our "cubic" filter. Notice in fig. 15 how the global shape of the initial PL is preserved, while small details are smoothed out.

Fig. 16 (a,b,c) show three PLs obtained using the "linear" filter on the initial PLs shown in fig. 7. This example illustrates the fact that our iterative algorithm prevents the PL from jumping over obstacles. It is not able to find shortest paths. The user can precisely choose where the PL should lie: if he/she wants the PL to traverse between the ear and the horn, he/she only

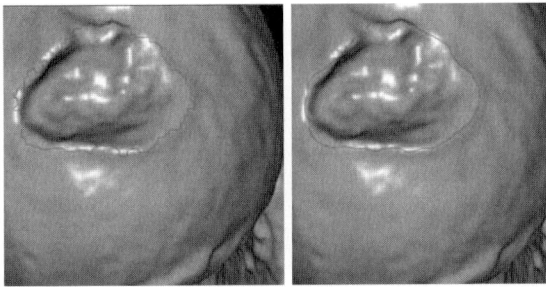

Fig. 14. *Ball joint mesh, 137062 vertices, 274120 faces. The initial PL shown left interpolates 30 user specified triangles. The smooth PL shown right is obtained using our cubic smoothing filter. This is a typical example where discrete geodesics can not be used.*

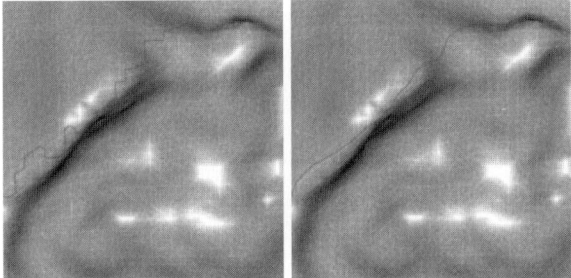

Fig. 15. *Local zoom on the top left corner of fig. 14. The mesh edges are shown in white.*

has to select one triangle in this area during the initial triangle strip construction, and the PL will remain in this area throughout the optimization process.

Fig. 17-left shows the PL at different stage of the optimization process. The initial PL is in red, and the final PL is in blue. Fig. 17-right illustrates the topological steps of the algorithm.

Fig. 18 illustrate a segmentation. It shows top a set of initial PLs, and bottom the final smooth PLs using the linear filter. Such a curve net can be useful for building a mesh parameterization.

8 Conclusion

In this paper a very simple and efficient algorithm for building smooth PLs on 2D-manifold triangulations has been introduced. By contrast with prior works, this algorithm is based purely on intrinsic geometric properties of the mesh and the PL, and in particular it does not rely on any parameterization.

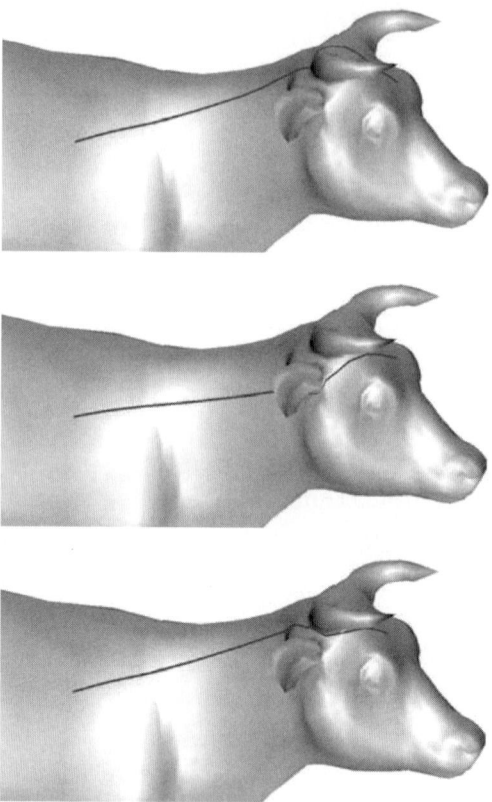

Fig. 16. *The three smooth PL are obtained using the linear filter with the initial PL shown in fig. 7. Our algorithms prevents the PL from jumping over obstacles.*

Two different smoothing filters were proposed, one that tends to straighten the PL, and the other that removes small details while preserving the overall shape of the PL.

In future works we will investigate other kind of filters, with larger support, and we will use the results of this paper as a tool for segmentation and visualization purposes.

Acknowledgments

This work was partially supported by the European Community 5-th framework program, with the Research Training Network MINGLE (Multiresolution IN Geometric modELing, HPRN-1999-00117). Special thanks to Alex Yvart for his Dijkstra code. Models are courtesy Cyberware.

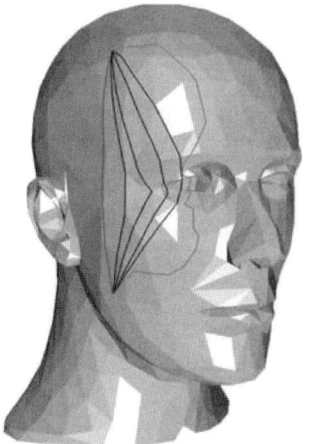 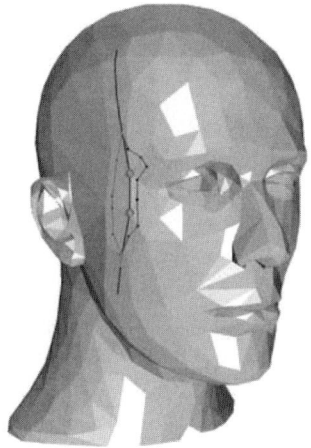

Fig. 17. *Iterative smoothing process (left). The topological modifications (right). Details are given in section 6 (see colorplate).*

References

1. AHO, A., AND ULLMAN, J. *Data structures and algorithms*. Addison-Wesley, 1979.
2. ALEKSANDROV, A. D., AND ZALGALLER, V. A. Intrinsic geometry of surfaces. *Translation of Mathematical Monographs 15* (1967).
3. CORMEN, T., LEISERSON, C., AND RIVEST, R. *Introduction to Algorithms*. MIT Press, 1990.
4. DIJKSTRA, E. A note on two problems in connection with graphs. *Numer. Math 1* (1959), 269–271.
5. ECK, M., DEROSE, T., DUCHAMP, T., HOPPE, H., LOUNSBERY, T., AND STUETZLE, W. Multiresolution analysis of arbitrary meshes. In *Proceedings of SIGGRAPH 1995* (1995), Computer Graphics Proceedings, Annual Conference Series, ACM, ACM Press / ACM SIGGRAPH, pp. 173–182.
6. FLOATER, M. Parameterization and smooth approximation of surface triangulations. *CAGD 14* (1997), 231–250.
7. KIMMEL, R., AND SEITHAN, J. Fast marching method on triangulated domains. *Proceedings of the National Academy of Science 95* (1998).
8. KRISHNAMURTHY, V. *Fitting Smooth Surfaces to Dense Polygon Meshes*. PhD thesis, Stanford University, 2000.
9. KRISHNAMURTHY, V., AND LEVOY, M. Fitting smooth surfaces to dense polygon meshes. In *Proceedings of SIGGRAPH 1996* (1996), Computer Graphics Proceedings, Annual Conference Series, ACM, ACM Press / ACM SIGGRAPH, pp. 312–324.
10. MA, K.-L., AND INTERRANTE, V. Extracting feature lines from 3d unstructured grids. In *Proceedings of Visualization 1997* (1997), IEEE, pp. 285–292.
11. MITCHELL, J., AND MOUNT, D. The discrete geodesic problem. *SIAM J. Copmput. 16* (1987), 647–668.

Fig. 18. *The top image shows the layout of 19 initial PL connecting 10 user specified triangles. The bottom image shows the final smooth PL using the linear filter. Such a net of smooth PL may be used for surface parameterization.*

12. PHAM-TRONG, V. *Détermination géométrique de chemins géodésiques sur des surfaces de subdivision.* PhD thesis, Université Joseph Fourier, Grenoble, 2001.
13. POLTHIER, K., AND SCHMIES, M. Straightest geodesics on polyhedral surfaces. In *Mathematical Visualization, 1998.* In H.C. Hege, K. Polthiers (eds.), Springer, 1998.
14. PRAUN, E., SWELDENS, W., AND SCHRÖDER, P. Consistent mesh parameterizations. *Computer Graphics Proceedings (SIGGRAPH 01)* (2001), 179–184.

Progressive Precision Surface Design

Mark A. Duchaineau[1] and Kenneth I. Joy[2]

[1] Lawrence Livermore National Laboratory
   duchaineau1@llnl.gov
[2] Center for Image Processing and Integrated Computing
   Computer Science Department
   University of California, Davis
   joy@cs.ucdavis.edu

Summary. We introduce a novel wavelet decomposition algorithm that makes a number of powerful new surface design operations practical. Wavelets, and hierarchical representations generally, have held promise to facilitate a variety of design tasks in a unified way by approximating results very precisely, thus avoiding a proliferation of undergirding mathematical representations. However, traditional wavelet decomposition is defined from fine to coarse resolution, thus limiting its efficiency for highly precise surface manipulation when attempting to create new non-local editing methods.

Our key contribution is the *progressive wavelet decomposition* algorithm, a general-purpose coarse-to-fine method for hierarchical fitting, based in this paper on an underlying multiresolution representation called *dyadic splines*. The algorithm requests input via a generic *interval query* mechanism, allowing a wide variety of non-local operations to be quickly implemented. The algorithm performs work proportionate to the tiny compressed output size, rather than to some arbitrarily high resolution that would otherwise be required, thus increasing performance by several orders of magnitude.

We describe several design operations that are made tractable because of the progressive decomposition. *Free-form pasting* is a generalization of the traditional control-mesh edit, but for which the shape of the change is completely general and where the shape can be placed using a free-form deformation within the surface domain. Smoothing and roughening operations are enhanced so that an arbitrary loop in the domain specifies the area of effect. Finally, the sculpting effect of moving a tool shape along a path is simulated.

1 Introduction

The process of designing geometric shapes via computation is a critical activity for the making of films, computer games, automobiles and many other ends. Underpinning this design activity are mathematical representations and associated algorithms that facilitate a wide variety of manipulations of shape, such as creating overall proportions, placing details, then deforming the shape or otherwise modeling various quasi-physical manipulations. Unfortunately, no single mathematical representation is known that will provide exact analytic results to all surface operations of interest. Rather than introduce more and more specialized mathematics, a recent trend has been to support many operations in a single, unified representation using approximation theory and hierarchical algorithms [5,6,14,20,24].

Wavelets have been used in surface modeling by Gortler and Cohen [13], who have introduced methods based upon an "oracle" which drives their adaptive refinement. Other non-local design techniques have been proposed by Biermann *et al.* [3] and by Litke *et al.* [21], who have extended this work to subdivision surfaces.

In this paper, we introduce a multiresolution framework that allows coarse-to-fine (i.e. *progressive*) computation of a broad set of non-local shape manipulations. This work formalizes and extend the hierarchical B-splines of Forsey and Bartels [11,12], creating new applications of this method.

The key technique we introduce is the *progressive wavelet decomposition*, whereby the usual fine-to-coarse filtering and truncation is replaced by coarse-to-fine selective refinement. This switch in orientation is generally not possible unless the input data are represented and operations are evaluated in a generic hierarchical fashion, which we term *interval queries*. The abstract input interface to the progressive wavelet decomposition is therefore in the form of an interval query oracle, which the transform calls in response to selective refinement requests on the operation output. The interval query mechanism is inspired by the methods of interval analysis [23], and the research into modeling systems built on those concepts [17,27]. A simple, hierarchical parametric representation, *dyadic splines* [8], is used at the lowest level. A dyadic spline is defined by alternately performing B-spline refinement and adding displacement vectors. The coarse-to-fine processing proceeds in the following phases:

1. Split a leaf of the domain-interval bintree in two, and put the (so far uncomputed) wavelet coefficients overlapping these intervals at that scale onto to the active coefficient list.
2. Invoke the interval-query oracle to the target function, which provides a local Bézier patch estimate and error bound. Do this on all the domain intervals that the newly-active wavelet coefficients depend on.
3. In this neighborhood, compute the estimated values and associated error bounds of the scaling function coefficients, dyadic spline displacements, and wavelet coefficients using the appropriate local weighted-average filters.
4. Propagate improved values up to coarser resolutions if warranted, using the local wavelet decomposition filters.

The split request can be made in any order that an application chooses. A good generic ordering of these requests involves placing the domain bintree leaves on a priority queue ordered by the size of the error bounds in the neighborhood. These phases are repeated over and over until a desired accuracy is achieved or a desired time limit is reached.

We evaluate our approach with respect to six criteria:

1. **Output-sensitive computation:** Our progressive decomposition algorithm performs work proportionate to the compressed (approximated) output size. This is similar to the best algorithms in more specialized settings such as view-dependent optimization [7,15], multiresolution surface editing [11,31], and multiresolution painting [2], yet provides a kind of generic "plug in" architecture that eases the addition of new manipulation operations.
2. **Guaranteed error bounds:** The formulation of our transform not only is guaranteed to converge, but provides strict error bounds at every step in the progressive sequence.
3. **Fixed memory footprint:** We provide a caching system for the interval queries that allows the transform to restrict the working memory footprint to a tiny subset of the total data accessed, traversed, evaluated or output.
4. **Rate-distortion curves:** Our coarse-to-fine processing produces accuracies comparable to traditional fine-to-coarse methods at higher refinement, but suffers somewhat at coarser resolutions because the selective refinement and local approximations are based on "fuzzy" knowledge of the underlying function. In a sense this is the price that must be paid to get progressive computation, but it does not appear affect overall convergence rates.
5. **Selective refinement:** the algorithm allows applications that know where and in what order they want detail in a function domain. Interestingly, this includes feeding the output of the progressive transform into other interval-query oracles and progressive transforms, leading to a closed system for progressive computation.
6. **High-level design tools:** We devised surface design applications that are interesting in their own right but make a large point: they show the possibility of quasi-physical operations that more closely match the intuition gained from non-digital model building, as opposed to the tedium of "pulling on the control net" by hand. In a sense this follows in the footsteps of the development of Computational Solid Geometry (CSG) [25], free-form deformations [26], and hyperpatch modeling [16].

2 Dyadic Spline Representation

This section will give a brief review of the dyadic spline representation, giving its formulation and the properties most critical the progressive decomposition algorithm. Complete details are available in [8].

The general idea is depicted in Figure 1. An initial coarse grid of control points is alternately split and perturbed until some limit function is produced. The common uniform B-spline weighted averaging is used, and the perturbations are simple vector additions. The set of functions represented in this way is dense in L_p, meaning that all functions of interest in practical situations can be accurately converted to a dyadic spline.

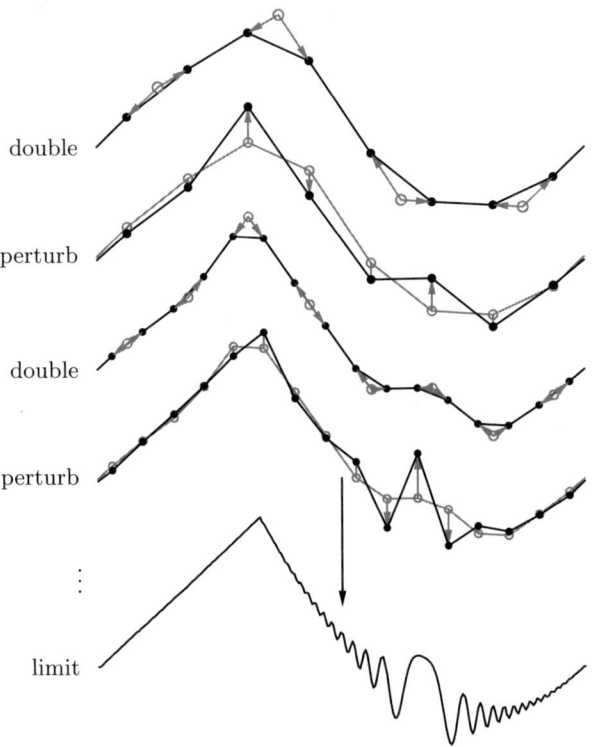

Fig. 1. A dyadic spline is the limit of a sequence of doubling (B-spline refinement) and perturbing (hierarchical displacement) operations. A broad class of functions can be stored this way, but more importantly this view of a function facilitates a general form of progressive evaluation and computation.

Bintrees are based on the dyadic rational numbers

$$\{i/2^\ell \mid i, \ell \in \mathcal{Z}\}$$

where a hierarchy of one-dimensional intervals is indexed by *level* ℓ and *position* i as

$$I_{\ell,i} = \left[i/2^\ell, (i+1)/2^\ell\right)$$

For higher dimensions, the hierarchy of one-dimensional intervals becomes a hierarchy of two- or three-dimensional intervals by splitting intervals in half along one axis at a time. These intervals have a level ℓ, current axis a, and m indices $i_1, \ldots, i_m$:

$$I_{\ell,a,i_1,\ldots,i_m} = I_{\ell+1,i_1} \times \cdots \times I_{\ell+1,i_{a-1}} \times I_{\ell,i_a} \times \cdots \times I_{\ell,i_m}$$

This hierarchy is important since it forms the fundamental spatial structure that all the various weighted averaging schemes use. *Displacements* and *range positions* associated with $I_{\ell,i}$ will be denoted by $D_{\ell,i}$ and $P_{\ell,i}$ respectively. B-spline subdivision can be expressed by a weighted-averaging formula

$$P_{\ell,2i} = \sum_j \alpha_{n,j} P_{\ell-1,i+j}$$
$$P_{\ell,2i+1} = \sum_j \beta_{n,j} P_{\ell-1,i+j}$$

where the $\alpha_{n,j}$ and $\beta_{n,j}$ are weights derived from the dyadic rationals (see [7]). This can be extended to include the displacements by the recurrence

$$P_{\ell,2i} = \sum_j \alpha_{n,j} P_{\ell-1,i+j} + D_{\ell,2i}$$
$$P_{\ell,2i+1} = \sum_j \beta_{n,j} P_{\ell-1,i+j} + D_{\ell,2i+1}$$

Note that in this simple form (without wavelets), the dyadic spline is defined by the base control mesh P_0 and displacements D_ℓ for $\ell = 1, \cdots, \infty$.

The 1-D formulation is extended to m dimensions by utilizing a tensor-product. Here, the one-dimensional filtering is applied along each of the axes.

A formulation of dyadic splines that is most useful in this work involves the specification of four linear operators (filters):

S = subdivide to obtain the next finer level
F = fit points to the next finer level
C = compact the displacements
E = expand displacements

These four filters are defined by the following fundamental relationship.

$$\begin{bmatrix} \mathbf{F} \\ \mathbf{C} \end{bmatrix} [\mathbf{S}|\mathbf{E}] = \begin{bmatrix} \mathbf{I} & 0 \\ 0 & \mathbf{I} \end{bmatrix}$$

In other words, the operation of fitting and compacting the differences from prediction should be the inverse of the operation of subdividing and expanding the compacted differences.

The subdivision operator **S** is defined by the dyadic spline recurrence

$$P_\ell = \mathbf{S} P_{\ell-1} + D_\ell$$

The subdivision filter, combined with the fit, compaction and expansion filters, form the usual wavelet decomposition bank depicted in Figure 2. (Note that the **C** operator in effect eliminates the factor of two redundancy in the displacement representation of a function.) The $P_{\ell,i}$ values are *scaling function coefficients*, and the compacted displacements $Q_{\ell,i}$ are *wavelet coefficients*.

The fit operator approximates the ideal least-squares fit operator

$$\mathbf{F}^\infty = (\mathbf{S}^T \mathbf{S})^{-1} \mathbf{S}^T$$

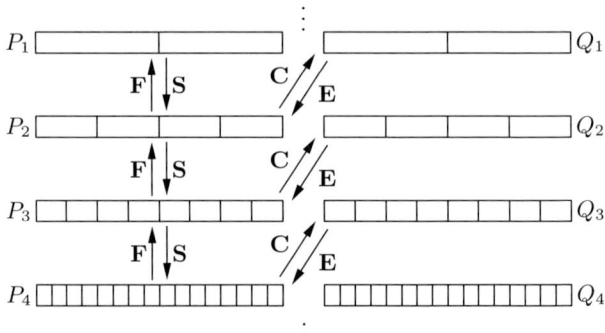

Fig. 2. The dyadic spline wavelet filter bank, showing the data flow dependencies and scale relationships of the four filters **F** (fit), **S** (subdivide), **C** (compact) and **E** (expand).

but with finite support. By setting the central (near diagonal) elements of **F** to $\mathbf{F}^\infty$, and leaving an appropriate number as available degrees of freedom, this inverse property can be maintained by solving a tiny linear system (see [8] or [28]).

The filters are applied in various orders depending on the operation desired. For traditional wavelet decomposition, one assumes some fine-level P_ℓ is given, and the computation proceeds as

$$\cdots$$
$$P_3 = \mathbf{F}P_4 \quad Q_3 = \mathbf{C}P_4$$
$$P_2 = \mathbf{F}P_3 \quad Q_2 = \mathbf{C}P_3$$
$$\cdots$$

For synthesis, this is reversed:

$$\cdots$$
$$P_3 = \mathbf{S}P_2 + \mathbf{E}Q_3$$
$$P_4 = \mathbf{S}P_3 + \mathbf{E}Q_3$$
$$\cdots$$

To convert from the simple dyadic spline representation (base-mesh plus displacements) to wavelets, while at the same time keeping a consistent and optimized version of the displacements, the following is used:

$$\cdots$$
$$D_3 \mathrel{+}= \mathbf{F}D_4 \quad Q_3 = \mathbf{C}D_4 \quad D_4 = \mathbf{E}Q_3$$
$$D_2 \mathrel{+}= \mathbf{F}D_3 \quad Q_2 = \mathbf{C}D_3 \quad D_3 = \mathbf{E}Q_2$$
$$\cdots$$

This level of redundancy is useful for the formulation and implementation of surface design operations. At the end of this process, only the base mesh and wavelet coefficients are stored to disk or sent over the network.

3 Progressive Wavelet Decomposition

We will assume that the ideal target function (the result of an editing operation, for example), is denoted $g(t)$, and that we have available an oracle that will return a local Bézier-curve estimate for $t \in I_{\ell,i}$ of $\tilde{g}_{\ell,i}(t) = \sum_j G_{\ell,i,j} B_{\ell,i,j}(t)$, where the $G_{\ell,i,j}$ are control points and $B_{\ell,i,j}(t)$ are the Bernstein basis functions of some desired polynomial degree [9]. In addition to the local polynomial, we also need an error estimate $E_{\ell,i}$ such that $g(t) \in [\tilde{g}_{\ell,i}(t) - E_{\ell,i}, \tilde{g}_{\ell,i}(t) + E_{\ell,i}]$ for $t \in I_{\ell,i}$.

Suppose in a progressive decomposition that we desire to have estimates for $P_{\ell,i}$ for some intermediate level of resolution ℓ, for example at the leaves of the current bintree refinement. Given the filters **F**, **C**, **E**, and **S** we can then compute all the positions $P_{\ell',i}$, displacements $D_{\ell',i}$ and wavelet coefficients $Q_{\ell',i}$ for levels $\ell' < \ell$ (values at or coarser than ℓ). So our problem is reduced to simulating what would happen if we were to perform the wavelet filtering on the infinitely resolved Bézier curves. Since the filtering process, even in this limit, is fundamentally just a linear operation of weighted averaging, we can separately precompute the infinite-limit wavelet decomposition of the Bernstein basis functions, and at runtime simply look these results up to directly compute estimated positions $\tilde{P}_{\ell,i}$ as a weighted average of the nearby estimate control points $G_{\ell,i',j}$:

$$\tilde{P}_{\ell,i} = \sum_{j,s} \beta_{s,j} G_{\ell,i+s,j}$$

for the precomputed Bernstein-basis limit fits $\beta_{s,j}$. Note that due to scale invariance these weights depend only on relative position s and basis function index j, not on the level ℓ. The estimate fit kernels $\beta_{s,j}$ are shown in Figure 3. Note that the nonzero weights are in a narrow local neighborhood.

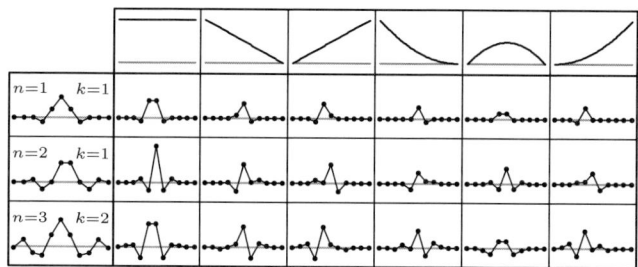

Fig. 3. Estimate-fit kernels $\beta_{s,j}$ for dyadic splines of degree $n = 1, \cdots, 3$ (with respective filter width parameter $k = 1, 1, 2$), and for Bernstein basis functions for degrees $0, 1, 2$.

The various positions, displacements and wavelet coefficients, $\tilde{P}_{\ell',i}$, $\tilde{D}_{\ell',i}$, and $\tilde{Q}_{\ell',i}$, can now be computed using the wavelet filters. It is straightforward to obtain strict error bounds on these values since error bounds on the inputs are known and the entire process is simple linear weighted averaging [23].

The wavelet decomposition algorithm proceeds to use this machinery to create a progressive sequence of increasingly accurate approximations to the target function $g(t)$. A pictorial example is shown in Figure 4. The target function in this case is a sequence of "bumps within bumps" defined as the sum of transcendental functions, specifically, translated and dilated versions of the "mother bump"

$$b(t) = \begin{cases} e^{-\tan^2(\frac{\pi}{2}t)} & \text{if } t \in (-1,1) \\ 0 & \text{otherwise} \end{cases}$$

These bump functions have closed forms for their derivatives of various degrees, and known monotonic regions, so it is straightforward to create local estimates with bounds.

The rate-distortion curve for the example is plotted in Figure 5 (in black), compared to the usual greedy algorithm that uses fine-to-coarse processing to throw away wavelet coefficients that contribute least to the error. Note especially that the accuracies are relatively worse for the progressive transform at low numbers of coefficients (due to it's fuzzy awareness of the target function), yet it "catches up" to the quality and convergence rates of the traditional greedy algorithm at higher counts.

The extension to the tensor-product setting is straightforward, as all the filtering operations just described can be performed on one axis at a time just as with subdivision. Whereas a univariate bintree decomposition $I_{\ell,i}$ was indexed by level ℓ and index i, the multivariate bintree requires an additional axis counter $a \in \{1, \ldots, m\}$ and multiple indices $i_1, \ldots, i_m$. To simplify the appearance of the multivariate bintree intervals

$$I_{\ell,a,i_1,\ldots,i_n} = I_{\ell+1,i_1} \times \cdots \times I_{\ell+1,i_{a-1}} \times I_{\ell,i_a} \times \cdots \times I_{\ell,i_n}$$

a shorthand of

$$I_{\mathcal{L},\mathbf{i}} = I_{\ell,a,i_1,\ldots,i_n}$$

will be used, where $\mathcal{L} = (\ell, a)$ and $\mathbf{i} = (i_1, \ldots, i_m)$. The composition $\mathcal{L} = (\ell, a)$ will be referred to as a *layer*, and is analogous to the level in the univariate case. Note that the intervals $I_{\mathcal{L},\mathbf{i}}$ still form a binary tree. The displacements are now denoted $D_{\mathcal{L},\mathbf{i}}$, and the positions $P_{\mathcal{L},\mathbf{i}}$.

An example progression for a 2-D domain with a few conical bumps is shown in Figure 6. Note how the progressive decomposition naturally adapts to the sharp features of this target function.

In the remainder of this paper we will describe four high-level surface design operations: smoothing, roughening, free-form pasting, and scraping. Smoothing has been done by several research groups, with Kobbelt *et al.* [19] the most recent. Both Kobbelt *et al.* and Ying *et al.* [30] have described roughening operations. Free-form pasting has been described previously by

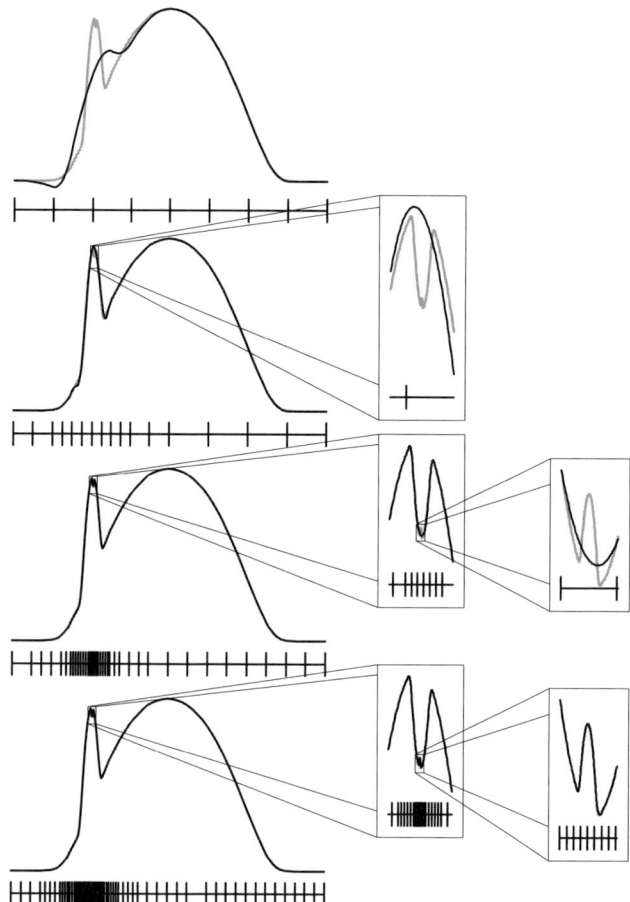

Fig. 4. A target function with extremely fine-scale features (shown as a sequence of insets) is progressively computed into a wavelet decomposition under the max error norm (L_∞). The progressive transform selectively refines where errors are not guaranteed to be low, leading to a natural adaptation of the refinement around the fine features. In this example, the transform is computed over a thousand times faster than if a sufficiently fine uniform sampling of the target function were used as the starting point.

Forsey and Bartels [11], while scraping was described by Khodakovsky and Schröder [18]. Our contribution is to use a common progressive wavelet transform methodology in the creation of these operations.

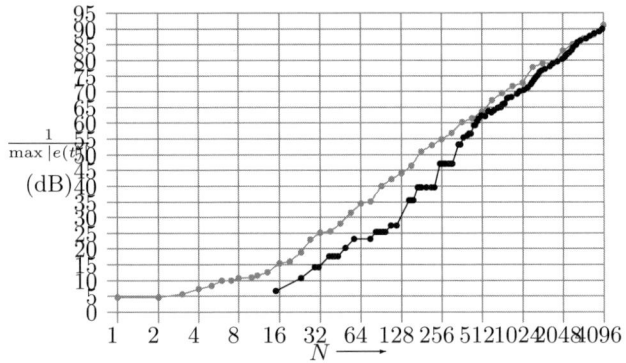

Fig. 5. Coarse to fine progression of our new transform (black) is relatively low accuracy compared to the traditional greedy algorithm at low coefficient counts, yet is nearly identical at higher counts.

4 Smoothing and Roughening Within a General Loop

This section will describe the implementation of operations within a restricted domain area defined by a closed loop of Bézier curves. The first notion is that of performing global smoothing, which is defined for smoothing parameter ℓ_s as

$$\bar{D}_{\mathcal{L},\mathbf{i}} = \begin{cases} D_{\mathcal{L},\mathbf{i}} & \text{if } \ell < \ell_s \\ (\ell_s - (\ell - 1))D_{\mathcal{L},\mathbf{i}} & \text{if } \ell - 1 < \ell_s \leq \ell \\ 0 & \text{otherwise } (\ell_s \leq \ell - 1) \end{cases}$$

This is similar to the smoothing defined in [10], but here extended to higher dimensions via tensor products.

For local smoothing, a generalization of the smoothing segment is needed. For this, a smoothing area is defined using the concept of a trim curve [4], previously used in the methods for trimmed surface patches. A trim curve $c(t)$ is a continuous, periodic mapping from $t \in [0, 1]$ to the surface domain $(u, v) \in \Re^2$. This curve encloses a domain area that will serve as the locality to be smoothed.

The smoothing operation blends between the original displacements $D_{\mathcal{L},\mathbf{i}}$ and the smoothed ones $\bar{D}_{\mathcal{L},\mathbf{i}}$. The blend factor q is defined as the fraction of $D_{\mathcal{L},\mathbf{i}}$'s interval of influence I_D that overlaps the area enclosed by $c(t)$. The computation of this overlap is discussed below. The blend factor q is then applied as:

$$\check{D}_{\mathcal{L},\mathbf{i}} = (1-q)D_{\mathcal{L},\mathbf{i}} + q\bar{D}_{\mathcal{L},\mathbf{i}}$$

Some results of local smoothing are shown in Figure 7.

It is nontrivial to compute the area of overlap of an interval I and the area enclosed by a trim curve $c(t)$. However, the well-known Warnock algorithm for polygon visibility [29] can be adapted to this problem. Although the concern here is only for determining the area of a single "polygon" $c(t)$ within a "view

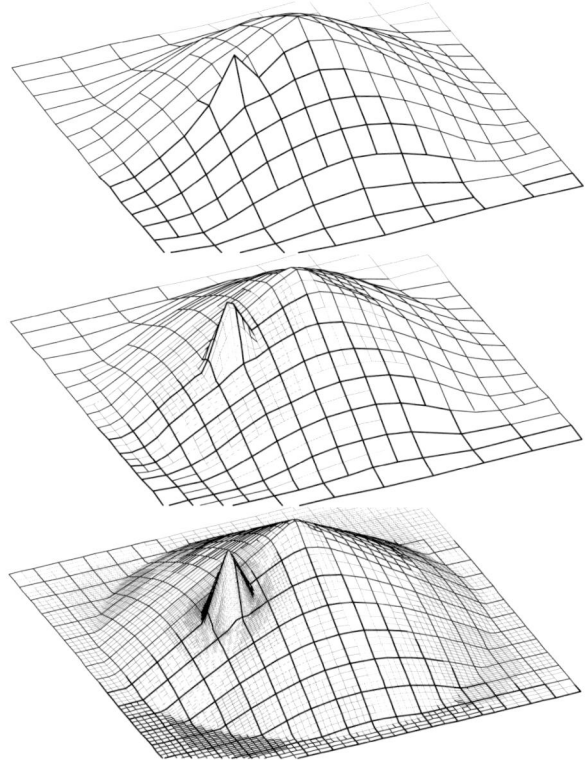

Fig. 6. A demonstration of the progressive wavelet decomposition of a function as a height with 2-D domain, consisting of two overlayed conical protrusions. The wavelet coefficients are built in a coarse-to-fine sequence, shown from top to bottom.

window" I, the Warnock algorithm has a useful property of dividing I into smaller intervals until each interval either misses $c(t)$, $c(t)$ crosses the interval in a simple way, or the interval is small. Winding number computations are used in this algorithm to determine which intervals (or which parts of crossed intervals) are inside the trim curve. To apply the polygon techniques to a curve, the curve must be approximated by a polygon. For the purposes of interactive editing, it is sufficient to ensure that the approximation error is within a small fraction of the width of the interval I. If $c(t)$ is a dyadic spline, or is in B-spline form, standard subdivision techniques can be applied to accomplish this [9]. In the implementation used here, "simple" crossings consist of two or fewer polygon edges, and a bintree decomposition of I is used. An example Warnock-style decomposition of a trim area is shown in Figure 8.

For roughening, we added random displacements in the manner of midpoint-displacement fractals [22]. The *direction* of the displacements is taken to be in the normal direction of a smoothed version of the surface, similar to what

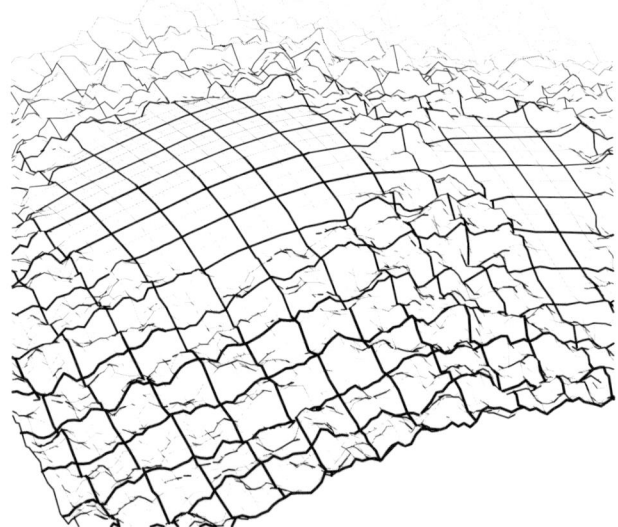

Fig. 7. A formerly rough surface is smoothed within the area enclosed by a trim curve.

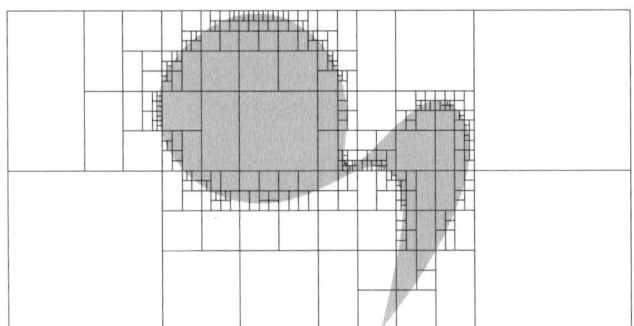

Fig. 8. Decomposition of the trim-curve domain area using a variant of the Warnock algorithm.

was done in the curve case in [10]. The offset frame for a surface displacement $D_{\mathcal{L},\mathbf{i}}$ is defined as the two unit tangents and unit normal at the point of maximum influence, taken with respect to the smoothed version of the surface.

Let $f(u,v)$ be the surface and $\bar{f}(u,v)$ be a smoothed version of the surface for smoothing parameter ℓ_s. Then the offset coordinate frame applied to offset displacement $\hat{D}_{\mathcal{L},\mathbf{i}}$ is then defined as

$$A_{\mathcal{L},\mathbf{i}} = [\mathbf{p}\ \mathbf{q}\ \mathbf{r}]$$

where

$$\mathbf{p} = \frac{\bar{f}_u(u_m, v_m)}{\|\bar{f}_u(u_m, v_m)\|}$$

$$\mathbf{q} = \frac{\bar{f}_v(u_m, v_m)}{\|\bar{f}_v(u_m, v_m)\|}$$

$$\mathbf{r} = \frac{\mathbf{p} \times \mathbf{q}}{\|\mathbf{p} \times \mathbf{q}\|}$$

and (u_m, v_m) is the domain point of maximum influence. Now the application of $A_{\mathcal{L},\mathbf{i}}$ to $\hat{D}_{\mathcal{L},\mathbf{i}}$ gives the standard displacement as

$$D_{\mathcal{L},\mathbf{i}} = A_{\mathcal{L},\mathbf{i}} \hat{D}_{\mathcal{L},\mathbf{i}}$$

This gives the effect that details track the position and orientation of the smooth underlying surface.

Global roughening is produced by adding random vectors in the local smoothed-normal direction to the fine-resolution wavelet coefficients. The localization of the roughening effect is accomplished in the same manner as local surface smoothing. An example of local roughening is shown in Figure 9.

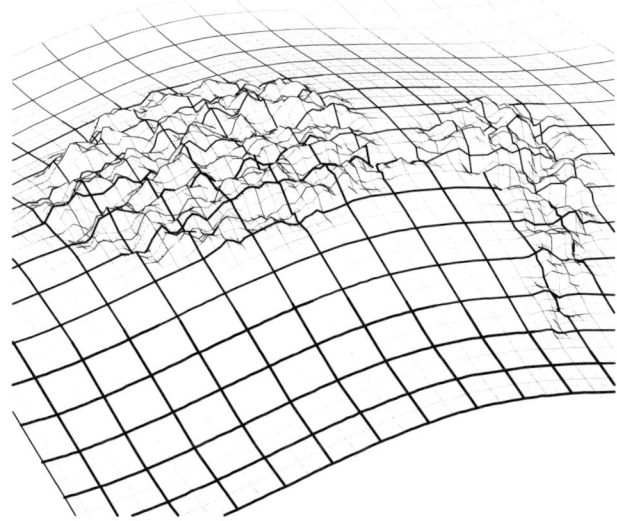

Fig. 9. Random fractal offsets are made in the local smoothed normal direction. The magnitude of the offset is modulated by the Warnock in/out overlap fractions.

5 Free-form Pasting

In this section we make a more flexible version of the *Pasting* operation introduced in [1]. The idea is to allow an arbitrary template shape to be offset from the input surface, where the placement of the template is given by a general domain-to-domain mapping. This placement strategy is akin to a free-form deformation [26] in the 2-D case. We similarly choose a bicubic Bézier patch to formulate the 2-D to 2-D mapping. The advantage over earlier pasting formulations is that we have somewhat more general template placement and shape control, but most importantly our results are computed in progressive order to any accuracy for the precise, continuous offset definition.

Let the input surface be $f(u,v)$. Let $g(s,t)$ be a scalar-valued template function and let $h(s,t)$ be an invertible domain-positioning function into (u,v). The basic template edit effect is defined as

$$\hat{f}(u,v) = f(u,v) + cG(u,v)$$

where c is a control vector for the *generalized basis function*

$$G(u,v) = g(h^{-1}(u,v))$$

Note that $h^{-1}(u,v)$ is only defined for $(u,v) \in h(J)$ where J is the interval domain of $h(s,t)$. When appropriate, assume that $G(u,v)$ is zero when $(u,v) \notin h(J)$. Also note that the control vector c may be derived from another control vector $\hat{c}$ that is defined in a local offset frame A, similar to the offset-frame displacements for roughening.

The result of a template edit, $\hat{f}(u,v)$, is approximated using the progressive transform. Local estimates are formed using interval-analytic techniques. This approximation process is described shortly. An example result of template editing is depicted in Figure 10.

The template edit result $\hat{f}(u,v)$ is approximated by using the progressive transform to approximate the generalized basis function $G(u,v) = g(h^{-1}(u,v))$. This approximation, denoted $\tilde{G}(u,v)$, scales a control vector c before added it to $f(u,v)$. The approximate template-edit result is $\tilde{f}(u,v) = f(u,v) + c\tilde{G}(u,v)$. Applying the progressive decomposition algorithm to approximate $G(u,v)$ reduces to finding local estimates. The remainder of this section will discuss the computation of suitable local estimates.

This discussion will use first-order interval estimates throughout. To develop an estimate for $g(h^{-1}(u,v))$, an estimate will first be constructed for $h^{-1}(u,v)$ based on an estimate of $h(s,t)$. This will be composed with an estimate of $g(s,t)$ to give the desired estimate of $g(h^{-1}(u,v))$.

Let $h(s,t)$ have the first-order interval estimate

$$\tilde{h}(s,t) = H \begin{bmatrix} s \\ t \end{bmatrix} + \begin{bmatrix} u_0 \\ v_0 \end{bmatrix} + \delta$$

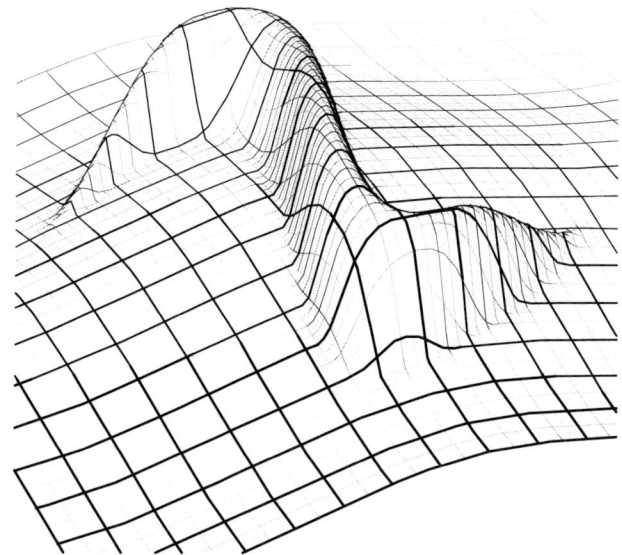

Fig. 10. A simple extruded hill shape is offset from the input surface in the continuous, smoothed local normal directions. The placement of the offset template is specified with an S-shaped bicubic Bézier patch controlled in the input-surface domain.

where H is an invertible 2×2 matrix, and δ is an interval in (u,v) space. Assume that this estimate holds for $h^{-1}(I)$, where I is an interval in (u,v) space. An interval estimate for $h^{-1}(u,v)$ is

$$\tilde{h}^{-1}(u,v) = H^{-1}\begin{bmatrix} u \\ v \end{bmatrix} + \begin{bmatrix} s_0 \\ t_0 \end{bmatrix} + \epsilon$$

where

$$\begin{bmatrix} s_0 \\ t_0 \end{bmatrix} = -H^{-1}\begin{bmatrix} u_0 \\ v_0 \end{bmatrix}$$

and where the error interval ϵ is chosen so that

$$\epsilon \supset -H^{-1}\delta$$

This estimate holds for $(u,v) \in I$. The error ϵ may be computed as the bounding box of the image of the four corners of δ under the transform $-H^{-1}$.

Now suppose $g(s,t)$ has the estimate

$$\tilde{g}(s,t) = [g_s \; g_t]\begin{bmatrix} s \\ t \end{bmatrix} + g_0 + \gamma$$

for error interval γ, and suppose this holds for $(s,t) \in h^{-1}(I)$. Then an estimate for $g(h^{-1}(u,v))$ over I is

$$[g_s \ g_t] \left(H^{-1} \begin{bmatrix} u \\ v \end{bmatrix} + \begin{bmatrix} s_0 \\ t_0 \end{bmatrix} + \epsilon \right) + g_0 + \gamma$$

For a surface $f(u,v)$, let $\bar{f}(u,v)$ be the smoothed version of the surface for smoothing parameter ℓ_{s1}. The tangents of this smoothed surface are normalized to give

$$\hat{\mathbf{p}}(u,v) = \frac{\bar{f}_u(u,v)}{\|\bar{f}_u(u,v)\|}$$

$$\hat{\mathbf{q}}(u,v) = \frac{\bar{f}_v(u,v)}{\|\bar{f}_v(u,v)\|}$$

These normalized tangents are approximated as dyadic splines (using the progressive transform) to allow the second stage of smoothing. Let $\tilde{\mathbf{p}}(u,v)$ and $\tilde{\mathbf{q}}(u,v)$ be the approximations to the normalized tangents, and $\bar{\mathbf{p}}(u,v)$ and $\bar{\mathbf{q}}(u,v)$ be the smoothed versions of these for smoothing parameter ℓ_{s2}. A final normalization and cross product gives the axis vectors of the desired offset frame

$$A(u,v) = [\mathbf{p}(u,v) \ \mathbf{q}(u,v) \ \mathbf{r}(u,v)]$$

where

$$\mathbf{p}(u,v) = \frac{\bar{\mathbf{p}}(u,v)}{\|\bar{\mathbf{p}}(u,v)\|}$$

$$\mathbf{q}(u,v) = \frac{\bar{\mathbf{q}}(u,v)}{\|\bar{\mathbf{q}}(u,v)\|}$$

$$\mathbf{r}(u,v) = \frac{\mathbf{p}(u,v) \times \mathbf{q}(u,v)}{\|\mathbf{p}(u,v) \times \mathbf{q}(u,v)\|}$$

The continuous offset-frame template edit becomes

$$\hat{f}(u,v) = f(u,v) + A^{-1}(u_m, v_m) A(u,v) c G(u,v)$$

where (u_m, v_m) is the domain point of maximum influence for $G(u,v)$. The transform $A^{-1}(u_m, v_m)$ is optional, but has the desirable effect that pulling the control vector c in (x,y,z) space causes the point $\hat{f}(u,v)$ to move in the same direction, as would happen when pulling the control vectors of conventional basis functions.

6 Precision Sculpting with Tool and Path

This section provides the interval-query mechanism for precisely sculpting a surface by moving a tool shape along a path in the surface domain.

A single surface "scrape" is defined by specifying tool depth in an offset-frame normal direction for each (u, v), where depth zero occurs at a smoothed version of the surface. The offset frame tangent and normal directions $\mathbf{p}(u, v)$, $\mathbf{q}(u, v)$ and $\mathbf{r}(u, v)$ are obtained from $A(u, v)$ as in the smoothing/roughening operations. The result of scraping is defined by the maximum of the tool depth and the depth of the original surface with respect to the smooth surface.

Let $f(u, v)$ be a given surface and $\bar{f}(u, v)$ be the smoothed surface for some smoothing parameter ℓ_s. Let $D_T(u, v)$ be the given tool depth function, and define the surface depth as

$$D_S(u, v) = -\mathbf{r}(u, v) \cdot (f(u, v) - \bar{f}(u, v))$$

The result depth will be

$$D(u, v) = \max\{D_T(u, v), D_S(u, v)\}$$

Since the surface position $f(u, v)$ does not generally reside on the line through $\bar{f}(u, v)$ in the normal direction $\mathbf{r}(u, v)$, some means of blending from the surface to the scrape boundary is needed. A scrape boundary occurs when $D_T(u, v) = D_S(u, v)$. A simple blending method is to linearly move the surface towards the normal line as $D_S(u, v) - D_T(u, v)$ goes from positive to zero. The blend factor is defined as

$$q = \begin{cases} 0 & \text{if } D_S(u,v) - D_T(u,v) < 0 \\ \frac{D_S(u,v) - D_T(u,v)}{H} & \text{if } 0 \leq D_S(u,v) - D_T(u,v) < H \\ 1 & \text{if } H \leq D_S(u,v) - D_T(u,v) \end{cases}$$

where H is a user-supplied blend distance. The blend factor is applied to define the scrape result as

$$\hat{f}(u, v) = \bar{f}(u, v) + D(u, v)\mathbf{r}(u, v) + \\ q((f(u, v) - \bar{f}(u, v)) \cdot \mathbf{p}(u, v))\mathbf{p}(u, v) + \\ q((f(u, v) - \bar{f}(u, v)) \cdot \mathbf{q}(u, v))\mathbf{q}(u, v)$$

Interval estimates are used so that the progressive transform may capture the scrape result as a dyadic spline. An example of a single scrape is shown in Figure 11.

Superimposing multiple scrapes as a simultaneous operation is performed by letting the tool depth function be defined as the maximum of the individual scrape tool depth functions

$$D_T(u, v) = \max_i D_i(u, v)$$

Otherwise the formulation above remains intact. The result of two simultaneous scrapes is shown in Figure 12.

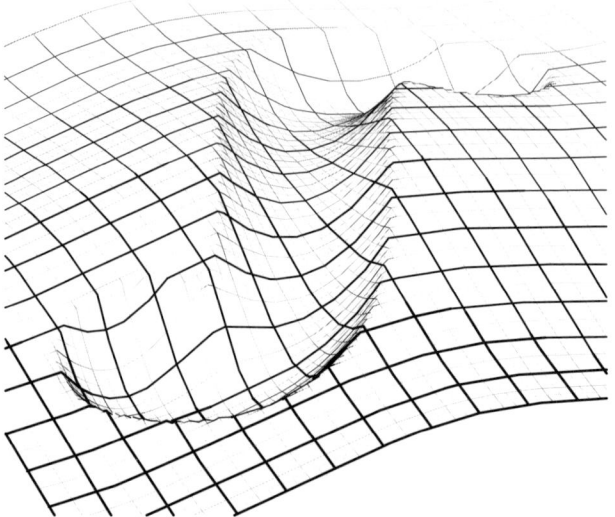

Fig. 11. A single "scrape" of a tool shape along a path.

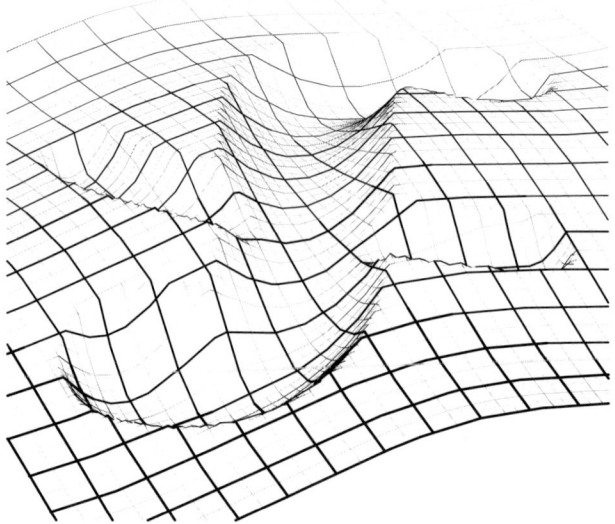

Fig. 12. Two overlayed surface scrapes.

7 Conclusion and Future Work

The main discovery, in reviewing this work, is that (a) it is not obvious how to efficiently perform wavelet compression directly to the results of mathematical surface operations, yet (b) it *is* possible to be efficient when an intermediate interval-query oracle supplies local Bézier estimates. We demonstrated by example that formulating these operations as oracle responses is tractable for a significant number of design modes that might be envisioned. We offer the following thoughts on future challenges and potential applications:

- usefulness for other wavelets
  The progressive decomposition algorithm should be applicable to wavelet representations other than the dyadic splines. Only two parts of the top-down algorithm have some sensitivity to the wavelets chosen: the comparison of the wavelet approximation versus the local estimate, and the incremental, sparse updates to the wavelet coefficients as more active wavelets are added during processing. It seems likely that these issues can be solved for many wavelet schemes, including those defined on subdivision surfaces and volumes.
- tuning for various norms
  The choices of which domain intervals to split and which intervals are "done" should be made with the desired norm in mind. This seems to be fairly straightforward, but has not been investigated so far.
- optimization of rate-distortion curves
  A major difficulty is trying to approach the optimal rate-distortion curves, especially early in the progressive approximation process. This is hard because the local estimates only give fuzzy knowledge of the target function. Perhaps an adaptive, recursive estimation strategy could be devised that would improve this knowledge.
- general techniques for providing local estimates
  In the discussions in this paper, the applications of the progressive decomposition algorithm used *ad hoc* techniques to provide local estimates to target functions. Current investigations are under way to find general, automatic methods for obtaining local estimates for a wide variety of target functions.

Acknowledgements

This work was performed under the auspices of the U.S. Department of Energy by University of California Lawrence Livermore National Laboratory under contract No. W-7405-Eng-48.

References

1. BARGHIEL, C., BARTELS, R., AND FORSEY, D. 1995. Pasting spline surfaces. In *Mathematical Methods for Curves and Surfaces*, Vanderbilt University Press, Nashville, TN, 31–40.

2. BERMAN, D. F., BARTELL, J. T., AND SALESIN, D. H. 1994. Multiresolution painting and compositing. In *Proceedings of SIGGRAPH 94*, ACM SIGGRAPH / ACM Press, Orlando, Florida, Computer Graphics Proceedings, Annual Conference Series, 85–90.
3. BIERMANN, H., MARTIN, I., ZORIN, D., AND BERNARDINI, F. 2001. Sharp features on multiresolution subdivision surfaces. In *Proceedings of the ninth Pacific Conference on Computer Graphics and Applications (PACIFIC GRAPHICS-01)*, IEEE Computer Society, Los Alamitos, CA, B. Werner, Ed., 140–149.
4. CASALE, M. S. 1987. Free-form solid modeling with trimmed surface patches. *IEEE Computer Graphics & Applications 7*, 1 (January), 33–43.
5. CIRAK, F., SCOTT, M. J., ANTONSSON, E. K., ORTIZ, M., AND SCHRÖDER, P. 2002. Integrated modeling, finite-element analysis, and engineering design for thin-shell structures using subdivision. *Computer-Aided Design 34*, 2 (February), 137–148.
6. DEROSE, T. D., KASS, M., AND TRUONG, T. 1998. Subdivision surfaces in character animation. In *Proceedings of SIGGRAPH 98*, ACM SIGGRAPH / Addison Wesley, Orlando, Florida, Computer Graphics Proceedings, Annual Conference Series, 85–94.
7. DUCHAINEAU, M. A., WOLINSKY, M., SIGETI, D. E., MILLER, M. C., ALDRICH, C., AND MINEEV-WEINSTEIN, M. B. 1997. ROAMing terrain: Realtime optimally adapting meshes. *IEEE Visualization '97* (November), 81–88.
8. DUCHAINEAU, M. A. 1996. *Dyadic Splines*. PhD thesis, Dept. of Computer Science, University of California, Davis. http://graphics.cs.ucdavis.edu/ duchaine/dyadic.html.
9. FARIN, G. 1999. *NURBS: From Projective Geometry to Practical Use*. A.K. Peters, Natick MA.
10. FINKELSTEIN, A., AND SALESIN, D. H. 1994. Multiresolution curves. In *Proceedings of SIGGRAPH 94*, ACM SIGGRAPH / ACM Press, Orlando, Florida, Computer Graphics Proceedings, Annual Conference Series, 261–268.
11. FORSEY, D. R., AND BARTELS, R. H. 1988. Hierarchical b-spline refinement. In *Computer Graphics (Proceedings of SIGGRAPH 88)*, vol. 22, 205–212.
12. FORSEY, D. R., AND BARTELS, R. H. 1995. Surface fitting with hierarchical splines. ACM Transactions on Graphics Systems 14, No. 2, 134–161.
13. GORTLER, S. J., AND COHEN, M. F. 1995. Hierarchical and variational geometric modeling with wavelets. In *1995 Symposium on Interactive 3D Graphics*, P. Hanrahan and J. Winget, Eds., ACM SIGGRAPH, 35–42. ISBN 0-89791-736-7.
14. GUSKOV, I., VIDIMCE, K., SWELDENS, W., AND SCHRÖDER, P. 2000. Normal meshes. *Proceedings of SIGGRAPH 2000* (July), 95–102.
15. HOPPE, H. 1997. View-dependent refinement of progressive meshes. In *Proceedings of SIGGRAPH 97*, ACM SIGGRAPH / Addison Wesley, Los Angeles, California, Computer Graphics Proceedings, Annual Conference Series, 189–198.
16. JOY, K. 1991. Utilizing parametric hyperpatch methods for modeling and display of free-form solids. In *SMA '91: Proceedings of the First Symposium on Solid Modeling Foundations and CAD/CAM Applications*, ACM Press / ACM, held June 5-7, 1991 in Austin, Texas, USA., 245–254.
17. KASS, M. 1992. Condor: Constraint-based dataflow. In *Computer Graphics (Proceedings of SIGGRAPH 92)*, vol. 26, 321–330.

18. KHODAKOVSKY, A., AND SCHRÖDER, P. 1999. Fine level feature editing for subdivision surfaces. In *Proceedings of the Fifth Symposium on Solid Modeling and Applications (SM-99)*, ACM Press, New York, W. F. Bronsvoort and D. C. Anderson, Eds., 203–211.
19. KOBBELT, L., CAMPAGNA, S., VORSATZ, J., AND SEIDEL, H.-P. 1998. Interactive multi-resolution modeling on arbitrary meshes. In *SIGGRAPH 98 Conference Proceedings*, Addison Wesley, M. Cohen, Ed., Annual Conference Series, ACM SIGGRAPH, 105–114. ISBN 0-89791-999-8.
20. LEE, A., MORETON, H., AND HOPPE, H. 2000. Displaced subdivision surfaces. In *Proceedings of SIGGRAPH 2000*, ACM Press / ACM SIGGRAPH / Addison Wesley Longman, Computer Graphics Proceedings, Annual Conference Series, 85–94.
21. LITKE, N., LEVIN, A., AND SCHRÖDER, P. 2001. Fitting subdivision surfaces. In *Proceedings Visualization 2001*, T. Ertl, K. Joy, and A. Varshney, Eds., IEEE Computer Society Technical Committee on Visualization and Graphics Executive Committee, 319–324.
22. MILLER, G. S. P. 1986. The definition and rendering of terrain maps. In *Computer Graphics (Proceedings of SIGGRAPH 86)*, vol. 20, 39–48.
23. MOORE, R. E. 1979. *Methods and Applications of Interval Analysis*. SIAM, Philadelphia.
24. PERRY, R. N., AND FRISKEN, S. F. 2001. Kizamu: A system for sculpting digital characters. In *Proceedings of SIGGRAPH 2001*, ACM Press / ACM SIGGRAPH, Computer Graphics Proceedings, Annual Conference Series, 47–56.
25. REQUICHA, A. A. G., AND VOELCKER, H. B. 1982. Solid modeling: a historical summary and contemporary assessment. *IEEE Computer Graphics & Applications 2* (March), 9–22.
26. SEDERBERG, T. W., AND PARRY, S. R. 1986. Free-form deformation of solid geometric models. In *Computer Graphics (Proceedings of SIGGRAPH 86)*, vol. 20, 151–160.
27. SNYDER, J. M., AND KAJIYA, J. T. 1992. Generative modeling: A symbolic system for geometric modeling. In *Computer Graphics (Proceedings of SIGGRAPH 92)*, vol. 26, 369–378.
28. STOLLNITZ, E. J., DEROSE, T. D., AND SALESIN, D. H. 1996. *Wavelets for Computer Graphics: Theory and Applications*. Morgann Kaufmann, San Francisco, CA.
29. WARNOCK, J. E. 1969. A hidden-surface algorithm for computer generated half-tone pictures. Tech. Rep. TR 4–15, NTIS AS-733 671, Computer Science Department, University of Utah.
30. YING, L., HERTZMANN, A., BIERMANN, H., AND ZORIN, D. 2001. Texture and shape synthesis on surfaces. In *Rendering Techniques 2001: 12th Eurographics Workshop on Rendering*, Eurographics, 301–312. ISBN 3-211-83709-4.
31. ZORIN, D., SCHRÖDER, P., AND SWELDENS, W. 1997. Interactive multiresolution mesh editing. In *Proceedings of SIGGRAPH 97*, ACM SIGGRAPH / Addison Wesley, Los Angeles, California, Computer Graphics Proceedings, Annual Conference Series, 259–268.

Access to Surface Properties up to Order Two for Visualization Algorithms

Helwig Hauser[1], Thomas Theußl[2], and Eduard Gröller[2]

[1] VRVis Research Center, http://www.VRVis.at/
[2] Institute of Computer Graphics and Algorithms, Vienna University of Technology, http://www.cg.tuwien.ac.at/home/

Summary. Elaborated visualization techniques which are based on surfaces often are independent from the origin of the surface data. Nevertheless, many of the previously presented visualization methods were developed for a specific type of surface, although principally applicable to generic surfaces. In this paper we discuss a model for a general access to surface properties up to order two, i.e., surface-point locations, normals, and curvature properties, (almost) regardless of the origin of the surface. Surface types and access algorithms are compared and summarized. At the end of this paper we shortly present an implementation of this model.

Key words: visualization, surfaces, surface properties.

1 Introduction

Surfaces are important geometric primitives for 3D visualization [20]. Useful techniques are available to render surfaces of various kind. For instance, scalar data volumes ($\mathbf{R}^3 \rightarrow \mathbf{R}$) from medical applications are represented using iso-surfaces [17, 19]. Three-dimensional vector fields ($\mathbf{R}^3 \rightarrow \mathbf{R}^3$) from flow analysis are visualized by the use of stream surfaces [12, 16, 28].

In the past years, advanced visualization techniques based on surfaces were proposed which use semi-transparency and local curvature properties to enhance the perceptability of surfaces in 3D. Gerstner, for example, demonstrated the use of multiple, semi-transparent iso-surfaces for visualizing very large datasets [6]. Interrante et al. [13–15] show how curvature-based techniques enhance the use of surfaces for the visualization of volumetric data. Surface curvature also plays an important role in surface design [5], surface fairing [11, 26], surface trimming [10], surface evaluation and analysis [2, 25], and surface visualization [4].

In this paper we develop an access model to surface properties up to degree two, (almost) regardless of the origin of the surface. Several algorithms which are necessary to deal with different types of surfaces are discussed. A C++ implementation called SMURF – short for SMart SURFace model – of such an abstract access layer between advanced visualization algorithms and surfaces of various origin (see Fig. 1) is described to demonstrate the ease-of-use of this approach.

One advantage of specifying a generic interface like SMURF is that visualization techniques are easily ported from one application to another. Algorithms like modulating the opacity of the surface according to its curvature properties are not

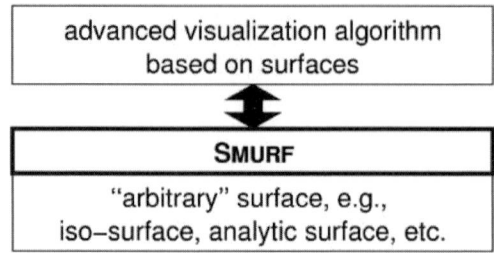

Fig. 1. SMURF is a generic interface between surface-based visualization and surface implementation.

bound to one application, but can be re-used for other surfaces as well. A similar approach in the area of mesh access is described by Rumpf et al. [22].

The remainder of this paper is organized as follows. First we give an overview of several surface types apparent in visualization (Sect. 2). We then discuss the access to surface properties up to degree two in terms of the previously mentioned surface types (Sect. 3). This section includes a review of algorithms which are necessary for accessing different surface types. An implementation of this model (SMURF) is presented in Sect. 4. Some results of SMURF applications are discussed in Sect. 5.

2 Surface Types

In the following we describe seven types of surfaces which are often used in computer graphics and visualization. Surfaces can be defined implicitly, for example, as an iso-surface of scalar volume data, or explicitly, i.e., analytically. Using SMURF the following surface types can be dealt with:

Implicitly defined iso-surfaces for discrete scalar data volumes (in the following case 1) – scalar data values $f_{\text{samp}}(\mathbf{x}_i)$ are given at certain discrete locations $\mathbf{x}_i$ in 3D, e.g., on a regular grid or as scattered data. A certain interpolant $f(\mathbf{x})$ of these values f_{samp} is considered to implicitly define an iso-surface $\mathbf{s}$ (corresponding to a certain iso-value $f_{\mathbf{s}}$): $\mathbf{s} = \{\,\mathbf{x}\,|\,f(\mathbf{x}) = f_{\mathbf{s}}\,\}$.

Implicitly defined iso-surfaces for analytic scalar data in 3D (case 2) – a scalar function $f(\mathbf{x})$ is given (as a "black box"), which can be evaluated at arbitrary locations $\mathbf{x}$ in 3D. A scalar continuum over 3D is assumed to be the application of the function to all points. A certain iso-value $f_{\mathbf{s}}$ specifies the iso-surface $\mathbf{s} = \{\,\mathbf{x}\,|\,f(\mathbf{x}) = f_{\mathbf{s}}\,\}$.

Implicitly defined stream surfaces for discrete vector fields (case 3) – vectorial data $\mathbf{v}_{\text{samp}}(\mathbf{x}_i)$ is given at certain discrete locations $\mathbf{x}_i$, for example, on a curvilinear grid. For a specific initial line segment or curve $\mathbf{s}_0(u)$ the corresponding stream surface $\mathbf{s}(u,t)$ is implicitly defined as $\mathbf{s}_0(u) + \mathbb{N}_0^t\,\mathbf{v}(\mathbf{s}(u,\tau))\,\mathrm{d}\tau$ where $\mathbf{v}(\mathbf{x})$ is an interpolant of the discrete values $\mathbf{v}_{\text{samp}}(\mathbf{x}_i)$.

Implicitly defined stream surfaces for analytically specified dynamical systems (case 4) – a vectorial function $\mathbf{v}(\mathbf{x})$ is given (as a "black box") to be evaluated at arbitrary locations $\mathbf{x}$ in 3D. A vectorial continuum over 3D is assumed as the application of the function to all points. A certain initial line segment or curve $\mathbf{s}_0(u)$ is implicitly integrated to define the stream surface $\mathbf{s}(u,t) = \mathbf{s}_0(u) + \mathbb{N}_0^t \mathbf{v}(\mathbf{s}(u,\tau)) \, d\tau$.

Explicitly defined parametric surfaces (case 5) – such a surface is defined by a parametric function $\mathbf{s}: \mathbf{R}^2 \to \mathbf{R}^3$.

Explicitly defined surfaces given in implicit form (case 6) – an equation $f(\mathbf{x}) = 0$ defines a surface in 3D (note that this case is similar to case 2).

Explicitly defined discrete surface approximations (case 7) – a mesh, i.e., a set of polygons is used to explicitly specify an approximation of a smooth surface.

There are other surface types as well, for example, explicitly expressing one coordinate in terms of the others, $\mathbf{s}(x,y) = (\, x \ y \ z(x,y) \,)^\mathrm{T}$. Usually they can be either transformed into one of the above mentioned cases, or appear rather rarely. Therefore, they are not considered separately in this paper.

3 Access to Surface Properties

Algorithms used for visualization of volumetric data can be broadly separated into two groups:

Image space techniques are usually based on *ray casting*. The data is intersected with a viewing ray, which is defined by an eye point and a viewing direction, to locate visible surface locations.

Object space techniques project the data onto the image plane to render the surface. In this case often incremental *surface curve traversal* is used to loop over the surface object.

Elaborated surface visualization methods usually are based on surface properties up to the order of two, i.e., the calculation of surface-point locations, surface normals, and surface curvature properties. The access to surface properties, i.e., the evaluation of these properties for certain points of the surface, involves a number of algorithms [8] which are dependent on the type of the surface. Examples are function reconstruction and gradient approximation. In the following we briefly summarize the most common approaches for the surface types described in Sect. 2.

3.1 Surface-point Location

When rendering a surface, the most important information about the data is where surface points lie. Based on the knowledge of surface point locations, higher-order properties, such as surface normals or curvature properties, can be computed.

Since a generic interface should be usable for image space and object space techniques, SMURF supports both *ray casting* and incremental *surface curve traversal*. In the following part we firstly discuss ray casting with respect to surfaces of various types.

Ray Casting. The intersection of a certain ray (given by a view-point **eye** and a viewing direction **dir**) and the surface to be shown yields a sorted list of surface-points (hit list). Usually just the first entry in the hit list is investigated as the (one

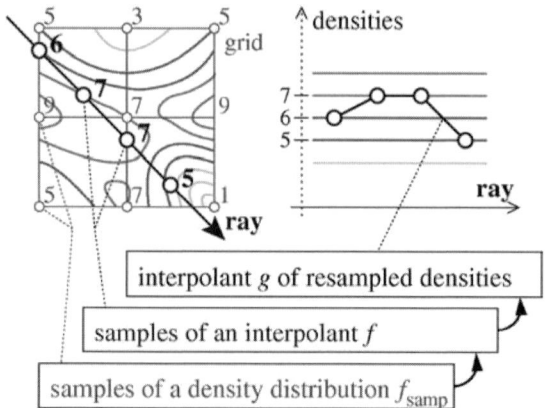

Fig. 2. 2D example of root finding along a ray for iso-surface ray tracing – often two-fold reconstruction is used (color version on color plate)!

and only) visible intersection. Also, advanced visualization algorithms also use semi-transparency of surfaces, thus requiring the computation of successive intersections. Therefore, a hit list of intersections should be returned by the "ray casting" surface interface (see Sect. 4). Depending on the type of surface, the intersection calculation is done differently:

Analytic solution (cases 5, 6, and 7) – in the case of an explicitly specified surface, the intersection between a ray and a surface usually can be expressed and sometimes also computed analytically. However, in the usual case the evaluation of this intersection expression has to be done using numerical methods like root finding (see below).

In the case of a parametric surface (case 5) the following two equations have to be solved in terms of u and v ($\mathbf{n}_1$ and $\mathbf{n}_2$ are two orthogonal vectors which are both normal to the viewing ray):

$$\mathbf{s}(u,v) \cdot \mathbf{n}_1 = \mathbf{eye} \cdot \mathbf{n}_1$$
$$\mathbf{s}(u,v) \cdot \mathbf{n}_2 = \mathbf{eye} \cdot \mathbf{n}_2$$
$$\text{where } \mathbf{n}_i \cdot \mathbf{dir} = 0 \text{ and } \mathbf{n}_i \cdot \mathbf{n}_j = \delta_{ij}$$

Depending on the complexity of $\mathbf{s}$, solving the above equations is usually not possible in closed form. There are numerical techniques to compute the list of intersections in terms of u and v [7].

In the implicit case (case 6) the following equation has to be solved in terms of λ:

$$f(\mathbf{eye} + \lambda \, \mathbf{dir}) = 0$$

Again, often numerical methods are required to solve the above equation.

In the polygonal case (case 7) theoretically all polygons have to be intersected with the ray to evaluate the hit list. However, spatial coherence can be exploited by using special data structures to speed up the intersection process [1, 9, 24]. Other simple but effective enhancements like back-face culling are available as well.

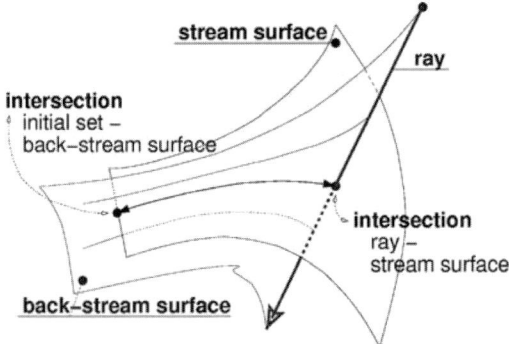

Fig. 3. Lazy evaluation ray casting of stream surfaces by the use of a "back-stream surface" (color version on color plate).

Root finding (cases 1 and 2) – considering a ray being cast into a density volume, e.g., a continuum that interpolates discrete data values (case 1) or the application of f to the entire domain (case 2), implicitly yields a scalar density function at all points of the ray. This function can be sampled along the ray to search for intersections with the iso-surface. An interpolant, for instance, linear interpolation, is used to approximate the function along the ray.

Figure 2 illustrates these steps in a 2D example. Discrete density values $f_{\text{samp}}(\mathbf{x}_i)$ are samples of a particular density distribution (arranged, for example, on a regular grid). One typical ray casting approach is to resample an interpolant f, often a tri-linear interpolation, at certain locations along the ray, i.e., $f(\mathbf{r}_i)$ with $\mathbf{r}_i = \mathbf{eye} + i\,\Delta\mathbf{dir}$. For the identification of the ray / iso-surface intersections $\mathbf{p}_k$ an interpolant $g(\mathbf{x})$ along the ray is taken (e.g., linear interpolation) with $g(\mathbf{r}_i) = f(\mathbf{r}_i)$. Equation $g(\mathbf{p}_k) = g(\mathbf{eye} + \lambda_k\,\mathbf{dir}) = f_\mathbf{s}$ is solved in terms of λ_k.

Note, that the use of a separate interpolant g yields a double reconstruction of the original function f. Instead, g can also be defined to be the projection of interpolant f onto the ray, which actually would be the more accurate solution. Unfortunately, this approach is rather complex as already the projection of a trilinear function f induces the interpolant g to be a cubic function in terms of λ_k [23].

Stream surface intersection (cases 3 and 4) – the most demanding problem within the task of locating surface-points is stream surface intersection. This is mainly due to the fact that stream surfaces are implicitly defined through an additionally required integration step of the underlying vectorial data. In flow visualization often pre-computed, i.e., pre-integrated, stream surfaces are used. Numerical techniques, like Euler or Runge-Kutta integration, are used to step-by-step generate a polygonal approximation of the stream surface, which afterwards is visualized using standard mesh rendering methods (compare to case 7).

Another approach exploits the reversibility of flow integration: instead of explicitly generating the stream surface itself, a "back-stream surface" is computed, considering the ray as an initial condition and performing flow integration backwards in time. Any intersection of this "back-stream surface" and the original initial set directly corresponds to an intersection of the investigated stream surface and the ray via a stream line (cf. Fig. 3).

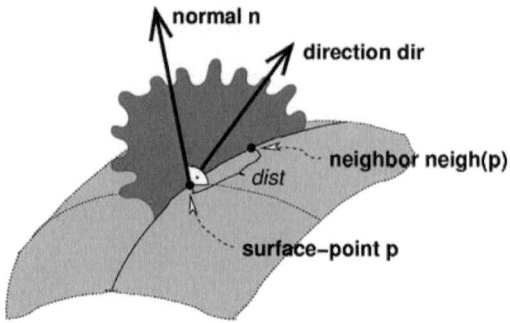

Fig. 4. Surface-curve traversal for iterative object-order surface rendering.

This approach is useful, for example, when lazy evaluation is used (see Sect. 4). Unfortunately this approach is rather expensive when many intersections should be computed. On the other hand, this duality (Fig. 3) can be exploited to increase numerical stability of the intersection computations – divergent flows are more accurately integrated backwards.

Surface-curve Traversal. In addition to ray casting, incremental surface-curve traversal was chosen as a complementary SMURF strategy to access surface-points. Starting with an initial surface-point **p**, a neighboring location separated by a specific distance *dist* is searched in a certain direction **dir**. Surface-point **p**, surface normal **n**, and direction **dir** define a plane which intersects the surface in a certain surface-curve. Out of both points on the curve which are *dist* (in terms of curve length) away from **p** the one which is mostly aligned with **dir** is considered to be the searched location. See Fig. 4 for an illustration of this procedure.

Iso-surfaces (cases 1 and 2) – for implicitly defined iso-surfaces a numerical approximation is used to loop over the surface-curve. First, a tangent vector **t** in surface point **p** is defined on the basis of the surface normal **n(p)** and direction **dir** by $\mathbf{t} = \mathbf{n(p)} \times \mathbf{dir} \times \mathbf{n(p)}$. Then, from a point $\mathbf{q} = \mathbf{p} + dist\,\mathbf{t}$ a local ray casting step is performed in direction $\pm \mathbf{n(p)}$. If no unique solution is found within a small distance from **q**, a sequence of smaller, iterative steps is performed instead of one with distance *dist*.

Stream surfaces (cases 3 and 4) – stepping along a stream surface is usually done either along the flow, i.e., in direction **v(p)**, or across the flow, i.e., along time lines. A step along the flow equals the integration of the underlying flow data from point **p** for a certain distance: $\mathbf{s_p}(t) = \mathbf{p} + \mathbb{N}_0^t \mathbf{v}(\mathbf{s_p}(\tau))\,d\tau$. A neighbor $\mathbf{neigh}(\mathbf{p}, \mathbf{v(p)}, dist)$ of point **p** therefore is computed as $\mathbf{s_p}(d)$ such that the curve-length between **p** and $\mathbf{s_p}(d)$ is *dist*. A surface neighbor of point $\mathbf{p} = \mathbf{s}(u_\mathbf{p}, t_\mathbf{p})$ along a time line in the stream surface is defined as $\mathbf{s}(u_\mathbf{p} \pm \Delta u, t_\mathbf{p})$ such that the length of the time line segment inbetween both points is *dist*. Usually, Δu not really approximates *dist* in a sufficiently accurate way, so an iterative approach is required like bisectioning.

Explicit surfaces (cases 5, 6, and 7) – in the case of a parametric surface (case 5), the tangent vector $\mathbf{t} = \mathbf{n} \times \mathbf{dir} \times \mathbf{n}$, which corresponds to direction **dir**, is first decomposed into parameters $u_\mathbf{t}$ and $v_\mathbf{t}$ such that $\mathbf{t} = u_\mathbf{t}\mathbf{s}_u + v_\mathbf{t}\mathbf{s}_v$ ($\mathbf{s}_\omega = \partial \mathbf{s}/\partial \omega$). Then, surface point $\mathbf{s}(u_\mathbf{p} + \frac{dist}{|\mathbf{t}|}u_\mathbf{t}, v_\mathbf{p} + \frac{dist}{|\mathbf{t}|}v_\mathbf{t})$ can be used as a first approximation in an iterative procedure for locating the surface neighbor. In case 6, the procedure for case 2 is used. In case 7, the intersection line between the polygon on which point **p** lies and the plane which is spanned by **n(p)** and **dir**, is computed. If the distance of **p** and the polygon edge (in the direction of the intersection line) is larger than *dist*, then the respective neighbor of **p** lies on the same polygon. Otherwise, the neighbor polygon has to be searched across the polgon edge. There, the procedure repeats with the intersection of polygon and plane, etc.

3.2 Surface Normal Computation

Various computer graphics algorithms use surface normals, e.g., for shading or backface culling. The acquisition of a normal corresponding to a certain surface-point again depends on the type of surface:

Analytic solution (cases 5 and 6) – if the surface is given explicitly, usually the surface normal at a certain point can be computed analytically. In the parametric case (case 5) the cross-product $\partial \mathbf{s}/\partial u |_\mathbf{p} \times \partial \mathbf{s}/\partial v |_\mathbf{p}$ of two tangent vectors yields a (not yet normalized) surface normal at point **p**. Of course, this is only possible if both tangents are not collinear. In the implicit case (case 6) the gradient $\nabla f|_\mathbf{p}$ is a surface normal of the iso-surface through **p** (not normalized).

Gradient reconstruction from densities (cases 1 and 2) – assuming a function $f(\mathbf{x})$ which can be evaluated at arbitrary points **x** (f is either the "black box", case 2, or an interpolant, case 1), surface normals (not normalized) can be computed using central differences, for example:

$$\mathbf{n}(\mathbf{p}) = \nabla f|_\mathbf{p} \approx \frac{1}{2} \begin{pmatrix} f(\mathbf{x}+\mathbf{e}_1) - f(\mathbf{x}-\mathbf{e}_1) \\ f(\mathbf{x}+\mathbf{e}_2) - f(\mathbf{x}-\mathbf{e}_2) \\ f(\mathbf{x}+\mathbf{e}_3) - f(\mathbf{x}-\mathbf{e}_3) \end{pmatrix}$$

$$\mathbf{e}_i = (\delta_{1i}\ \delta_{2i}\ \delta_{3i})^\mathrm{T}$$

Higher-order approximations of the gradient are possible as well. In general, an arbitrarily complex derivative filter can be applied for gradient reconstruction [3, 18].

Note, that in case 2 central differences easily are evaluated at arbitrary surface locations, whereas in case 1 (when dealing with data samples on a regular grid) normals are usually approximated at grid locations and then interpolated within cells, for example, using tri-linear interpolation.

Normal reconstruction from polygons (case 7) – a standard procedure for reconstructing normals within polygons is used for Phong shading [21]: at the vertices of a polygon a weighted sum of all the normals of adjacent polygons is computed. These vertex normals are interpolated within the polygon to approximate the normals of the surface which is approximated by the polygons.

Stream surface normals (cases 3 and 4) – In the case of a pre-computed stream surface in the form of a set of polygons, again techniques for case 7 can be used. In the other case, the (not yet normalized) surface normal **n(p)** can be computed as the cross-product of two tangent vectors:

$$\mathbf{n}(\mathbf{p}) = \mathbf{n}(\mathbf{s}(u_\mathbf{p}, t_\mathbf{p})) = \mathbf{v}(\mathbf{p}) \times (\mathbf{s}(u_\mathbf{p}+\Delta u, t_\mathbf{p}) - \mathbf{s}(u_\mathbf{p}-\Delta u, t_\mathbf{p}))$$

One tangent, **v(p)**, equals the vectorial data in the surface point whereas the other tangent is numerically approximated as (central) difference from neighboring stream lines in the stream surface.

3.3 Surface Curvature

Second-order surface properties, i.e., curvature information, is used to enhance surface-based visualization. Shape and location of a surface can be better perceived, for example, if curvature directed strokes are applied to the surface [14].

Surface curvature usually is expressed in several terms, e.g., Gaussian or mean curvature. Both curvature properties depend on a surface-curvature definition [5] which is dependent on a specific tangent direction. Principal directions are those tangent directions which yield either maximum or minimum curvature.

Curvature calculation for parametric surfaces (case 5) – The first and second fundamental coefficents (with the usual abbreviations) of a parametric surface $s(u, v)$ are defined as

$$E = s_u \cdot s_u, \quad F = s_u \cdot s_v, \quad G = s_v \cdot s_v$$
$$L = s_{uu} \cdot n, \quad M = s_{uv} \cdot n, \quad N = s_{vv} \cdot n$$

with $n = (s_u \times s_v)/|s_u \times s_v|$ being the unit normal vector and $s_u = \partial s/\partial u$, $s_v = \partial s/\partial v$. The normal curvature in tangent direction $u' : v'$ is

$$\kappa = -\frac{Lu'^2 + 2Mu'v' + Nv'^2}{Eu'^2 + 2Fu'v' + Gv'^2}$$

For κ being extremal it must satisfy the equation [5]

$$\det \begin{bmatrix} \kappa E - L & \kappa F - M \\ \kappa F - M & \kappa G - N \end{bmatrix} = 0$$

The extreme values κ_1 and κ_2 are the principal curvatures of the surface at x and

$$\kappa_1 \kappa_2 = (LN - M^2)/(EG - F^2)$$
$$\kappa_1 + \kappa_2 = (NE - 2MF + LG)/(EG - F^2)$$

are the Gaussian and mean curvature, respectively.

Curvature calculation for implicit surfaces (case 6) – In a surface-point p of interest we consider $n(p)$ to be a unit normal of the plane which is tangent to the surface through p, i.e., $n(p) = \nabla f|_p / |\nabla f|_p|$. Assuming e_1 to be an arbitrary vector of unit length contained in the tangent plane, we construct a local Frenet frame:

$$\Phi = \begin{pmatrix} e_1 & e_2 & n(p) \end{pmatrix}$$
$$e_1 \cdot n(p) = 0, \quad e_2 = n(p) \times e_1$$

Searching for the principal curvature of the surface through p, we have to investigate the changes of $n(x)$ near p with respect to changes of x within the tangent plane, i.e., $x = p + r e_\varphi$ with e_φ being a unit length vector orthogonal to $n(p)$, i.e., lying in the tangent plane.

Direction e_φ, where $\nabla n|_p \cdot e_\varphi$, i.e., the directional derivative of n near p into direction e_φ, is greatest in terms of length (called $e_{\tilde{\varphi}}$ in the following), is then the first principal direction of the surface through p. The second principal direction is orthogonal to both $e_{\tilde{\varphi}}$ and $n(p)$. The related curvatures are the lengths of the directional derivatives along the principal directions.

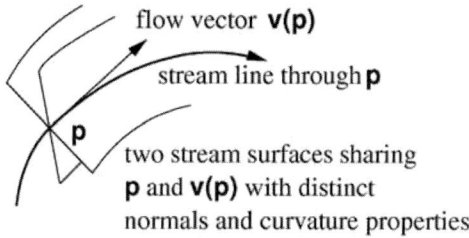

Fig. 5. The definition of a stream surface through a point **p** is ambiguous – depending on the seeding structure, a variety of different stream surfaces are possible. Therefore, normals and curvature properties of stream surfaces are of limited use for the visualization of flow data.

As derivation is a linear operator, the directional derivative of **n** into some direction $\mathbf{e}_\varphi = \cos\varphi\,\mathbf{e}_1 + \sin\varphi\,\mathbf{e}_2$ can be written in terms of the directional derivative of **n** into directions $\mathbf{e}_1$ and $\mathbf{e}_2$:

$$\nabla\mathbf{n}|_\mathbf{p} \cdot \mathbf{e}_\varphi = \left(\nabla\mathbf{n}|_\mathbf{p} \cdot \mathbf{e}_1 \quad \nabla\mathbf{n}|_\mathbf{p} \cdot \mathbf{e}_2\right) \cdot \begin{pmatrix} \cos\varphi \\ \sin\varphi \end{pmatrix}$$

Since $\nabla\mathbf{n}|_\mathbf{p} \cdot \mathbf{e}_\varphi$ is orthogonal to $\mathbf{n}(\mathbf{p})$ also, we can express it in terms of $\mathbf{e}_1$ and $\mathbf{e}_2$ by the use of decomposition with $(x\ y)^T = (\cos\varphi\ \sin\varphi)^T$:

$$\begin{pmatrix} x' \\ y' \end{pmatrix} = \underbrace{\begin{pmatrix} \mathbf{e}_1^T \\ \mathbf{e}_2^T \end{pmatrix} \cdot \begin{pmatrix} \nabla\mathbf{n}|_\mathbf{p} \cdot \mathbf{e}_1 \\ \nabla\mathbf{n}|_\mathbf{p} \cdot \mathbf{e}_2 \end{pmatrix}^T}_{\mathbf{A} = (\omega_{ij}),\ \omega_{ij} = \mathbf{e}_i \cdot \nabla\mathbf{n}|_\mathbf{p} \cdot \mathbf{e}_j} \cdot \begin{pmatrix} x \\ y \end{pmatrix}$$

Searching for the greatest eigenvector $(x_{\bar\varphi}\ y_{\bar\varphi})^T$ of matrix **A**, directly yields the corresponding first principal direction via $\mathbf{e}_{\bar\varphi} = (\mathbf{e}_1\ \mathbf{e}_2) \cdot (x_{\bar\varphi}\ y_{\bar\varphi})^T$.

Curvature reconstruction from densities (cases 1 and 2) – to reconstruct curvature properties of iso-surfaces obtained from scalar volume data essentially the same procedure as for implicit surfaces can be used. A function $f(\mathbf{x})$ is assumed, which can be evaluated at arbitrary points **x** (see Sect. 2), as well as a function $\mathbf{n}(\mathbf{p}) = \nabla f|_\mathbf{p}\,/\,\left|\nabla f|_\mathbf{p}\right|$ which yields the unit normal at an arbitrary point **p** (see Sect. 3.2). Again the eigenvalue decomposition of matrix $\mathbf{A} = (\mathbf{e}_i \cdot \nabla\mathbf{n}|_\mathbf{p} \cdot \mathbf{e}_j)_{ij}$ in terms of a local Frenet frame gives the searched curvature properties.

Stream surface curvature (cases 3 and 4) – in the case of a pre-computed stream surface techniques described for case 7 (see below) are used. In the case of a stream surface on demand curvature properties could be derived by investigating the changes of a normal with respect to changes within the tangent plane. It must be noted here that stream surface curvature is rarly used for visualization since it easily might be misinterpreted as a property of the underlying vector field. Figure 5 illustrates why stream surface curvature – even stream surface normals – usually lack importance in visualization. Both properties heavily depend on the choice of the initial condition.

Curvature reconstruction from polygons (case 7) – To reconstruct curvature properties from polygons, one obvious procedure would be to construct an

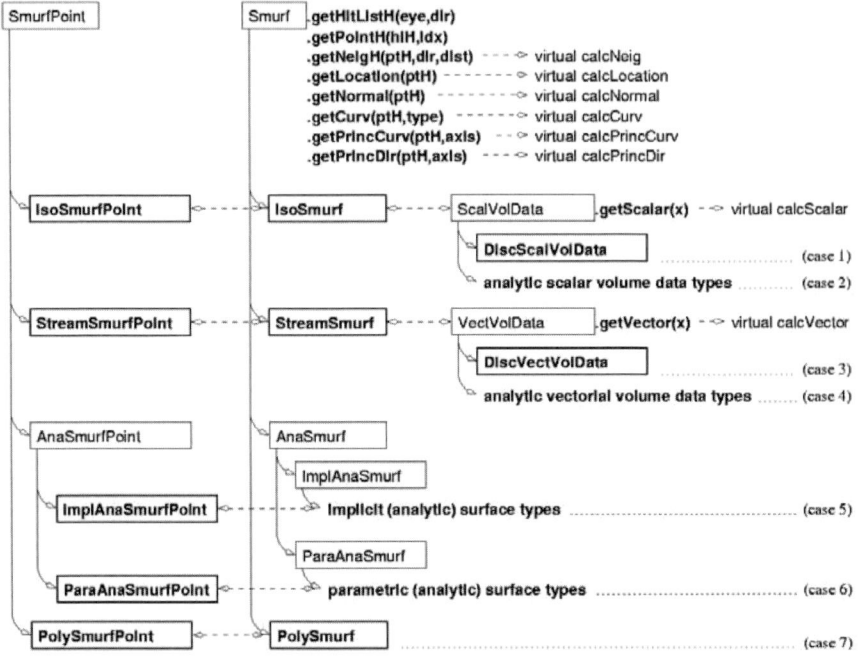

Fig. 6. SMURF class hierarchy and interface.

interpolant and calculate analytically the curvature of the interpolant. Todd and McLeod [27], however, report that this approach yields in general completely unsatisfying results.

Therefore, they propose to approximate the Dupin indicatrix from the vertices of the polygons, by exploiting Meusniers theorem [5], estimating normal curvatures in particular directions (which requires to estimate the normal, for example, with the approach used in Phong shading, as described in Sect. 3.2) and finally fitting a central conic to that data.

4 SMURF Classes

After having identified the different surface types (Sect. 2) and access schemes (Sect. 3) the integration in a C++ class hierarchy called SMURF (see Fig. 6) is straight-forward. An abstract base class provides all common properties of the different surface types and the access interface to surface properties as virtual functions. Sub-classes, corresponding to the surface types, are derived from this abstract base class and redefine the access schemes accordingly.

To distinguish discrete scalar volume data-sets (case 1) from analytic scalar volume functions (case 2) a class ScalVolData hides the interface. Therefore, class IsoSmurf can treat these two cases the same way. The same holds for discrete vector fields (case 3) and analytically specified dynamical systems (case 4) via class VectVolData. This means, that all the seven cases of surface types which were presented in Sect. 2 are mapped to four sub-classes of SMURF, i.e., ISOSMURF, STREAMSMURF, ANASMURF, and POLYSMURF. See Fig. 6 for the relations between these classes.

An important concept in the implementation is the one of lazy evaluation, i.e., computing not more than necessary at a certain point in time. For example, ray casting can be terminated after finding the desired intersection point. Further surface properties are evaluated on demand, and stored for future use. Therefore, a class hierarchy `SmurfPoint` is introduced which mirrors the `Smurf` class hierarchy and serves as memory element for the specific surface types, i.e., it stores all relevant information already computed which can be re-used. Again, Fig. 6 illustrates the relations between these classes.

5 Results, Applications

By hiding the intrinsic differences between the surface types identified in Sect. 2 SMURF supports the user with the following tasks:

Implementation of advanced visualization techniques – SMURF eases this task by providing an interface for obtaining surface properties independently of the surface type. Figure 7 shows an iso-surface with crosses aligned to the principal directions (similar to Beck et al. [2]). Figure 8 was generated using the code depicted in Fig. 9 – any other surface type could be rendered using the same code by just changing the very first line.

Comparison of algorithms – for instance, reconstruction schemes can be easily compared by sampling an analytic function and applying a visualization algorithm to both the scalar data volume and the analytic function. In Fig. 10 this concept was used to compare linear and cubic interpolation with respect to function reconstruction and the computation of the Gaussian curvature with the corresponding analytic function (a quadratic in 3D). Linear reconstruction of densities is clearly seen in Fig. 10(a), whereas there is no perceivable difference between Fig. 10(b) and (c). Comparing the curvature plot (color was used to visualize the Gaussian curvature), subtle differences can be obtained even between Fig. 10(b) and (c).

Figure 11 compares the quality of curvature reconstruction (by calculating principal curvature lines of a cylinder) depending on linear and cubic density reconstruction. In Fig. 11(a) small errors accumulated during numerical integration of the curvature lines are clearly visible.

6 Future Work

One obvious disadvantage of a general scheme like SMURF is that it principally suffers from inefficiency. Performance can be improved by including, e.g., intelligent caching strategies and implementation short-cuts. Furthermore, it should be possible to exploit ray-to-ray coherence of visualization algorithms. Thus, again, caching and addressing of external data must be allowed by the scheme.

Another idea is to extend the SMURF concept to a 'set of SMURFs' class with a similar interface. Consecutive intersections along a ray are reported in correct order

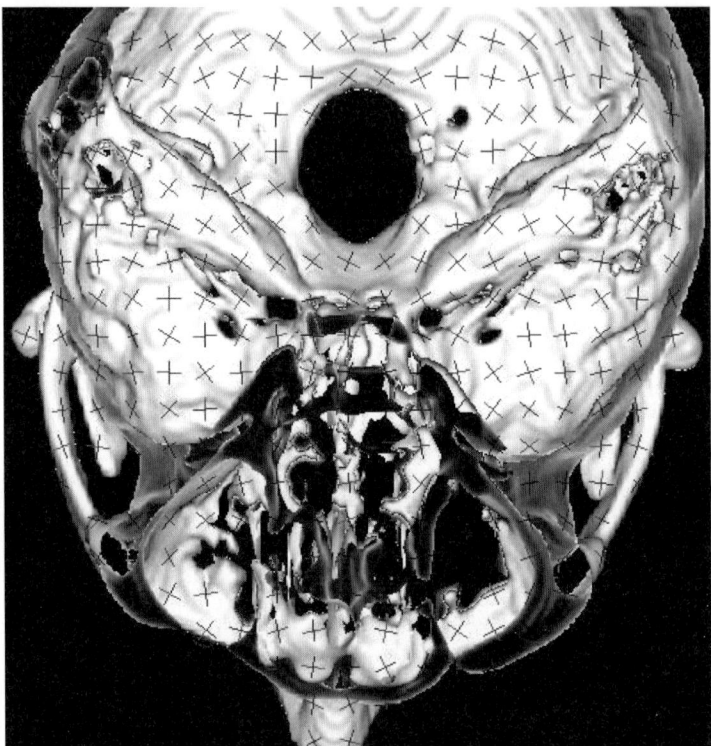

Fig. 7. Iso-surface computed for ten slices, scanned from a human head with curvature crosses.

from different surfaces, e.g., stacked iso-surfaces or multiple stream surfaces. Even surfaces of different type could be easily combined using this concept, for example, patient data together with objects from (virtual) surgery planning.

7 Conclusions

Our general purpose surface interface SMURF allows to easily re-use elaborated visualization techniques that are based on surfaces in 3D. Examples are curvature-directed strokes or plotting curvature lines, with surfaces originating from various applications like iso-surfaces from medical applications or stream surfaces from flow visualization. Surface properties up to the order of two, i.e., curvature information, are available to the user in a transparent way. Ray casting as well as incremental surface-curve traversal are provided as surface access strategies. Thus, advanced surface visualization techniques can be developed without having to care about specific algorithms for calculating particular surface properties. Their portability to other surface types is another advantage of the SMURF concept.

For the realization of this concept we first identified the most often used surface types and compared various algorithms for accessing surface properties in general

Access to surface properties up to order two for visualization algorithms   119

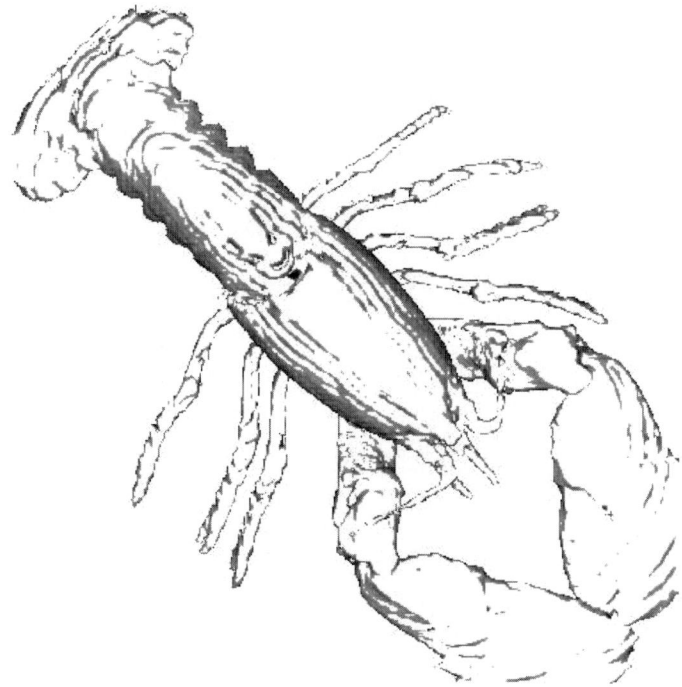

Fig. 8. Contour display of a lobster CT scan – see Fig. 9 for the SMURF-code used to render this image.

```
Smurf *pSmurf = new IsoSmurf("lobster.dat",threshold);

for (p=pFirstPixel(); p!=NULL; p=pNextPixel())
{
  VEC3 dir                = normalize(*p-eye);
  SmurfHitListHandle HLH  = pSmurf->getHitListH(eye,dir);
  SmurfPointHandle   PH   = pSmurf->getPointH(HLH,0);
  VEC3 normal             = pSmurf->getNormal(PH);

  if (-dir*normal < 0.6)
    p->set(1-(-dir*normal));
  else
    p->set(0);
}
```

Fig. 9. Code used for drawing lobster contours depicted in Fig. 8.

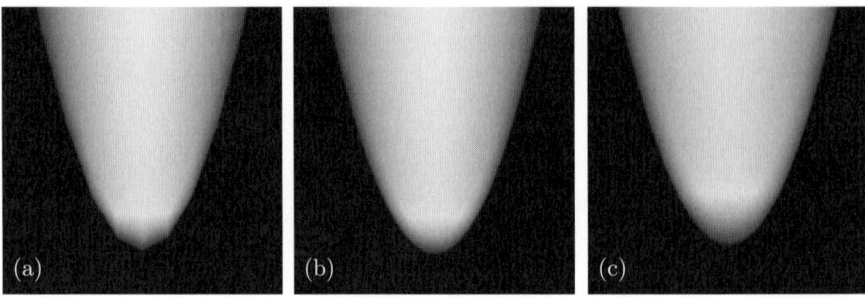

Fig. 10. Gaussian curvature plot using linear (a) and cubic (b) function reconstruction vs. analytic computation (c) – see the color version on the color plate.

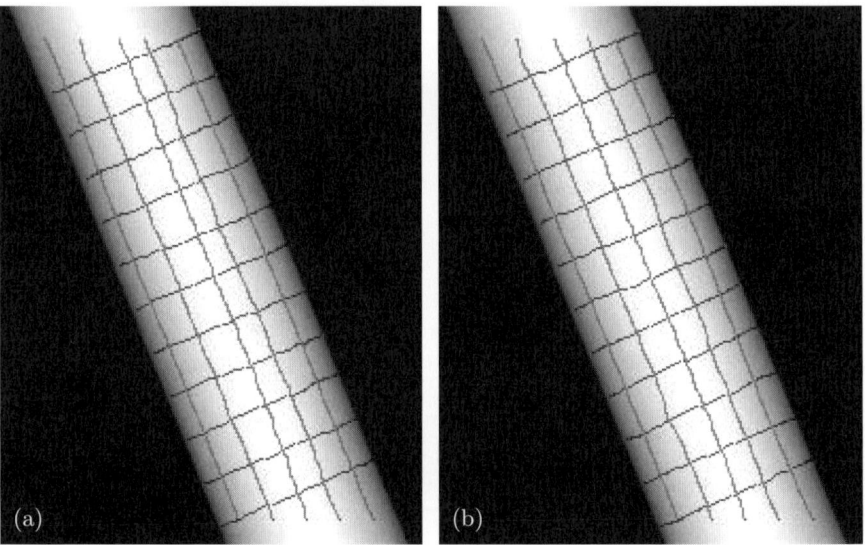

Fig. 11. Quality of curvature calculations depending on the function reconstruction scheme – linear (a) vs. cubic (b) reconstruction – color version on color plate.

before we unified them in a unique interface. As most visualization applications have to deal with sampled data, analytic evaluation is discussed as well as various reconstruction schemes. The usefulness of this approach is demonstrated by several results we obtained with our actual implementation.

Acknowledgments

The work presented in this publication has been supported by $V^{is}M^{ed}$. Please refer to http://www.vismed.at/ for further information on this project. Parts of this work have been done in the VRVis Research Center (http://www.VRVis.at/) in Vienna, Austria, which is funded by an Austrian research program called K plus.

References

1. James Arvo and David Kirk. A survey of ray tracing acceleration techniques. In Andrew Glassner, editor, *An introduction to ray tracing*, pages 201–262. Academic Press, 1989.
2. James Beck, Rida Farouki, and John Hinds. Surface analysis methods. *IEEE Computer Graphics and Applications*, 6(12):18–36, 1986.
3. Mark Bentum, Barthold Lichtenbelt, and Tom Malzbender. Frequency Analysis of Gradient Estimators in Volume Rendering. *IEEE Transactions on Visualization and Computer Graphics*, 2(3):242–254, 1996.
4. John Dill. An application of color graphics to the display of surface curvature. *Computer Graphics*, 15(3):153–161, August 1981.
5. Gerald Farin. *Curves and Surfaces for Computer Aided Geometric Design*. Morgan Kaufmann, 4th edition, 1997.
6. Thomas Gerstner. Multiresolution extraction and rendering of transparent isosurfaces. *Computers and Graphics*, 26(2):219–228, April 2002.
7. Andrew Glassner, editor. *An Introduction to Ray Tracing*. Acad. Press, 1989.
8. Hans Hagen, Stefanie Hahmann, Thomas Schreiber, Yasuo Nakajima, Burkard Wordenweber, and Petra Hollemann-Grundstedt. Surface interrogation algorithms. *IEEE Computer Graphics and Applications*, 12(5):53–60, Sept. 1992.
9. Vlastimil Havran. *Heuristic Ray Shooting Algorithms*. PhD thesis, Czech Technical University, Praha, Czech Republic, April 2001. Available from http://www.cgg.cvut.cz/~havran/phdthesis.html.
10. Josef Hoschek and Franz-Josef Schneider. Spline conversion for trimmed rational Bézier- and B-spline surfaces. *Computer Aided Design*, 22(9):580–590, 1990.
11. Andreas Hubeli and Markus Gross. Fairing of non-manifolds for visualization. In *Proceedings IEEE Visualization*, pages 407–414, 2000.
12. Jeff Hultquist. Constructing stream surfaces in steady 3D vector fields. In *Proceedings IEEE Visualization*, pages 171–177, 1992.
13. Victoria Interrante. Illustrating surface shape in volume data via principal direction-driven 3D line integral convolution. *Computer Graphics*, 31(Annual Conference Series):109–116, August 1997.
14. Victoria Interrante, Henry Fuchs, and Stephen Pizer. Conveying the 3D shape of smoothly curving transparent surfaces via texture. *IEEE Transactions on Visualization and Computer Graphics*, 3(1):98–117, 1997.
15. Victoria Interrante, Henry Fuchs, and Steven Pizer. Illustrating transparent surfaces with curvature-directed strokes. In *Proceedings IEEE Visualization*, pages 211–218, 1996.
16. Helwig Löffelmann, Lukas Mroz, Eduard Gröller, and Werner Purgathofer. Stream arrows: Enhancing the use of streamsurfaces for the visualization of dynamical systems. *The Visual Computer*, 13:359–369, 1997.
17. William Lorensen and Harvey Cline. Marching cubes: A high resolution 3D surface construction algoritm. *Computer Graphics*, 21(4):163–168, July 1987.
18. Torsten Möller, Raghu Machiraju, Klaus Müller, and Roni Yagel. Evaluation and Design of Filters Using a Taylor Series Expansion. *IEEE Transactions on Visualization and Computer Graphics*, 3(2):184–199, 1997.
19. Gregory Nielson and Bernd Hamann. The asymptotic decider: Removing the ambiguity in marching cubes. In *Proceedings IEEE Visualization*, pages 83–91, 1991.

20. Gregory Nielson, Bruce Shriver, and Lawrence Rosenblum. *Visualization in Scientific Computing*. IEEE Computer Society Press, 1990.
21. Bui-Tuong Phong. Illumination for computer generated pictures. *Communications of the ACM*, 18(6):311–317, 1975.
22. Martin Rumpf, Alfred Schmidt, and Kunibert Siebert. Functions defining arbitrary meshes - A flexible interface between numerical data and visualization. *Computer Graphics Forum*, 15(2):129–142, 1996.
23. Georgios Sakas, Marcus Grimm, and Alexandros Savopoulos. Optimized maximum intensity projection (MIP). In *Proceedings* EUROGRAPHICS *Rendering Workshop*, pages 51–63, 1995.
24. Hanan Samet. *Applications of Spatial Data Structures: Computer Graphics, Image Processing, and GIS*. Addison-Wesley, 1989.
25. Lee Seidenberg, Robert Jerad, and John Magewick. Surface curvature analysis using color. In *Proceedings IEEE Visualization*, pages 260–267, 1992.
26. Gabriel Taubin. A signal processing approach to fair surface design. *Computer Graphics*, 29(Annual Conference Series):351–358, November 1995.
27. Philip Todd and Robin McLeod. Numerical estimation of the curvature of surfaces. *Computer-Aided Design*, 18(1):33–37, January 1986.
28. Jarke van Wijk. Implicit stream surfaces. In *Proceedings IEEE Visualization*, pages 245–252, 1993.

Modeling Rough Surfaces

Yootai Kim[1], Raghu Machiraju[1], and David Thompson[2]

[1] Department of Computer and Information Science, The Ohio State University
2015 Neil Avenue, Columbus, OH 43210
{*yootai, raghu* }*@cis.ohio-state.edu*
[2] Computational Simulation and Design Center, Mississippi State University
Box 9627, Mississippi State, MS 39762
dst@erc.msstate.edu

1 Introduction

Many surfaces in nature are rough. A rough surface can be defined as a surface that has a fractal dimension. In fact, one measure of roughness is the fractal dimension [1, 14]. Rough surfaces are observed at all scales independent of their origin; for example, a microscopic view of metal substrate, a cauliflower, ice, and mountains are all rough at some level. Additionally, rough surfaces are frequently generated by various technical processes, such as molecular beam epitaxy (MBE).

Rough surfaces are also common in synthetic environments. Techniques for realistic image synthesis have improved dramatically. However, there is no easy solution to the problem of generating rough surfaces. The recent movie *Ice Age* produced by Blue Sky Studios is a good example. Although the synthetic ice world in the feature film was visually appealing, the computer generated ice models were not realistic enough to match the visual richness of natural ice. Furthermore, the production process still requires much labor and somewhat *ad hoc* methods.

In materials science, numerous models have been developed to study surface growth phenomenon [1]. In general, it is difficult to develop a viable continuum model of surface growth phenomenon and then solve the resulting differential equations. Therefore, discrete models play an important role in the prediction of surface growth phenomena. A discrete model is defined as a system with discrete variables and update rules. Since many rough surfaces in nature are formed by deposition and diffusion processes, discrete models simulating these processes can reasonably reproduce natural rough surfaces. Typical results from these simulations consist of point clusters with nontrivial topologies.

For instance, consider the results of the two-dimensional simulation using the diffusion limited aggregation (DLA) cluster growth model shown in Fig 1. The distribution of the islands and dendrites is unusual and cannot be generated by traditional means. Also, in addition to fractal like microstructures, prominent large-scale structures are produced. Therefore, a sophisticated methodology is needed to capture and represent these complex surfaces. Explicit methods of representation using triangulation are unable to capture the intricacies of these surfaces in their entirety. Implicit methods do possess the capability to represent surfaces with complex topologies. However, it is necessary to consider methods that provide variable amounts of smoothness. Level set methods, through appropriate choice of initial conditions and front velocities, can extract surfaces of differing smoothness while preserving the underlying topology and are routinely used for reconstruction of smooth surfaces. We consider their use for reconstruction of rough surfaces.

Fig. 1. A fractal surface growth simulation: DLA

In this paper, we propose a method for generating rough surfaces using discrete surface growth models. Our goal is to develop easily controllable methods for generating rough surfaces for computer graphics applications. We employ a two-pass method. A point set is generated using a discrete model based on surface growth and evolution. Then, the resulting surface is extracted by a level set method. This two-pass process provides more flexibility to users by separating the surface extraction step from the data generation step. The simple rule-based discrete simulations we employ here have several advantages. First, the results are convincing since they are derived from physical processes. Second, implementation is easy and the computations using the methods are not expensive. Lastly, users can exercise control by simply changing the discrete update rules. After generating an initial data set from the application of a discrete model, a level set method is used to obtain an implicit surface representation. This approach allows us to easily handle complex topologies and compute intrinsic geometric properties of the surface. In addition, it is easy to deform the shape for animation and to combine several objects to generate an elaborate model.

Our paper is organized as follows. In Section 2, we review related efforts for rough surface generation. We then provide an overview of our techniques in Section 3. Later, in Section 4 and 5, we describe the details of several surface growth models and our level-set-based surface extraction method. In Section 6, we provide results that demonstrate the potential of our approach and in the final section we draw conclusions and present a discussion of future work.

2 Related Work

The morphology of rough surfaces can be described by fractal models and concepts. To date, fractal models have been the primary method to model natural surfaces. They can be classified as one of the following five approaches: Poisson faulting [14,25], Fourier filtering [14,15,25], midpoint displacement [7, 11, 16, 20], successive random additions [25], and noise synthesis [8, 16]. The Poisson faulting process is a sum of randomly placed step functions with random heights, which generates a Brownian surface. Musgrave [17] compares these methods in a comprehensive way. However, all produce surfaces defined in terms of height fields and cannot generate arbitrary surfaces. In particular, these techniques cannot generate surfaces of arbitrary genus as required by rough and amorphous materials. These methods generate random fractals which are scale invariant in a statistical sense. Gross and large structures are hard to obtain through the deployment of these techniques.

Procedural textures [4] can also be used to simulate a rough surface. Lewis [12] suggested a solid noise synthesis algorithm for surface texturing and stochastic modeling. Worley [27] obtained good results using a cellular texture basis function for organic skin and tiled stone. Fleischer [6] proposed a cellular development simulation to model organic surface details such as scales, feathers, or thorns. The results are very promising; however, it is hard to devise a cellular automata simulation and conversion functions to obtain desired results.

Computer graphics researchers in the past have used deposition concepts for different purposes. Musgrave performed random deposition followed by surface relaxation to emulate thermal weathering processes [17]. Fearing's accumulation model [5] also employed similar ideas for modeling fallen snow. Finally, Dorsey used fractal growth models such as random and ballistic depositions to model weathering of metallic surfaces [3].

Implicit surface methods have been used to represent rough surfaces. Relevant efforts include Hart's implicit representation of rough surfaces [10] and Greene's voxel space automata [9]. Hart derived implicit formulae for fractal representations. He generated wooden surfaces and blended them to demonstrate the power of implicit representation. Greene simulated the growth of plants using simple relationship rules in discrete volumes.

The efforts we report here are different in several ways. The surfaces we produced are not just height fields. Additionally, our techniques produce both microscopic and macroscopic structural variations. For example, DLA models allow the development of larger gross structures. Further, we employ level set formulations to extract the final surfaces. It should be noted that, while researchers in the physical sciences have used growth models for some time, their results are mostly based on two-dimensional models. Thus, the extraction of a three-dimensional growth surface is certainly novel as reported here.

3 Overview of Rough Surface Generation

Our rough surface generation process consists of two modules: the surface growth simulator and the surface reconstructor. The surface growth simulator generates a point set based on user-specified initial conditions. The initial conditions depend on the surface growth model and include parameters for update rules and the initial configuration of the seed points. In this paper, we use two surface growth models: random deposition with surface relaxation (RDSR) and diffusion limited aggregation (DLA). While RDSR is a simple local growth model, DLA is a non-local growth model. It was previously demonstrated that RDSR and DLA surfaces are fractal in nature [1]. These fractals belong to the class of self-affine fractals which are invariant under anisotropic transformation. The surface reconstructor captures surfaces from the results of the growth simulator and generates a surface representation such as a polygonal model (See figure 2).

Thus, our two-pass approach divides the problem into two well-separated sub-problems. Each sub-problem has been studied extensively and many mature technologies can be brought to bear on its solution. In addition to surface growth models derived from materials science, any rule-based method can be used.

A level set method is used to obtain an implicit surface representation. It is a computational technique for tracking evolving interfaces and is used in a wide range of areas such as physics, materials science, and computer vision [23]. Our surface reconstruction method is based on a level set method proposed by Zhao [30]. In [30], an initial surface is continuously deformed toward a final surface in a potential flow direction. The final surface can be extracted as a polygonal model using the marching cubes method [13] and then rendered with standard graphics software.

4 Surface Growth Simulation

We now describe two fractal surface growth models: random deposition with surface relaxation (RDSR) and diffusion limited aggregation (DLA). While

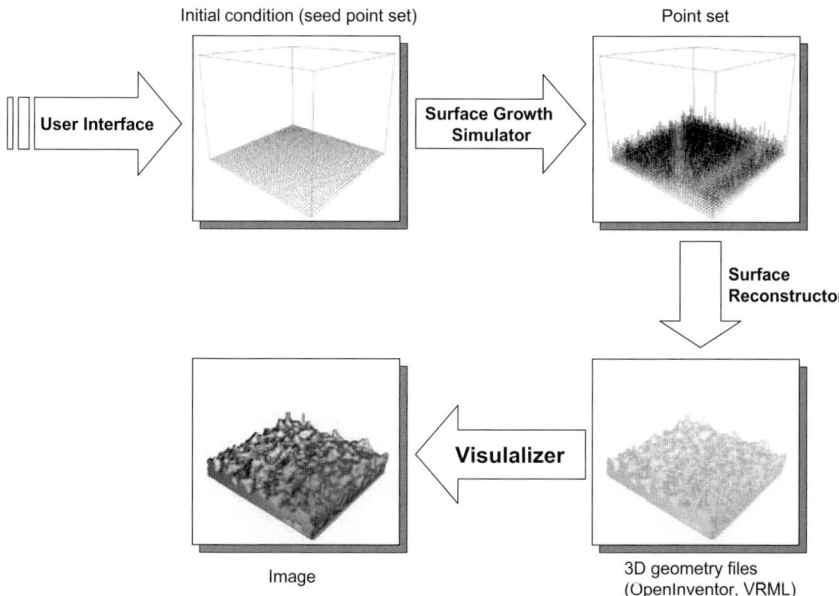

Fig. 2. Rough surface generator pipeline

RDSR is a local growth model, DLA is a nonlocal growth model. DLA generates more diverse surfaces than RDSR. More detailed descriptions of these and other methods can be found in [1].

4.1 Random Deposition with Surface Relaxation (RDSR)

We first explain the random deposition model (RD) because it easily leads to RDSR. RD is the simplest local growth model. From a randomly chosen site over the surface, a particle drops vertically until it reaches the top of the column under it, whereupon it is deposited (see Fig. 4.1). In RDSR, the deposited particle diffuses along the surface up to a finite distance, stopping when it finds the position with the lowest height (see Fig. 4.1). Due to the relaxation process, the final surface will be smoother than one generated by RD [1].

The most important difference between RD and RDSR is that an RDSR-generated surface is correlated through the relaxation process. The interface width, another measure of surface roughness, is defined by the root mean square fluctuation in the surface height. The interface width grows indefinitely for RD surfaces, but saturates for RDSR surfaces. RD surfaces look very rough and protrusive while RDSR surfaces appear smoother and more natural.

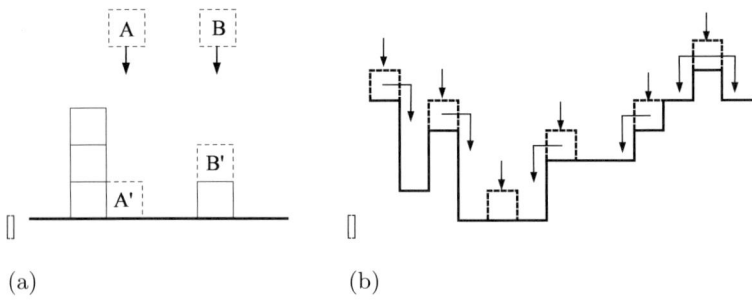

Fig. 3. (a) RD model, (b) RDSR model

4.2 Diffusion Limited Aggregation (DLA)

DLA is the most widely-known nonlocal cluster growth model. The working of the model is illustrated in Figure 4. A seed particle is fixed at a site in the bottom plane. A second particle is then released from a random position distant from the seed. It moves following a Brownian trajectory or a random walk until it reaches one of the four neighbor sites of the stationary seed whereupon it sticks with some probability forming a two-particle cluster. Then, a new particle is released which can stick to any of the five perimeter sites of the two-particle cluster. This process is then repeated. The diffusive effect is achieved through the use of Brownian motion of particles and the release of particles from clusters.

The nonlocality of DLA is due to the shadowing effect generated by the branches of the cluster. There is a much higher probability that a new released particle will be captured by the outlying portions of the cluster than in the interior regions. In other words, the interior region is shadowed by the branches on the perimeter (see Fig. 1). Hence, the growth rate depends not only on the local morphology, but also on the global geometry of the cluster. DLA can generate various surfaces from dendritic structures to a moss-like structure depending on the sticking probability and the nonlocal growth effect [21].

5 Surface Reconstruction

We now explain our level-set-based surface reconstruction algorithm. We follow the level set formulation in [30]. Since the method of [30] is targeted to the reconstruction of smooth surfaces, we employ a different potential flow for the reconstruction of rough surfaces in the level set formulation.

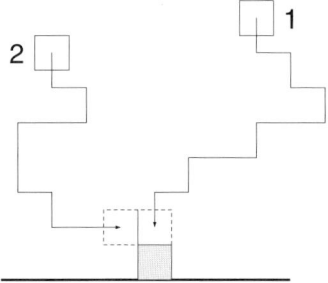

Fig. 4. Growth model for DLA

5.1 Level Set Formulation

In general, the surface growth simulators produce surfaces that do not have simple topologies. This makes explicit surface representation almost impossible to implement. The level set method is a powerful numerical technique for the deformation of implicit surfaces. The level set formulation works in any number of dimensions. The data structure is very simple and topological changes are handled easily.

The level set method was originally introduced by Osher and Sethian in [19] to capture evolving surfaces by curvature flow and has been successfully used to track interfaces for wide variety of problems. See [18, 23] for a comprehensive review. The two key steps of the level set method are described below.

Embed the Surface A co-dimension one surface Γ is defined as the zero isosurface of a scalar (level set) function $\varphi(\mathbf{x})$, i.e., $\Gamma = \{\mathbf{x} : \varphi(\mathbf{x}) = 0\}$. $\varphi(\mathbf{x})$ is negative inside Γ and positive outside Γ. In practice, the signed distance function is preferred as a level set function. Geometric properties of the surface Γ, such as the normal and mean curvature can be easily computed from $\varphi(\mathbf{x})$ using:

$$\text{outward unit normal:} \quad \mathbf{n} = \frac{\nabla\varphi}{|\nabla\varphi|} \quad (1)$$

$$\text{mean curvature:} \quad \kappa = \nabla \cdot \frac{\nabla\varphi}{|\nabla\varphi|}. \quad (2)$$

Embed the Motion The time evolution PDE for the level set function is obtained by differentiating $\varphi(\Gamma(t), t)$ to obtain

$$\varphi_t + \frac{d\Gamma(t)}{dt} \cdot \nabla\varphi = 0 \quad \Longleftrightarrow \quad \varphi_t + v_n |\nabla\varphi| = 0. \quad (3)$$

Here, v_n is the normal velocity of $\Gamma(t)$ which may depend on external physics or global and local geometric quantities.

To develop the level set PDE, one needs to extend the velocity, v_n in Eq. (3), which is given by the motion of the original surface. Let S denote a point set. Define $d(\mathbf{x}) = distance\_function(\mathbf{x}, S)$ to be the closest distance between the point $\mathbf{x}$ and S. We use the convection model of a surface Γ in a velocity field $\mathbf{v}(\mathbf{x})$ described by the PDE

$$\frac{d\Gamma(t)}{dt} = \mathbf{v}(\Gamma(t)). \tag{4}$$

Then, we can naturally extend the convection to all level sets of $\varphi(\mathbf{x}, t)$ to obtain

$$\frac{d\varphi}{dt} = -\mathbf{v}(\mathbf{x}) \cdot \nabla \varphi. \tag{5}$$

While Zhao used $\mathbf{v}(\mathbf{x}) = -\nabla d(\mathbf{x})$ in [30], we use $\mathbf{v}(\mathbf{x}) = -d(\mathbf{x})$ because the computation of $\nabla d(\mathbf{x})$ on a highly rough surface is very unstable. Thus, the level set formulation of our convection model is

$$\frac{d\varphi}{dt} = d(\mathbf{x})|\nabla \varphi|. \tag{6}$$

5.2 Numerical Implementation

There are three key numerical elements in our surface reconstruction. First, a fast algorithm is required to compute the distance function to an arbitrary data set on a rectangular grid. Second, we are required to find a good initial surface for our level set PDE to reduce the computational cost of solving the PDE. Third, we need a fast and stable solver for the PDE. As shown in Fig. (5), we obtain an initial surface, Γ_i by deforming the bounding surface Γ_0 following an approximate normal flow of Γ_0. Then, we deform the offset surface Γ_i to get the final surface Γ_f by solving Eq. (6).

Computing the Distance Function The distance function $d(\mathbf{x})$ to an arbitrary data set S is computed by solving the following Eikonal equation:

$$|\nabla d(\mathbf{x})| = 1, \quad d(\mathbf{x}) = 0, \; \mathbf{x} \in S. \tag{7}$$

We use the algorithm in [30] that combines upwind differencing with Gauss Seidel iterations of alternating sweeping orders to solve the differential equation (7). In two dimension, the following upwind differencing is used to discretize Eq. (7),

$$[(d_{i,j} - x_{min})^+]^2 + [(d_{i,j} - y_{min})^+]^2 = h^2 \tag{8}$$

where h is the grid size, n is the total number of grid points, $i = 1, \ldots, n, j = 1, \ldots, n$,

$$(x)^+ = \begin{cases} x & x > 0 \\ 0 & x \leq 0, \end{cases}$$

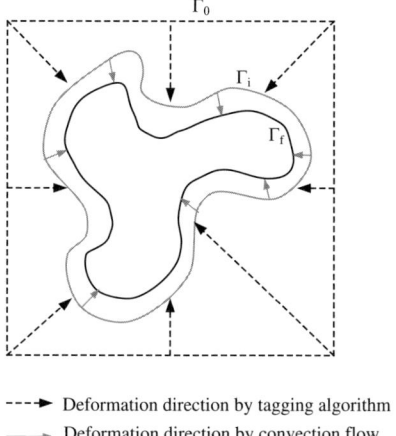

Fig. 5. Deformation methods (Γ_0: an exterior bounding surface, Γ_i: an initial surface for a level set solver, Γ_f: a final surface obtained by a level set solver)

and
$$x_{min} = \min(d_{i-1,j}, d_{i+1,j}) \quad y_{min} = \min(d_{i,j-1}, d_{i,j+1}).$$

The solution for Eq. (8) satisfies
$$\min(x_{min}, y_{min}) < d_{i,j} \leq \min(x_{min}, y_{min}) + h.$$

Hence, the exact solution for the nonlinear Eq. (8) is given by:

$$d_{i,j} = \begin{cases} \min(x_{min}, y_{min}) + h & if \ |\delta| \geq h \\ \dfrac{x_{min} + y_{min} + \sqrt{2h^2 - \delta^2}}{2} & if \ |\delta| < h \end{cases} \quad (9)$$

with $\delta = x_{min} - y_{min}$. Then, the distance function is obtained by solving Eq. (8) on every grid cell in the following four sweeping orderings:

$$(1) i = 1 : n, j = 1 : n \quad (2) i = 1 : n, j = n : 1$$
$$(3) i = n : 1, j = n : 1 \quad (4) i = n : 1, j = 1 : n.$$

Usually, the solution converges within five or six sweeps in two dimension, and nine sweeps in three dimension. See [28] for details and proofs.

Finding an Initial Surface In our approach, we continuously deform an initial surface to the final surface by following the convection flow direction. If we start with an initial surface that is too far from the real shape, it will take a long time to evolve the PDE. A good guess for the initial surface helps to speed convergence to the final surface. To find an initial surface such that

$\{\mathbf{x}: d(\mathbf{x}) = \epsilon\}$ where ϵ is an offset distance specified by the user, we employ a simple tagging algorithm based on a region growing method in discrete space.

We start from any initial exterior region such as a bounding box. Every grid cell is initially tagged as interior, boundary, or exterior. We denote the interior, boundary, and exterior region as Ω, $\partial\Omega$, and $\overline{\Omega}$ respectively. Let $d_{ij} = d(\mathbf{x}_{ij})$ be the unsigned distance of $\mathbf{x}_{ij}$ to the data set S. We say $\mathbf{x}_{ij} > \mathbf{x}_{kl}$ or $\mathbf{x}_{ij}$ is farther than $\mathbf{x}_{kl}$ or $\mathbf{x}_{ij}$ is larger than $\mathbf{x}_{kl}$ if $d_{ij} > d_{kl}$. We deform the initial tagged boundary $\partial\Omega$ to the final tagged boundary using the tagging algorithm in Algorithm 1.

Require: $S \in \Omega$
1: ϵ: offset distance, $\partial\Omega$: a tagged boundary
2: **while** maximum distance of $\partial\Omega \geq \epsilon$ **do**
3:    Pick the most distant point $\mathbf{x}_{<ij} \in \partial\Omega$
4:    **if** All interior neighbors of $\mathbf{x}_{ij}$ are closer to S **then**
5:       Add $\mathbf{x}_{ij}$ into $\overline{\Omega}$ and Put its interior neighbors into $\partial\Omega$.
6:    **end if**
7: **end while**
8: The final $\partial\Omega$ is the offset surface.

Algorithm 1: Tagging algorithm to find an initial offset surface, $d(\mathbf{x}) = \epsilon$

We maintain a priority queue for $\partial\Omega$ so that the most distant point can be identified quickly. After tagging, we recompute the distance function for the tagged boundary. We obtain the signed distance function by negating the distance function at all interior cells.

Solving the Level Set PDE We can continuously deform the initial signed distance function, $\varphi(\mathbf{x})$, by solving the level set PDE given in Eq. (3). If we solve the PDE in a brute force way, the computational cost is $O(N^3)$ at each time step for the grid size N. The computational cost reduces to $O(N^{2/3})$ using the fast local level set method [22]. Instead of computing on every grid cell, the computation is restricted to a narrow tube around the zero level set (see Fig. 6). Since the solution of Eq. (3) often becomes very flat or steep at the front $\Gamma(t)$, a redistancing algorithm is needed to keep $\varphi(\mathbf{x}, t)$ a signed distance function and smooth in a neighborhood of the front. An upwind scheme is used for space discretization of Eq. (3), and an essentially non-oscillatory Runge-Kutta scheme is used for time approximation. Details for the discretization scheme can be found in [19, 29].

We outline the main algorithm.

1. Update tubes, T and N, where

$$T = \{\mathbf{x} : |\varphi(\mathbf{x})| < \gamma\}$$
$$N = \{(x_i, y_i) : \min_{-1 \leq \nu, \mu \leq 1} |\varphi_{i+\nu, j+\mu}| \leq \gamma\}.$$

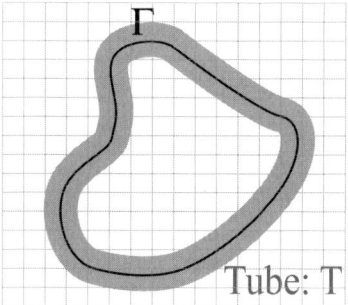

Fig. 6. Computation is only performed on the gray region (Tube: T) around the zero level set Γ

The level set is advanced in time in tube T while the redistancing step is performed in tube N.

2. Advance: Update φ in tube T for one time step to obtain $\tilde{\varphi}$ by an ODE time stepping method. Instead of using v_n in Eq. (3), $c(\varphi)v_n$ is used to prevent numerical oscillations at the tube boundary, where the cut-off function, $c(\varphi)$, is defined by:

$$c(\varphi) = \begin{cases} 1 & if \ |\varphi| \leq \beta \\ \dfrac{(|\varphi| - \gamma)^2(2|\varphi| + \gamma - 3\beta)}{(\gamma - \beta)^3} & if \ \beta < |\varphi| \leq \gamma \\ 0 & if \ |\varphi| > \gamma \end{cases} \quad (10)$$

3. Redistance: Apply the redistancing step to $\tilde{\varphi}$ on the tube N. Evolve the following Hamilton-Jacobi equation until $d(\mathbf{x}, \tau)$ reaches a steady state solution $d_s(\mathbf{x})$:

$$d_\tau + S(d)(|\nabla d| - 1) = 0,$$

$$d(\mathbf{x}, 0) = d_0(\mathbf{x}) = \tilde{\varphi}(\mathbf{x}, t), \quad (11)$$

$$S = \dfrac{d}{\sqrt{d^2 + |\nabla d|^2 h^2}}.$$

If d_0 is already close to a distance function, the redistancing operation usually takes only one or two iterations within the tube N.

4. Update the new φ by:

$$\varphi(\mathbf{x}) = \begin{cases} -\gamma & if \ d_s(\mathbf{x}) < -\gamma \\ d_s(\mathbf{x}) & if \ |d_s(\mathbf{x})| \leq \gamma \\ \gamma & if \ d_s(\mathbf{x}) > \gamma \end{cases} \quad (12)$$

6 Results

We use the *vispack* library [26] to implement parts of the surface reconstructor. The zero level set or the required implicit surface is extracted using suitable vispack routines that invoke the marching cubes method [13]. The result is available as VRML output. Since VRML is a common three-dimensional format supported by many free and commercial renderers, it provides a flexible choice to users depending on their needs. The computations were conducted on a SGI workstation with a 225MHZ MIPS R10000 processor and 1 GByte of memory. For a 100^3 computational grid, it takes 3–4 minutes to compute a distance function and an initial surface respectively. It takes about 20 seconds for the first order solver, and a minute with the second order solver for one time stepping. We use the second order solver to obtain the presented results.

Fig. 7 shows the deformation of the bunny model by the convection flow given by Eq. (6). The initial surface, Fig. 7(b) is the approximate offset surface from the true surface, i.e. $\{\mathbf{x} : d(\mathbf{x}) = h\epsilon\}$, which is obtained by the tagging algorithm. The initial surface displays aliasing artifacts since the tagging algorithm is a procedural rather than a numerical method. It should be noted that our convection flow is good enough for rough surface characterization, though the result is not as smooth as the one using the weighted minimal surface model in [30]. In the weighed minimal surface model, an additional curvature term regularizes the surface, which is not desired for rough surfaces.

In Fig. 8, we show a rough surface generated by a RDSR simulation on a sphere. An initial surface with an offset of $\epsilon = 3$ is used for all rough surface examples. The final surface in Fig. 8(b) is naturally rough.

Fig. C.4 shows a rough terrain. It is generated from a DLA simulation which is performed on a 64^3 grid and the computational grid size is $115 \times 115 \times 49$. This example illustrates how well the level set method captures complex geometries and topologies including arches. The arches are clearly seen in Fig. C.5. Fig. C.6 is another example of the DLA simulation with different initial conditions. It demonstrates that our method can produce visually appealing natural scenery.

7 Conclusions and Future Work

We presented a new rough surface modeling technique using a fractal surface growth model and a level-set-based method for surface extraction. Our method is flexible because of its modular design. A fractal surface growth model guarantees the surface is natural and rough. It also provides the user some control over the shape of the resulting surface. The implicit representation obtained using a level set method handles the resulting complex topologies naturally. It is simple and easy to implement as well. We generated very promising results using these two methods in combination.

An immediate problem is the control of the roughness of the surface, perhaps using the interface width, fractal dimension, or some other measure. One approach is to vary the input conditions. For example, by simulating DLA on a complex object surface, we can generate more interesting results than a typical displacement mapping method. We may use an image as the distribution of the initial seeds for DLA. Another alternative would be to employ environmental fields. Two issues of concern associated with the level set method are the computational cost and the stability of the PDE solver. It may be possible to employ fast and robust methods that are potentially less accurate for computer graphics applications. We are planning to investigate the adaptive, semi-Lagrangian method presented in [24]. Finally, it would be interesting to add roughness directly to the level set function grid by applying a physics-based velocity function such as the dendritic growth in a Stefan problem [2].

References

1. A. L. Barabási and H. E. Stanley. *Fractal Concepts In Surface Growth*. Cambridge University Press, Cambridge, 1995.
2. S. Chen, B. Merriman, S. Osher, and P. Smereka. A simple level set method for solving stefan problems. *J. Comput. Phys.*, 135:8–29, 1997.
3. J. Dorsey and P. Hanrahan. Modeling and rendering of metallic patinas. In Holly Rushmeier, editor, *Proceedings of ACM SIGGRAPH 1996*, Computer Graphics Proceedings, Annual Conference Series, pages 387–396, August 1996.
4. D. Ebert, K. Musgrave, D. Peachey, K. Perlin, and S. Worley. *Texturing and Modeling: A Procedural Approach*. Academic Press, New York, 1994.
5. P. Fearing. Computer modelling of fallen snow. In Kurt Akeley, editor, *Proceedings of ACM SIGGRAPH 2000*, Computer Graphics Proceedings, Annual Conference Series, pages 37–46, New York, July 2000. ACM, ACM Press / ACM SIGGRAPH.
6. K. W. Fleischer, D. H. Laidlaw, B. L. Currin, and A. H. Barr. Cellular texture generation. In Robert Cook, editor, *Proceedings of ACM SIGGRAPH 1995*, Computer Graphics Proceedings, Annual Conference Series, pages 239–248, August 1995.
7. A. Fournier, D. Fussell, and L. Carpenter. Computer rendering of stochastic models. *Communications of the ACM*, 25(6):371–384, June 1982.
8. G. Y. Gardner. Functional modeling of natural scenes. *SIGGRAPH Course Notes: 28 Functional Based Modeling*, 28:41–49, 1988.
9. N. Greene. Voxel space automata: Modeling with stochastic growth processes in voxel space. *Computer Graphics(Proceedings of ACM SIGGRAPH 89)*, 23(3):175–184, July 1989.
10. J. Hart. Implicit representation of rough surfaces. *Implicit Surfaces'95*, pages 33–44, April 1995.
11. J. P. Lewis. Generalized stochastic subdivision. *ACM Transactions on Graphics*, 6(3):167–190, July 1987.
12. J. P. Lewis. Algorithms for solid noise synthesis. *Computer Graphics(Proceedings of ACM SIGGRAPH 89)*, 23(3):263–270, July 1989.

13. W. E. Lorensen and H. E. Cline. Marching cubes: A high resolution 3D surface construction algorithm. *Computer Graphics(Proceedings of ACM SIGGRAPH 87)*, 21(4):163–169, July 1987.
14. B. B. Mandelbrot. *The Fractal Geometry of Nature*. W. H. Freeman and Co., San Francisco, 1982.
15. G. A. Mastin, P. A. Watterberg, and J. F. Mareda. Fourier synthesis of ocean scenes. *IEEE Computer Graphics and Applications*, 7(3):16–23, March 1987.
16. G. S. P. Miller. The definition and rendering of terrain maps. *Computer Graphics(Proceedings of ACM SIGGRAPH 86)*, 20(4):39–48, August 1986.
17. K. Musgrave, C. E. Kolb, and R. S. Mace. The synthesis and rendering of eroded fractal terrains. *Computer Graphics(Proceedings of ACM SIGGRAPH 89)*, 23(3):41–50, July 1989.
18. S. Osher and R. Fedkiw. Level set methods: an overview and some recent results. *Journal of Computational Physics*, 169:463–502, 2001.
19. S. Osher and J. Sethian. Fronts propagating with curvature dependent speed: Algorithms based in hamilton-jacobi formulations. *Journal of Computational Physics*, 79:12–49, 1988.
20. H. O. Peitgen and S. Dietmar, editors. *The Science of Fractal Images*. Springer-Verlag, New York, 1988.
21. H. O. Peitgen, H. Jürgens, and S. Dietmar, editors. *Chaos and Fractals*. Springer-Verlag, New York, 1992.
22. D. Peng, B. Merriman, H. Zhao, S. Osher, and M. Kang. A pde based fast local level set method. *Journal of Computational Physics*, 155:410–438, 1999.
23. J. Sethian. *Level Set Methods and Fast Marching Methods*. Cambridge University Press, 1999.
24. J. Strain. A fast modular semi-lagrangian method for moving interfaces. *Journal of Computational Physics*, 161:512–528, 2000.
25. R. F. Voss. Random fractal forgeries. In R. A. Earnshaw, editor, *Fundamental Algorithms for Computer Graphics*. Springer-Verlag, Berlin, 1988.
26. R. T. Whitaker. Vispack:a c++ object oriented library for processing volumes, images, and level set surface models. 2002.
27. S. Worley. A cellular texture basis function. In H. Rushmeier, editor, *Proceedings of ACM SIGGRAPH 1996*, Computer Graphics Proceedings, Annual Conference Series, pages 291–294, August 1996.
28. H. Zhao. Fast sweeping method for eikonal equations. *preprint*, 2002.
29. H. Zhao, T. Chan, B. Merriman, and S. Osher. A variational level set approach to multiphase motion. *Journal of Computational Physics*, 127:179–195, 1996.
30. H. Zhao, S. Osher, and R. Fedkiw. Fast surface reconstruction using the level set method. In *Proceedings of IEEE Workshop on Variational and Level Set Methods in Computer Vision (VLSM 2001)*, pages 194–201, Jul 2001.

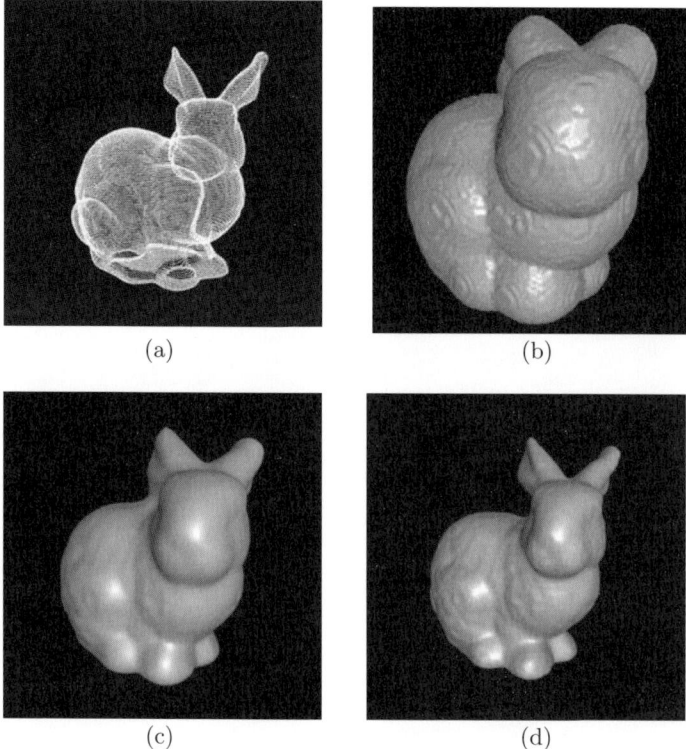

Fig. 7. Bunny deformed by the convection flow, $135 \times 134 \times 112$ grid: (a) point set, (b) initial surface with $\epsilon = 17$, (c) 100 iterations, (d) 200 iterations

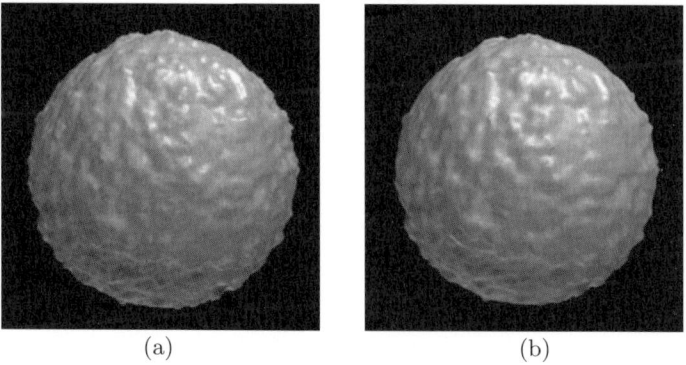

Fig. 8. RDSR simulation, $112 \times 115 \times 114$ grid: (a) initial surface with $\epsilon = 3$, (b) 30 iterations

A Feature Based Method for Rigid Registration of Anatomical Surfaces

Georgios Stylianou

Arizona State University Tempe AZ 85287 *stylianou@asu.edu*

Summary. The problem is: given two three-dimensional (3D) triangulated surfaces A, B find the best rigid transformation that brings surface A as close as possible to surface B. The most commonly used method in the literature is iterative closest point (ICP). ICP uses either the whole surface description or feature points on the surfaces. The drawback of ICP is that it requires a good initial estimation to achieve convergence. This paper proposes a variation of ICP, randomized iterative closest curve (RICC). The input of RICC is feature curves of the input surfaces. The input surfaces, in our examples, are 3D triangulated brain structures. The feature curves are mathematically based shape features, crest lines that have anatomical significance on brain structures. The advantage of RICC versus ICP is that it does not need an initial estimation to converge to a global minimum. Also, RICC exploits the structurally meaningful feature lines instead of using just the points that are structurally meaningless. Randomization is used to boost its performance in hard cases. Experimental results show that RICC achieves fast convergence in all cases, regardless of input surfaces' pose.

1 Introduction

In the field of medical imaging, the complex 3D structures of multisectional images or their 3D reconstructions are sometimes difficult to understand and process. Especially when the anatomy is highly varied or if abnormality is present the correlation of visible structure information prior and during a disease provides even more challenges for diagnosis and treatment. For example, in the clinical diagnosis or prognosis of Alzheimer's disease it is useful to be able to evaluate the local or global deformations of the brain for different time frames. Precise methods for fully automated surface registration are required such that qualitative and quantitative studies can be done on brain structures.

Many methods exist for rigid registration of 2D images or 3D surfaces and most of them are based on the Iterative Closest Point (ICP) method first introduced by Besl and McKay [2] and Zhang et al. [19] independently. Feldmar and Ayache [8] perform rigid and affine registration by minimizing the combined distance between positions, surface normals and curvatures. Sharp et al. [15] use invariant features for correspondence selection in their modified ICP algorithm that works well under noise-free conditions. Hsu et al. [9] use edge and surface features and implement classical ICP to perform rigid registration of brain surfaces. Maurer et al. [12] register 3-D images by implementing a variance of ICP by using weighted geometrical features. Hsu and Maurer do not even give the initial estimation they use. Declerck et al. [5] use feature points and ICP. Dalley [4] and Rusinkiewicz [13] give good surveys that compare all ICP variants. Other ICP based methods are [3, 10, 11].

A deficiency of ICP based algorithms is that an initial estimation is necessary to bring the source surface as close as possible to the target surface prior to running the algorithm [2, 19]. This drawback of getting an initial estimation complicates the ICP method significantly because usually initial estimations require more information than just feature points or curves. For initial estimation, Feldmar and Ayache [8] use principal curvature and principal frame, Sharp [15] uses moment invariants in conjuction with a modified distance metric and Declerck [5] assumes the images to be registered are approximately aligned. In addition the algorithm may have to be executed several times prior to convergence [10, 12]. Furthermore, feature points on 2D images or 3D surfaces used for the registration, even though are significant, are structurally meaningless. Finally, ICP has slow monotonic convergence when and if it converges.

In this article, we propose a method for rigid registration of 3D brain structures conceptually similar to ICP but with the advantage of fast convergence. The algorithm is fully automated and does not need initial estimation to be performed. We name this method Randomized Iterative Closest Curve (RICC). The input of RICC is feature curves instead of feature points that are the input of ICP. We use as input curves feature lines known as crest lines. Crest lines are significant feature lines on the brain structure. They have specific names and they are known to be located in areas of the brain that control a human's cognitive and body functions. In other words, crest lines can alone partially describe the brain structure. The brain is the main but not the only application of crest lines. In previous work, we have developed a fully automatic algorithm for crest line extraction from 3D triangulated surfaces [16]. Since then we have improved the algorithm's robustness to noise [17]. The proposed algorithm prior to randomization is already faster and has better convergence than ICP but it still suffers the problem that when the source and target surfaces are not close to each other it does not converge. Therefore, we added the randomization and boosted its performance.

2 Crest Lines

Crest lines extraction was addressed in prior work [16, 17]. Here we give the crest lines definition.

Definition. Crest lines are defined as the set of of all points satisfying

$$D_{\mathbf{t}_1} k_1(u, v) = 0 \tag{1}$$

where D_{t_1} is the directional derivative in direction t_1, k_1 is the largest principal curvature and $\mathbf{t}_1$ is its domain direction on a point (of the surface) with domain coordinates (u, v).

They are shape features with the main characteristic of using local information to yield a global description of the surface. Crest lines are local shape features by definition. But when all the crest lines are viewed together, they partially describe the surface.

Figure 1 shows an example of a crest line. Because a crest point has maximum largest curvature in its corresponding direction, a crest line naturally follows the direction of the smallest curvature of its composing crest points.

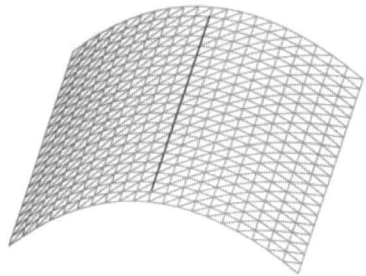

Fig. 1. A crest line example.

Also, figures 2 and 3 show parts of brain surface with crest lines that are identical to data used in the experiments.

3 Randomized Iterative Closest Curve

The proposed method is very similar to Iterative Closest Point (ICP) as first stated by Besl and McKay [2] and Zhang [19] independently in 1992. The input of the algorithm is two sets of curves A, B. Every curve is an ordered sequence of points in R^3.

1. Make a random rotation.
2. Repeat
3. Create pairs (a, b) where $a \in A$, $b \in B$, such that $H(a, b)$ is minimum where H is the Hausdorff distance.
4. Create pairs $(\mathbf{p}_i, \mathbf{q}_i)$, $\{\mathbf{p}_i\} \in a$, $\{\mathbf{q}_i\} \in b$.
5. Compute rigid motion $(R, \mathbf{t})$ such that $\frac{1}{N}\sum_{i=1}^{N} ||R\mathbf{p_i}+\mathbf{t_i}-\mathbf{q_i}||$ is minimum where N is the number of pairs, R is a 3×3 rotation matrix, and $\mathbf{t}$ is a translation vector in R^3, and apply the transformation.
6. If $(R, \mathbf{t})$ does not bring A closer to B then undo step 5 and apply a random rotation R.
7. Until the termination criterion is reached.

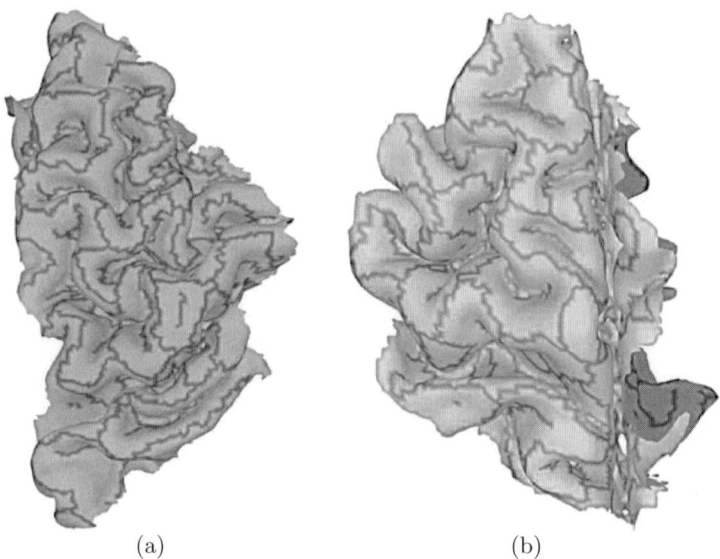

Fig. 2. Crest Lines on parts of a brain surface.

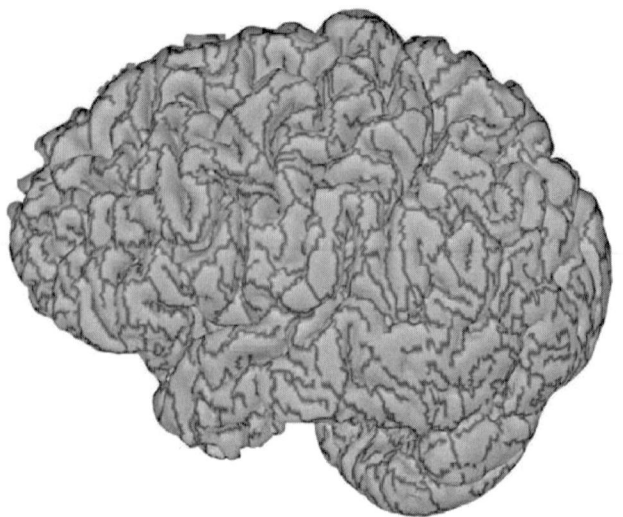

Fig. 3. Crest lines on the cortical surface.

3.1 Initial Random Rotation

The initial random rotation is performed to bring the rotation angles between the input surfaces A and B in general position. The reason is that if the two surfaces are not in general position, the convergence rates can be in certain cases significantly higher than if they were in general position. This was suggested by the experimental results, as shown in the results section. In addition, this step can be regarded as using a random rotation to give the surfaces an unknown distance instead of a predefined one.

3.2 Curve Pairing

Curve pairing is the first step of the algorithm. It is a function $f : A \to B$ where $f(a) = b$ if $H(a,b)$ is minimum, $a, b \subset R^3$ and $a \in A$, $b \in B$. That is every curve $a \in A$ is paired with its closest curve $b \in B$, and their distance $H(a,b)$ is calculated using the Hausdorff distance. The Hausdorff distance of two curves $a, b \subset R^3$ is:

$$H(a,b) = \max(\mathbf{H}(a,b), \mathbf{H}(b,a)) \qquad (2)$$

$$\mathbf{H}(a,b) = \max_{\mathbf{p} \in a} \min_{\mathbf{q} \in b} \| \mathbf{p} - \mathbf{q} \| \qquad (3)$$

Function f must be one-to-one but does not need to be onto. As it is described it is not one-to-one but it is onto. To make it one-to-one, we delete all the curves $a_i \in A$ mapped to the same curve $b \in B$ but one. We keep a pair (a_i, b) only if it has smaller Hausdorff distance than the pairs (a_j, b), where $i\ n(e^-)j$. The reason we delete the rest of the pairs is that a rigid motion cannot match many curves to one curve.

3.3 Point Pairing

The input of step 2 is a pair (a, b) where a, b are two curves. The objective is to create point pairs $(\mathbf{p}, \mathbf{q})$ such that $\mathbf{p} \in a$ and $\mathbf{q} \in b$. Suppose $|a| = n$ and $|b| = m$ then we have three cases:

case $n = m$: The point pairs are $(\mathbf{p}_1, \mathbf{q}_1), (\mathbf{p}_2, \mathbf{q}_2), ..., (\mathbf{p}_n, \mathbf{q}_n)$.
case $n < m$: The point pairs are $(\mathbf{p}_1, \mathbf{q}_k), (\mathbf{p}_2, \mathbf{q}_{k+1}), ..., (\mathbf{p}_n, \mathbf{q}_{k+n-1})$ where k is such that $\sum_{j=1}^{n} \| \mathbf{p}_j - \mathbf{q}_{k+j-1} \|$ is minimum for some $k \in [1, m-n+1]$.
case $n > m$: The point pairs are $(\mathbf{p}_k, \mathbf{q}_1), (\mathbf{p}_{k+1}, \mathbf{q}_2), ..., (\mathbf{p}_{k+m-1}, \mathbf{q}_m)$ where k is such that $\sum_{j=1}^{m} \| \mathbf{p}_{k+j-1} - \mathbf{q}_j \|$ is minimum for some $k \in [1, n-m+1]$.

3.4 Calculating the Rigid Motion

After creating all the point pairs, we calculate the rigid motion that consists of a rotation and a translation. A rotation is a 3×3 matrix and a translation is a vector in R^3. Several methods exist to calculate the rigid motion including singular value decomposition [1], quaternion [7] and dual quaternion [18]. Eggert [6] compared those methods and reported that they have little differences. Here, we use the dual quaternion method.

3.5 Random Rotation

The method as described so far is already more powerful than ICP but still has the problem of convergence to a local minimum, when dealing with hard cases, for more details see the results. The iterative algorithm, similarly to ICP, converges to some minimum. If that minimum is not satisfactory, then the only way to jump out of it is to apply some rigid motion. The rigid motion must be large enough to jump out of the local minimum (and not get back again) but should also be small enough such that we do not miss a better minimum (or global minimum). The only way to achieve that is to apply a random rotation. By random I mean that a random number generator is used to give the rotation angles used to build the random rotation matrix. The matrices used for random rotations are created the same way. In the results, we show the number of iterations until convergence for different random rotation ranges.

3.6 Termination Criteria

Termination criteria can be one of the following:

- When we reach a certain threshold.
- When the minimum does not improve during a certain number of iterations.
- When we exceed a certain number of iterations.

In our examples we already know the global minimum. Thus we terminate when we reach that minimum. But in the general case the preferred stopping criterion would usually be when the minimum does not improve during a certain number of iterations.

4 Computational Complexity

The computational complexity of ICP is $O(knm) + c$. k is the number of iterations till convergence and n, m are the numbers of feature points of the source and target surfaces, respectively. c is the time complexity for the initial estimation, which in some cases needs up to the same time needed for convergence.

The computational complexity of RICC is $O(kNM)$. k is the number of iterations till convergence and N, M are the numbers of feature curves of the source and target surfaces, respectively. Also, Hausdorff distance requires $O(n_i m_j)$ time, where n_i is the number of points of source surface's feature curve i and m_j is the number of points of target surface's feature curve j.

Data structures such as k-d trees and range trees can be used to make both algorithms more efficient. After analyzing the computational complexities of ICP and RICC we observe that they are quite similar. RICC is slightly faster because it does not have initial estimation and exhibit faster convergence. In addition, it can be spedup after utilizing the structure of the feature curves to reduce the total number of points, as it is shown below.

To make the algorithm faster, we have approximated the crest lines with linear splines. We use linear splines to calculate the Hausdorff distance. Because the calculation of the Hausdorff distance of two curves a, b takes $O(nm)$ time, where n, m is the number of points of curves a, b respectively, reducing the number of points of a, b speeds up the calculation of the Hausdorff distance. After approximating the curves with linear splines, we have reduced the number of points of each curve by $65\% - 90\%$. This yields a reduction of operations to get the Hausdorff distance by $88\% - 99\%$. Eventually this amounts to a total reduction of points of about 80%.

The reduction is dramatic. In conjuction, with RICC for fast rigid registration this result is even better. Furthermore, if we did not have feature curves, which have structure incorporated, we could not apply this method.

5 Results

We have experimented on various different implementations of RICC. RICC is compared with ICP, a randomized implementation of ICP and a derandomized implementation of RICC.

For the experiments real brain data (cortical surface) have been used. I have experimented using the same surfaces because the global minimum is known to be zero. I did not use data where the global minimum is unknown because due to the complexity of the data and the fact that they are located in the three-dimensional space, it is extremely hard to evaluate whether ICP or RICC have reached the global minimum. The cortical surfaces had approximately 20000 vertices, and the valley crest lines consist of about 1000 points which is 5% of the initial data.

Table 1 shows the convergence rates for ICP, ICC, $RICC^1$, $RICC^2$. Entries with 'x' mean that the algorithm did not converge. The experiments are based on exact and approximate matching. The first column shows the rotation angles of surface B vs. surface A. The rotation angles are manually selected to check the performance of the algorithms in these (easy and hard) cases. ICP is the standard iterative closest point algorithm but without initial estimation. ICC is the iterative closest curve algorithm without the random step. $RICC^1$ is the randomized ICC which performs a random rotation about the x and z axes. $RICC^2$ performs random rotation about the z axis. $RICC^1$ and $RICC^2$ do not make an initial random rotation.

ICP exhibits extremely slow convergence when and if it converges. The randomized implementation of ICP does not behave any better. The reason is the monotonicity of ICP. Because ICP is strictly monotone, in every iteration it moves closer to a minimum than the previous iteration, until it converges to some minimum. On the other hand, ICC is not strictly monotone. ICC is much faster than ICP but even though it converges for some harder cases, it still suffers from the same problem. The enhancement is the randomization. $RICC^1$ usually converges and it is observed that it is the same as ICC with random choices. When RICC converges monotonically to a minimum it is the same as ICC; thus they need the same number of iterations to converge. But when ICC has trouble the random choice boosts the performance of the algorithm to achieve - usually fast - convergence.

The problem of RICC is that its behavior is based on the random rotation in conjunction with the input surfaces. A random rotation about the x, y, z axis or any of them combined affects the convergence rate significantly. For instance, when the input surface B is surface A rotated about the z axis, performing a random rotation about the z axis we have a very fast convergence and random rotations about the x axis results in a very slow convergence. This is not shown very clearly in table 1 but when surface B is rotated about the x or y axis the convergence rates go up to 150 iterations for $RICC^1$. But when RICC was doing random rotations about the x or y axis then the convergence rates were going only up to 30 iterations.

Table 1. Convergence rates for iterative closest point (ICP), iterative closest curve (ICC), random ICC with random rotation about the x and z axes ($RICC^1$) and random ICC with random rotation about the z axis ($RICC^2$).

Rotation	ICP	ICC	$RICC^1$	$RICC^2$
(0,0,45)	18	4	4	4
(0,0,60)	34	11	22	10
(0,0,90)	83	12	14	9
(0,0,180)	x	111	32	20
(30,30,30)	x	11	11	11
(45,45,45)	x	x	8	11
(60,60,60)	x	x	13	56
(180,0,90)	x	x	23	x

The solution to this drawback was to add an initial random rotation; thus putting the initial surface in general position. The random rotation is performed always about the x and z axes and the results are shown in table 2. Now, in some cases there is slower convergence than prior the initial random rotation but in general the algorithm exhibits better, more balanced convergence rates.

So far, we have observed that the algorithm works very well when dealing with exact matching. We also have experimented with approximate matching to evaluate the robustness of the algorithm. Approximate matching means that feature lines extracted from surface A are a subset of the feature lines extracted from surface B. Surfaces A and B are exactly the same but they are represented by a different number of feature lines. These results are in table 3.

Table 2. Convergence rates for exact matching in the general case RICC.

Rotation	RICC	Rotation	RICC
(0,0,45)	12	(0,90,0)	34
(0,0,60)	24	(0,180,0)	62
(0,0,90)	30	(45,0,0)	39
(0,0,180)	12	(60,0,0)	12
(45,45,45)	21	(90,0,0)	28
(60,60,60)	12	(180,0,0)	61
(0,45,0)	36	(180,0,90)	58
(0,60,0)	49		

They show that the algorithm is robust. Even though the convergence rate is slower, the algorithm converges in those more realistic cases.

Table 3. Convergence rates for approximate matching in the general case RICC.

Rotation	RICC
(45,0,0)	15
(180,0,0)	61
(0,45,0)	88
(0,0,45)	9
(0,0,180)	71
(45,45,45)	34
(60,60,60)	42
(180,0,90)	133

The experimental results suggest that when the source and target surfaces differ by a rotation about only one axis (x, y or z) then the best random rotation axis is that axis, meaning we will have faster convergence if we use that axis only. But unless we know a priori that they differ only about one axis, we must cover all the cases by using a random rotation about at least two axes.

Finally, figure C.7 shows the 12 iterations of RICC till convergence in the case where the target surface is rotated 45 degrees about z axis. In addition, it shows the bounding box of the source surface that is recalculated in every iteration.

6 Conclusions

The problem addressed in this article was the rigid registration of two surfaces. The standard method is iterative closest point that suffers from slow convergence, does not always converge and uses points on the surfaces which are, in essence, structurally meaningless. Instead, we have proposed a feature-based method, iterative closest curve, that has faster convergence than ICP and uses meaningful line features. We enhanced ICC with randomization and generalized it such that it can get fast convergence for any rotation. By conducting experiments on real data, we showed that RICC is robust and efficient. Finally, we have further sped it up by approximating every curve by a linear spline.

Acknowledgments

This work was supported in part by the Arizona Alzheimer's Disease Research Center.

References

1. Arun, K., Huang, T., Blostein, S. (1987): Least-squares fitting of two 3-D point sets. IEEE Trans. PAMI, **9**, 698–700
2. Besl, P.J., McKay, N.D. (1992): A method for Registration of 3-D Shapes. IEEE Trans. Pattern Analysis and Machine Intelligence, **14**, **2**, 239–256
3. Chen, C.-S., Hung, Y.-P., Cheng, J.-B. (1998): A Fast Automatic Method for Registration of Partially-Overlapping Range Images. IEEE Sixth Int'l Conf. on Computer Vision, 242–248
4. Dalley, G., Flynn, P. (2001): Range Image Registration: A Software Platform and Empirical Evaluation. Proceedings of IEEE Third Int'l Conf. on 3-D Digital Imaging and Modeling, 246–253
5. Declerck, J., Feldmar, J., Goris, M.L., Betting, F. (1997): Automatic Registration and Alignment on a Template of Cardiac Stress and Rest Reoriented SPECT Images. IEEE Transactions on Medical Imaging, **16**, **6**, 727–737
6. Eggert, D.W., Lorusso, A., Fisher, R.B. (1997): Estimating 3-D Rigid Body Transformations: A Comparison of Four Major Algorithms. Machine Vision and Applications, **9**, 272–290
7. Faugeras, O., Hebert, M. (1986): The representation, recognition, and locating of 3D shapes from range data. Int'l J. Robotics Res., **5**, **3**, 27–52
8. Feldmar, J., Ayache, N.J. (1996): Rigid, Affine and Locally Affine Registration of Free-Form Surfaces. Int'l J. Computer Vision, **18**, **2**, 99–119
9. Hsu, L.-Y., Loew, M.H., Ostuni, J. (1999): Automated Registration of Brain Images Using Edge and Surface Features. IEEE Engineering in Medicine and Biology Magazine, **18**, **6**, 40–47
10. Kapoutsis, C.A., Vavoulidis, C.P., Pitas, I. (1999): Morphological Iterative Closest Point Algorithm. IEEE Transactions on Image Processing, **8**, **11**, 1644–1646
11. Luck, J., Little, C., Hoff, W. (2000): Registration of Range data using a Hybrid Simulated Annealing and Iterative Closest Point Algorithm. Proceedings of the 2000 IEEE Int'l Conf. on Robotics and Automation, **4**, 3739–3744

12. Maurer Jr, C.R., Aboutanos, G.B., Dawant, B.M., Maciunas, R.J., Fitzpatrick, J.M. (1996): Registration of 3-D Images Using Weighted Geometrical Features. IEEE Transactions on Medical Imaging, **15**, **6**, 836–849
13. Rusinkiewicz, S., Levoy, M. (2001): Efficient Variants of the ICP Algorithm. Proceedings of the IEEE Third Int'l Conf. on 3-D Digital Imaging and Modeling, 145–152
14. Rössl, C., Kobbelt, L., Seidel, H.-P. (2000): Extraction of feature lines on triangulated surfaces using morphological operators. AAAI 2000 Spring Symposium Series "Smart Graphics", March 20–22, Stanford, CA, USA
15. Sharp, G.C., Lee, S.W., Wehe, D.K. (2002): ICP Registration Using Invariant Features. IEEE Transaction on Pattern Analysis and Machine Intelligence, **24**, **1**, 90–102
16. Stylianou, G., Farin, G. (2003): Shape Feature Extraction. In: Farin, G., Hammann, B., Hagen, H. (ed) Hierarchical Approximation and Geometrical Methods for Scientific Visualization. Springer Berlin Heidelberg
17. Stylianou, G. (2003): Automatic Crest Line Extraction from Anatomical Surfaces. Proceedings of SPIE Medical Imaging Conference, to appear
18. Walker, M.W., Shao, L., Volz, R.A. (1991): Estimating 3-D location parameters using dual number quaternions. CVGIP: Image Understanding, **54**, 358–367
19. Zhang, Z.Y. (1994): Iterative Point Matching for Registration of Free-Form Curves and Surfaces. Int'l J. Computer Vision, **13**, **2**, 119–152
20. Zhang, Z.Y. (1992): Iterative Point Matching for Registration of Free-Form Curves and Surfaces. technical report RR-1658, Inria, France

Part III

Wavelets and Compression on Surfaces

Lifting Biorthogonal B-spline Wavelets

Martin Bertram

University of Kaiserslautern, P.O. Box 3049, D-67653 Kaiserslautern
bertram@informatik.uni-kl.de

Summary. Multiresolution data representations provide an indispensable tool for the compression, progressive transmission, and visualization of scientific data. Wavelet transforms based on B-spline scaling functions are frequently used to obtain continuous surface- and volume representations at multiple levels of resolution. Starting with a fine-resolution data set, wavelet decomposition provides a sequence of coarser approximations based on B-spline scaling functions and a set of wavelet coefficients containing the geometric differences with respect to the finer levels. The inverse transform can be used to reconstruct the finer levels of resolution from these wavelet coefficients within linear computation time. Biorthogonal wavelet transforms facilitate local computation of decomposition and reconstruction. We survey the most relevant results regarding wavelets and present different biorthogonal wavelet constructions based on highly efficient and simple-to-use lifting operations. Our approaches are suitable for lossy and lossless compression of large-scale data sets.

1 Introduction

The discrete wavelet transform [6, 16, 18, 23] is a useful tool for data compression, progressive transmission, and multiresolution modeling. It provides a sparse representation for highly detailed functions. The corresponding basis functions are dilated and translated versions of one function that, due to its shape, is called *wavelet*. Sparse representations are extremely useful in image compression [14] and for solving difficult mathematical problems, like integrating radiosity kernels [5, 21] and partial differential equations (PDEs) [7, 8, 27]. The discrete wavelet transform is computed in linear time, which makes it superior to most other multiresolution techniques. Biorthogonal wavelet transforms facilitate local computation of analysis and synthesis, in contrast to semi-orthogonal wavelet constructions [12, 20] where a banded linear system of equations needs to be solved for global fitting.

In this paper, we proviode a survey of the most important results and applications for the wavelet transform, particularly focusing on lossy and lossless data compression. We use the lifting scheme [25] to construct biorthogonal wavelet transforms requiring few arithmetic operations for for efficiently modeling large volumetric data sets. Our constructions are applicable to both lossless and lossy data compression.

2 Data Compression

Classical image compression algorithms consist of the following three steps [14]:

(i) Compute a linear, bijective transform of an image.
(ii) Quantize the resulting coefficients (for lossy compression only).
(iii) Apply an entropy coding scheme compressing the coefficients.

The first step de-correlates the information present in an image or data set. Exploiting spacial *correlation*, *i.e.*, similarity of local function values, like RGB color or greyscale values, leads to a sparse representation in a Fourier or wavelet basis, see Figure 1. For lossless compression, the transform is computed in integer arithmetic to obtain integer coefficients that can be encoded without quantization. In the case of lossy compression, the coefficients resulting from the transform are rounded to closest numbers on a certain grid to obtain values in a discrete set of symbols represented by integers. Coefficients of small absolute value can be replaced by zero for high lossy compression rates, see Figure 2. The integer coefficients are compressed by a coding scheme, for example by arithmetic coding [19, 28], exploiting that small absolute values appear more frequently than greater ones.

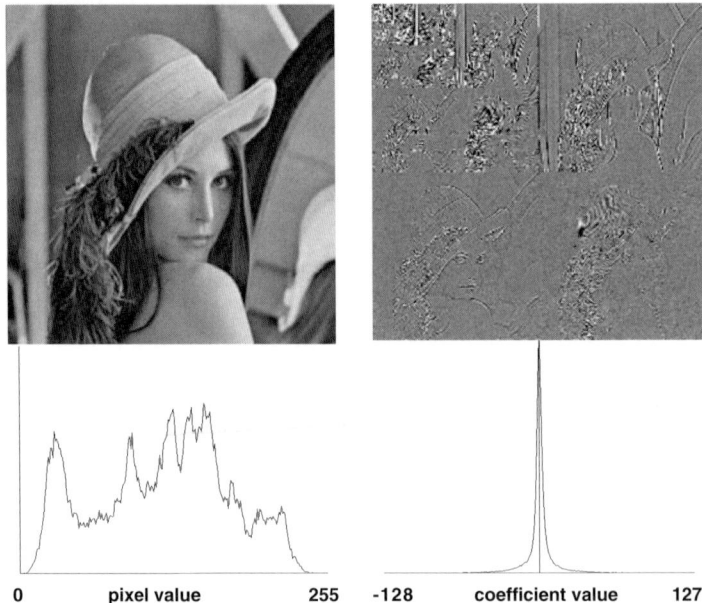

Fig. 1. Wavelet transform (using a linear spline wavelet) and histogram of pixel and coefficient values.

To reconstruct an image, the compressed coefficients are decoded. If they have not been quantized in the compression step, the original coefficient values are restored and the inverse transform exactly reproduces the original image. In the case of quantization, the reconstructed image contains an error. In many cases, however,

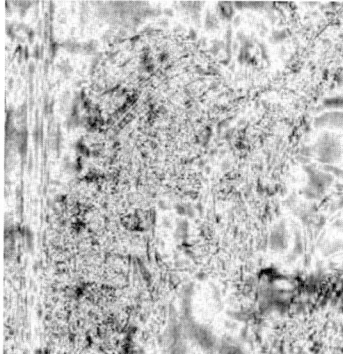

Fig. 2. Image reconstructed from ten percent of coefficient values (left) and reconstruction error scaled by a factor of ten (right).

the human eye does not recognize this quantization error since it has been introduced in the range of a transform and is smoothly distributed across the image domain by the inverse transform. When using the wavelet transform, high compression rates in the order of 64:1 [14] are obtained by lossy compression with small visible artifacts.

2.1 Choosing a Transform

Early compression standards, like JPEG, are based on the *discrete cosine transform* (DCT) using a basis of different-frequency cosine functions. Due to their global support, these basis functions are not well-suited for representing local image detail. Therefore, JPEG subdivides an image into frames of 8×8 or 16×16 pixels before transforming it. This fragmentation, however, may become visible in reconstructed images at high compression rates. Newer standards, like JPEG2000, are based on wavelet-like techniques to overcome these problems.

When applying the wavelet transform, fragmentation of images is not necessary, since wavelets have local support. The wavelet transform recursively separates high-frequency details from the remaining lower frequencies by band-pass and low-pass filtering. For images and higher-dimensional data sets, these one-dimensional filtering operations are applied once to every dimension (rows and columns). The coefficients resulting from low-pass filtering with respect to *all* dimensions provide a coarse representation of the data set. (The size of this coarse data set is $1/2^d$ of the original data set, where d is the dimensionality.) Only these low-pass filtered coefficients are recursively transformed, using the same set of filters, see Figure 3.

Modeling scientific volume data for compression, progressive transmission, and visualization can be accomplished with the same techniques used for image compression [26]. Generalization to multiple dimensions is straight forward, since the one-dimensional filtering operations are applied independently in every canonical direction. The basis functions of the multi-dimensional transform are tensor products of the one-dimensional functions (*wavelets* and *scaling functions*).

For compression and visualization of scientific data sets the representation needs to satisfy tight prescribed error bounds. In many cases, it is not acceptable to

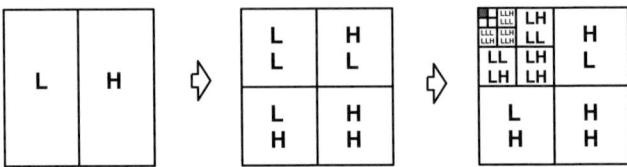

Fig. 3. The wavelet transform separates details of different frequency bands by recursive band-pass and low-pass filtering.

introduce an unpredictable quantization error. Instead, the samples are rounded to a prescribed precision and represented as integer numbers. These are then processed by a bijective integer-to-integer transform and encoded without loss [3].

2.2 Coefficient Coding

In the coding step, the small range of coefficient values and the fact that small coefficients occur more often than larger ones are exploited to obtain an actual compression. For a range of n possible coefficient values (symbols) with occurance probabilities $p_1, \ldots, p_n$, the minimal average number of bits per symbol is

$$\sum_{i=1}^{n} -p_i \, log_2\,(p_i). \qquad (1)$$

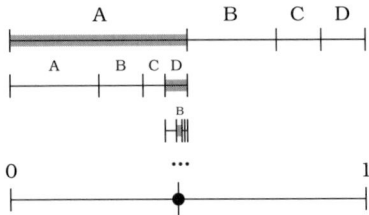

Fig. 4. Arithmetic coding of the symbol list "ADBAAABC." The average number of bits per symbol is 1.75 for this example.

This optimum is asymptotically reached by *arithmetic coding*. This scheme encodes a list of symbols by subdividing an interval, say $[0, 1]$, according to the probabilities of all symbols. The subinterval corresponding to the current symbol in the list is selected and recursively subdivided according to the succeeding symbols, see Figure 4. Finally, the encoded list of symbols is represented as a single, highly precise number identifying the final subinterval within the original interval. From this number, the list of symbols can be uniquely recovered.

Despite of its linear computation time, arithmetic coding requires a lot of implementation to avoid numerical problems, since an entire data set is encoded into one single number. Other coding schemes, like *Huffman coding* [16], provide a more efficient implementation, but they do not obtain optimal compression rates. An implementation for an arithmetic coder is provided by Witten *et al.* [28].

A major advantage of arithmetic coding is that the probabilities for symbols can be changed adaptively. The local distribution of coefficient values is typically much different in dense and sparse regions of a data set and for coefficients of different types. Locally adapting the distribution based on the most recently coded coefficients can further increase compression rates. An approach exploiting correlation between different levels of a wavelet transform, useful for sparse data, are zerotrees described by Shapiro [22].

3 Wavelet Transform

3.1 Continuous Wavelet Transform

The continuous wavelet transform [6,9,16] is a signal processing tool analyzing the local frequency spectrum of a function f (for fundamentals in signal processing, see Haykin [13]). This spectrum is provided by the Fourier transform

$$\widehat{f}(\omega) = \int_{\mathbb{R}} f(x)\, e^{-i\omega x}\, dx \quad \text{where} \quad i = \sqrt{-1}. \tag{2}$$

Band-pass filtering corresponds to "selecting" a part of the spectrum $\widehat{f}$ by multiplying it with $\widehat{\psi}$, where ψ is a continuous band-pass filter:

$$\widehat{f * \psi} = \widehat{f}\, \widehat{\psi}. \tag{3}$$

The *convolution operator* "$*$" is defined as

$$f * \psi(t) = \int_{\mathbb{R}} f(s)\, \psi(t-s)\, ds. \tag{4}$$

To create time-frequency localization tools, we introduce two parameters a and b that dilate and translate a given band pass ψ in the following way:

$$\psi_{a,b}(x) = \frac{1}{\sqrt{|a|}}\, \psi\left(\frac{x-b}{a}\right), \quad a \in \mathbb{R} \neq 0,\ b \in \mathbb{R}. \tag{5}$$

b translates the filter ψ and a translates its spectrum.

Convolution with this set of translated and dilated filters defines a new transform providing the desired localization property (in the limits of the uncertainty principle). For the existence of the inverse transform, ψ must satisfy the *wavelet condition*:

Definition 2. *[16] A function $\psi \in L^2(\mathbb{R})$ is called wavelet, if there exists a constant $c_\psi \in \mathbb{R} > 0$ with*

$$c_\psi = 2\pi \int_{\mathbb{R}} \frac{|\widehat{\psi}(\omega)|^2}{|\omega|}\, d\omega.$$

The continuous wavelet transform and its inverse are defined as

$$\mathbf{W}_\psi f(a,b) = \frac{1}{\sqrt{c_\psi}} \langle f, \psi_{a,b}\rangle_{L^2} = \frac{1}{\sqrt{c_\psi}} \int_{\mathbb{R}} f(x)\, \overline{\psi_{a,b}(x)}\, dx,$$
$$f(x) = \frac{1}{\sqrt{c_\psi}} \int_{\mathbb{R}^2} \mathbf{W}_\psi f(a,b)\, \psi_{a,b}(x) \frac{da\, db}{a^2}. \tag{6}$$

We note that the inner product in the wavelet transform is equivalent to the convolution operator. The wavelet condition implies that $\widehat{\psi}$ decreases quickly approaching $\omega = 0$ and that the average (direct current) of ψ is zero. Examples for wavelets and their Fourier transforms are shown in Figure 5.

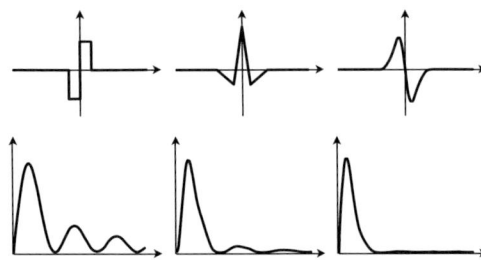

Fig. 5. Three different wavelets (top) and amplitudes of their Fourier transforms (bottom).

3.2 Discrete Wavelet transform

The continuous wavelet transform is highly redundant. A discrete sampling of the transform on a dyadic grid $\{(a,b) = (2^i, 2^i j) \mid i, j \in \mathbb{Z}\}$, see Figure 6, is sufficient for reconstruction, if the corresponding wavelet basis $\{\psi_{2^i, 2^i j}\}$ is a *frame* [16]:

Definition 3. *[16] A system of functions $\{\psi_{i,j} \mid i, j \in \mathbb{Z}\}$ is called frame for $L^2(\mathbb{R})$, if constants $0 < A, B < \infty$ exist so that for every $f \in L^2(\mathbb{R})$*

$$A\|f\|_{L^2}^2 \leq \sum_{i,j \in \mathbb{Z}} \|\langle f, \psi_{i,j}\rangle_{L^2}\|^2 \leq B\|f\|_{L^2}^2. \tag{7}$$

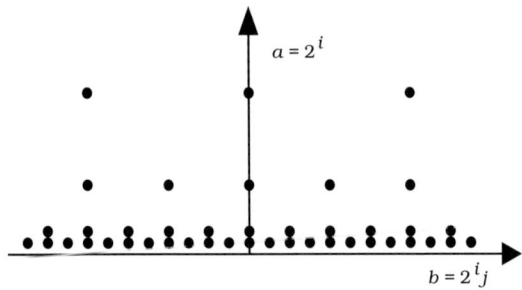

Fig. 6. Dyadic grid samples that are sufficient for reconstruction.

The linear-time algorithm for the discrete wavelet transform (DWT), known as *fast wavelet transform* [17], is based on the concept of *multi-scale analysis* [9, 23] defining a sequence of nested spaces V_j, spanned by *scaling functions*. Every space V_j provides the basis functions for one particular level of resolution. The differences (details) between two levels of resolution are represented in spaces W_j spanned by wavelets, see Figure 7.

Definition 4. *[9, 16] A multi-scale analysis is defined by a set of nested spaces V_j and a scaling function φ with the following properties:*

1. $V_j \subset V_{j+1} \quad \forall j \in \mathbf{Z}$,

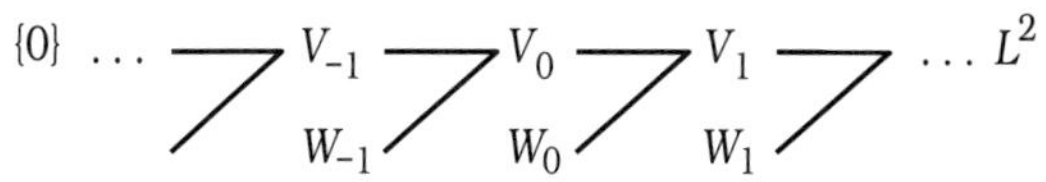

$$\{0\} \cdots \nearrow V_{-1} \nearrow V_0 \nearrow V_1 \nearrow \cdots L^2$$
$$\phantom{\{0\} \cdots \nearrow\ } W_{-1} W_0 W_1$$

Fig. 7. Nested spaces V_j and their duals W_j of a multi-scale analysis.

2. $\bigcup_{j \in \mathbf{Z}} V_j = L^2(\mathbb{R})$,
3. $\bigcap_{j \in \mathbf{Z}} V_j = \{0\}$.
4. $f \in V_j \Leftrightarrow f(2 \cdot) \in V_{j+1} \quad \forall j \in \mathbf{Z}$, and
5. $V_0 = span\{\varphi_k^0 := \varphi(\cdot - k) \mid k \in \mathbb{Z}\}$ is a Riesz basis, i.e., there exist constants $0 < A, B < \infty$ such that

$$A \sum_{k \in \mathbb{Z}} c_k^2 \leq \| \sum_{k \in \mathbb{Z}} c_k \, \varphi_k^0 \|_{L^2} \leq B \sum_{k \in \mathbb{Z}} c_k^2 \quad \forall c \in l^2(\mathbb{Z}). \tag{8}$$

Starting with a certain multi-scale analysis, we construct wavelet spaces W_j that are complements of V_j in V_{j+1}, i.e.,

$$V_j \oplus W_j = V_{j+1}.$$

All spaces V_j and W_j are dilated versions of V_0 and W_0, spanned by translates of a scaling function φ and a wavelet ψ, respectively:

$$V_j = \overline{span\{\varphi_k^j := \varphi(2^j \cdot - k) \mid k \in \mathbb{Z}\}}, \quad \text{and}$$
$$W_j = \overline{span\{\psi_k^j := \psi(2^j \cdot - k) \mid k \in \mathbb{Z}\}}.$$

A good choice for W_j is the orthogonal complement of V_j in V_{j+1}. In this case, all spaces W_j are mutually orthogonal. If the function ψ is constructed so that all ψ_k^0 are mutually orthogonal, for example by using Gram-Schmidt orthonormalization, then the set of all wavelets ψ_k^j forms an orthogonal basis for $L^2(\mathbb{R})$. Orthogonal bases generally lead to small coefficients for representing functions. An example for an orthogonal wavelet basis is defined by the *Haar wavelet*, shown in Figure 8.

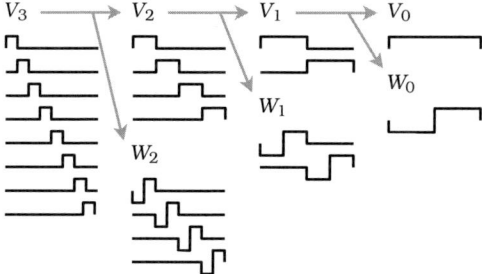

Fig. 8. Basis functions for multi-scale analysis based on the Haar wavelet.

Orthogonality, symmetry, smoothness, and compact support are conflicting goals, however. The discontinuous Haar wavelet is actually the only real-valued wavelet that is orthogonal, symmetric, and has compact support [9]. For geometric modeling purposes, however, symmetry, compact support, and certain degrees of smoothness are essential. These considerations lead to wavelet constructions where the spaces V_j and W_j are not orthogonal. In this case, there exist sets of *dual spaces* $\tilde{V}_j$ and $\tilde{W}_j$ defined as

$$\tilde{V}_j \perp W_j, \quad \tilde{W}_j \perp V_j, \quad \text{and}$$
$$\tilde{V}_j \oplus W_j = \tilde{W}_j \oplus V_j = V_{j+1}.$$

The dual spaces are spanned by dilates and translates of a dual scaling function $\tilde{\varphi}$ and a dual wavelet $\tilde{\psi}$ defining also a multi-scale analysis. Wavelet constructions of this type are thus called *biorthogonal* [6, 9].

The DWT is simply a basis transform from V_{j+1} into V_j and W_j that is recursively applied. Since this basis transform is regular, *i.e.*, it has an inverse, neither ψ, nor $\tilde{\psi}$ need to satisfy the wavelet condition defined in the context of the continuous wavelet transform.

The basis transform uniquely defines sequences h, l, $\tilde{h}$, and $\tilde{l}$ with the properties

$$\varphi_k^{j+1} = \sum_{i \in \mathbf{Z}} (l_{k-2i}\, \varphi_i^j + h_{k-2i}\, \psi_i^j), \tag{9}$$

$$\varphi_k^j = \sum_{i \in \mathbf{Z}} \tilde{l}_{i-2k}\, \varphi_i^{j+1}, \quad \text{and} \tag{10}$$

$$\psi_k^j = \sum_{i \in \mathbf{Z}} \tilde{h}_{i-2k}\, \varphi_i^{j+1}. \tag{11}$$

If the functions φ and ψ have compact support, then all four sequences are finite. These sequences represent discrete high-pass and low-pass filters that are used in the fast algorithm for the DWT.

Projections of an analytical function f into spaces V_j and W_j are represented by sets of coefficients c^j and d^j, respectively,

$$f_{V_j}(x) = \sum_{k \in \mathbf{Z}} c_k^j\, \varphi_k^j(x) \quad \text{and} \tag{12}$$

$$f_{W_j}(x) = \sum_{k \in \mathbf{Z}} d_k^j\, \psi_k^j(x). \tag{13}$$

Transforming the coefficients c^{j+1} into a coarser representation defined by coefficients c^j plus a representation for the differences (details) by coefficients d^j is called *decomposition* or *analysis*. The inverse operation, *i.e.*, obtaining coefficients c^{j+1} from c^j and d^j, is called *reconstruction* or *synthesis*.

Starting with a coefficient representation c^n for a function f at fine resolution, its DWT is computed by successive decomposition steps, resulting in sets of coefficients d^{n-1}, $d^{n-2}, \ldots, d^0$, and c^0, see Figure 9. The wavelet coefficients d^j are sparse since they capture the small differences between two levels of resolution. The inverse DWT reconstructs the coefficient sets $c^1, c^2, \ldots, c^n$ again.

The decomposition and reconstruction formulae are based on the four discrete filters defined above. The filter h is a band pass separating the highest frequency band (the details) from a function, l is a low-pass filter generating a coarser approximation without the highest frequency band, and $\tilde{h}$ and $\tilde{l}$ are necessary to invert the transform. These are the standard decomposition and reconstruction formulae:

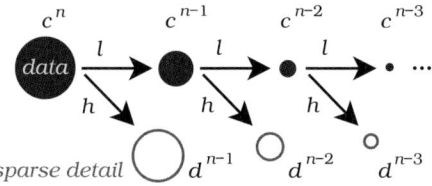

Fig. 9. Decomposition process.

Decomposition:

$$c_i^j = \sum_{k \in \mathbb{Z}} l_{k-2i}\, c_k^{j+1}, \tag{14}$$

$$d_i^j = \sum_{k \in \mathbb{Z}} h_{k-2i}\, c_k^{j+1}. \tag{15}$$

Reconstruction:

$$c_k^{j+1} = \sum_{i \in \mathbb{Z}} (\tilde{l}_{k-2i}\, c_i^j + \tilde{h}_{k-2i}\, d_i^j). \tag{16}$$

The algorithm known as fast wavelet transform applies this decomposition formula recursively starting with N coefficients c^n that are considered as samples of a function f. Every decomposition step is computed in linear time with respect to the number of coefficients, since the filters l and h have finite length. The number of transformed coefficients decreases by one half for every level of resolution, since the wavelet coefficients d^j are not transformed again. Thus, the overall complexity is proportional to $N + \frac{1}{2}N + \frac{1}{4}N + \frac{1}{8}N + \cdots = 2N = O(N)$.

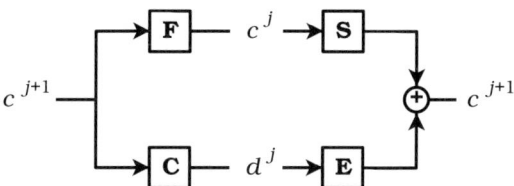

Fig. 10. Wiring diagram for decomposition (left) and reconstruction (right). We note that the coefficients c^j are recursively transformed, which is not shown in the diagram.

In geometric modeling, the discrete filters h, l, $\tilde{h}$, and $\tilde{l}$ are associated with the linear operations **F** (fitting), **C** (compaction of detail), **S** (subdivision), and **E** (expansion of detail), respectively. Using matrix notation, we obtain $\mathbf{F}_{ij} = l_{j-2i}$, $\mathbf{C}_{ij} = h_{j-2i}$, $\mathbf{S}_{ij} = \tilde{l}_{i-2j}$, and $\mathbf{E}_{ij} = \tilde{h}_{i-2j}$. The decomposition and reconstruction process is illustrated in Figure 10.

We recall that the operators **F** and **C** compute a regular basis transform from a space V_{j+1} into complementary spaces V_j and W_j, and that the inverse basis transform is computed by **S** and **E**. This imposes the following compatibility constraints on the construction of these operators:

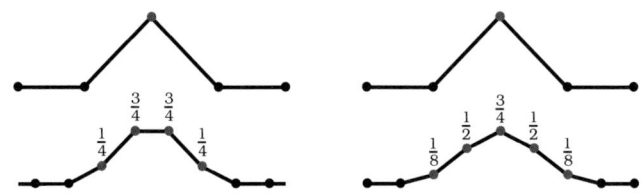

Fig. 11. Dyadic refinement for quadratic and cubic B-splines.

$$\begin{aligned} \mathbf{FS} &= \mathbf{CE} = I \quad \text{and} \\ \mathbf{FE} &= \mathbf{CS} = 0, \end{aligned} \quad (17)$$

where I is the identity matrix.

The operators $\mathbf{S}$ and $\mathbf{E}$ define the shape of the scaling functions and wavelets based on recursive subdicvision:

$$\begin{aligned} \varphi_i^j &= \mathbf{S}^\infty \, \delta_i^j \quad \text{and} \\ \psi_i^j &= \mathbf{S}^\infty \, \mathbf{E} \, \delta_i^j, \end{aligned} \quad (18)$$

where δ_i^j is the Kronecker symbol serving as coefficient sequence (or control polygon). The subdivision process for B-splines uses the filter $\tilde{l}$ summarized in Table 1, see Figure 11.

degree	$\tilde{l}_{0,1}$	$\tilde{l}_{-1,2}$	$\tilde{l}_{-2,3}$	degree	$\tilde{l}_0$	$\tilde{l}_{\pm 1}$	$\tilde{l}_{\pm 2}$	$\tilde{l}_{\pm 3}$
0	1			1	1	$\frac{1}{2}$		
2	$\frac{3}{4}$	$\frac{1}{4}$		3	$\frac{3}{4}$	$\frac{1}{2}$	$\frac{1}{8}$	
4	$\frac{5}{8}$	$\frac{5}{16}$	$\frac{1}{16}$	5	$\frac{5}{8}$	$\frac{15}{32}$	$\frac{3}{16}$	$\frac{1}{32}$

Table 1. Sequence $\tilde{l}$ defining subdivision operator $\mathbf{S}$ for B-spline scaling functions of different polynomial degrees [4].

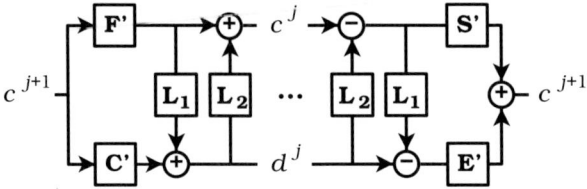

Fig. 12. Lifting Scheme: an arbitrary number of lifting operations can be applied to shape scaling functions and wavelets.

3.3 The Lifting Scheme

Lifting [25] subdivides the filtering operations into multiple smaller operations, simplifying the construction of wavelets and often reducing computation times.

The individual lifting operations can often be approximated by bijective integer-to-integer operations, providing a DWT that is entirely computed in integer arithmetic for lossless compression.

The general lifting scheme for the decomposition formula is illustrated in Figure 12, see also [15, 24]. In the simplest case, the operators **F**′ and **C**′ defining the lazy wavelet transform are just down-sampling operators. The inverse DWT is defined by the same lifting operations, applied with negated signs in reverse order.

4 Wavelet Construction

We now provide some examples for factoring wavelets into lifting operations. A rigorous mathematical treatment of this topic was done by Daubechies/Sweldens [10]. Consider the wavelets defined in Table 2. ψ_H is the piecewise constant Haar wavelet, ψ_Q is a quadratic B-spline wavelet with small support [1], and ψ_D is a wavelet constructed by Duchaineau [11].

wavelet	h_{-1}	h_0	h_1	h_2	l_{-3}	l_{-2}	l_{-1}	l_0	l_1	l_2	l_3	l_4
ψ_H		1	-1					$\frac{1}{2}$	$\frac{1}{2}$			
ψ_Q	$-\frac{1}{4}$	$\frac{3}{4}$	$-\frac{3}{4}$	$\frac{1}{4}$			$-\frac{1}{4}$	$\frac{3}{4}$	$\frac{3}{4}$	$-\frac{1}{4}$		
ψ_D	$-\frac{1}{4}$	$\frac{3}{4}$	$-\frac{3}{4}$	$\frac{1}{4}$	$\frac{1}{12}$	$-\frac{1}{4}$	0	$\frac{2}{3}$	$\frac{2}{3}$	0	$-\frac{1}{4}$	$\frac{1}{12}$

wavelet	$\tilde{h}_{-3}$	$\tilde{h}_{-2}$	$\tilde{h}_{-1}$	$\tilde{h}_0$	$\tilde{h}_1$	$\tilde{h}_2$	$\tilde{h}_3$	$\tilde{h}_4$	$\tilde{l}_{-1}$	$\tilde{l}_0$	$\tilde{l}_1$	$\tilde{l}_2$
ψ_H				$\frac{1}{2}$	$-\frac{1}{2}$					1	1	
ψ_Q			$\frac{1}{4}$	$\frac{3}{4}$	$-\frac{3}{4}$	$-\frac{1}{4}$			$\frac{1}{4}$	$\frac{3}{4}$	$\frac{3}{4}$	$\frac{1}{4}$
ψ_D	$-\frac{1}{12}$	$-\frac{1}{4}$	0	$\frac{2}{3}$	$-\frac{2}{3}$	0	$\frac{1}{4}$	$\frac{1}{12}$	$\frac{1}{4}$	$\frac{3}{4}$	$\frac{3}{4}$	$\frac{1}{4}$

Table 2. Examples for discrete filters defining wavelet transforms.

The decomposition formula for the Haar wavelet ψ_H is being evaluated independently for every pair of adjacent coefficients,

$$\begin{pmatrix} c_i^j \\ d_i^j \end{pmatrix} = \begin{pmatrix} \frac{1}{2} & \frac{1}{2} \\ 1 & -1 \end{pmatrix} \begin{pmatrix} c_{2i}^{j+1} \\ c_{2i+1}^{j+1} \end{pmatrix}. \tag{19}$$

The matrix in this equation can be split into two operations updating only one coefficient at a time, as illustrated in Figure 13:

$$\begin{pmatrix} \frac{1}{2} & \frac{1}{2} \\ 1 & -1 \end{pmatrix} = \begin{pmatrix} 1 & -\frac{1}{2} \\ 0 & 1 \end{pmatrix} \begin{pmatrix} 1 & 0 \\ 1 & -1 \end{pmatrix}. \tag{20}$$

In general, every regular 2×2-matrix M with non-zero entries on the diagonal can be split in this way, based on four constants a_1, b_1, a_2, and b_2 satisfying

$$M = \begin{pmatrix} b_2 & a_2 \\ 0 & 1 \end{pmatrix} \begin{pmatrix} 1 & 0 \\ a_1 & b_1 \end{pmatrix} = \begin{pmatrix} b_2 + a_2 a_1 & a_2 b_1 \\ a_1 & b_1 \end{pmatrix}. \tag{21}$$

For the lifted decomposition formula for the Haar wavelet, we first re-label the coefficients $c_i \leftarrow c_{2i}^{j+1}$ and $d_i \leftarrow c_{2i+1}^{j+1}$. Then, we compute the lifting operations

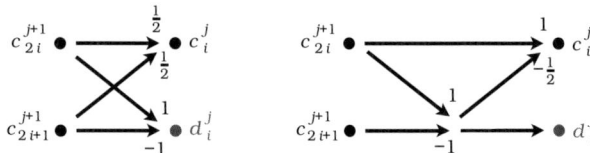

Fig. 13. Decomposition for Haar wavelet (left) and equivalent lifting operations (right).

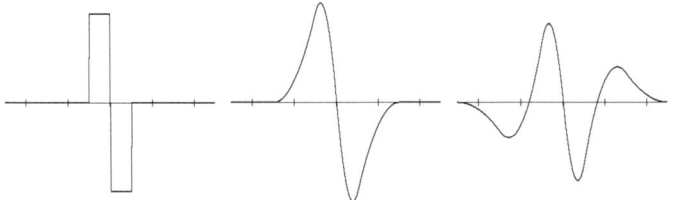

Fig. 14. Wavelets ψ_H, ψ_Q, and ψ_D.

$$d'_i \leftarrow c_i - d_i$$
$$c'_i \leftarrow c_i - \tfrac{1}{2}d'_i,$$

providing the new coefficients $c_i^j \leftarrow c'_i$ and $d_i^j \leftarrow d'_i$.

Decomposition for the quadratic B-spline wavelet ψ_Q is defined according to the entries in Table 2 as

$$\begin{pmatrix} c_i^j \\ d_i^j \end{pmatrix} = \begin{pmatrix} -\tfrac{1}{4} & \tfrac{3}{4} & \tfrac{3}{4} & -\tfrac{1}{4} \\ -\tfrac{1}{4} & \tfrac{3}{4} & -\tfrac{3}{4} & \tfrac{1}{4} \end{pmatrix} \begin{pmatrix} c_{i-1}^{j+1} \\ c_i^{j+1} \\ c_{i+1}^{j+1} \\ c_{i+2}^{j+1} \end{pmatrix}. \qquad (22)$$

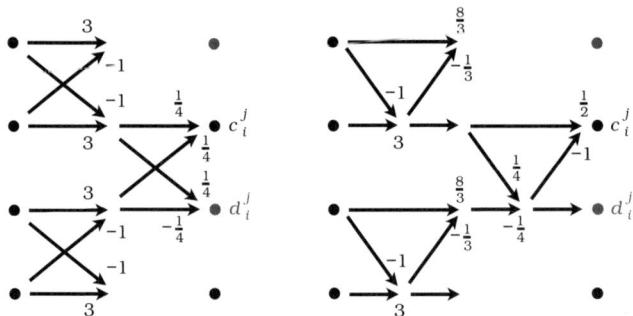

Fig. 15. Factoring the decomposition for wavelet ψ_Q into two matrix multiplications (left), which can be split into four lifting operations (right).

As illustrated in Figure 15, this formula can be factored into two 2×2-matrix multiplications with unique matrix entries (except for scaling the intermediate coefficients). We have computed these matrix entries satisfying equation (22) by solving a system of non-linear equations. These matrix multiplications are

$$\begin{pmatrix} d'_{i-1} \\ c'_i \end{pmatrix} \leftarrow \begin{pmatrix} 3 & -1 \\ -1 & 3 \end{pmatrix} \begin{pmatrix} d_{i-1} \\ c_i \end{pmatrix} \quad \text{and}$$
$$\begin{pmatrix} c''_i \\ d''_i \end{pmatrix} \leftarrow \begin{pmatrix} \frac{1}{4} & \frac{1}{4} \\ \frac{1}{4} & -\frac{1}{4} \end{pmatrix} \begin{pmatrix} c'_i \\ d'_i \end{pmatrix}. \tag{23}$$

Both matrices can be split into two lifting operations each, resulting in the lifting scheme

$$\begin{aligned} c'_i &\leftarrow -d_{i-1} + 3c_i \\ d'_{i-1} &\leftarrow \tfrac{8}{3} d_{i-1} - \tfrac{1}{3} c'_i \\ d''_i &\leftarrow \tfrac{1}{4} c'_i - \tfrac{1}{4} d'_i \\ c''_i &\leftarrow \tfrac{1}{2} c'_i - d''_i. \end{aligned} \tag{24}$$

Combining the second and third lifting operations results in the more compact scheme,

$$\begin{aligned} c'_i &\leftarrow -d_{i-1} + 3c_i \\ d''_i &\leftarrow \tfrac{1}{4} c'_i - \tfrac{2}{3} d_i + \tfrac{1}{12} c'_{i+1} \\ c''_i &\leftarrow \tfrac{1}{2} c'_i - d''_i. \end{aligned} \tag{25}$$

For lifting the wavelet ψ_D, we exploit that it is based on the same compaction-of-difference filter h as the wavelet ψ_Q. We observe that the fitting filters of both wavelets, denoted as l^D and l^Q, satisfy the relation

$$l_i^D = l_i^Q + \tfrac{1}{3}(h_{i-2} - h_{i+2}). \tag{26}$$

The lifting scheme for ψ_D is thus identical to the one for ψ_Q, with one additional lifting operation, given by

$$c'''_i \leftarrow c''_i - \tfrac{1}{3} d''_{i-1} + \tfrac{1}{3} d''_{i+1}. \tag{27}$$

We note that the last two lifting operations can again be combined. Despite of its long filters, the wavelet transform for ψ_D requires only one more lifting operation as the transform for the compact Haar wavelet ψ_H.

It can be verified that computing the lifted wavelet transforms requires fewer operations than computing the original decompositon and reconstruction schemes. Another advantage of lifting is that the boundary treatment becomes much simpler. The local lifting operations can individually be adapted to the boundaries.

The reconstruction schemes for our transforms are simply obtained by computing the inverse of every lifting operation in reverse order. For example, the inverse of equation (27) is defined as

$$c''_i \leftarrow c'''_i + \tfrac{1}{3} d''_{i-1} - \tfrac{1}{3} d''_{i+1}. \tag{28}$$

When using integer arithmetic for lossless compression purposes, the term added to a coefficient is rounded to a closest integer. For the inverse operation, the same rounded term is subtracted, providing perfect reconstruction [3].

5 Numerical Results

5.1 Lossless Compression Example

We have applied the lossless compression algorithm to a *Rayleigh-Taylor instability* (turbulent flow) simulation, courtesy of Lawrence Livermore National Laboratory. This data set represents a three-dimensional, time-varying scalar field in a resolution of 512^3 byte samples for each of 301 time steps. Since the data set is sparse in the first time steps, a high lossless compression in the order of 1:20 can be achieved.

Due to the large size of the data set, the compression algorithm is applied locally to smaller blocks fitting into main memory. For numerical examples, we have chosen a horizontally aligned 512^2 slice and a 64^3 brick in the middle of the data set at the last, most turbulent time step. We have used the folowing four integer wavelet transforms to the slice:

- ψ_H: the Haar wavelet.
- ψ_Q: the quadratic B-spline wavelet with small support.
- ψ_D: the quadratic wavelet constructed by Duchaineau [11].
- ψ_1: a linear B-spline wavelet constructed in [2].

The slice and its wavelet transforms (scaled by ten with a 50 percent grey level added) are shown in Figure 16. It can be observed that the discontinuous Haar wavelet represents high frequency parts well but leads to large coefficients on coarse levels of resolution.

The numerical compression results for the slice of the Rayleigh-Taylor instability data set are summarized in Table 3. Computation times were measured on a 194 MHz MIPS R10000 processor on an SGI Onyx system. The arithmetic code length is shown in percent of the size of the uncompressed slice (262144 bytes). The code lengths and the corresponding compression rates are computed from equation (1). We note that the size of the coefficient distribution is not included. It is possible to recover the distribution adaptively from the decoded coefficients resulting in even higher compression rates. We have also counted the number of zero coefficients and recorded the range of coefficient values for every transform.

Numerical compression results for a 64^3 brick of the Rayleigh-Taylor data set are provided in Table 4. The total number of samples is the same as for the slice, but the transforms are computed in three dimensions. Higher compression rates are obtained, due to the correlation of function values in the third dimension.

wavelet	comput. time [sec]	arith. code length [%]	compr. rate	number of zeros [%]	coeff. range min	max
ψ_H	0.055	51.0	1.96	20.7	−67	112
ψ_Q	0.097	35.6	2.81	35.5	−49	46
ψ_D	0.115	35.8	2.79	35.6	−59	50
ψ_1	0.078	39.5	2.53	37.3	−94	110

Table 3. Lossless compression results for 512^2 slice of Rayleigh-Taylor data set.

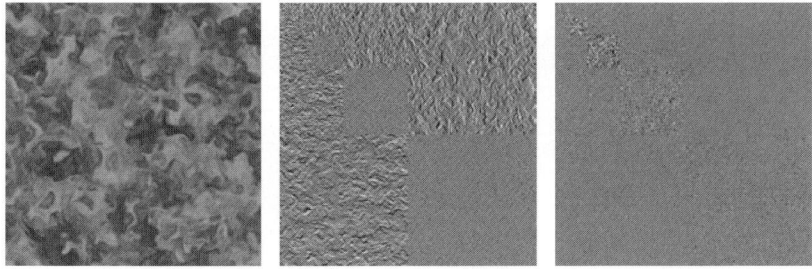

Fig. 16. 512^2 slice of Rayleigh-Taylor data set (left), Haar transform (middle), and quadratic B-spline wavelet transform (right). The coefficients are scaled by ten, and a grey level (0.5 on the interval $[0,1]$) is added.

wavelet	comput. time [sec]	arith. code length [%]	compr. rate	number of zeros [%]	coeff. range min	max
ψ_H	0.103	34.5	2.90	41.7	−43	60
ψ_Q	0.177	29.4	3.40	43.3	−89	62
ψ_D	0.222	29.0	3.45	44.3	−95	62
ψ_1	0.151	26.7	3.75	52.6	−83	103

Table 4. Lossless compression results for 64^3 brick of Rayleigh-Taylor data set.

5.2 Lossy Compression Examples

As an example for lossy compression based on floating-point data, we have chosen the Crater Lake data set, courtesy of U.S. Geological Survey. We have re-sampled this data set to a regular grid of 320 × 448 samples, which is about 90 percent of the original data set (159272 samples).

Table 5 shows the maximal and L^2 quantization errors in percent of the amplitude for different wavelet transforms at compression rates of ten and 100. The compression rates are obtained by scaling the wavelet coefficients of each transform by an individual factor before rounding them to integers. Some of the reconstructed terrain data sets are depicted in Figure 17.

wavelet	compression rate 10		compression rate 100	
	L^∞-error	L^2-error	L^∞-error	L^2-error
ψ_H	3.55	0.89	16.34	4.01
ψ_Q	1.18	0.24	9.66	2.03
ψ_D	1.53	0.29	12.21	2.49
ψ_1	1.41	0.26	7.94	1.67

Table 5. Lossy compression results for Crater Lake.

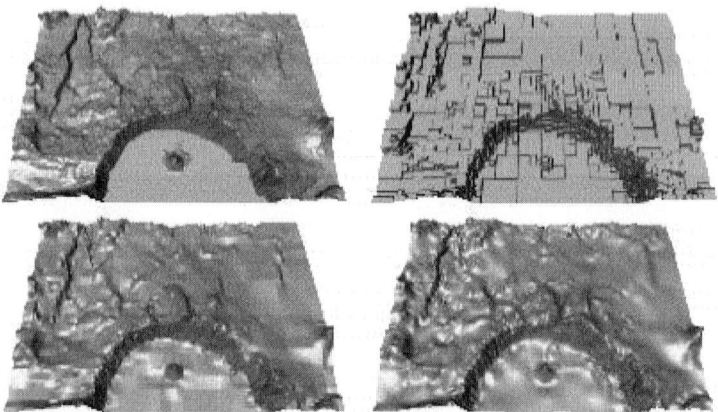

Fig. 17. Crater Lake data set and reconstructions from a 1:100 compression. Full resolution (top left), Haar wavelet ψ_H (top right), linear B-spline wavelet ψ_1 (bottom left), and Duchaineau's wavelet ψ_Q (bottom right).

Acknowledgements

Parts of this work were performed under the auspices of the U.S. Department of Energy (DOE) by University of California Lawrence Livermore National Laboratory (LLNL) under contract No. W-7405-Eng-48. We thank Mark Duchaineau, Ken Joy, and Bernd Hamann for their helpful discussions improving the present work.

References

1. M. Bertram, *Multiresolution Modeling for Scientific Visualization*, Ph.D. Thesis, University of California at Davis, July 2000. http://daddi.informatik.uni-kl.de/ bertram/
2. M. Bertram, M.A. Duchaineau, B. Hamann, and K.I. Joy, *Generalizing lifted tensor-product wavelets to irregular polygonal domains*, Data Visualization: The State of The Art, Kluver Academic Publishers, 2003, pp. 289–300.
3. R. Calderbank, I. Daubechies, W. Sweldens, and B.-L. Yeo, *Wavelet transforms that map integers to integers*, Applied and Computational Harmonic Analysis, Vol. 5, No. 3, Academic Press, July 1998, pp. 332–369.
4. A.S. Cavaretta , C.A. Micchelli , and W. Dahmen, *Stationary Subdivision*, American Mathematical Society, Boston, MA, 1991.
5. P. Christensen, E. Stollnitz, D. Salesin, T. DeRose, *Wavelet radiance*, Proceedings of the Fifth Eurographics Workshop on Photorealistic Rendering Techniques, Springer-Verlag, Berlin, Germany, 1995, pp. 295–309 & 432.
6. C.K. Chui, *An Introduction to Wavelets*, Academic Press, 1992.
7. W. Dahmen, *Wavelet and multiscale methods for operator equations*, Acta Numerica, Vol.6, Cambridge University Press, 1997, pp. 55–228.
8. W. Dahmen, *Wavelet methods for PDEs – some recent developments*, IGPM report No. 183, RWTH Aachen, Germany, Dec. 1999.

9. I. Daubechies, *Ten Lectures on Wavelets*, Society for Industrial and Applied Mathematics (SIAM), Philadelphia, 1992.
10. I. Daubechies and W. Sweldens, *Factoring wavelet transforms into lifting steps*, Journal of Fourier Analysis and Applications, Vol. 4, No. 3, CRC Press Inc., 1998, pp. 245–267.
11. M. Duchaineau, *Dyadic splines*, Ph.D. thesis, Department of Computer Science, University of California, Davis, 1996. http://graphics.cs.ucdavis.edu/
12. A. Finkelstein and D.H. Salesin, *Multiresolution curves*, Computer Graphics, Proceedings of Siggraph '94, ACM, 1994, pp. 261–268.
13. S. Haykin, *Digital Communication*, Wiley, New York, 1988.
14. M.L. Hilton, B.D. Jawerth, A. Sengupta, *Compressing still and moving images with wavelets*, Multimedia Systems, Vol. 2, Springer, 1994, pp. 218–227.
15. L. Kobbelt and P. Schröder, *A multiresolution framework for variational subdivision*, ACM Transactions on Graphics, Vol. 17, No. 4, Oct. 1998, pp. 209–237.
16. A.K. Louis, P. Maaß, A. Rieder, *Wavelets: Theorie und Anwendungen* (in German), B.G. Teubner, Stuttgart, Germany, 1994.
17. S.G. Mallat, *A theory for multiresolution signal decomposition: the wavelet representation*, IEEE Transactions on Pattern Analysis and Machine Intelligence, Vol. 11, No. 7, July 1989. pp.674–93.
18. Y. Meyer, *Wavelets, Algorithms & Applications*, Society for Industrial and Applied Mathematics (SIAM), Phiadelphia, 1993.
19. A. Moffat, R.M. Neal, and I.H. Witten, *Arithmetic coding revisited*, ACM Transactions on Information Systems, Vol. 16, No. 3, July 1998, pp. 256–294.
20. E.G. Quak and N. Weyrich, *Decomposition and reconstruction algorithms for spline wavelets on a bounded interval*, Applied and Computational Harmonic Analysis 1, 1994, pp. 217-231.
21. P. Schröder and P. Hanrahan, *Wavelet methods for radiance computations*, Proceedings of the Fifth Eurographics Workshop on Photorealistic Rendering Techniques, Springer Berlin, Germany, 1995, pp. 310–326 & 433.
22. J.M. Shapiro, *Embedded image coding using zerotrees of wavelet coefficients*, IEEE Transactions on Signal Processing, Vol. 41, No. 12, Dec. 1993.
23. E.J. Stollnitz, T.D. DeRose, D.H. Salesin, *Wavelets for Computer Graphics– Theory and Applications*, Morgan Kaufmann Publishers Inc., 1996.
24. W. Sweldens and P. Schröder, *Building your own wavelets at home*, Wavelets in Computer Graphics, ACM Siggraph Course notes, 1996, pp. 15-87.
25. W. Sweldens, *The lifting scheme: a custom-design construction of biorthogonal wavelets*, Applied and Computational Harmonic Analysis, Vol. 3, No. 2, 1996, pp. 186–200.
26. H. Tao, R.J. Moorehead, *Progressive transmission of scientific data using biorthogonal wavelet transform*, IEEE Visualization 1994, pp. 93–99.
27. O.V. Vasilyev, D.A. Yuen, S. Paolucci, *Solving PDEs using wavelets*, Computers in Physics, Vol. 11, No. 5, AIP, 1997, pp. 429–435.
28. I.H. Witten, R.M. Neal, and J.G. Cleary, *Arithmetic coding for data compression*, Communications of the ACM, Vol. 30, No. 6, June 1987, pp. 520–540.

Tree-based Data Structures for Triangle Mesh Connectivity Encoding

Ioannis Ivrissimtzis, Christian Rössl, and Hans-Peter Seidel

Max-Planck-Institut für Informatik,
Stuhlsatzenhausweg 85, Saarbrücken, 66123, GERMANY
{*ivrissim,roessl,hpseidel*} *@mpi-sb.mpg.de*

1 Introduction

Triangle meshes have recently emerged as the de facto standard in many Computer Graphics applications, generating a research interest in finding data structures able to represent them efficiently. This is not a trivial task given that the size of a typical meshes can vary from few hundreds triangles, up to hundreds of millions of triangles for some very detailed models.

A typical data set describing a 3D triangle mesh model consists of connectivity, geometry and some external attributes. The connectivity describes the way the vertices of the mesh are connected with edges and faces. It captures intrinsic topological properties of the mesh, containing all the information related to the genus and the existence and size of boundaries. The geometry of the mesh describes the actual positions of the vertices in the 3-dimensional Euclidean space $\mathbf{R}^3$. Finally, there are the other attributes of the mesh, like color or normals. In this paper we deal with the encoding of connectivity.

The choice of the most suitable data structure for connectivity encoding usually depends on the characteristics of the application. If the mesh is going to be manipulated heavily, then the focus is on the efficient traversal and access of the vertices, edges and triangles. Therefore, the most suitable data structures, like for example the winged-edge data structure, have a lot of redundant information. For rendering purposes more compact representations are preferred. The usual choice is a simple shared vertex representation encoding triangles as triplets of indices into a vertex table. The *ply*-format, supported also by the Stanford 3D Scanning Repository, is a well known example of such a representation. For storage and efficient transmission of the mesh over low-bandwidth networks as the Internet the focus is on the further compression of the data.

In this paper we propose a novel data structure for encoding triangle mesh connectivity, consisting of a binary tree with positive integers attached on its nodes. It is simple enough so that it can be implemented easily, and compact enough so that it can be used for compression purposes. It arises naturally from a Divide and Conquer Algorithm which we will describe in detail. Notice that in the compression literature the focus is almost solely on the algorithms themselves, and some of the most widely used methods do not use any data structure other than a stream of symbols.

The main advantage of introducing and studying this data structure is that we can find a simple criterion for deciding whether a member of the data structures represents a valid mesh or not. This criterion can be used directly to check the validity of a code without going into the reconstruction of the mesh, or indirectly for enhancing the compression ratio. Criteria of this kind are rather rare in the literature, because if the encoding is very compact usually the corresponding algorithm is very involved, while otherwise it is very difficult to deal with the redundant information.

1.1 Overview

In Section 2 we give some basic standard terminology. In Section 3 we give a brief overview of the existing encoding techniques. In Section 4 we describe in detail the Divide and Conquer Algorithm, mainly following [12]. Finally, in Section 5 we study the data structure acquired by the Divide and Conquer Algorithm, showing that we can achieve very good compression ratios and also obtain an interesting new insight into the complex mathematics of triangle meshes.

2 Terminology

Some basic terminology we use throughout the paper.

The *Rooted Triangle Mesh* is a triangle mesh, with one directed edge on the boundary marked. The marked edge is called the *gate*. If the mesh has no boundary we create one by removing one triangle. Usually we will refer to the rooted triangle meshes simply as meshes, and we will make the distinction between rooted and unrooted meshes only when it is necessary. Also, notice that the same distinction can be carried over to the trees. A *tree* is defined as an acyclic graph and the *rooted tree* as a tree with one node, the *root*, marked. Again, it is customary to refer to the rooted trees simply as trees if there is no room for confusion.

A *zig-zag strip* of *length n* is a mesh with $n+2$ vertices $(v_0, v_1, \ldots, v_{n+1})$ and n triangles $v_i v_{i+1} v_{i+2}$ with $0 \leq i \leq n-1$ (cf. Fig. 1). In the rooted zig-

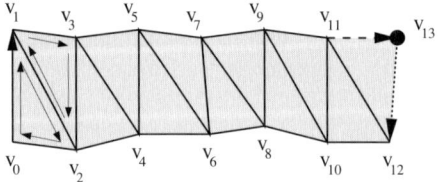

Fig. 1. A zig-zag strip of length 12. The gate is marked with a solid thick arrow, the left leading directed edge is dashed, the right is dotted, and the leading vertex is marked by a black dot.

zag strips we use here the gate will always be the directed edge (v_0, v_1). The vertex v_{n+1} will be called the *leading vertex* and the triangle $v_{n-1}v_nv_{n+1}$ the leading triangle. Assuming that all the triangles have the same orientation, as indicated by the gate, the oriented leading triangle is (v_{n-1}, v_n, v_{n+1}) for odd n and (v_n, v_{n-1}, v_{n+1}) for even n. For odd n, (v_n, v_{n+1}) and (v_{n+1}, v_{n-1}) are the left and the right directed leading edges, respectively. For even n the left and right leading directed edges are (v_{n-1}, v_{n+1}) and (v_{n+1}, v_n), respectively. In both cases the left leading directed edge is the one pointing at the leading vertex.

3 Overview of Mesh Encoding Algorithms

In the last years the increasing popularity of triangle meshes has rapidly accelerated the pace of research in the area of mesh encoding. As a result, many diverse techniques have emerged for the encoding of triangle mesh connectivity, each one with some advantages over all the others when a particular class of meshes is considered. Some of the earlier techniques include the encoding of the connectivity as a permutation of the vertices [5], the topological surgery method [21], where a mesh is encoded as a vertex tree together with the dual face tree, and the pioneering Cut Border Machine [10], [8], which was the first recursive method based on a traversal of the triangles of the mesh.

More recently two major branches of encoding techniques have started to emerge. The valence driven techniques, where the algorithm traverses the vertices of the mesh, and the EdgeBreaker like methods, where the algorithm traverses the faces of the mesh. In the valence driven methods, each vertex transmits its valence together with some special symbols. The method was initiated by Touma-Gotsman in [22], with a non-adaptive traversal of the mesh vertices, while Alliez-Desbrun used an adaptive traversal of the vertices [1], reporting the best compression ratios in practice. Alliez and Desbrun's algorithm has been expanded to more complex problems such as the progressive transmission of a 3D model including the geometry [2], or the encoding of polygonal meshes [14], giving again the best reported results when compared with any other similar method.

The other major branch of recently developed techniques is based on the EdgeBreaker algorithm. The original EdgeBreaker was proposed in [19]. It traverses a tree of the faces of a mesh and for each face returns one symbol from an alphabet of five, determining the adjacencies of that face with the not yet conquered part of the mesh. Its main advantage over the valence driven methods is the existence of a sharp guaranteed worst-case bound, which originally was reported at 4 bits per vertex, and later was improved to 3.67 bits per vertex [15], and to 3.55 bits per vertex [9]. Numerous other improvements in the efficiency of the technique followed. For example, [18] make the encoding and decoding process linear in time, [11] simplify further the decoding algorithm especially for meshes of arbitrary topology, while [20] is an adaptation of the EdgeBreaker for highly regular meshes. The Divide and Conquer algorithm we study here traverses the faces of the mesh and can be classified into the family of the EdgeBreaker like schemes.

At a more theoretical level, we notice that some of the mathematical foundations of the triangle mesh connectivity encoding were laid much earlier, in 1962, with the results of Tutte [23] on the enumeration of the planar rooted triangulations. Tutte found generating functions for the number of distinct rooted triangulations over a plane, and studied their asymptotic behavior. His results established a theoretical upper bound of 3.24 bits per vertex for the encoding of sufficiently large triangle meshes. In [4], Tutte's initial condition that two boundary vertices cannot be connected with a non-boundary edge is removed and it is proven that the generating functions have the same asymptotic behavior.

Finally, in [6], a method employing an approach very similar to ours was reported for graph compression. There, a graph is decimated with the use of graph separators, that is, subgraphs whose removal separates the graph into components of roughly similar size, and in a recursive process they encode these components in a tree data structure. The different setting, especially the nature of the graph as a combinatorial rather than a geometric object, makes much more difficult conclusive answers, something that, as we see in this paper, is not the case with triangle meshes.

4 The Divide and Conquer Algorithm

In this section we describe in detail the Divide and Conquer algorithm. The algorithm was introduced in [12] where some additional details can be found. We first describe the algorithm for planar meshes, then we study some basic characteristics of its behavior, and show the necessary modifications for the handling of arbitrary topology. We conclude the section with a brief discussion of the internal similarities between our algorithm and the EdgeBreaker.

4.1 The Divide and Conquer Algorithm for Planar Meshes

In the beginning of the encoding process we mark the boundary vertices of the mesh as conquered. We randomly choose a directed edge on the boundary of the mesh as the initial gate and we build a zig-zag strip conquering its vertices. We stop when we arrive at an already conquered vertex. The latter happens when the leading vertex of the zig-zag strip reaches either the boundary of the mesh or another vertex of the strip. In both cases the original rooted mesh splits into two rooted submeshes: The left submesh with the left leading directed edge of the strip as gate, and the right submesh with the right leading edge of the strip as gate. Any of these submeshes or both can be empty. We continue recursively, encoding separately the two submeshes and the encoding process terminates when all the submeshes are empty. A typical situation is shown in Fig. 2. We organize the data acquired in this process in the form

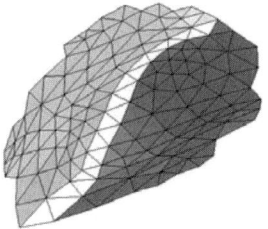
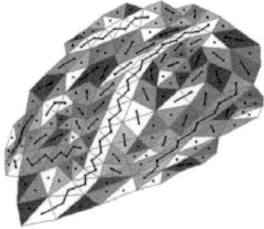

Fig. 2. A triangle strip divides each mesh into two submeshes (left). The left and the right submeshes are processed the same way recursively defining a binary tree. The result is shown on the right side, the black lines denote the strip connectivity. We can encode a planar triangle mesh as the resulting binary tree with only the strip lengths stored in its nodes.

of a binary tree with the strip lengths stored in its nodes. The length of the initial zig-zag strip is stored in the root of the tree, the encoding of the left submesh is the left branch of the tree, and the encoding of the right submesh is the right branch of the tree.

Encoding a triangle mesh can be written in pseudo code as:

```
tree = encode(edge) {
  node.length=0;
  while (!conquered(oppositeVertex(edge)) { // GROW
    conquer(oppositeVertex(edge),triangle(edge));
    edge=nextStripEdge(edge);
    ++node.length;
  }
  if (node.length==0)
    return NULL;                            // STOP
  node.left =encode(leftLeading(edge));     // RECURSION
  node.right=encode(rightLeading(edge));
  return node;
}
```

The encoding starts with the enter gate as argument to `encode()`, e.g. (v_0, v_1) in Fig. 1. `oppositeVertex()` returns the "leading" vertex opposite of the edge in the same triangle, e.g. $v_2 \in \triangle(v_0, v_1, v_2)$. `nextStripEdge()` finds the next edge in the next triangle of the strip, e.g. $(v_2, v_1) \in \triangle(v_2, v_1, v_3)$, and `leftLeading()`/`rightLeading()` return the adjacent edges in the same triangle, e.g. $(v_1, v_2), (v_2, v_0)$. Navigation in the mesh can be reduced to several calls to `next(edge,orientation)` and `neighbor(edge)` operations returning the next edge and the neighboring directed edge in the adjacent triangle.

Conversely, in one recursive step of the decoding process we have a zig-zag strip of specified length and two rooted meshes, and we glue them together. We identify the gate of the left rooted mesh with the left leading edge of the strip, and we glue the left boundary of the strip with the boundary of the mesh stopping at the gate of the strip. Then we repeat the same for the right rooted mesh and the right boundary of the strip. The gate of the new mesh is the gate of the strip.

Assume that the tree encoding the mesh is traversed in preorder. Then the recursive decoding can be sketched as follows:

```
decode() {
  length=getNextNode();     // READ, preorder traversal
  if (length>0) {           // else: leaf, STOP
    [ enter,leading[2] ]=createStrip(length);
    enterSub[0]=decode();   // RECURSION
    enterSub[1]=decode();
    for (i=0;i<2;++i)       // GLUING
      if (enterSub[i]!=EMPTY)
        glue(leading[i],enterSub[i]);
    return enter;
  }
  return EMPTY;
}
```

Here, getNextNode() returns the values of the tree nodes, createStrip() creates a triangle strip and returns its enter gate/edge, e.g. (v_0, v_1), in Fig. 1, and the two leading[] edges, e.g. $(v_{11}, v_{13}), (v_{13}, v_{12})$. The recursive call to decode() creates the two submeshes and returns their enter gates. Finally, the strip boundaries are glued to the submeshes. The number of boundary edges for glue() can easily be calculated from the strip's length and parity, e.g. 6 and 7 for the strip of length 12. Note that some modification is needed to handle the self intersection, valence 3 and non-planar topology cases.

E.g. in Fig. 2 (left) we glue starting from the leading edges (top right in the picture) first the blue and then the green submesh to the dividing strip (red). In both cases gluing stops at the gate of the strip (bottom left). Fig. 3 shows an example run of the algorithm for a simple mesh.

Note that a strip may intersect not only with the boundary of a submesh but also with itself resulting in a loop (cf. Fig. 4, left). This is the same situation as before but with an inner and an outer submesh. The leading vertex now induces some kind of singularity for gluing: When we start gluing from the inner leading edge (red-blue border) we cannot decide which direction to take once we arrive back at the leading vertex as the outer boundary has not been glued yet. This problem can be resolved by gluing both leading edges first (glueSingleEdges(leading[1],enterSub[1]) removes the singularity) before gluing along the whole strip (glue(leading[0],enterSub[0])).

Another fine point of the algorithm is the occurrence of empty submeshes. A strip of length 1, that is a single triangle, can have one or both submeshes empty if one or both leading edges are boundary edges of the mesh. A strip of length 2 can only have the left submesh empty because the non-gate vertex of the first triangle of the strip is not boundary. Otherwise the process of growing the strip would stop there. In a strip of length greater than 2 both the leading directed edges are not boundary and an empty submesh occurs only when the strip intersects itself and the internal submesh is empty. This happens precisely when the strip passes through a vertex of valence 3. The

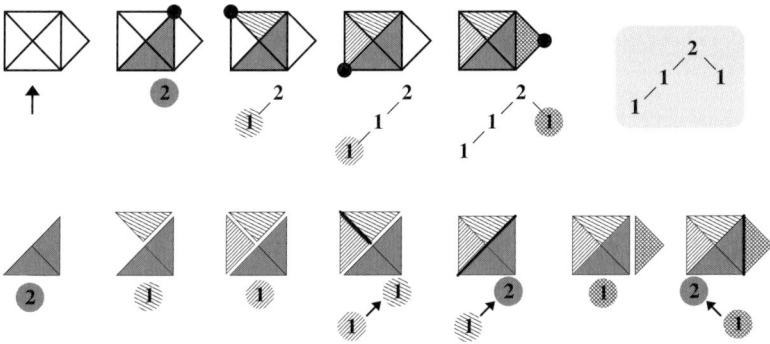

Fig. 3. Example run of the algorithm. *Top row*: The mesh (top left) is recursively encoded as a binary tree (top right) starting from the gate shown by the arrow. The four resulting strips are filled with different patterns. The first strip of length 2 partitions the mesh into two submeshes, all other strips have one or two empty submeshes. The dot denotes the leading vertex of the current strip, i.e. the point where the strip touches a conquered/boundary vertex. For every new strip the corresponding tree state is shown with the new node highlighted. *Bottom row*: For decoding, the tree is traversed in preorder. The strips are created during the top-down traversal of the tree nodes, and submeshes are glued (thick line) when an edge of the tree is followed in bottom-up direction. In the figure the corresponding tree nodes and edges are shown below the state of the mesh.

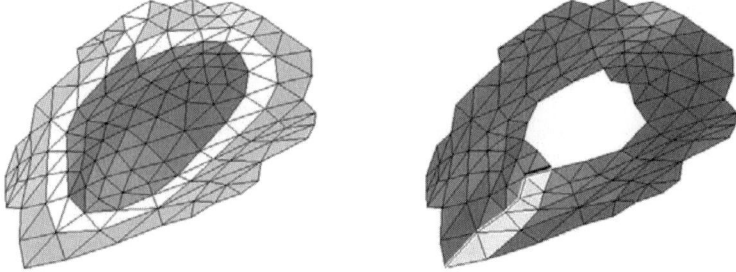

Fig. 4. Left: A strip may intersect itself resulting in a loop. Right: Handles or holes prevent the strip from splitting the mesh into two parts. A special *split* code is output instead of the empty left submesh. It references the corresponding edge in the right submesh drawn black. With this information the remaining boundary (white) can be glued.

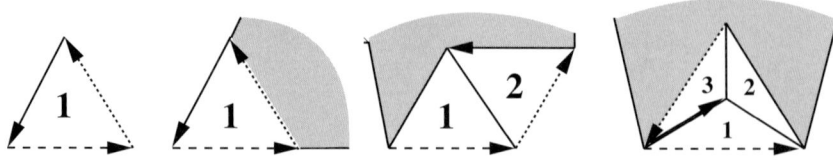

Fig. 5. From left to right. All strips are entered through the dashed gate (bottom line). The first strip of length 1 has both submeshes empty, while the next strip of length 1 has only the right submesh empty (solid gate). The strip of length 2 has the left submesh empty (dotted gate). The strip of length 3 has the right submesh empty (solid gate). Notice that in the case of a strip with length greater than 2 the empty submesh is in the interior of the strip and not on the boundary (valence 3 case).

parity of the length of the strip determines if the right or the left mesh is empty (cf. Fig. 5).

This case has to be checked separately, as the empty submesh does not correspond to boundary. If the strip `length>2` and the submesh `k` is empty (`enterSub[k]==EMPTY`) then an extra glue operation is needed: `glue(stripEdge,enterSub[k])`, where `stripEdge` is the corresponding edge of the strip (cf. triangle 1 in Fig. 5, right) which can be referenced easily by a fixed navigation path back through the strip.

4.2 Analysis of the Divide and Conquer Algorithm

Next, we give two basic propositions concerning the behavior of the Divide and Conquer Algorithm. The first proposition is about the size of the obtained the binary tree, and the second describes the relationship between the structure of the tree and the strip length information stored in its nodes.

The number of vertices of the mesh will be denoted by v, the number of triangles of the mesh, which is equal to the sum of the strip lengths, will be denoted by t, and the number of the strips, which is equal to the number of the nodes of the tree, will be denoted by n. We have

Proposition 5. *Let n be the number of the strips obtained from the divide and conquer algorithm. We have $n = v - 2$, where v is the number of the vertices of the mesh.*

Proof: We notice that a strip of length l conquers $l - 1$ vertices, and thus n strips of total length t conquer $t - n$ vertices. The number of vertices to be conquered is v_I, the number of interior vertices of the mesh, hence, we have $t - n = v_I$, giving $n = t - v_I$. The latter can also be written $n = v - 2$ as we immediately see by applying Euler's formula. □

The binary tree and the strip length information can be combined in a single data structure, namely a binary tree with positive integer weights assigned to its nodes. The next question is when such a data structure is valid, i.e. when it represents a planar triangle mesh. We have

Proposition 6. *Let T be a tree with n nodes and let the sum of the strip lengths be t. Then, T represents a planar mesh if and only if*

(i) $t + 1 \leq 2n$ holds for T and all its subtrees.
(ii) The strip lengths of the nodes with only left child are odd, and the strip lengths of the nodes with only right child are either 1 or even.

Proof: Let b be the number of the boundary vertices of a planar mesh. Applying Euler's formula gives

$$b = 2v - t - 2 \qquad (1)$$

which from Proposition 5 becomes

$$b = 2n - t + 2 \qquad (2)$$

Then, the inequality $3 \leq b$, holding for the boundary of the mesh, gives $t + 1 \leq 2n$. Notice that the inequality $3 \leq b$ is a standard assumption in the literature of meshes as $b = 2$ would give two boundary vertices connected between them with two different edges, while $b = 1$ would give a vertex connected with itself.

Condition (ii) describes the exceptional cases where one of the submeshes is empty and two edges of the dividing strip are glued together.

Conversely, if the condition (i) of the proposition holds, there is enough free boundary to perform all the gluing operations and we get a valid triangle mesh, while condition (ii) guarantees that we can perform the gluing in the exceptional cases as well. □

4.3 Arbitrary Topology

In the case of arbitrary topology the main difference is that a dividing strip can have the same non-empty submesh on the left and the right. The simplest example is a planar mesh with a hole in it, which is topologically equivalent to a cylinder (cf. Fig. 4, right). In this case we need to encode only one branch of the tree and give some additional information on how the boundary of the corresponding submesh is glued to the other side of the strip.

If the genus of the mesh is g and there are h holes in it, it is a simple topological fact that the number of the strips which do not separate the mesh is at most $2g + h$. For each such strip we need $O(2 \log v)$ bits to identify the directed edge for gluing, and in the worst case $O(\frac{1}{2} \log v)$ bits for the extra symbol.

4.4 The Connection with the EdgeBreaker

Before the study of the obtained data structure, it worth having a look at the algorithm in a more general setting, clarifying some aspects which might be obscured in a very concrete exposition. A first observation is that the algorithm, like the EdgeBreaker, implicitly induces a traversal of the triangles of the mesh. This traversal of the triangles can be seen at two levels, the first level is the traversal of the triangles of a zig-zag strip, from the root of the strip to the leading edges, and the second level is the traversal of the tree that stores these zig-zag strips.

Also, we notice that the algorithm works not only with zig-zag strips but with general strips as well. In fact, any bitstream the encoder and the decoder would agree on, defines a way of building general strips, and thus a variation of the method. Here we use the zig-zag strips as the most natural choice for making a strip.

The fan strip is another important class of strips. In this case, assuming a preorder traversal of the tree, we get the same traversal of the triangles as the EdgeBreaker, and our approach differs only in the interpretation of the obtained data. In fact, we can translate the encoding of the binary tree and the strip lengths into the familiar C, L, E, R, S string of the EdgeBreaker and vice-versa. Each strip length n can be written as a string of $n-1$ C's, and for each node of the tree we use one of the L, R, S, E symbols, depending on whether it has only left child, only right child, two children or no child. Notice that because our data structure is non-linear, namely, a binary tree rather than a symbol stream, it is not necessary to assume any particular traversal of the tree to interpret the data. Although this has advantages in the theoretical analysis of the algorithm, nevertheless, in the implementation we usually assume a traversal to make things simpler.

5 Tree-based Data Structures

With the Divide and Conquer Algorithm, the encoding of triangle mesh connectivity encoding can be split into two separate but closely related subtasks. The encoding of the binary tree, and the encoding of the strip lengths stored in its nodes. We study these two encodings, first separately and then in their interrelation. For simplicity we deal with planar meshes only. But, taking into account the special output symbols and the changes of the Euler's characteristic they introduce, the arbitrary topology can be treated in a similar manner.

5.1 Binary Tree Encodings

The binary tree is one of the most popular structures for storage and maintenance of data, and thus, many basic related problems have been widely studied. For a brief comparative study of different binary tree encodings see [13] and [16].

In our context, the relevance of the tree encoding methods becomes apparent with the observation that any binary tree can correspond to a planar mesh. Indeed, if all the strip lengths are equal to 1, then by Proposition 6 every binary tree gives a valid planar mesh. The number of all the binary trees with n nodes is given by the Catalan number C_n

$$C_n = \frac{(2n)!}{n!(n+1)!} \sim \frac{4^n}{\sqrt{\pi}n^{3/2}} \qquad (3)$$

see [7] Exercise 9.8. Thus, there is an asymptotic bound of 2 bits per node.

Some of the existing methods for binary tree encoding enumerate all the trees with n nodes so that each tree can be represented by an integer number. Obviously, such methods achieve optimal compression ratios, but the encoding and decoding cost is very high, and especially for the large trees we use here this cost is prohibitive. Other encoding methods traverse the tree, transmitting one letter for each node. These methods are separated into two categories. The methods which use a fixed alphabet and those using an alphabet depending on the number of the nodes n.

The most common fixed alphabet encodement uses 4 letters, let say $\{L, R, S, E\}$ to make the analogy with the EdgeBreaker clearer. Each letter determines if the corresponding node has only left, only right, both or none children. The cost is $2n$ bits, which, asymptotically achieves the bound of 2 bits per node given by (3). By Proposition 5 this is also equal to 2 bits/v.

Another well-known fixed alphabet encoding is the Zaks' sequences which use the two letters alphabet $\{0, 1\}$. The encoding process first transforms the binary tree into a complete binary tree by appending new leaves wherever possible, that is, two new leaves at any old leaf and one new leaf at any single child node. Then we traverse the nodes of the complete binary tree in preorder, transmitting a 1 if the node is internal and a 0 if the node is a leaf. There are standard algorithms deciding when a given sequence of 0's and 1's is the Zaks' sequence of a tree.

The size of the file encoding the tree can be further reduced with Arithmetic Coding [17]. In this case the compression ratio also depends on the traversal of the tree. Notice that there are traversals, like the inorder, which do not work with the above four or two letter encodings, because we can not reconstruct the tree without some additional information.

The variable alphabet encodements now, most often use the letters of the set $\{0, 1, \ldots, n\}$. When it comes to compression issues they have the problem that it is more difficult to find a sharp guaranteed upper bound which can be trivially found for a fixed alphabet. Nevertheless, many times they are more flexible, and there are standard algorithms determining the validity of a code, making it easy to eliminate the transmission of redundant information.

An example of variable length encoding is the weight sequence, see [16]. There, the letter corresponding to a node is the number of the nodes of its left subtree. The inductive argument showing that the method works is that if we know the number of the nodes of the tree, and the number of the nodes of the left subtree we can find the number of nodes of the right subtree by subtraction, and we can continue recursively this process until reaching the leaves. The weight sequence encoding of a tree has a special interest because by Proposition 5 the number of the nodes n is related to v the number of the vertices of the mesh. For this reason, the weight sequence is a useful intermediate representation of a binary tree corresponding to a mesh.

5.2 Strip Lengths Encodings

The encoding of the strip lengths is equivalent to the encoding of $v - 2$ numbers summing up to t. The range of the numbers is from 1 up to the length of the largest strip we create. Although the largest strip has always length less than $\lfloor \frac{v}{2} \rfloor$, still this is not a sharp bound with any practical use.

The simplest way to encode such a sequence of numbers is to represent the number k as a word of k bits consisting of $k - 1$ 1's followed by a 0. The total cost in this case is t, and the total cost of encoding the mesh, including the tree, is bounded by $t + 2v - 4 < 4v$. It worth noticing here that the guaranteed performance of 4 bits/v of this very coarse encoding method, can not be improved without going deeper into the study of the relationship between the encoding tree and the corresponding strip-lengths.

Another, very popular method to encode a sequence of numbers is the Huffman coding. Each number is assigned a unique code and the number of bits we spend on each code depends on the probability of each number to appear in the sequence. After the Huffman coding we can use Arithmetic Coding to further exploit any existing entropy. Notice that the compression achieved with the Arithmetic Coding depends on the order the strip lengths are transmitted, that is, on the particular traversal of the tree. Also, notice that because now the tree is given, and unlike the situation in Subsection 5.1, any traversal works. For a survey of different tree traversals, see [3].

If the mesh has a relatively small boundary the number of vertices, and thus, the number of tree nodes is about half the number of triangles. Therefore, the average strip length is near 2, and the entropy of strip lengths largely depends on the number of strips with length 2. We have noticed by experiment that the more regular a mesh the more strips of length 2 occur in its encoding. For an intuitive explanation of the last, see Figure 2: a large strip passing through a regular area of the mesh creates a regular boundary with vertices of valence 4, and this creates a lot of strips with length 2. This observation partly justifies the choice of the leading directed edges of the dividing strip as the gates of the two submeshes. Another deterministic choice of the gates, for example near the middle of the dividing strip, would increase the length of the strips near the root of the tree but the result would be worse compression ratios, because the total entropy would decrease.

The Table 1 shows the experimental results for preorder and postorder encoding of the tree with a 4 and a 2 letter alphabet, and preorder, inorder, postorder and level traversal in the encoding of the strip lengths, for a variety of meshes. In [12] we used preorder traversals for both the tree and the strip lengths and we transmitted them in an interwoven fashion, that is, at each node the code of the strip length followed immediately after the tree code. Comparing the results there, even with the most favorable combination of separate transmission of tree and strip lengths file, we see that they are better. That means that there is a lot of entropy in the blend of tree codes and strip lengths, and this entropy is exploited by the Arithmetic Coder.

Mesh	#V	#F	$t_{pre,1}$	$t_{pre,2}$	$t_{post,2}$	$s_{pre,H}$	s_{pre}	s_{post}	s_{in}	s_{level}	IRS02
david1	315	586	2.46	2.46	2.46	1.40	2.08	2.08	2.03	2.08	3.94
david2	1512	2924	2.11	2.10	2.11	1.43	1.66	1.67	1.66	1.67	3.56
david3	6035	11820	2.03	2.03	2.03	1.46	1.58	1.60	1.53	1.62	3.39
david4	24085	47753	2.05	2.06	2.06	1.46	1.51	1.53	1.39	1.55	3.18
dinosaure	14070	28136	2.04	2.06	2.05	1.45	1.52	1.54	1.37	1.55	3.13
fandisk	6475	12946	1.52	1.50	1.50	1.51	1.23	1.24	0.75	1.57	1.95
mannequin1	428	839	2.34	2.34	2.34	1.36	1.89	1.91	1.87	1.91	3.79
mannequin2	11703	23402	0.92	0.91	0.90	1.30	0.84	0.83	0.34	1.21	1.06
venus	8268	16532	2.04	2.05	2.05	1.49	1.60	1.61	1.55	1.63	3.46
max-planck	100086	199996	1.20	1.16	1.16	1.52	1.08	1.06	0.67	1.39	1.42

Table 1. Compression ratios in bits/vertex for 10 models from [1]. The trees (t) are encoded separately from the strip lengths (s). The suffixes indicate pre-, post-, in-, and level-oder traversal, the 1-alphapbet (Zaks) or 2-alphabet. For $s_{pre,H}$ only Huffman coding is applied, Arithmetic Coding and a fixed alphabet is used in all other columns. The last column shows the results in [12].

Next we assume that the tree code and the strip length code of a node are sent one after the other. That means that we assume the same traversal for both the tree code and the strip lengths code. We study the relationship between the two encodings, and we see how we can save information from the tree code using information from the strip-lengths and vice versa.

5.3 Tree First Transmission

Here we assume that the code of the tree is sent first, using the $\{L, R, S, E\}$ alphabet, and the code of the strip length follows. Some simple observations can reduce the amount of information we have to send for the strip lengths.

A first observation is that because of Proposition 6, condition (ii), the weight of an R node is either 1 or an even number, while the weight of an L node is always odd. Another observation, from the same Proposition, condition (i), is that all the leaves have strip length 1, and do not need encoding. Going one step further, we notice that if an R is just above a leaf then, by condition (i), the corresponding strip length is either 1 or 2 and can be encoded with a single bit. Similarly, an L node just above a leaf can only have a strip length 1, and we do not need to transmit any information. Fig. 6 shows the encoding near the leaves and the corresponding meshes.

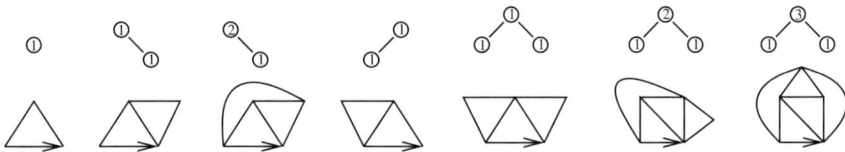

Fig. 6. The encoding trees and the corresponding meshes near the leaves. Notice that some of the rooted meshes are isomorphic as unrooted meshes.

The above observations are the simplest instances of a more general feature of our approach, namely that every node of the tree represents a gluing operation and Proposition 6 gives a simple criterion to determine when such a gluing operation is legal. Therefore, we can treat these instances in a unified way by determining at each node all the legal gluing operations, and sending only the necessary information for the decoder to distinguish between them. This assumes a postorder reconstruction of the mesh.

Checking the criterion of the special gluings is straightforward. By Proposition 6 condition (ii), if the node is an L then the set of possible strip lengths is restricted to the odd numbers, while if the node is an R the set of possible strip lengths consists of the even numbers and the 1. For a criterion checking the regular gluings, i.e. Proposition 6 condition (i), for each node we need the number of nodes of its subtree, and the sum of the strip lengths of that subtree. This information can be held in an auxiliary data structure, where each node $\mathbf{n}_j$ of the tree is assigned two integers (n_j, w_j), the number of nodes and the sum of strip lengths of its subtree. Notice that the integers w_j are essentially the weight sequence of the tree, see Subsection 5.1. When we process a new node $\mathbf{n}_j$, we first find its children, let say $\mathbf{n}_{j_l}, \mathbf{n}_{j_r}$, and if the integers assigned to them are (n_{j_l}, w_{j_l}) and (n_{j_r}, w_{j_r}), then

$$n_j = n_{j_l} + n_{j_r} + 1 \quad \text{and} \quad w_j = w_{j_l} + w_{j_r} + s \qquad (4)$$

where s is the strip length corresponding to $\mathbf{n}_j$. Then, Proposition 6, condition (i), gives

$$w_{j_l} + w_{j_r} + s + 1 \leq 2(n_{j_l} + n_{j_r} + 1) \qquad (5)$$

giving

$$s \leq 2n_{j_l} + 2n_{j_r} - w_{j_l} - w_{j_r} + 1 \qquad (6)$$

This way we find explicitly the set of all the strip lengths giving a legal gluing operation at a node, and instead of transmitting the actual strip length we send an offset determining its position in the set of all legal values. If the set contains only one element, that is, when the set is the {1} we send no information. If the set has two elements, that is, if it is either the {1, 2} or the {1, 3} we send only one bit. In all the other cases it is better to resort to Huffman encoding of the offsets rather than using an ad hoc code for every particular set, because of the significantly higher frequencies of the short strips.

The algorithm we just described can also be used in conjunction with the original Divide and Conquer algorithm, where we transmit the actual strip lengths, as a test checking the validity of a code. In this case, proceeding as above, we find the set of all legal strip lengths corresponding to a node and we check if the actual value of s lies into that set. If this happens for every node then all the gluing operations are legal and the code describes a valid mesh. If there is a node with strip length outside the set, then the decoding process will break at that point. Of course the testing algorithm must be coupled with an algorithm checking the legality of the corresponding tree code, which can be found in the literature.

5.4 Strip-lengths First Transmission

Suppose now that we first transmit the strip-length of a node and then the tree code corresponding to it. From Proposition 6, condition (ii), even strip lengths correspond to an S or an R node, while odd strip lengths greater than 1 correspond to an S or an L node. Therefore, we need a single bit for the tree code of the nodes with strip length greater than 1.

Table 2 shows some more results obtained with the encoding techniques described in Subsections 5.3, 5.4. An additional column compares to the Edgebreaker-like traversal when fan-strips are used instead of zig-zag strips. We also show the results obtained the by the Alliez-Desbrun method [1], which currently gives the best compression ratios.

5.5 The Valence 3 Vertices

From Proposition 6, and the above discussion of its implications, it is apparent that the valence 3 vertices are very characteristic. Their peculiarity arises from the fact that a zig-zag strip collapses to itself only when passing through a valence 3 vertex. Equivalently, it is the only case when a strip of length greater than 2 can correspond to an R or an L node. Therefore, in many cases it may pay off to have an initial preprocessing step clearing the mesh from its valence 3 vertices. Such a strategy to improve the efficiency of an algorithm was also proposed in [2].

After the clearance step we work as in Subsections 5.3, 5.4 separating the case of tree code transmission first from the case of strip length transmission first. Sending the tree code first, we know that an L node can only store a strip length equal to 1, because any greater strip length would create a valence 3

Mesh	pre	post,H	post	fan	A&D
david1	3.94	3.30	3.94	3.86	2.96
david2	3.53	3.35	3.53	3.57	2.88
david3	3.36	3.40	3.36	3.37	2.70
david4	3.13	3.39	3.14	3.19	2.52
dinosaure	3.09	3.38	3.10	3.13	2.25
fandisk	1.90	3.42	1.94	1.94	1.02
mannequin1	3.79	3.29	3.78	3.79	2.51
mannequin2	1.02	3.27	1.07	1.05	0.37
venus	3.42	3.40	3.42	3.45	2.37
max-planck	1.38	3.43	1.42	1.42	n/a

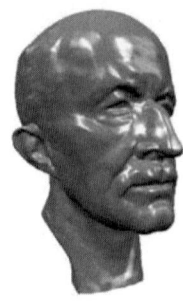

Table 2. Compression ratios in bits/vertex. The table shows results for transmitting the tree (2-alphabet) and indices of valid strip lengths in an interwoven fashion in pre- and post- oder with Huffman coding only resp. Arithmetic Coding. The *fan* column gives results from using not zig-zag strips but fan-strips and a preorder traversal, fixed alphabeth which corresponds to the bitstream obtained from the Edgebreaker (with Arithmetic Coding applied). The compression ratios are similar. The right most column shows the results of Alliez-Desbrun [1] for comparison. In this paper renderings of the corresponding meshes can be found. The Max-Planck mesh with 100086 vertices has also been used for testing.

vertex, and therefore, we do not need to send any strip length information. Similarly, a strip length corresponding to an R is either 1 or 2, and is encoded in a single bit. On the other hand, if we first transmit the strip length of a node, then any length greater than 2 corresponds to an S node and we do not need any extra tree code. A strip length equal to 2, corresponds to either an S or an R node and we need a single bit for the tree code.

6 Conclusion

We described a Divide and Conquer algorithm for the encoding of triangle mesh connectivity. The naturally arising data structure for the storage of the obtained information is a binary tree with positive integer numbers assigned to its nodes. We studied this data structure, showing that there is a deep correlation between the structure of the tree and the assigned integers, which can benefit the performance of the algorithm.

References

1. P. Alliez and M. Desbrun. Valence-Driven connectivity encoding for 3D meshes. In *EUROGRAPHICS 01 Conference Proceedings*, pages 480–489, 2001.
2. Pierre Alliez and Mathieu Desbrun. Progressive compression for lossless transmission of triangle meshes. In *SIGGRAPH 01, Conference Proceedings*, pages 195–202, 2001.

3. Alfs Berztiss. A taxonomy of binary tree traversals. *BIT*, 26:266–276, 1986.
4. W.G. Brown. Enumeration of triangulations of the disk. *Proc. Lond. Math. Soc., III. Ser.*, 14:746–768, 1964.
5. M. Denny and C. Sohler. Encoding a triangulation as a permutation of its point set. In *Proceedings of the 9th Canadian Conference on Computational Geometry*, pages 39–43, May 15–17 1997.
6. N. Deo and B. Litow. A structural approach to graph compression. In *MFCS Workshop on Communications*, pages 91–101, 1998.
7. Ronald L. Graham, Donald E. Knuth, and Oren Patashnik. *Concrete mathematics: a foundation for computer science.2nd ed.* Addison-Wesley, 1994.
8. Stefan Gumhold. Improved cut-border machine for triangle mesh compression. In *Erlangen Workshop '99 on Vision, Modeling and Visualization, IEEE Signal Processing Society*, 1999.
9. Stefan Gumhold. New bounds on the encoding of planar triangulations. Technical Report WSI-2000-1, Wilhelm-Schikard-Institut für Informatik, Tübingen, March 2000.
10. Stefan Gumhold and Wolfgang Straßer. Real time compression of triangle mesh connectivity. In *SIGGRAPH 98 Conference Proceedings*, pages 133–140, July 1998.
11. Isenburg and Snoeyink. Spirale reversi: Reverse decoding of the edgebreaker encoding. *CGTA: Computational Geometry: Theory and Applications*, 20, 2001.
12. I. Ivrissimtzis, C. Rössl, and H-P. Seidel. A divide and conquer algorithm for triangle mesh connectivity encoding. In *Pacific Graphics 02, Conference Proceedings*, 2002.
13. Jyrki Katajainen and Erkki Mäkinen. Tree compression and optimization with applications. *Int. J. Found. Comput. Sci.*, 1(4):425–447, 1990.
14. Andrei Khodakovsky, Pierre Alliez, Mathieu Desbrun, and Peter Schröder. Near-optimal connectivity encoding of 2-manifold polygon meshes. *Graphical Models, special issue on compression*, 2002.
15. Davis King and Jarek Rossignac. Guaranteed 3.67V bit encoding of planar triangle graphs. In *Proceedings of the 11th Canadian Conference on Computational Geometry*, pages 146–149, 1999.
16. E. Mäkinen. A survey in binary tree codings. *Comput. J.*, 34(5):438–443, 1991.
17. Moffat, Neal, and Witten. Arithmetic coding revisited. *ACMTOIS: ACM Transactions on (Office) Information Systems*, 16, 1998.
18. Rossignac and Szymczak. Wrap&zip decompression of the connectivity of triangle meshes compressed with edgebreaker. *CGTA: Computational Geometry: Theory and Applications*, 14, 1999.
19. Jarek Rossignac. Edgebreaker: Connectivity compression for triangle meshes. In *IEEE Transactions on Visualization and Computer Graphics*, volume 5 (1), pages 47–61. 1999.
20. Szymczak, King, and Rossignac. An edgebreaker-based efficient compression scheme for regular meshes. *CGTA: Computational Geometry: Theory and Applications*, 20, 2001.
21. Gabriel Taubin and Jarek Rossignac. Geometric compression through topological surgery. *ACM Transactions on Graphics*, 17(2):84–115, April 1998.
22. Costa Touma and Craig Gotsman. Triangle mesh compression. In *Proceedings of the 24th Conference on Graphics Interface (GI-98)*, pages 26–34, June 18–20 1998.
23. W.T. Tutte. A census of planar triangulations. *Can. J. Math.*, 14:21–38, 1962.

Compression of Normal Meshes

Andrei Khodakovsky[1] and Igor Guskov[2]

[1] Caltech *akh@cs.caltech.edu*
[2] University of Michigan *guskov@eecs.umich.edu*

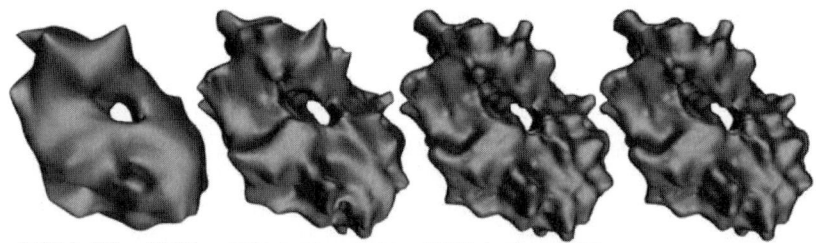

263B (e:131, p:38dB)   790B (e:37, p:49dB)   2788B (e:10, p:60dB)   35449B (e:2, p:74dB)

Fig. 1. Partial reconstructions from a progressive encoding of the molecule model. File sizes are given in bytes, errors in multiples of 10^{-4} and PSNR in dB (model courtesy of The Scripps Research Institute).

Summary. Normal meshes were recently introduced as a new way to represent geometry. A normal mesh is a multiresolution representation which has the property that all details lie in a known normal direction and hence the mesh depends only on a *single scalar per vertex*. Such meshes are ideally suited for progressive compression. We demonstrate such a compression algorithm for normal meshes representing complex, arbitrary topology surfaces as they appear in 3D scanning and scientific visualization. The resulting coder is shown to exhibit gains of an additional 2-5dB over the previous state of the art.

1 Introduction

The growth of computing power of personal computers and recent progress in shape acquisition technology facilitate the wide use of highly detailed meshes in industry and entertainment. Similarly, scientific visualization applications tend to produce ever finer meshes, such as iso-surfaces. In their raw, irregular form, acquired meshes are complex and often unmanageable due to their sheer size and irregularity [23]. It is therefore important to find more efficient and compact representations. Algorithms for efficient encoding of such meshes have been described in both single rate [2, 15, 29, 30, 35, 37] and progressive settings [1,3,6,18,24,27,28,34]. For an overview of 3D geometry compression see [36].

One should recognize the fact that compression is always a trade-off between size and accuracy. That is especially true for meshes that come from shape acquisition or iso-surface extraction, and which always carry sampling error and acquisition noise. Compression methods for such meshes can be lossy as long as the approximation error of the reconstruction stage is comparable with the sampling error. In the past eight years a number of efficient "remeshing" techniques have appeared that replace the original mesh with a mesh consisting of a number of "regular" pieces, such as B-spline [12], NURBS [20], or subdivision connectivity [11,21,22] patches. Recently, even more regular resampling techniques were introduced that use a single regular patch for the whole model [14]. Naturally, one should expect that the remeshed model should behave much better with regards to compression algorithms. This expectation was confirmed in [19], where the MAPS algorithm [22] was included as part of a progressive geometry coder. In particular, the paper [19] makes it clear that any mesh representation can be considered as having three components: geometry, connectivity, and parameterization; moreover, the last two components are not relevant for the representation of the geometry. For semi-regular mesh hierarchies, one can make a reasonable assumption that the normal component of the detail coefficients stores the geometric information, whereas the tangential components carry the parametric information (see also [9,16]).

Most of the existing remeshing algorithms purposefully remove almost all of the connectivity information from the mesh, and also reduce the parametric information. One may wonder: *how much of the parametric information can be removed from the surface representation?* The answer is *almost all of it* as demonstrated in [17]. In fact, by fixing the transform algorithm to use unlifted Butterfly wavelets it is possible to build a semi-regular mesh whose details lie almost exclusively in the local normal direction. Consequently, the geometry of "normal" meshes is fully represented by a single scalar per vertex, instead of the usual three. Therefore it is natural to use normal meshes for compression as we do in this paper.

Contribution The goal of this paper is to demonstrate the additional coding gains possible when employing normal semiregular meshes rather than standard semiregular remeshes (such as those produced by MAPS [22]).

2 Compression Algorithms

The progressive geometry coding described in [19] requires three necessary components: a remeshing algorithm, a wavelet transform, and a zerotree coder. In [19] the MAPS remesher [22] was used, followed by the Loop or Butterfly wavelet transform. In this paper, we are using the normal remesher of [17] which was specifically designed to produce detail coefficients with no tangential components when an *unlifted Butterfly* wavelet transform is applied to the produced semi-regular mesh. We use the same zerotree coder as in [19]. The following sections will briefly overview the algorithms used for compression. For a more detailed exposition, the reader is referred to [19] and [17].

2.1 Normal Remesher

The normal remesher starts with a closed irregular mesh, and proceeds to build a semi-regular mesh hierarchy approximating the original model. The algorithm is described in [17] and consists of two stages. Using mesh simplification, a base mesh is chosen that is topologically equivalent to the original mesh. The connectivity of this base mesh will eventually become the connectivity of the coarsest level in the semi-regular hierarchy. Also at this stage a net of surface curves is initialized that splits the irregular mesh into a number of non-intersecting regions (these regions are in a one-to-one correspondence with the coarsest level faces of the semi-regular mesh being constructed). To complete the first stage of the algorithm, the net of surface curves is propagated to the finest level of the original irregular mesh, and a relaxation of global vertex positions within the surface is performed to even out their distribution and improve aspect ratios of the base mesh triangles.

In the second stage of the algorithm, a "piercing procedure" is applied recursively to obtain positions of finer level points of the semi-regular mesh. Thus, the semi-regular mesh is refined and, to maintain the status quo, the corresponding regions of the irregular mesh are split into smaller subregions. A global adaptive parameterization is maintained on the original mesh in order to keep the piercing process under control and to enable fast construction of surface curves. Surface patches are parameterized with Floater's parameterization scheme [13]. Every time a patch is split into smaller ones, a new parameterization is computed for the resulting sub-patches.

The described two-stage process produces a semi-regular mesh that has mostly normal detail coefficients except for a small number of locations where the piercing did not find any "valid" intersection point with the original surface. For such exceptional vertices, the recorded detail is not scalar, and will have three components. The percentage of non-normal coefficients varies depending on the geometric properties of a given mesh and the corresponding coarse level points chosen in the first stage of the algorithm. Typically, the number of non-normal coefficients is below 10% for adaptive meshes that have the same number of vertices as the original irregular mesh. It may be possible to decrease the percentage of the non-normal details by a non-linear optimization approach, however we observed that the impact of non-normal coefficients on the compression results is not significant.

The interpolating character of the Butterfly subdivision scheme makes the construction of the normal remesh relatively straightforward since once a vertex is introduced during refinement its position stays the same on finer levels of the hierarchy. In contrast, the construction of normal meshes corresponding to any non-interpolating subdivision scheme (such as Loop) would have to be much more involved both algorithmically and computationally. In both interpolating and non-interpolating cases the normality condition introduces a system of constraints on the surface point sampling. However, for a non-interpolating scheme in order to satisfy these constraints one would need to use a global nonlinear optimization procedure. The metamesh approach to the normal mesh construction used for Butterfly normal meshes would become unwieldy for such an optimization, and a different, possibly parameterization-based approach needs to be designed for Loop normal mesh construction. We leave it as a direction for future work.

Displaced subdivision surfaces A similar goal of achieving a "scalar" geometry description was addressed in the work on displaced subdivision surfaces [21]. That approach represents the original shape as a single resolution normal displacement from a base Loop subdivision surface, which is achieved by a careful construction of the base mesh. While the DSS representation is purely scalar, the typical sizes of the DSS base meshes are on the order of magnitude higher than the ones obtained in the normal remesher even for geometrically simple models [17] [21].

2.2 Wavelet Transform

A semi-regular surface representation is a sequence of approximations at different levels of resolution: the corresponding nested sequence of linear spaces is given as $V_0 \subset V_1 \subset \ldots \subset V_n$. Here V_0 is the space of coarse base meshes, and V_n is where the finest level meshes live. The wavelet transform maps this mesh hierarchy into a representation that consists of the base mesh and the sequence of wavelet coefficients that express differences between successive levels of the semi-regular mesh hierarchy. Thus, the wavelet transform decomposes the surface with respect to the hierarchy $V_n = V_0 + W_1 + \ldots + W_n$, where $V_j = V_{j-1} + W_j$ for $j = 1, \ldots, n$. [26]

Typically, an appropriate wavelet transform achieves a better statistical distribution of the coefficients in the surface representation. This works well because vertex coordinates of a smooth surface are correlated. This correlation can be exploited through the prediction of finer level geometry based on the coarser level geometry (low-pass filtering). The difference between the prediction and the actual geometry is represented by wavelet coefficients (high-pass filtering). For smooth surfaces we expect to achieve good prediction, which leads to smaller wavelet coefficients. Figure 2 shows the histograms of the distribution of original geometry coordinates as well as the distribution of the corresponding wavelet coefficients. Note that there are very few large coefficients (these actually contain important information) while the remain-

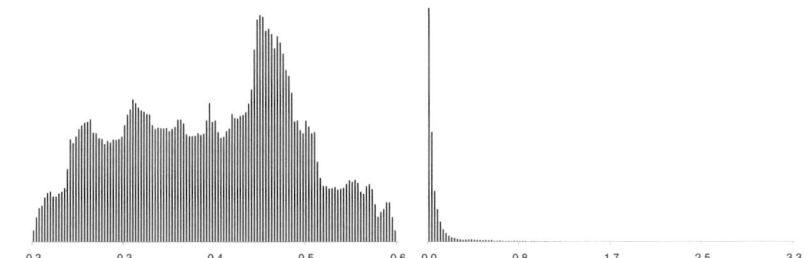

Fig. 2. Histograms of distribution of the original geometry and Loop wavelet coefficients for Venus head model.

ing majority of the coefficients are almost zero. Therefore, this representation is much more suitable for compression.

We start our wavelet construction by fixing a predictor, *i.e.* the subdivision scheme. The subdivision defines an embedding $V_j \subset V_{j+1}$.

$$\begin{pmatrix} s_e^{j+1} \\ s_o^{j+1} \end{pmatrix} = \begin{pmatrix} P_e \\ P_o \end{pmatrix} (s^j).$$

P_o block computes positions of of new (*odd*) vertices and P_e is an update on the old (*even*) points. There are two classes of subdivision schemes: *interpolating* (e.g. Butterfly) and *approximating* (e.g. Loop or Catmull-Clark) [38]. Interpolating schemes insert odd points as weighted averages of even points but never update the positions of even points (P_e is the identity matrix). Approximating schemes not only compute odd points but also update positions of even points.

The wavelet space expresses the difference between finer and coarser spaces. The problem of designing the wavelet transform can be stated as a matrix completion problem: find such two matrices Q_e and Q_o that the whole transform $V_j + W_{j+1} \mapsto V_{j+1}$ is invertible.

$$\begin{pmatrix} s_e^{j+1} \\ s_o^{j+1} \end{pmatrix} = \begin{pmatrix} P_e & Q_e \\ P_o & Q_o \end{pmatrix} \begin{pmatrix} s^j \\ w^{j+1} \end{pmatrix}. \tag{1}$$

Of course, the properties of the wavelet transform greatly depend on the choice of the completion. In signal processing, the wavelets are typically required to define a stable transform, to have some number of vanishing moments, and to satisfy some orthogonality conditions. Additionally, the transforms with sparse Q_e and Q_o are preferable, since they can be implemented more efficiently. For surfaces, the wavelet theory is far from being fully developed, and the importance of some of the above mentioned qualities still has to be established.

In the remainder of this section we describe the particular wavelet transforms that we use for the geometric compression of semi-regular meshes.

Linear subdivision. First, we consider a simple case of the linear prediction (subdivision) scheme and the corresponding wavelet transform. The linear subdivision scheme computes the predicted geometry position of an odd point as the average of the two endpoints of the edge associated with the new point (Figure 4). The details (wavelet coefficients) are the differences between predicted and original positions of the odd points. Note, that this is an interpolating scheme, that is the even vertices do not change their values between levels of the hierarchy. Therefore, details are associated only with odd points. The wavelet transform preserves the number of degrees of freedom. This is true in general, though for approximating schemes the construction gets more complex [19].

After the above transform we adjust the scaling between the coarser coefficients and the details. Namely, the details on level j are divided by 2^j. This scaling arises from the L_2 normalization of the subdivision basis functions. Linear subdivision basis functions are hat functions centered at a vertex and supported on the one-ring of that vertex. In the pure subdivision setting, basis functions are normalized to have the same maximal value at all the levels of the hierarchy. However, for compression purposes it is beneficial to decompose the surface with respect to an L_2 normalized basis. Since the area of the support shrinks by the factor of 4 between successive levels, the finer basis functions must be multiplied by 2, which accumulates so that the detail coefficients on level j must be divided by 2^j. Note that because of such a normalization the coarser details will be encoded with higher precision than the finer ones. The same scaling strategy applies not only for linear subdivision wavelets but also for all higher order schemes described below.

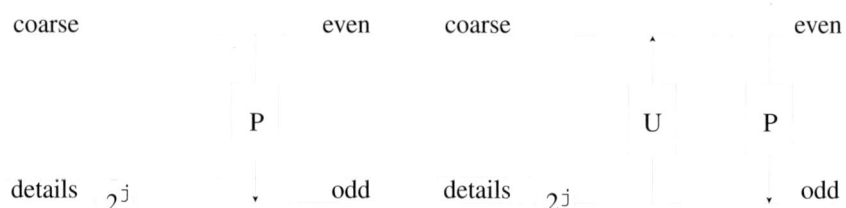

Fig. 3. The diagram illustrates the wavelet transform in the lifting framework. On the left the initial construction, on the right an update step is added.

A two-level transform can be illustrated with the following diagram (Figure 3). The wires on the left correspond to the coarse model (top) and details (bottom). The right hand side corresponds to the finer level geometry. The forward transform (analysis) is a transition from right to left: the predictor is evaluated on even vertices and subtracted from the odd vertex positions. Finally, the top wire is divided by 2^j. Reconstruction is a transition in the opposite direction: the top wire is rescaled back, then the predictor, computed using coarser coefficients (bottom wire) is added to the details. Equivalently, the transforms are expressed by the following formulas:

$$\begin{pmatrix} s^{j-1} \\ w^j \end{pmatrix} = \begin{pmatrix} I & 0 \\ -2^{-j} P & 2^{-j} I \end{pmatrix} \begin{pmatrix} s_e^j \\ s_o^j \end{pmatrix}, \quad \begin{pmatrix} s_e^j \\ s_o^j \end{pmatrix} = \begin{pmatrix} I & 0 \\ P & 2^j I \end{pmatrix} \begin{pmatrix} s^{j-1} \\ w^j \end{pmatrix}.$$

For stability and good approximation properties of a wavelet transform, it is important that the wavelet basis has some number of vanishing moments. In our construction, primal wavelets do not have even the zeroth vanishing moment. This problem can be fixed using the lifting construction: During the analysis step we modify coarse coefficients by subtracting a specially defined linear combination of just computed details (Figure 3). For example, using only the closest details with weight -0.125 enforces a zero moment of the primal wavelets. The same approach can be used to improve other properties of the wavelets. For instance, instead of enforcing moments, one can improve an angle between the wavelet space W_i and the coarser space V_{i-1}.

Higher order schemes. The linear scheme is very simple and easy to implement but it does not produce smooth surfaces. Several other methods for building wavelet transforms on semi-regular meshes exist [4,26,32]. For compression of smooth surfaces it is advantageous to use higher order schemes. Such schemes can achieve better approximation leading to smaller detail coefficients and therefore to better compression performance. In this work we used two particular wavelet transforms: Loop wavelets and non-lifted Butterfly wavelets.

The Loop [25] wavelets we use were first described in [19]. These have the advantage of using the Loop subdivision scheme in its inverse transform and perform quite well. However, a more natural wavelet transform to use in this work is the unlifted version of Butterfly wavelets [10, 39] because the *exact* same transform is used to produce normal meshes. A detailed description of the construction of Butterfly wavelets can be found in [32].

Figure 4 shows the Butterfly subdivision stencils in the regular case. The scheme uses an eight-point stencil and produces smooth surfaces. Similar to the linear case, an update step can be added to improve the wavelet transform. In general, such an update step can improve coding performance, but it is not suitable for normal mesh compression. The reason is that the normal mesh is constructed with respect to "pure" butterfly transform. Therefore, the wavelet coefficients will not be normal to the surface if any update step is added.

Note that the Butterfly wavelet transform uses finite filters for both analysis and reconstruction, and is therefore faster than the forward Loop transform which requires the solution of a sparse linear system. The Loop reconstruction filter has support of the same size as the Butterfly filter. On the other hand, we found that for non-normal meshes the Loop wavelet transform typically yields better visual appearance than the Butterfly transform, with comparable error measures (Figure 9).

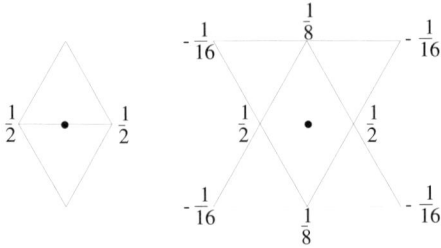

Fig. 4. Stencils for the linear (left) and Butterfly (right) subdivision in a regular case.

Local frames. Typically, the x, y, and z wavelet components are correlated [19]. We exploit this correlation by expressing wavelet coefficients in local frames induced by the coarser level. This is especially true for normal meshes. Almost all wavelet coefficients computed for normal meshes have only a normal component. Figure 5 shows histograms of the latitude angles θ (the angle from the normal axis) of the Butterfly and Loop wavelet coefficients in a local coordinate frame. Since the normal mesh is built using Butterfly subdivision almost all Butterfly coefficients have only normal components (those are seen as two peaks at 0 and 180 degrees of the histogram).

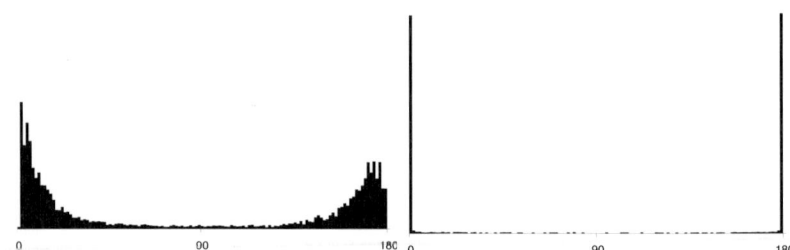

Fig. 5. Histograms of the Loop (left) and Butterfly (right) wavelet coefficients latitude θ angle for the Venus head model in a local frame.

2.3 Zerotree Coding

We encode components of wavelet coefficients separately, that is, our coder essentially consists of three independent zerotree coders. The bits from the three coders are interleaved to maintain progressivity.

A general principle of wavelet coefficient encoding is to send the highest order bits of the largest magnitude coefficients first. They will make the most significant contributions towards reducing error. The zerotree algorithm [8,

31,33] groups all coefficients in hierarchical zerotree sets such that with high probability all coefficients in a given set are simultaneously below threshold.

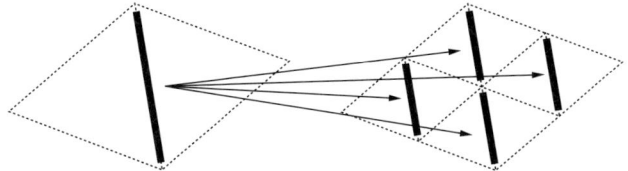

Fig. 6. A coarse edge (left) is parent to four finer edges of the same orientation (right).

The main distinction of our setting from the image case is the construction of the zerotrees. For images, one associates the coefficients with a quadrilateral face and the trees follow immediately from the face quadtree. For semi-regular mesh hierarchies, the main insight is that while scale coefficients are associated with vertices, wavelet coefficients have a one to one association with edges of the coarser mesh. Vertices do not have a tree structure, but edges do. Each edge is the parent of four edges of the same orientation in the finer mesh as indicated in Figure 6. Hence, each edge of the base domain forms the root of a zerotree; it groups all the wavelet coefficients of a fixed wavelet subband from its two incident base domain triangles. We found that the hierarchies, which have such a subband separation property lead to better compression performance (up to 0.5dB in some examples)

As was established in [19], the normal components of wavelet coefficients have more effect on geometric error than the tangential components. We therefore scale the normal components by four before quantization, so that they are effectively encoded with two more bits of precision than their tangential counterparts.

The scale coefficients from the coarsest level are uniformly quantized and progressively encoded. Each bitplane is sent as the zerotree descends another bit plane. Finally, the output of the zerotree algorithm is encoded using an arithmetic coder leading to about 10% additional reduction in the file size.

3 Adaptive Reconstruction

There is a trade-off between the remeshing error and the size of the resulting semi-regular mesh. With adaptive remeshing this trade-off is local. We refine the mesh where the error is maximal and leave it coarse where the remeshing error is small (Figure 7). Note, that the coarse remesh and its refinement give the same compression performance at low bit-rates, since refinement usually introduces small details at finer levels. These details are ignored by the zerotree coder at low bit-rates.

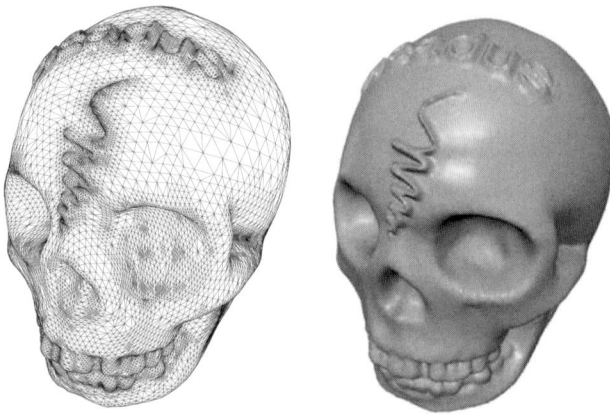

Fig. 7. Adaptive normal mesh for the skull (19138 vertices) with relative L^2 error of 0.02% and relative L^∞ error of 0.09%. The base mesh is a tetrahedron (4 vertices) while the original mesh has 20002 vertices (model courtesy of Headus). For compression results see Figure 11.

The Butterfly based wavelet transform is fully adaptive: both analysis and reconstruction filters are finite. Therefore, it can take any adaptively remeshed semi-regular mesh as input. All regions that are not subdivided to the finest level define zero wavelet coefficients. These coefficients "virtually" exist as an extension of non-uniform zerotrees. The situation is more complicated for Loop wavelets: the analysis step requires solving a sparse linear system and its extension to adaptively refined meshes is not currently supported.

The zerotree coder step is a separate procedure that is naturally adaptive: there is no difference between non-uniform and uniform zerotrees with zeros attached to extra nodes of the latter.

During the reconstruction (for both Butterfly and Loop compression) we subdivide faces only when we decode wavelet coefficients that belong to those faces. After we decode all the coefficients we subdivide all the faces until we meet an adaptive flatness threshold [40]. This step is important since we want to produce smooth surfaces even at low bit-rates. The flatness threshold is controlled by the user and can be chosen depending on the reconstruction bit-rate and the performance of the end-user graphics system.

4 Results and Discussion

We measured the performance of our coder [19] with normal meshes using both Butterfly and Loop wavelet transforms. For comparison we use the CPM coder of Pajarola and Rossignac [27] and the single-rate coder of Touma and Gotsman [37]. We plotted rate-distortion curves for the Touma-Gotsman coder by changing the coordinate quantization between 8 and 12 bits. We also compare our results with the performance of the coder of [19] for the Venus, horse and rabbit models.

Error was measured using the publicly available METRO tool [5]. All graphs show error between the reconstruction and the irregular original model as a function of the number of bits per vertex of the original mesh. All errors are given as PSNR (PSNR= $20\log_{10}($ BBoxDiag$/L^2$-error)).

	Venus	rabbit	horse	dinosaur	skull	molecule	David
V_{or}	50K	67K	48K	14K	20K	10K	275K
V_b	42	71	112	128	4	53	283
$E_r, 10^{-4}$	0.47	0.47	0.51	1.7	2.3	6.3	1.04
r_n	1580	759	754	2973	1861	794	7500

Table 1. Number of vertices in original models (V_{or}), base domain of remeshes (V_b), relative L^2 error (E_r) in units of 10^{-4}, and the number of non-normal coefficients (r_n).

We found that the coding with normal meshes has better performance compared to MAPS meshes. Even using a Loop wavelet transform we observe an improvement. For example, for the horse model (Figure 10) Loop wavelets on the normal mesh allow 1.5 times (3.5dB) smaller distortion than on the MAPS mesh. Note, that an improvement in the high bit-rate limit is due to the smaller remeshing error. More important is that we have better distortion for all bit-rates which happens because of the normality of the mesh. The additional improvement of about a factor of 1.2 (1.5dB) comes from using Butterfly wavelets summing up to a total of 5dB improvement compared to [19]. For the rabbit and Venus models the coders using MAPS and normal meshes have closer performance. However we still observe an improvement of at least 2dB.

Better compression rate of normal meshes obtained using Butterfly wavelets is not surprising given the fact that almost all the wavelet coefficients have a single nonzero component instead of three. A more remarkable result is a better compression rate of the normal meshes versus other remeshes when the Loop wavelets are used. Indeed, in that case one does not see scalar wavelet coefficients, and the reason for better performance is less explicit. We believe that this better performance can be explained by the smoothness of normal parameterization across the patch boundaries. One indication why this can be true is given by the following invariance property of the normal remesher: *given a Butterfly subdivision mesh as the input (together with extraordinary vertices as base vertices), the normal remesher will exactly reproduce the original mesh.* In practice, we need to also assume that all the normal piercing

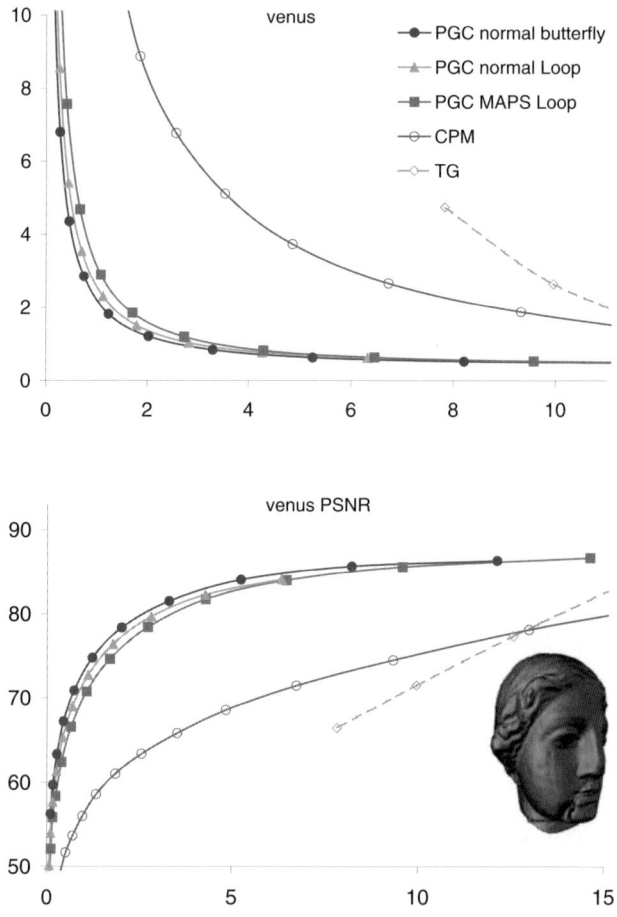

Fig. 8. Rate distortion for our coder with Butterfly and Loop wavelets using MAPS and normal meshes, TG, and CPM coders for the Venus model. At the top relative L^2 error in multiples of 10^{-4} as a function of bits/vertex. At the bottom the PSNR in dB.

steps of the remesher succeeded which is usually true for meshes with good triangle aspect ratios. It is clear then that given a mesh well approximated by a subdivision surface one can hope to produce the normal mesh with small well compressible wavelet details. Since the Butterfly subdivision surface parameterization is smooth the normal mesh parameterization can be expected to be smooth as well. The corresponding result for the one-dimensional case of normal curves has been recently proven in [7].

The above discussion makes explicit the fact that the Floater parameterization is employed in the normal remesher in an auxiliary role – it is only

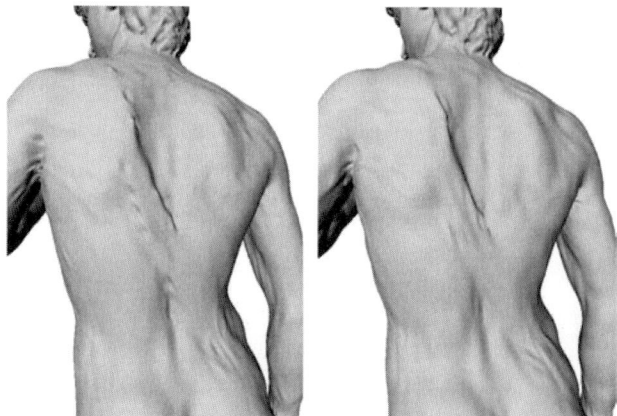

Fig. 9. Comparison of partial reconstructions using butterfly (left) and Loop (right) wavelets for approximately the same PSNR value (compressed files are around 10KB). Note the bumps on the butterfly surface.

used when the normal piercing fails. Otherwise, the local point sampling is fully independent of the particular parameterization scheme.

The fact that encoding with Loop wavelets benefits from Butterfly normal meshes is remarkable. Experimentally we found that if we treat wavelet components completely separately, Loop normal components compress better than Butterfly normal components. Given this evidence we should expect further improvement with Loop based normal meshes.

Figure 11 shows rate-distortion curves for the dinosaur, skull, and molecule models. These have coarser original meshes, therefore we allow a larger remeshing error (see the discussion on the natural choice of remeshing error in [19]). Note, that the adaptive remesh of the skull model has less vertices than the original, and has a reasonably small remeshing error. The base domain for the skull is a tetrahedron. In order to perform comparison with Loop based compression we uniformly subdivided the adaptive remesh with the Butterfly scheme.

For all the meshes mentioned in this paper (except for the David model), the remeshing step of our algorithm takes less than 10 minutes to complete. The forward Loop transform of the coder requires the solution of a sparse linear system, which takes about 30 seconds. Loop reconstruction and both Butterfly forward and backward transforms take 2-3 seconds. Zerotree encoding and decoding take about 1 second.

The largest model we were able to transform into the normal representation was a simplified model of the full David statue from the Digital Michelangelo project [23] (see Figure 12). Both the original and the remeshed model had on the order of 300,000 vertices. Our implementation of the normal remesher took thirty minutes to produce the normal mesh. It is clear that in

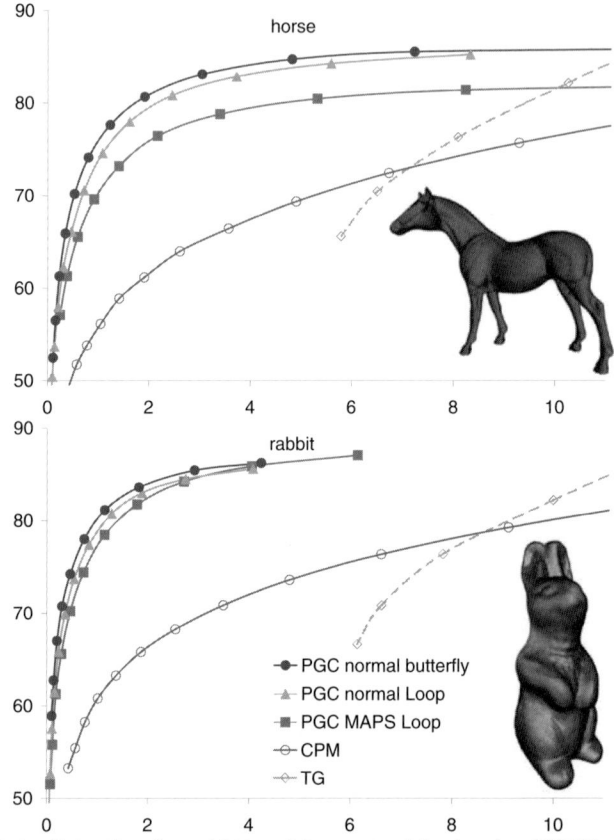

Fig. 10. Rate-distortion for rabbit and horse models showing PSNR as a function of bits/vertex.

order to remesh larger models, one needs to implement an out-of-core variant of the remeshing algorithm. Additionally, more efficient multigrid techniques for the patch parameterization can also be beneficial for better performance. We leave these improvements as future work.

5 Conclusions

In this paper we show that normal meshes improve performance of progressive geometry compression. We observe improvement of 2-5 dB depending on the model. We also describe adaptive compression which allows finer control on the number of vertices in the reconstructed model. The comparison between the Loop and Butterfly based compression methods is performed: in both cases compression is improved with normal meshes, and while the meshes compressed with Butterfly wavelets exhibit smaller mean-square error, the

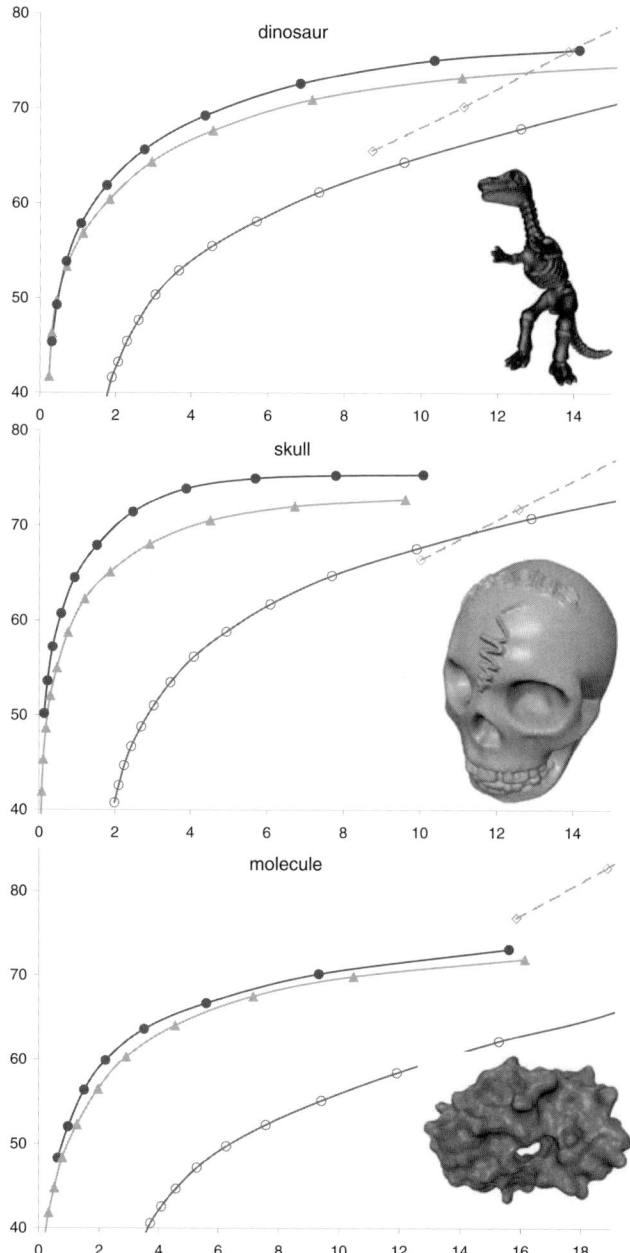

Fig. 11. Rate-distortion for our coder with Butterfly and Loop wavelets using normal meshes, TG and CPM coders for dinosaur, skull, and molecule models in PSNR as a function of bits/vertex.

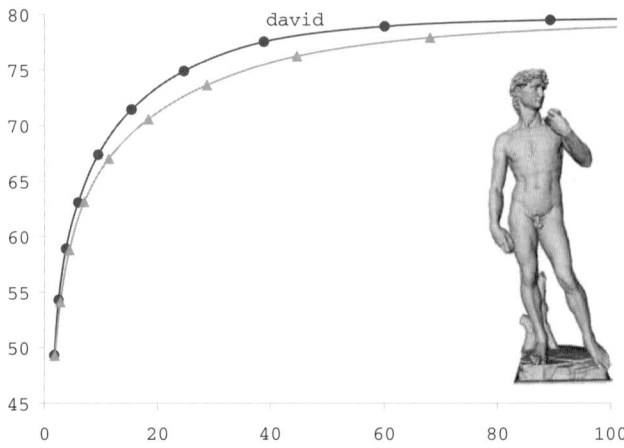

Fig. 12. Rate-distortion for our coder with Butterfly and Loop wavelets using normal meshes for a david model in PSNR as a function of file size (in 10^3 Bytes).

Loop based compression gives better visual quality of the reconstructed surfaces.

Directions for future work include design of normal meshes using Loop wavelets, as well as generalization of the normal meshing to more topologically complex and dynamically changing surfaces.

Acknowledgments This work was supported in part by NSF (ACI-9624957, ACI-9721349, DMS-9874082, CCR-0133554). The paper was written with great encouragement and invaluable help from Wim Sweldens and Peter Schröder. Kiril Vidimče contributed to the current implementation of the normal remesher. Datasets are courtesy Cyberware, Headus, The Scripps Research Institute, University of Washington, and the Digital Michelangelo project at Stanford University [23].

References

1. ALLIEZ, P., AND DESBRUN, M. Progressive compression for lossless transmission of triangle meshes. In *SIGGRAPH 2001 Conference Proceedings* (2001), pp. 198–205.
2. ALLIEZ, P., AND DESBRUN, M. Valence-driven connectivity encoding of 3d meshes. In *Eurographics 2001 Conference Proceedings* (Sept. 2001), pp. 480–489.
3. BAJAJ, C. L., PASCUCCI, V., AND ZHUANG, G. Progressive compression and transmission of arbitrary triangular meshes. *IEEE Visualization '99* (1999), 307–316.
4. BERTRAM, M., DUCHAINEAU, M. A., HAMANN, B., AND JOY, K. I. Bicubic subdivision-surface wavelets for large-scale isosurface representation and visualization. In *Proceedings of IEEE Visualization 2000* (2000), pp. 389–396.

5. CIGNONI, P., ROCCHINI, C., AND SCOPIGNO, R. Metro: Measuring error on simplified surfaces. *Computer Graphics Forum 17*, 2 (1998), 167–174.
6. COHEN-OR, D., LEVIN, D., AND REMEZ, O. Progressive compression of arbitrary triangular meshes. *IEEE Visualization '99* (1999), 67–72.
7. DAUBECHIES, I., RUNBORG, O., AND SWELDENS, W. Normal multiresolution approximation of curves. Preprint, Department of Mathematics, Princeton University, 2002.
8. DAVIS, G., AND CHAWLA, S. Image coding using optimized significance tree quantization. In *Prodeedings Data Compression Conference* (1997), IEEE, pp. 387–396.
9. DESBRUN, M., MEYER, M., SCHRÖDER, P., AND BARR, A. H. Implicit fairing of irregular meshes using diffusion and curvature flow. *Proceedings of SIGGRAPH 99* (1999), 317–324.
10. DYN, N., LEVIN, D., AND GREGORY, J. A. A butterfly subdivision scheme for surface interpolation with tension control. *ACM Transactions on Graphics 9*, 2 (1990), 160–169.
11. ECK, M., DEROSE, T., DUCHAMP, T., HOPPE, H., LOUNSBERY, M., AND STUETZLE, W. Multiresolution analysis of arbitrary meshes. *Proceedings of SIGGRAPH 95* (1995), 173–182.
12. ECK, M., AND HOPPE, H. Automatic reconstruction of b-spline surfaces of arbitrary topological type. *Proceedings of SIGGRAPH 96* (1996), 325–334.
13. FLOATER, M. S. Parameterization and smooth approximation of surface triangulations. *Computer Aided Geometric Design 14* (1997), 231–250.
14. GU, X., GORTLER, S., AND HOPPE, H. Geometry images. *ACM SIGGRAPH 2002* (2002), 355–361.
15. GUMHOLD, S., AND STRASSER, W. Real time compression of triangle mesh connectivity. *Proceedings of SIGGRAPH 98* (1998), 133–140.
16. GUSKOV, I., SWELDENS, W., AND SCHRÖDER, P. Multiresolution signal processing for meshes. *Proceedings of SIGGRAPH 99* (1999), 325–334.
17. GUSKOV, I., VIDIMČE, K., SWELDENS, W., AND SCHRÖDER, P. Normal meshes. *Proceedings of SIGGRAPH 2000* (2000), 95–102.
18. HOPPE, H. Efficient implementation of progressive meshes. *Computers & Graphics 22*, 1 (1998), 27–36.
19. KHODAKOVSKY, A., SCHRÖDER, P., AND SWELDENS, W. Progressive geometry compression. *Proceedings of SIGGRAPH 2000* (2000), 271–278.
20. KRISHNAMURTHY, V., AND LEVOY, M. Fitting smooth surfaces to dense polygon meshes. *Proceedings of SIGGRAPH 96* (1996), 313–324.
21. LEE, A., MORETON, H., AND HOPPE, H. Displaced subdivision surfaces. In *Proceedings of the Computer Graphics Conference 2000 (SIGGRAPH-00)* (2000), pp. 85–94.
22. LEE, A. W. F., SWELDENS, W., SCHRÖDER, P., COWSAR, L., AND DOBKIN, D. Maps: Multiresolution adaptive parameterization of surfaces. *Proceedings of SIGGRAPH 98* (1998), 95–104.
23. LEVOY, M. The digital michelangelo project. In *Proceedings of the 2nd International Conference on 3D Digital Imaging and Modeling* (Ottawa, October 1999).
24. LI, J., AND KUO, C. Progressive coding of 3-d graphic models. *Proceedings of the IEEE 86*, 6 (1998), 1052–1063.
25. LOOP, C. Smooth subdivision surfaces based on triangles. Master's thesis, University of Utah, Department of Mathematics, 1987.

26. LOUNSBERY, M., DEROSE, T. D., AND WARREN, J. Multiresolution analysis for surfaces of arbitrary topological type. *ACM Transactions on Graphics 16*, 1 (1997), 34–73. Originally available as TR-93-10-05, October, 1993, Department of Computer Science and Engineering, University of Washington.
27. PAJAROLA, R., AND ROSSIGNAC, J. Compressed progressive meshes. Tech. Rep. GIT-GVU-99-05, Georgia Institute of Technology, 1999.
28. PAJAROLA, R., AND ROSSIGNAC, J. SQUEEZE: Fast and progressive decompression of triangle meshes. In *Proc. CGI* (2000).
29. ROSSIGNAC, J. Edgebreaker: Connectivity compression for triangle meshes. *IEEE Transactions on Visualization and Computer Graphics 5*, 1 (1999), 47–61.
30. ROSSIGNAC, J., AND SZYMCZAK, A. Wrap&zip: Linear decoding of planar triangle graphs. Tech. Rep. GIT-GVU-99-08, Georgia Institute of Technology, 1999.
31. SAID, A., AND PEARLMAN, W. A new, fast, and efficient image codec based on set partitioning in hierarchical trees. *IEEE Transaction on Circuits and Systems for Video Technology 6*, 3 (1996), 243–250.
32. SCHRÖDER, P., AND SWELDENS, W. Spherical wavelets: Efficiently representing functions on the sphere. *Proceedings of SIGGRAPH 95* (1995), 161–172.
33. SHAPIRO, J. Embedded image-coding using zerotrees of wavelet coefficients. *IEEE Transactions on Signal Processing 41*, 12 (1993), 3445–3462.
34. TAUBIN, G., GUEZIEC, A., HORN, W., AND LAZARUS, F. Progressive forest split compression. *Proceedings of SIGGRAPH 98* (1998), 123–132.
35. TAUBIN, G., AND ROSSIGNAC, J. Geometric compression through topological surgery. *ACM Transactions on Graphics 17*, 2 (1998), 84–115.
36. TAUBIN, G., AND ROSSIGNAC, J., Eds. *3D Geometry Compression*. No. 21 in Course Notes. ACM Siggraph, 1999.
37. TOUMA, C., AND GOTSMAN, C. Triangle mesh compression. *Graphics Interface '98* (1998), 26–34.
38. ZORIN, D., AND SCHRÖDER, P., Eds. *Subdivision for Modeling and Animation*. Course Notes. ACM SIGGRAPH, 1999.
39. ZORIN, D., SCHRÖDER, P., AND SWELDENS, W. Interpolating subdivision for meshes with arbitrary topology. *Proceedings of SIGGRAPH 96* (1996), 189–192.
40. ZORIN, D., SCHRÖDER, P., AND SWELDENS, W. Interactive multiresolution mesh editing. *Proceedings of SIGGRAPH 97* (1997), 259–268.

New Results in Signal Processing and Compression of Polygon Meshes

Gabriel Taubin

IBM T.J. Watson Research Center, P.O. Box 704, Yorktown Heights, NY 10530
taubin@computer.org

Polygon models, which are used in most graphics applications, require considerable amounts of storage, even when they only approximate precise shapes with limited accuracy. To support internet access to 3D models of complex virtual environments or assemblies for electronic shopping, collaborative CAD, multi-player video games, and scientific visualization, representations of 3D shapes must be compressed by several orders of magnitude. Furthermore, several closely related methods have been proposed in recent years to smooth, de-noise, edit, compress, transmit, and animate very large polygon meshes, based on topological and combinatorial methods, signal processing techniques, constrained energy minimization, and the solution of diffusion differential equations. This paper is an overview of some of our recent results in this area: Dual Mesh Resampling, Linear Anisotropic Mesh Filtering, Bi-Level Isosurface Compression, Space-Optimized Texture Maps, and Volume Warping for Adaptive Isosurface Extraction.

1 Introduction

A polygonal mesh is defined by the association between the faces and their sustaining vertices (connectivity), by the vertex positions (geometry), and by optional colors, normals and texture coordinates (properties). These properties can be bound to the vertices, faces, or corners of the mesh. In general we look at vertex positions and the optional properties as signals defined on the mesh connectivity. In this paper we concentrate on signals defined on mesh vertices to simplify the exposition.

Polygon Mesh Resampling and Smoothing. The problem of smoothing or de-noising large irregular polygon meshes of arbitrary topology has motivated most of the recent work in geometric signal processing [42, 44, 49]. For example, mesh smoothing algorithms are used to remove measurement noise from range data in [43], to remove systematic noise from isosurfaces in [47], and as a global geometry predictor to compress multiresolution triangle meshes in [48]. From a signal processing point of view, a polygon mesh can be regarded as the result of sampling a smooth surface, where the mesh connectivity plays the role of an irregular sampling space, and the vertex positions are the values produced by the sampling process. In this framework sometimes it is necessary to change the sampling rate, i.e., to transfer a signal from one mesh connectivity to another mesh connectivity with the same topology. In the classical problem of uniform sampling rate conversion in signal processing [8, 54], under conditions determined by Shanon's sampling theorem, when the sampling rate is reduced (fewer faces than vertices in the primal mesh) the frequency content of the signal determines whether loss of information (due to aliasing) occurs or not, and when the sampling rate is increased (more faces than vertices in the primal mesh), no loss of information occurs because the resampled signal is of low frequency. Similarly, in the case of irregular meshes a resampling process should work directly in the digital domain without reconstructing the analog signal, should minimize the loss of information, and the resampled signal should be a linear function of the original signal. In section 3 we discuss the problem of resampling with respect to the connectivity of the dual mesh, addressed by the algorithm introduced in [45].

Mesh smoothing operators are also used as design tools in interactive geometric modelling systems in conjunction with other operators that modify the connectivity, the topology and/or the geometry of the mesh. Amongst these other operators, connectivity refinement schemes based on recursive subdivision have become popular to design multi-resolution geometric models [60]. In this context it is important to be able to apply smoothing operators locally, adaptively, and also with more general constraints. In addition to being able to specify the position of certain vertices, it is also desirable to be able to set the values of some normal vectors, particularly along curves defined by mesh edges. With normal vector constraints surface patches can be joined with tangent plane continuity, and the sharpness of ridge curves embedded in the surfaces can be controlled. In the now classical Laplacian smoothing algorithm and its derivatives, imposing interpolatory vertex position constraints is easy, but imposing interpolatory normal constraints is not. A new smoothing algorithm is introduced in [46] where both vertex position and face normal interpolatory constraints are easy to apply. We discuss this algorithm, which also solves the problem of denoising without tangential drift, in section 4.

Polygon Mesh Compression. A number of general purpose polygon mesh compression algorithms have been introduced in recent years. Deering [9] developed a mesh compression scheme for hardware acceleration. Taubin and Rossignac [50], Touma and Gotsman [53], Rossignac [39], Gumhold and Strasser [14], and others, introduced methods to encode the connectivity of triangle meshes with no loss of information. King et. al. [23] developed a method to compress quadrilateral meshes. Isenburg and Snoeyink [18], Konrod and Gotsman [26], and Khodakovsky et.al. [22] invented methods to encode the connectivity of polygon meshes. These algorithms focus on compressing the connectivity information very efficiently, and are all based on a traversal of the primal or dual graph of the mesh. Some of them compress connectivity of very regular meshes to a small fraction of a bit per vertex, and all to 2-4 bits per vertex in the worst case. When the geometry information (vertex coordinates, and optionally normals, colors, and texture coordinates) is also taken into account, the cost per vertex increases considerably. For example, adding only vertex coordinates quantized to 10 bits per vertex lifts the cost to typically 8-16 bits per vertex. Based on these considerations, it makes sense to consider special purpose algorithms to compress certain families of polygon meshes, such as isosurface meshes, particularly because all these approaches are incompatible with the out-of-core nature of isosurface extraction algorithms that traverse the voxels in scan order. In section 5 we describe the isosurface compression algorithm introduced in [47]. Other meshes not well compressed with general purpose mesh compression algorithms are textured meshes, where the compression of the textured images is not addressed. In section 6 we discuss a new algorithm to compress textured meshes introduced in [2]. Adaptive mesh simplification algorithms can be regarded as lossy connectivity compression techniques. In section 7 we describe the method to generate adaptive isosurfaces introduced in [1], and based on the same relaxation algorithm as the texture optimization algorithm introduced in [2].

2 Signal Processing on Meshes

Linear time and space algorithms are desirable to operate on large data sets, particularly for applications such as surface design and mesh editing, where interactive rates are a primary concern. The simplest smoothing algorithm that satisfies the linear complexity requirement is Laplacian smoothing, a well established iterative algorithm introduced in the mesh generation literature to improve the quality of meshes used for finite element computations [12,15]. In this context boundary vertices of the mesh are constrained not to move, but internal vertices are simultaneously moved in the direction of the barycenter of their neighboring vertices. And then the process is iterated a number of times. When Laplacian smoothing is applied to the noisy vertex positions of a 3D polygon mesh without constraints, noise is removed, but significant shape distortion may be introduced. This problem is also called *shrinkage*. When a large number of Laplacian smoothing steps are iteratively performed, the shape undergoes significant deformations, eventually converging to the centroid of the original data. The $\lambda|\mu$ algorithm introduced in [43] solves the shrinkage problem with a simple alternating sign modification that converts the Laplacian smoothing algorithm into a low-pass filter, but not the tangential drift problem. In the same paper [43] Discrete Fourier Analysis is extended to signals defined on meshes as a tool to understand and predict the behavior of these linear smoothing processes. This work was followed by a number of extensions, improvements, applications, and closely related algorithms, which addressed the following existing and/or open problems with mesh filters to some extent: enhancement and prevention of tangential drift, boundary and crease curve detection and enhancement, and introduction and

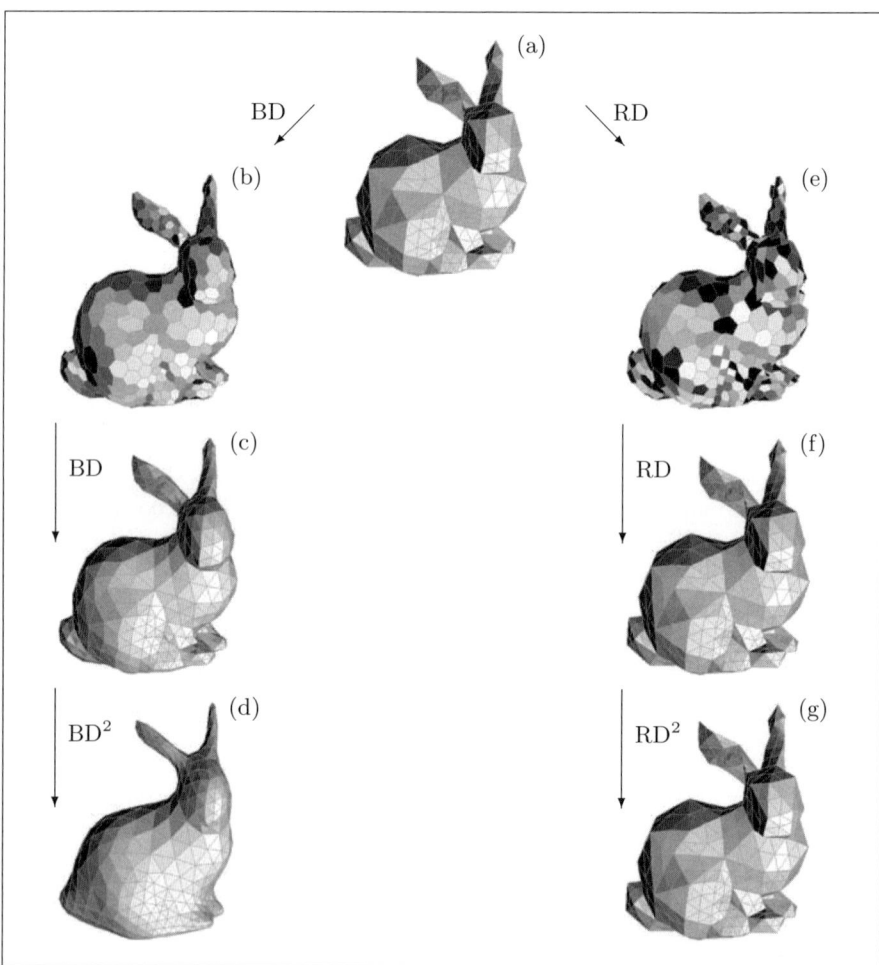

Fig. 1. Barycenter dual mesh operator (BD) vs. resampling dual mesh operator (RD). (a): primal mesh; (b) dual mesh; (c) D^2 applied to primal mesh has same connectivity but different geometry; (d) D^4 applied to primal mesh displays evident shrinkage; (e): resampling dual mesh; (f): primal geometry is recovered when RD^2 is applied to primal mesh; (g): RD^4 applied to primal mesh also recovers primal geometry. In general, some signal loss may occur when RD^2 is applied to the primal mesh, but the sequence RD^{2n} converges fast.

intuitive control of position and normal constraints. We review some of these contributions here.

Taubin et al. [51] introduced efficient and robust algorithms to evaluate any finite impulse response (FIR) linear filter on mesh signals. Zorin et al. [61] showed an application to multi-resolution interactive shape design. Kuriyama and Tachibana [27] used an anisotropic diffusion mechanism to generate smooth interpolatory subdivision surfaces with vertex position and normal constraints, and tension control. Hausler and Karbacher [16, 21] described a non-linear method for smoothing polygon meshes based on circular arc approximations. Kobbelt *et al.* [24, 25] introduced a multi-resolution approach to denoising and shape design based on linear filters and constrained minimization. Desbrun *et al.* [11] addressed the tangential drift problem observed in meshes with irregular edge length and angle distribution with a non-linear Laplacian operator and an infinite impulse response (IIR) filter with rational transfer function. And more recently, several authors [7, 10, 34, 36] have proposed closely related non-linear algorithms that extend to meshes the anisotropic diffusion approach to image segmentation introduced by Perona and Malik [38].

The ability to impose constraints to the smoothing process, such as specifying the position of some vertices, values of some normal vectors, the topology and geometry of ridge curves, or the behavior of the smoothing process along the boundaries of the mesh, is needed in the context of free-form interactive shape design. In the Laplacian smoothing algorithm and its linear derivatives, imposing interpolatory vertex position constraints is easy, but imposing interpolatory normal constraints is not. Taubin [43] showed that by modifying the neighborhood structure, i.e., the weights that control the diffusion process, certain kind of constraints can be imposed without any modification of the algorithm, while other constraints require minor modifications and the solution of small linear systems. Bierman *et al.* [5] showed how to construct subdivision surfaces with vertex normal constraints. Kobbelt et al. [24, 25] formulated the problem as an energy minimization problem, and solved it efficiently with a multi-resolution approach on levels of detail hierarchies generated by decimation. Kuriyama [27] and Yamada et al. [57] introduced soft vertex normal constraints through a spring-based approach, but reported slow convergence.

Ohtake, Belyaev, and Bogaevski [35] proposed a non-linear modification of Laplacian smoothing which increases the mesh regularity and reduces the risk of over-smoothing. Ohtake and Belyaev proposed a related algorithm to detect ridges and ravines in poltgon meshes [4, 33]. Along this path, Ohtake, Belyaev, and Bogaevski [34, 36] then introduced a non-linear crease-enhancing algorithm where a coupled diffusion process simultaneously processes face normals and vertex positions. Simplifying and speeding-up this algorithm were some of our motivations in the development of the algorithm introduced in this paper. Another related algorithm by Ohtake, Horikawa, and Belyaev [37] can be used to smooth tangential vector fields.

3 Dual Mesh Resampling

The dual of a manifold polygonal mesh without boundary is commonly defined as another mesh with the same topology (genus) but different connectivity (vertex-face incidence), in which faces and vertices occupy complementary locations and the position of each dual vertex is computed as the center of mass (barycenter or centroid) of the vertices that support the corresponding face. This process defines the *barycenter dual* mesh operator. Other dual operators can be constructed based on new rules to compute the position of the dual vertices. Signal loss is measured by comparing the original signal with the output of the square of a dual mesh operator, which are two signals defined on the same mesh connectivity. The barycenter dual operator produces significant signal loss. In fact, the linear operator defined by the square of the barycenter dual mesh operator on the primal vertex positions is a second order smoothing operator that displays the same kind of shrinkage behavior as Laplacian smoothing [43], always producing shrinkage when applied to non-constant vertex positions. A new *resampling dual* mesh operator is constructed in [45]. The square of this new operator reconstructs the original signal for the largest possible linear subspace of vertex positions. The construction of the dual vertex positions is regarded as a resampling process, where the primal vertex positions, which are signals defined on the primal mesh connectivity, are linearly resampled (transferred) according to the dual mesh connectivity. The problem is how to define resampling rules *as a function of the connectivity* so that loss of information is minimized.

We consider vertex, edge, and face *signals* defined on the vertices, edges, and faces of a mesh. These signals define vector spaces. For example, primal vertex positions are three-dimensional vertex signals, and dual vertex positions are three-dimensional face signals (vertex signals on the dual mesh). The role of the edge signals will become evident later. Since all the computations in this algorithm are linear and can be performed on each vertex coordinate independently, it is sufficient to consider one-dimensional signals. We arrange these one-dimensional signals as column vectors X_V, X_E, and X_F, of dimension V, E, and F, respectively. The element of X_V corresponding to a vertex v is denoted x_v, the element X_E corresponding to an edge e is denoted x_e, and the element of X_F corresponding to a face f is denoted x_f.

In [45] it is shown that the dual vertex positions of the Platonic solids [55] circumscribed by a common sphere can also be defined as the solution of a least-squares problem with the quadratic energy function

$$\varphi(X_V, X_F) = \sum_{e \in E} \|x_{v_1} + x_{v_2} - x_{f_1} - x_{f_2}\|^2 \qquad (1)$$

linking primal and dual vertex positions, and that this energy function is well defined for any manifold polygonal mesh without boundary. As a result the resampling dual operator is defined as the minimizer of this energy function with respect to X_F, while X_V is kept fixed, and the sum is taken over all the edges of the mesh. It is not difficult to verify that the square of the resampling dual operator produces the same input signal as output, if and only if the minimum value of the function $\varphi(X_V, X_F)$ with respect to X_F with

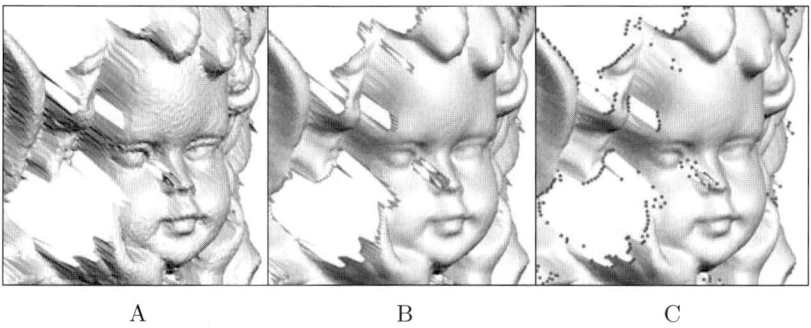

A   B   C

Fig. 2. Application to smoothing without tangential drift. (A) Noisy mesh with irregular edge length distribution. (B) After 4 steps of unit-length isotropic Laplacian smoothing of face normals with $\lambda = 0.5$, constrained boundary faces (blue), and fix vertices. (C) After 10 subsequent steps of face normal integrating anisotropic Laplacian smoothing on vertex positions with $\lambda = 0.5$, constrained boundary vertices (red), and fix face normals. Flat-shading used to enhance the faceting effect. Connectivity is kept constant.

fixed X_V is zero, because in that is the case, for each edge $e = \{v_1, v_2, f_1, f_2\}$ connecting two vertices and two faces, the segments joining the corresponding vertex positions and face positions intersect at their midpoints, i.e.,

$$\frac{1}{2}(x_{v_1} + x_{v_2}) = \frac{1}{2}(x_{f_1} + x_{f_2}) \, .$$

Since a general signal does not satisfy this property, the conditions under which loss of information occurs and is prevented are established in the paper, and the asymptotic behavior of iterative dual mesh resampling is studied, in fact defining the *space of low frequency signals*, i.e. the largest linear subspace of signals that can be resampled with no loss of information. In general the linear operator defined by the square of the resampling dual mesh operator is a smoothing operator that prevents shrinkage as in Taubin's $\lambda|\mu$ smoothing algorithm [43]. Explicit expressions for the new resampling dual vertex positions as linear functions of the primal vertex positions are derived in [45], as well as the formula for the resampling dual vertex positions as a function of the primal and dual Laplacian operators. This formula for the resampling dual vertex positions with the Laplacian operator leads to an efficient algorithm to compute the resampling dual vertex positions. This algorithm, which can be implemented as a minor modification of the Laplacian smoothing algorithm, converges very fast.

4 Linear Anisotropic Mesh Filtering

The mesh smoothing algorithm with positions and normal interpolatory constraints introduced in [46] can be decomposed into four main steps. In the first step a vector field of face normals is computed from the initial vertex positions and specified face normal constraints. In the second step the face normal field is low-pass filtered independently of the vertex positions while the unit-length nature of the vectors and the face constraints are preserved. In the third step the matrix weights for a linear anisotropic Laplacian operator are computed from the filtered face normals. Finally, in the fourth step the vertex positions are low-pass filtered with the linear anisotropic filter. Interpolatory face normal constraints are imposed as in the classical case by setting the normals associated with the constrained faces to the specified values. Similarly, the constrained vertices are set to the specified constrained values in the second phase of the algorithm. Since the rest of the machinery is the same as in other linear filters, the hierarchical approach to boundaries and creases [43] can be applied in conjunction with the two new Laplacian operators as well here. This algorithm constitutes a simpler and unified solution to the problems of smoothing mesh signals with position and normal interpolatory constraints, and of denoising without tangential drift. It is based on two new Laplacian operators, and on the existing machinery of linear filters.

The first new Laplacian operator is for signals defined on graphs and values on the unit sphere. These signals are called *sphere signals*, and the operator *spherical Laplacian*. In this operator the displacements between neighboring face normals are three-dimensional rotations, and the weighted averaging and scaling of these rotations is done in the domain of a local parameterization of the group of three-dimensional rotations $SO(3)$. This approach is stable, robust, and guaranteed to produce vector fields composed of unit-length vectors. Also, by carefully choosing the local parameterization of the group of rotations, the evaluation of this operator is almost as efficient as the classical isotropic Laplacian operator. Detailed analysis for three popular parameterizations is provided: the exponential map, quaternions, and Cayley's rational parameterization, which in fact define three different spherical Laplacian operators. A low-pass filter based on the classical isotropic Laplacian operator can be used to smooth the unit length face normal vectors defined on the dual graph, but the resulting vector field is no longer composed of unit-length vectors. These vectors can be normalized if non zero, but there is no guarantee that all the values will be non-zero, particularly if large angles exist between normals associated to neighboring faces. The new spherical Laplacian operator solves these problems.

The second new Laplacian operator is anisotropic, and defined for signals with values in Euclidean space. In fact this is a family of operators, which includes the classical isotropic Laplacian operator as a special case. These signals are called *Euclidean signals*, and the operators *Euclidean Laplacian operators*. In this new operators the weighting of displacements to neighbor-

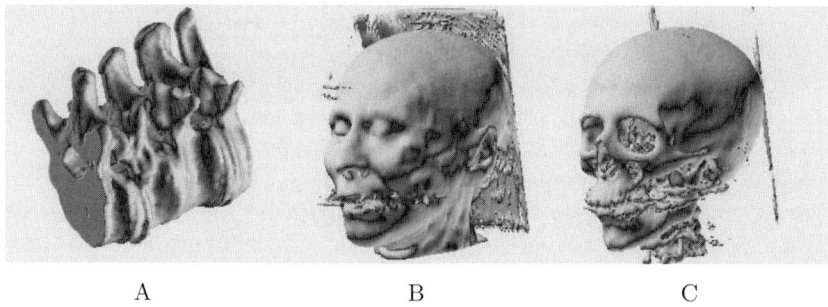

A   B   C

Fig. 3. Bi-Level Isosurface Compression. Compressed Cuberille isosurfaces as rendered by the Java decoder. A: spine set, $91 \times 512 \times 512$ voxels, level 1500, 381,278 faces, 381,667 vertices, compressed to **0.6182 bits per face**. B: UNC CThead data set, $113 \times 256 \times 256$ voxels, level 600, 294,524 faces, 294,018 vertices, compressed to **0.7437 bits per face**. C: UNC CThead data set, level 1160, 312,488 faces, 312,287 vertices, compressed to **0.8081 bits per face**.

ing vertices is given by matrices. These matrices defined on the vertices of the mesh can be regarded as a *discrete diffusion tensor field* that reinforces certain directions and attenuates others. The classical isotropic Laplacian operator with scalar weights is the special case where the matrix weights are diagonal matrices. To prevent tangential drift the weights are computed as functions of a field of face normals, in such a way that the tangential components of the displacement vectors are attenuated, and the normal components enhanced. The degree of anisotropy can be controlled even locally by using weights computed as affine combinations of the weights obtained from the face normals, and the isotropic weights. The resulting Laplacian smoothing algorithm with these matrix weights integrates the face normals in the least squares sense. Other anisotropic Laplacian operators can be constructed to produce different effects.

In both filters the weights that control the diffusion of information into a vertex from its neighbors are computed only once, and kept constant during the iterations. Filters where the weights are constant during the iterations are called *linear filters*. Non-linear filters re-compute the weights between iterations as a function of the data.

5 Bi-Level Isosurface Compression

Isosurface extraction algorithms construct polygon mesh approximations to level sets of scalar functions specified at the vertices of a 3D regular grid. The most popular isosurface algorithms [20] are *Cuberille* [6] and *Marching Cubes* [28]. We refer to the polygon meshes produced by these and related algorithms as *isosurface meshes*. Despite the widespread use of these meshes in scientific visualization and medical applications, and their very large size, special purpose algorithms to compress them for efficient storage and fast download have not been proposed until very recently [30,32,41,58,59]. A new, simple, and very efficient algorithm to compress isosurface data was introduced in [47]. This is the data extracted by isosurface algorithms from scalar functions defined on volume grids, and used to generate polygon meshes or alternative surface representations. In this algorithm the mesh connectivity and a substantial proportion of the geometric information are encoded to a fraction of a bit per Marching Cubes vertex with a context based arithmetic coder closely related to the JBIG binary image compression standard [19]. JBIG is short for *Joint Bi-level Image experts Group*, which is also the name of a standards committee. JBIG is one of the best available schemes for lossless image compression based on *context based arithmetic coding*, and can also be used for coding gray scale and color images with limited numbers of bits per pixel. The remaining optional geometric information that specifies the location of each Marching Cubes vertex more precisely along its supporting intersecting grid edge, is efficiently encoded in scan-order with the same mechanism. Vertex normals are optionally computed as normalized gradient vectors by the encoder and included in the bitstream after quantization and entropy encoding, or computed by the decoder in a postprocessing smoothing step. These choices are determined by trade-offs associated with an in-core vs. out-of-core decoder structure. The main features of this algorithm are its extreme simplicity and high compression rates

Isosurface algorithms take as input very large volume data files, and produce polygon meshes with very large number of vertices and faces. For remote visualization, we can transmit either the volume data and run the isosurface algorithm in the client, or compute the isosurface in the server and transmit the resulting polygon mesh. In both cases the transmission time constitutes a major bottleneck because of the file sizes involved, even using general purpose mesh compression schemes in the second case. And this is true without even considering the computational resources of the client. A third approach is followed in [47]. The only information from the volume data the isosurface algorithm uses to construct the polygon mesh is: which grid edges cross the desired level set, and where these intersection points are located within the edges. As a result the isosurface algorithm can be decomposed into two processes: the *server* or *encoder* process, which scans the volume data, determines intersecting edges, and computes locations of intersection points; and the *client* or *decoder* process, which reconstructs the polygon mesh from the data transmitted by the server process. The main contribution in this paper is a very simple scheme to efficiently encode these data. In addition, the tradeoffs associated with optionally computing normal vectors (used mainly for shading) in the server or the client are also considered.

The polygon mesh generated by Marching Cubes is the dual mesh of the quadrilateral mesh generated by the Cuberille algorithm [6]. The mesh generated by the Cuberille algorithm is the regularized (converted to manifold) boundary surface of the solid defined by the set of voxels corresponding to grid vertices with scalar value above the isovalue. Without regularization, in general this mesh is highly singular (non-manifold). The conversion to manifold requires duplication of vertices and edges, so that in the resulting mesh every edge has exactly two incident faces. Which vertices to duplicate and how to connect the faces can be determined by virtually *shrinking* the solid, moving the faces in the direction of the inside. The multiplicity of each dual grid vertex in the regularized mesh only depends on the local connectivity of the eight incident voxels. As in the Marching Cubes algorithm, the regularization can be done by table look-up while the volume data is being scanned, with a table of size $2^8 = 256$. What is important to note that the Cuberille algorithm can construct the isosurface mesh from the same information as the Marching Cubes algorithm. The edge intersections in the primal mesh specify the location of the face centroids of the Cuberille mesh. The location of the cuberille vertices can then be computed by local averaging, or by using more accurate schemes [13, 45]. In addition, the client can apply a number of subsequent smoothing algorithms to improve the mesh appearance [43, 46].

The situation is similar for normals. If computed in the server as the gradient of the scalar function at the edge intersection points [29, 31, 56], and included in the compressed data, the Marching Cubes decoder will treat them as vertex normals, and the Cuberille decoder as face normals. If the normals are not included in the compressed data, then it is up to the client to decide how to estimate them from the vertex coordinates and the connectivity information.

The implication of these observations is that there is considerable freedom in the implementation of the decoder, making absolutely no changes to the encoder or the compressed bitstream. It is not even necessary for the decoder to produce a polygon mesh as output. For visualization purposes, and in particular if normals are included in the compressed data, a point-based approach [40] could be very effective.

6 Space-Optimized Texture Maps

Texture mapping is a common operation to increase the realism of three-dimensional meshes at low cost, particularly in computer games. Image compression techniques can be used to address the problems created by small texture memory in graphics adapters. However, specialized hardware may be required [3, 17, 52] to decode the compressed texture on the fly to obtain interactive frame rates. Downsampling the texture image does not solve the problem either because except for simple textures containing repetitive patterns, details are typically not uniformly distributed across the image, and details are lost. A new texture optimization algorithm based on a nonlinear reduction of the physical space allotted to the texture image is introduced in [2]. This algorithm can be used in conjunction with general purpose image and mesh compression algorithms to compress textured meshes at lower distortion rates. This algorithm optimizes the use of texture space by warping the image and then computing new texture coordinates that compensate for the warping. Neither the mesh geometry nor its connectivity are modified. The method uniformly distributes frequency content of the image in the spatial domain by stretching the image in high frequency areas and shrinking it in low frequency regions. The warped image can be downsampled preserving the original details. The unwarping is performed by the texture mapping process. Hence, from the point of view of the rendering algorithms, the space-optimized textured model is no different from the original model. The texture map optimization process is an operator that transforms a textured model into another textured model.

In contrast with previous approaches, this method requires minimal user input. The frequency map automatically captures the relative importance of different regions in the image. The map is efficiently and robustly computed using a wavelet packet decomposition technique and a denoising filter. The effect of metric distortions is incorporated by analyzing the Jacobian of the parameterization and combining it with the frequency map. The image warping is obtained with a simple multi-grid relaxation algorithm. The final optimized image does not require special decompression hardware, and artifacts typical of compression techniques are not present.

7 Volume Warping for Adaptive Isosurface Extraction.

Methods for mesh simplification can be regarded as lossy connectivity compression techniques. Although many solutions have been proposed to reduce the number of polygons generated by isosurface algorithms, most require the creation of isosurfaces at multiple resolutions or the use of additional data structures, often hierarchical, to represent the volume. A new technique for adaptive isosurface extraction is proposed in [1], closely related to the approach used in [2] for texture map optimization. This new algorithm is easy to implement and allows the user to decide the degree of adaptivity as well as the choice of isosurface extraction algorithm. The method optimizes the extraction of the isosurface by warping the volume. In a warped volume areas of importance (e.g. containing significant details) are inflated while unimportant ones are contracted. Any isosurface extraction algorithm can be applied to the warped volume. The extracted mesh is subsequently unwarped such that the warped areas are re-scaled to their initial proportions. The resulting isosurface is represented by a mesh that is more densely sampled in regions decided as important.

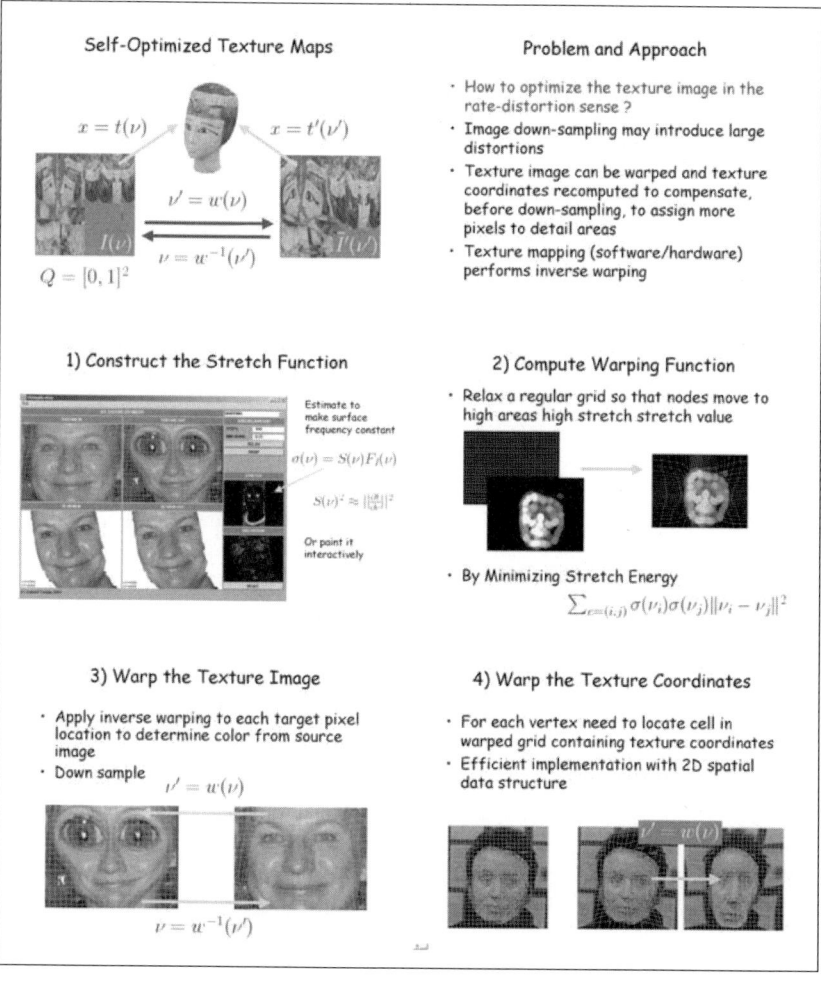

Fig. 4. Space-Optimized Texture Maps.

References

1. L. Balmelli, C. Morris, G Taubin, and F. Bernardini. Volume Warping for Adaptive Isosurface Extraction. In *IEEE Visualization 2002 Conference Proceedings*, Boston, Massachusetts, October 2002.
2. L. Balmelli, G. Taubin, and F. Bernardini. Space-Optimized Texture Maps. In *Eurographics 2002 Conference Proceedings*, Saarbrücken, Germany, September 2002.
3. A. C. Beers, M. Agrawala, and N. Chaddha. Rendering from compressed textures. *Proc. of SIGGRAPH*, pages 373–378, August 1996.
4. A. Belyaev, Y. Ohtake, and K. Abe. Detection of ridges and ravines on range images and triangular meshes. In *Vision Geometry IX Conference Proceedings*, number 4117 in SPIE Annual Meeting, pages 146–154, San Diego, July 2000.

5. H. Bierman, A. Levin, and D. Zorin. Piecewise smooth subdivision surfaces with normal control. In *Siggraph'2000 Conference Proceedings*, 2000.
6. L.S. Chen, G.T. Herman, R.A. Reynolds, and J.K. Udupa. Surface shading in the cuberille environment. *IEEE Computer Graphics and Applications*, 5(12):33–42, 1985.
7. U. Clarenz, U. Diewald, and M. Rumpf. Anisotropic geometric diffusion in surface processing,. In *Proceedings of IEEE Visualization 2000*, October 2000.
8. R.E. Crochiere and L.R. Rabiner. *Multirate Digital Signal Processing*. Signal Processing. Prentice-Hall, Inc., Englewood Cliffs, New Jersey, 1983.
9. M. Deering. Geometric compression. In *Siggraph'95 Conference Proceedings*, pages 13–20, August 1995.
10. M. Desbrun, M. Meyer, P. Schroder, and A. Barr. Anisotropic feature-preserving denoising of height fields and bivariate data. In *Proceedings of Graphics Interface 2000*, May 2000.
11. M. Desbrun, M. Meyer, P. Schröder, and A.H. Barr. Implicit fairing of irregular meshes using diffusion and curvature flow. In *Siggraph'99 Conference Proceedings*, pages 317–324, August 1999.
12. D.A. Field. Laplacian smoothing and delaunay triangulations. *Communications in Applied Numerical Methods*, 4:709–712, 1984.
13. S. Gibson. Constrained elastic surface nets: generating smooth surfaces from binary segmented data. In *Medical Image Computation and Computer Assisted Interventions, Conference Proceedings*, pages 888–898, 1998.
14. S. Gumhold and W. Strasser. Real time compression of triangle mesh connectivity. In *Siggraph'98 Conference Proceedings*, 1998.
15. K. Ho-Le. Finite element mesh generation methods: A review and classification. *Computer Aided Design*, 20(1):27–38, 1988.
16. G. Häusler and S. Karbacher. Reconstruction of smoothed polyhedral surfaces from multiple range images. In *Proceedings of 3D Image Analysis and Synthesis '97*, pages 191–198, Sankt Augustin, 1997. Infix Verlag.
17. S3 Inc. S3TC DirectX 6.0 standard texture compression. http://www.s3.com/savage3d/s3tc.html, 1998.
18. M. Isenburg and J. Snoeyink. Face fixer: Compressing polygon meshes with properties. In *Siggraph'2000 Conference Proceedings*, pages 263–270, July 2000.
19. ITU-T T.82 Information technology - Coded representation of picture and audio information - Progressive bi-level image compression, March 93. http://www.itu.int.
20. A.D. Kalvin. A survey of algorithms for constructing surfaces from 3d volume data. Technical Report RC 17600, IBM Research Division, January 1992.
21. S. Karbacher and G. Häusler. A new aproach for modeling and smoothing of scattered 3d data. In R.N. Ellson and H. Nurre, editors, *Proceedings of the SPIE on Three-Dimensional Image Capture and Applications*, volume 3313, pages 168–177, 1998.
22. A. Khodakovsky, P. Alliez, M. Desbrun, and P. Schröder. Near-optimal connectivity encoding of 2-manifold polygon meshes. *Geometric Models*, 2002. Special Issue on Processing of Large Polygonal Meshes (to appear).
23. A. King, D. Szymczak and J. Rossignac. Connectivity compression for irregular quadrilateral meshes. Technical Report GIT-GVU-99-36, Georgia Tech GVU, 1999.

24. L. Kobbelt, S. Campagna, J. Vorsatz, and H.-P. Seidel. Interactive multi-resolution modeling on arbitrary meshes. In *Siggraph'98 Conference Proceedings*, pages 105–114, July 1998.
25. L. Kobbelt, J. Vorsatz, and H.-P. Seidel. Multiresolution hierarchies on unstructured triangle meshes. *Computational Geometry Theory and Applications*, 1999. special issue on multi-resolution modeling and 3D geometry compression.
26. B. Konrod and C. Gotsman. Efficient coding of non-triangular meshes. In *Proceedings of Pacific Graphics*, Hong-Kong, 2000.
27. S. Kuriyama and K. Tachibana. Polyhedral surface modeling with a diffusion system. In *Eurographics'97 Conference Proceedings*, pages C39–C46, 1997.
28. W.E. Lorensen and A.V. Cline. Marching cubes: a high resolution 3d surface construction algorithm. *ACM Computer Graphics (Siggraph Conference Proceedings)*, 21(4):163–196, 1987.
29. T. Möller, R. Machiraju, K. Müller, and R. Yagel. A comparison of normal estimation schemes. In *IEEE Visualization'97, Conference Proceedings*, pages 19–26, 1997.
30. L. Mroz and H. Hauser. Space-Efficient Boundary Representation of Volumetric Objects. In *Proceedings of the Joint Eurographics-IEEE TCVG Symposium on Visualization (VisSym01)*, Ascona, Switzerland, May 2001.
31. L. Neumann, B. Csébfalvi, A. König, and E. Gröller. Gradient estimation in volume data using 4d linear regression. In *Eurographics 2000, Conference Proceedings*, pages 351–358, 2000.
32. K. G. Nguyen and D. Saupe. Rapid high quality compression of volume data for visualization. In *Eurographics'2001, Conference Proceedings*, 2001.
33. Y. Ohtake and A. Belyaev. Nonlinear diffusion of normals for stable detection of ridges and ravines on range images and polygonal models. In *IAPR Workshop on Machine Vision Applications MVA2000, Proceedings*, pages 497–500, Tokyo, Japan, November 2000.
34. Y. Ohtake, A. Belyaev, and I. Bogaevski. Mesh regularization and adaptive smoothing. *Computer-Aided Design*, 33:789–800, 2001.
35. Y. Ohtake, A. Belyaev, and I.A. Bogavevski. Polyhedral surface smoothing with modified laplacian and curvature flows. *The Journal of Three Dimensional Images*, 13(3):19–24, 1999.
36. Y. Ohtake, A.G. Belyaev, and I.A. Bogaevski. Polyhedral surface smoothing with simultaneous mesh regularization. In *Geometric Modeling and Processing 2000 Conference Proceedings*, pages 229–237, Hong Kong, April 2000.
37. Y. Ohtake, M. Horikawa, and A. Belyaev. Adaptive smoothing tangential direction fields on polygonal surfaces. In *Pacific Graphics 2001 Conference Proceedings*, pages 189–197, Tokyo, Japan, October 2001.
38. P. Perona and J. Malik. Scale-space and edge detection using anisotropic diffusion. *IEEE Transactions on Pattern Analysis and Machine Intelligence*, July 1990.
39. J. Rossignac. Edgebreaker: Connectivity compression for triangular meshes. *IEEE Transactions on Visualization and Computer Graphics*, 5(1):47–61, January-March 1999.
40. S. Rusinkiewicz and M. Levoy. Qsplat: A multiresolution point rendering system for large meshes. In *Siggraph'2000, Conference Proceedings*, 2000.
41. D. Saupe and J.-P. Kuska. Compression of isosurfaces. In *Proceedings of IEEE Vision, Modelling and Visualization (VMV 2001)*, Stuttgart, Germany, November 2001.

42. P. Shröder and W. Sweldens. Course 50: Digital Geometry Processing. Siggraph'2001 Course Notes, August 2001.
43. G. Taubin. A Signal Processing Approach To Fair Surface Design. In *Siggraph'95 Conference Proceedings*, pages 351–358, Los Angeles, California, August 1995.
44. G. Taubin. Geometric Signal Processing on Polygonal Meshes. In *Eurographics 2000 State of The Art Report (STAR)*, September 2000.
45. G. Taubin. Dual Mesh Resampling. In *Pacific Graphics 2001, Conference Proceedings*, Tokyo, Japan, October 2001.
46. G. Taubin. Linear Anisotropic Mesh Filtering. Technical Report RC-22213, IBM Research, October 2001.
47. G. Taubin. BLIC: Bi-Level Isosurface Compression. In *IEEE Visualization 2002 Conference Proceedings*, Boston, Massachusetts, October 2002.
48. G. Taubin, A. Guéziec, W. Horn, and F. Lazarus. Progressive forest split compression. In *Siggraph'98 Conference Proceedings*, pages 123–132, July 1998.
49. G. Taubin and L. Kobbelt. Course 17 : Geometric Signal Processing on Large Polygonal Meshes . Siggraph'2001 Course Notes, August 2001.
50. G. Taubin and J. Rossignac. Geometry Compression through Topological Surgery. *ACM Transactions on Graphics*, 17(2):84–115, April 1998.
51. G. Taubin, T. Zhang, and G. Golub. Optimal surface smoothing as filter design. In *Fourth European Conference on Computer Vision (ECCV'96)*, 1996. Also as IBM Technical Report RC-20404, March 1996.
52. J. Torborg and J. T. Kajiya. Talisman: Commodity realtime 3d graphics for the pc. *Proc. of SIGGRAPH*, pages 353–363, August 1996.
53. C. Touma and C. Gotsman. Triangle mesh compression. In *Graphics Interface Conference Proceedings*, Vancouver, June 1998.
54. P.P. Vaidyanathan. *Multirate Systems and Filter Banks*. Signal Processing. Prentice-Hall, Inc., Englewood Cliffs, New Jersey, 1993.
55. E. Weisstein. http://mathworld.wolfram.com/DualPolyhedron.html.
56. R. Yagel, D. Cohen, and A. Kaufman. Normal estimation in 3d discrete space. *The Visual Computer*, pages 278–291, 1992.
57. A. Yamada, T. Furuhata, K. Shimada, and K. Hou. A discrete spring model for generating fair curves and surfaces. In *Proceedings of the Seventh Pacific Conference on Computer Graphics and Applications*, pages 270–279, 1998.
58. S.N. Yang and T.S. Wu. Compressing isosurfaces generated with marching cubes. *The Visual Computer*, 18(1):54–67, 2002.
59. X. Zhang, C. Bajaj, and W. Blanke. Scalable Isosurface Visualization of Massive Datasets on COTS-Cluster. In *Proceedings of IEEE Symposium on Parallel Visualization and Graphics*, San Diego, CA, October 2001.
60. D. Zorin and P. Schröder. Course 23: Subdivision for Modeling and Animation. Siggraph'2000 Course Notes, July 2000.
61. D. Zorin, P. Schröder, and W. Sweldens. Interactive multiresolution mesh editing. In *Siggraph'97 Conference Proceedings*, pages 259–268, August 1997.

Part IV

Topology, Distance Fields, and Solid Modeling

Adaptively Represented Complete Distance Fields of Polygonal Models

Jian Huang[1] and Roger Crawfis[2]

[1] Computer Science, The University of Tennessee, Knoxville, TN
[2] Computer and Information Science, The Ohio State University, Columbus, OH

Summary. Distance fields are an important volume representation. However, distance fields constructed by straightforward discrete sampling can neither provide a quantitative measure of error nor fully capture detailed distance fields caused by corners in 3D geometries. We discuss here a complete distance field representation (CDFR) of polygonal models that does not rely on Nyquist sampling theory. In a CDFR volume, each voxel has a complete description of all surface polygons that affect the local distance field. CDFR can be adaptively represented without compromising accuracy. The adaptively represented complete distance field is shorted for ARCDF. For any desired distance, we can extract a surface contour in Euclidean distance, at any levels of accuracy, from the same CDFR or ARCDF representation. We further show any example of applying CDFR to a cutting edge CAD application involving high-complexity parts at un-precedented accuracy using only commonly available computational resources. Finally, although the general concepts presented here may be extended for parametric models as well, our current method can only handle polygonal models.

1 Introduction

A volume holds discrete spatial data samples of 3D fields, such as density, entropy, pressure or even vector fields. In an alias-free volume discretization, only frequency components below half the Nyquist sampling rate could be present. In order to utilize volume technologies, it is common practice to convert surface models, such as a polygonal mesh exported by a CAD package, to a volume representation. To do this, first, one needs to voxelize the surface model into a hollow volume representing the surface shape [11] [15]. Second, a distance transform is computed to construct a solid volume that stores a distance or thickness field recording distances to the surface. A number of distance metrics have been used in the distance transform. Examples range from the less accurate but efficient Manhattan and Chamfer [5] distances, to accurate but more expensive Euclidean distance.

Distance fields are scalar fields, with each element in the 3D volume representing the minimal distance from that location to the surface under study. It is common to use signed values to distinguish interior from exterior of the shape. All traditional distance fields are based on the Nyquist sampling theory. Compared to a surface, which is a sharp spatial impulse, a distance field is much smoother. It is more plausible to correctly sample a distance field using an affordable spatial resolution than to sample a surface shape directly. Surface shapes without sharp corners and edges can be reconstructed rather accurately from distance fields of relatively low resolutions [1] [9]. However, when corners and sharp edges are present in the surface shape, high frequency components are introduced. In this case, to correctly construct a discrete distance volume, super-sampling with exceptionally high volume resolution, as well as low-pass prefiltering, is necessary.

Different applications of distance fields have quite contrasting requirements of accuracy. On the one hand, general graphics applications using distance volume for morphing or rendering do not require a very tight bound of error as long as there are no visual artifacts in the rendered images. Hence, a qualitative measure of accuracy is enough. On the other hand, a majority of CAD applications, such as tooling feasibility, diecastibility studies, etc., have to address accuracy, or tolerances, using stringent quantitative metrics. Because there are no practically usable definition of geometric error after band-limiting filtering, conventional distance volumes are not widely used in CAD applications. One can always use a blindly chosen high spatial resolution, but it is not quite affordable by most applications. We hereby describe a possible approach to address this impasse.

In [13], to accurately represent distance fields of polygonal models discretely a complete distance definition disparate from Nyquist's theory was proposed. The new distance field representation was named a "complete distance field representation (CDFR)", because once a distance volume is constructed, one can extract or volume render any distance contour at any error tolerance directly from the distance volume. The advantage of being able to reuse a CDFR distance volume for whatever error tolerance, without having to redo the construction process, distinguishes the new approach from traditional distance volumes.

Hierarchical data structures [18] representing distance fields efficiently have been reported in [7], where adaptively sampled distance fields (ADF) were introduced. ADFs reduce volume resolution where fewer details are present locally. The specific ADF implementation described in [7] relies the Nyquist sampling theory and discards all geometric details beyond the cut-off bandwidth supported by the leaf level in an Octree [18]. Since the term detail was not quantitatively defined in [7], ADF could not be used in CAD/CAM applications. But the hierarchical concept in ADF can be directly utilized by CDFR, producing an Adaptively Represented Complete Distance Fields (ARCDF). ARCDF does not compromise accuracy in anyway while providing a way to trade additional storage for efficiency in distance contour extraction. Before discussing the details of CDFR and ARCDF in Section 3 through Section 5, we first survey the background literature of distance fields in general, in Section 2.

2 Introduction to Distance Fields

Traditionally, distance fields are defined as spatial fields of scalar distances to a surface geometry or shape. Each element in a distance field specifies its minimum distance to the shape. As long as the shape is represented by an oriented manifold, positive and negative distances can be used to distinguish outside and inside of the shape, for instance, using negative values on the outside and positive on the inside. Distance fields have a number of applications in constructive solid geometry [1] [7], surface reconstruction and normal estimation [9] and morphing [1] [3] [5]. Distance fields are also applied to concurrent engineering [16] where simulations and analysis involving the interior of geometries, such as die-casting simulation or thickness analysis of parts [19], are routine.

For an alias-free sampling of a signal, Nyquist's Law dictates that the sampling rate must be at least two times the highest frequency component in the signal. In spatial domain, surface geometry, not considering the interior of a geometrical object, is infinitesimally thin, and has an infinitely wide frequency spectrum. The sharp details on the surface, such as corners and edges, also reside on the high ends in the spectrum. Even with an overwhelmingly large volume resolution, one still needs extensive low-pass filtering to limit the bandwidth of the geometric shape. These low-pass filtering operations, with either simple box filters [11] [15] or specifically designed higher order filters, inevitably cause a loss of the exact surface details. Converting the surface shape to a distance field, which is smoother, provides a way to exactly locate the surface [9] during reconstruction. But the underlying assumption of having a completely smooth surface that is free from sharp corners and edges is unrealistic for most scenarios, because in most industrial designs, corners, edges, holes and tunnels are key features of any practical part. Of course one can focus on smooth objects, like a vase or teapot, but that type of models is not the norm in CAD/CAM.

Frisken et. al [7] developed a well analyzed framework for adaptively sampled distance fields (ADF), by which one can build hierarchies of distance fields, by varyng sampling rates according to the amount of local geometric details They used tri-linear interpolation to reconstruct distances, and were able to demonstrate a suite of applications with impressive rendering quality. However, ADF [7] does not fundamentally solve the problem of losing surface details in discrete representations. Further, after an ADF is constructed, the loss in geometric details is irreversible. When the primary goal of an application shifts from qualitative visual quality to quantitative accuracy, ADFs with trilinear interpolation may not satisfy the need.

A high quality distance field algorithm should address two core issues: (i) to accurately represent a continuous domain distance field discretely; and (ii) to provide a data structure supporting efficient processing of the distance volume. The first question can be answered by CDFR as a fundamental fix that preserves all geometric details in the distance field. The answer to the second question could be to use a hierarchical data structure [7] to efficiently store CDFR, providing a smooth transition between resolution levels.

3 Complete Distance Definition (CDD) & Complete Distance Field Representation (CDFR)

We now introduce a complete distance definition (CDD) for polygonal mesh. The CDD distances are Euclidean distances. Before discussing CDD, we will discuss a few observations that motivated the CDFR research.

3.1 Some Observations

Distance fields are very smooth in some simple scenarios. For instance, suppose in a 1-dimensional space, we have an impulse. It's frequency components extend to infinity. There is no way to use a finite sampling frequency to sample the impulse without aliasing. But on the other hand, as illustrated in Fig. 1, the signed distance field of the impulse is a linear function which extends from negative infinity to positive infinity. Sampling this linear function can be accurate with a relatively low sampling rate.

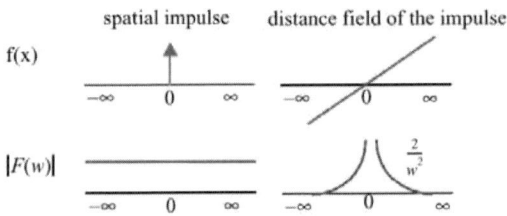

Fig. 1. It's impossible to accurately to sample the impulse function (left) without band limiting. But the corresponding distance field (right) can be sampled at low sampling rates. The spatial functions and the corresponding frequency spectrums are depicted on the top and bottom rows, respectively.

Unfortunately, this feature does not hold in higher dimensions where corners are present. The non-linear distance fields around the corners make it impossible to accurately recover the correct distance distribution from discrete samples. According to Nyquist's Law, to sample such complicated distance fields, one must low-pass filter the corners and smooth out the sharpness. But to capture the exact location of the impulse in Fig. 1, we do not have to use Nyquist sampling. Alternatively, all one needs is to place an anchor point at some location, and record the signed distance from the anchor to the impulse. This observation motivated complete distance definition.

3.2 Complete Distance Definition (CDD)

CDD is a set of parameters describing both the distance from a 3D point to a surface geometry primitive and the geometry primitive itself. Specifically, when the shape is represented as a mesh of triangles, CDD reduces to a tuple that consists of a scalar canonical distance value, and a description of the triangle with a vertex list and an edge list:

$$\langle \text{distance}, \langle v_1, v_2, v_3 \rangle, \langle e_1, e_2, e_3 \rangle \rangle \tag{1}$$

The value *distance* is the true Euclidean distance from the voxel center to a triangle.

For a triangle, *tri*, and a 3D point, *pnt*, if *pnt* orthogonally projects into *tri* (case C_1), the distance is the orthogonal distance from *pnt* to the plane where *tri* lies. Otherwise, if *pnt* orthogonally projects onto any of the three edges (case C_2), then the distance value is the shortest distance from *pnt* to an edge that *pnt* projects orthogonally onto. In case neither C_1 or C_2 applies (case C_3), the distance is the minimal distance from *pnt* to the three vertices.

We still use positive and negative signs to distinguish inside and outside. We term the triangle that is the closest to *pnt* as the *base triangle* of *pnt*. If *pnt* is closest to a triangle and the distance is of case C_1, then this triangle is *pnt*'s base triangle. If *pnt*'s distance is not case C_1, rather, it's case C_2 or C_3, looking for *pnt*'s *base triangle* is more complicated. For C_2 cases, let's label the projection point of *pnt* on that corresponding edge as, p' , and we record the vector pointing from p' to *pnt* as V. Between the two triangles sharing that edge, the triangle with a normal direction closer to V's direction, i.e. larger absolute dot product value, $|V \cdot Normal|$, is *pnt*'s *base triangle*. Very similarly, in C_3 cases, among the triangles sharing that closest vertex, we can easily find out the *base triangle* of *pnt* by comparing dot product values. We are interested in finding out *pnt* s base triangle, because by using the outward normal direction of the *base triangle* and the relative position of *pnt*, we can determine the sign of the distance at *pnt* without ambiguity. Please note that, if more than one triangles have their $|V \cdot Normal|$ values being the same, there is an ambiguous case. If the $V \cdot Normal$ values are all of the same sign, then one can randomly pick one of triangles as the *base triangle*; otherwise, one should propagate the sign from a neighbor of this voxel, whose sign is already determined without ambiguity. A second point of notice is that this method assumes all polygons in the mesh have a consistent vertex order, which may not be true with some industrial grade polygonal models due to artifacts or bugs during the exporting process. This problem would cause our sign discovery method to fail.

To better illustrate the process in determining the distance sign, in Fig. 2a, we show several 3D examples, shown in 2D. p_1 through p_6 are 2D points. t_1 through t_6 are triangles that form the surface mesh. $p_{2,3,4 and 5}$ are all case C_1. From the normal direction of t2, we can tell p_2 is outside, p_3 is inside. Similarly, using the normal of t_3 and t_5, one can tell that p_4 is inside, and p_5 is outside, respectively. p_1 and p_6 are both C_2. We show an enlarged view of these two cases in Fig. 2b. By comparing the dot products, we can tell p_1's sign is determined by t_2, and for p_6, it is decided by t_1.

Finally, to save storage, we store the description of all triangles in a separate array and only keep a triangle index in a CDD tuple.

3.3 A Complete Distance Field Representation (CDFR)

In this section, we show the process that uses CDD to build a complete distance field representation (CDFR), allowing exact capture of all geometric details, e.g. sharp corners and edges, to any level of accuracy.

Given a surface mesh, in the voxelization step, we store CDD tuples with each surface voxel, rather than single valued distances.

For each triangle touching a surface voxel, a CDD tuple is stored with that voxel. The end result of the voxelization step leaves all surface voxels with a list of CDD tuples, sorted in ascending order by distance values. Fig. 3 provides an example of voxelizing a single surface voxel, *Vox*. There are three triangles touching *Vox*. T_2 is case C_1, with T_1 and T_3 being case C_2 or C_3. The minimal distance of *Vox*, measured from the center of *Vox* is d_2. As a result, *Vox* has a sorted list of 3 CDD tuples. After voxelization, each voxel intersected by the surface contains a list of polygons cutting through it.

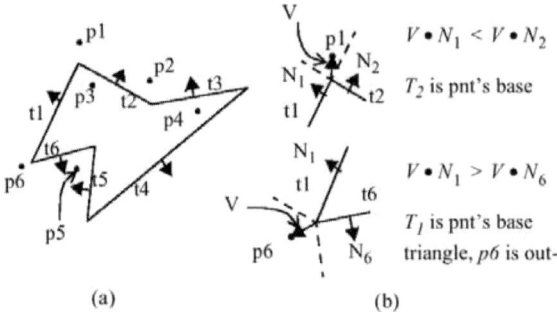

Fig. 2. (a) 2D illustrations of the process to determine the sign of the distance of a point. The solid black arrows depict the outward normal direction of each triangle. Points p_2 through p_5 project into the triangles, i.e. case C_1. The signs of the distances of p_2 through p_5 are determined by evaluating the normal direction of each point s base triangle. p_1 and p_6 are examples of C_2 cases. (b) Enlarged view of p_1 and p_6. Both p_1 and p_6 are outside.

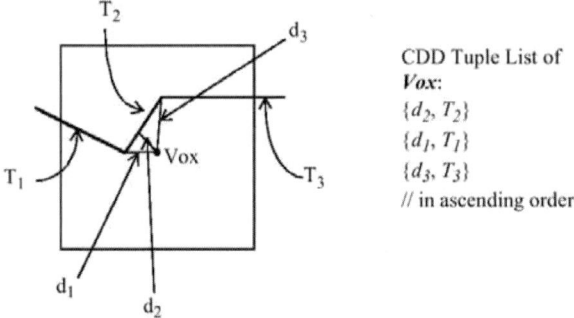

Fig. 3. A 2D illustration of building a CDD tuple list for a surface voxel, Vox. There are 3 triangles intersecting Vox. The CDD tuple list is organized in ascending distance order, with the minimal distance of Vox being d_2.

3.4 Distance Transform

For a distance transform, initially, we use an flooding algorithm to either eliminate all outside or inside voxels from our computation, depending on user's choice. For example, in most CAD/CAM applications, engineers care exclusively about the interior of a part. Of course, one can always distance transform both sides of a model by transforming each side separately, one by one. For the remaining voxels, a contour by contour CDD propagation is performed from the surface voxels to the interior. During this process, (i) each voxel propagate its CDD tuples to all its 26 neighbors [11]; (ii) each voxel then combines all newly received CDD tuples into a single triangle list, and computes the true Euclidean distances to these triangles from its own position; (iii) the resulting distance array is merged with the voxel s original CDD tuples in ascending order. The first CDD tuple in the list contains the new current distance, *cur_distance*, of this voxel. All the CDD tuples that contain a distance value within the range:

$$[cur\_distance, cur\_distance + \sqrt{3} * voxelsize] \qquad (2)$$

are stored with that voxel. This is a sufficient range to guarantee correctness in the distance transform, as we will prove in Section 4. The CDD tuples out of this range are discarded. This process, (i) through (iii), of distance transform iterates until no voxels find new CDD tuples from its 26-neighbors reduces its current CDD tuples list.

The distance transformation method discussed above is not as efficient as some others reported in the literature [2] [14]. However, those methods cannot be directly adopted by the CDFR approach. The main reason being accuracy. In CDFR, the chief goal is to guarantee accuracy. A proof of the sufficiency of the CDFR representation is the key. The best proof to date, devised in Section 4, requires that this straight-forward way of distance transform be used.

3.5 Extracting A Distance Contour

One way to use a distance field in CAD/CAM is to reconstruct or extract a distance contour. For instance, a user may ask to examine a certain distance contour within a part at an error tolerance of 0.05mm. The conventional way of reconstructing sub-voxel distance is to trilinearly interpolate in-between voxels [7]. Often times this reconstruction step is embedded in ray-casting procedures during rendering. In CDFR, it is done via exact computations. The extracted distance contours are stored as point-based models [6], which can be rendered at high interactive rates with splatting [12].

The extraction procedure works as following. Given a requested interior thickness, t ($t > 0$), we traverse these voxels with a distance value in the following range:

$$\left(t - \frac{\sqrt{3}}{2} voxelsize, t + \frac{\sqrt{3}}{2} voxelsize\right) \qquad (3)$$

The requested distance contour will pass through the span of these voxels. Unlike marching cubes [10], We do not use conditions like $minimalthickness \leq t$ and $maximalthickness \geq t$, because the underlying assumption of having a locally linear function is not true here. There could be cases where all 8 voxels surrounding a spatial location, where the thickness is the maximal, have thickness values lower than the maximal.

After identifying the relevant voxels based on (3), we subdivide each voxel into sub-voxels, or points [4]. In order to support user chosen error tolerance, E, the size of each sub-voxel must be:

$$\frac{\sqrt{3}}{2} subvoxelsize < E \qquad (4)$$

For each sub-voxel, or point, we compute the signed distance for all the CDD tuples resident on all 8 voxels surrounding that point location. Only points that have the minimal distance value within the range $[t - E/2, t + E/2]$ are extracted into the point-based distance contour.

3.6 High Quality Gradients

Besides using the distance contour for analysis, visualizations of the distance contours are also highly desired in applications. For point-based models, having high quality normal information on each point is essential for high image quality.

Our CDFR offers an additional advantage in this perspective. When extracting the distance contour from the base triangle of each sub-voxel, the normal of this point is computed. If this point is of case C_1 to its base triangle, then the normal of the base triangle is this point s true gradient. If the point is one of the cases C_2 or C_3, the gradient is the vector V in Fig. 2. For instance, in a C_2 case, the 3D point, P, first gets projected onto the closest edge. The gradient is the vector connecting P and its projection. In C_3 cases, the gradient direction is obtained by connecting P and the closest vertex. Therefore, the normal vectors computed for the whole point-based model is continuous and accurate. High quality per point shading is thus supported.

4 Proof of Sufficiency

To prove the correctness of CDFR, we need a proof of sufficiency. That is, when we need to reconstruct the local distance field in the span of any voxel, all the surface primitives affecting this local area are present on that voxel.

A surface primitive, such as a triangle, affects a local field in 3D space by being the closest surface triangle to at least one position in this local area. Based upon this observation, we devise our proof of sufficiency with a proof by contradiction:

Suppose in the CDFR, R, there exists a local voxel, V, in whose span there exists at least one point, $P(x, y, z)$, whose base triangle, T, is not resident on the voxel, V.

Without loss of generality, we write the distance from P to T as D. All distance fields are continuous functions, although they may not have continuous derivatives. For a point, $P(x + dx, y + dy, z + dz)$, that is closely neighboring P, the minimal distance from P to T is bounded by:

$$[D - \sqrt{dx^2 + dy^2 + dz^2}, \quad D + \sqrt{dx^2 + dy^2 + dz^2}] \tag{5}$$

Due to deduction, when P incrementally moves from P towards V's center point, it logically follows that the distance from V's center point to T is bounded by:

$$[D - \mathbb{N}_p^v ds, \quad D + \mathbb{N}_p^v ds] \tag{6}$$

Equation (6) can be rewritten as:

$$[D - dist(V, P), \quad D + dist(V, P)] \tag{7}$$

However, the minimum distance to P, which is D, must also be smaller than $minD + dist(V, P)$, with $minD$ denoting the minimum distance of the surface to V. Therefore, the distance of T to V, must be within the following range:

$$[minD, \quad minD + dist(V, P)] \tag{8}$$

Since P is in the span of V, the maximum possible distance is $\sqrt{3}/2 voxelsize$, the range in (8) is actually a subset of:

$$[minD, \quad \sqrt{3}voxelsize] \qquad (9)$$

Contradiction. Since during our distance propagation process, Equation (9) is exactly the range that we maintain on each voxel. Hence, triangle T must be resident on voxel V. The assumed case can not exist. Proof completed.

We do not claim our storage is minimal. We might have kept more CDD tuples on each voxel than necessary. However, an improved proof is necessary to guarantee correctness and save more storage.

5 Adaptively Represented Complete Distance Fields

CDFR stores on each voxel the indices of all surface triangles that affect the distance field in the local neighborhood around that voxel. These triangles are called "resident triangles". The accuracy of CDFR is guaranteed no matter what resolution is the CDFR volume. However, the resolution of the volumetric grid does affect the efficiency of computation when a distance contour is extracted, subdividing each voxel and computing the minimal distance of each sub-voxel to all resident triangles. Both having less sub-voxels within each voxel or less resident triangles would improve efficiency.

Utilizing a hierarchical data structure, it is more efficient to use different spatial resolutions in areas of different levels of complexity. For the purpose of adaptively representing CDFR, complexity can be straightforwardly defined as the number of triangles resident on a voxel. Similar to ADF, ARCDF uses octree as the data structure and construct the octree in a top-to-bottom fashion, after a CDFR is built, subdividing all voxels of high complexity. In this regard, an ARCDF could be considered as a forest.

The average number of resident triangles per voxel is quite low for models of a modest complexity. However, a small percentage of voxels may much more resident triangles. We partition these high complexity voxels into 8 octree subdivided children. Each octree child covers a smaller spatial span and inherits only those resident triangles that affects child's own spatial span from the parent. To do this, we compute the distance from each subvoxel center to all the resident triangles, and then discard all triangles with distances larger than:

$$minimal\_distance + \sqrt{3} * subvoxelsize \qquad (10)$$

This subdivision process is recursively until a certain stopping criteria is met. Our stopping criteria is for a voxel to have less than a user defined resident triangle count. Our results show that, although subdivision always speeds up the distance contour extraction process, subdivision does not consistently reduce the number of resident triangles significantly on all subdivided voxels. In areas where a large number of triangles affect the local distance field, no matter how small that region is, the number of resident triangles does not reduce greatly as a result of subdivision. Therefore, a smallest voxel size that can be used is also included in the stopping criteria. The subdivision process stops as soon as the smallest voxel size is reached.

ARCDF provides a level of quantitative accuracy that is not affordable using any traditional Nyquist based discrete volumes. As a result, the overhead of storage and computation incurred on each voxel in ARCDF is much higher than those of an ADF node. At the same time, ADF was developed for general graphics applications and can produce images at striking visual quality. This fact prevents any reasonable comparisons of efficiency or accuracy between ADF and ARCDF, although both algorithms utilize an octree-type hierarchy. In sum, ARCDF and ADF are designed for very different purposes.

6 Results and Analysis

The resolution of CDFR volumes do not affect the accuracy of the distance field. Also, the CDFR construction step is independent from the step that reconstructs iso-distance contours. Before we analyze the performance of our approach, we show images of distance contours on a few sample parts to demonstrate the accurate Euclidean distance fields obtained. All point-based models are rendered with splatting [12]. All results are collected on a SGI Octane with a 300MHz processor and 512MB memory. Table 1 provides a full description of the models used for test and analysis of our algorithm. We have also applied our algorithm to a very complicated part, Engine Cylinder Head, for heavy section detection. The details of heavy section detection and the Engine Cylinder Head model are described in Section 6.6.

Table 1. Physical Information of Test Models.

Model	No. Triangles	Size: x,y,z (inch)	Maximal Thickness(inch)
Cube	12	(5, 5, 5)	2.5
Tetrahedron	4	(1, 1, 1)	0.2
1-Tooth	16	(1, 2, 2)	0.48
6-Star	48	(1, 3, 3.46)	0.49
Connector	242	(6.9, 2.0, 2.9)	0.50
Brevi	1812	(38.1, 34.9, 96.0)	13.00

6.1 Proof-of-Concept Experiments

In convex parts, distance contours show the exact shape of the original surfaces, including sharp corners in Fig. 4(a)(b). Concavities cause complexity in distance fields. Two concave examples, a one-ended tooth and a six-pointed star are shown in Fig. 4(c)(d). In concave models, small thickness contours closer to the surfaces retain more detail of the surface, while deeper contours manifest more global features of the shape (Fig. 4c,d).

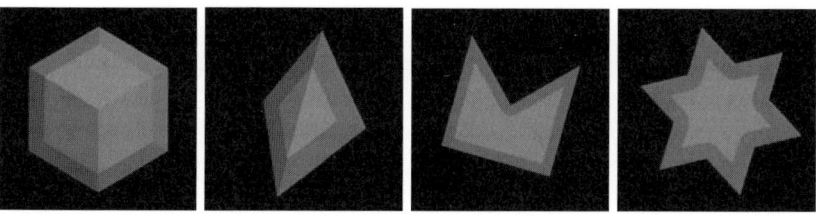

Fig. 4. A cube and tetrahedron, with the surface mesh shown in semi-transparency. The distance contours (shown in red, perpoint shaded) are of thickness (a) 0.6 inch and (b) 0.1 inch. Two concave examples, a 6-pointed star and a one-ended tooth, (c) for small thickness of 0.2 inch, and (d) increased thickness at 0.35 inch.

6.2 Real-World Models

We tested our approach on two industry production models, "connector" and "brevi". All contours in Fig. 5 are extracted to an accuracy of 1/1024 of the physical length of each part. In Fig. 5(left two) and Fig. 5(right two), respectively, we show the thickness contours within the "connector" and "brevi" parts, demonstrating sharp surface features affecting the core of the distance fields. The hole on the lower right corner of Fig. 5(right most) is a discontinuous point in the distance field.

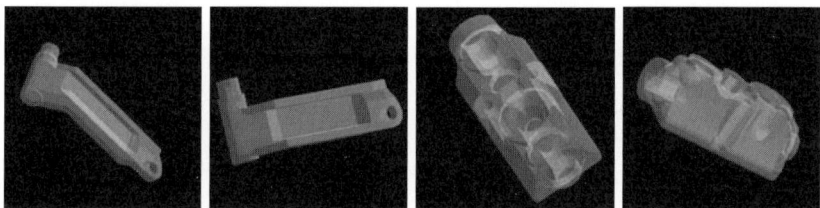

Fig. 5. Results of "connector', with contours of thicknesses 0.2 inch and 0.35 inch, and those of "brevi", at thicknesses of 10 inches and 4 inches. (in left to right order)

6.3 CDFR Size and Construction Time

The detailed storage structure on each voxel is shown in Fig. 6. Each voxel contains a 1-byte flag, $vCnt$. Empty voxels, i.e. outside voxels, have $vCnt$ set to zero. Of course, if a very complicated part is expected, one can always use a longer number, such as an integer, as $vCnt$. Surface voxels have a $vCnt$ value in the range between 1 and 127, denoting the number of triangles present on this voxel. Voxels entirely in the interior are distinguished by having a $vCnt$ larger than 127. The value $(vCnt-127)$ is the count of triangles present. For very complicated models at very low CDFR resolution, $vCnt$ may overflow. In that case, a larger number is needed for $vCnt$, or an adaptive subdivision scheme is required. By subdividing each voxel into an even number of sub-voxels, we also break up areas having a central curvature point,

```
CDD_voxel
{
  unsigned char vCnt; // in/suf/out, and counter of triangles
  float cur_distance; // current minimal distance on this voxel
  int triangles[triangle count]; // dynamic array of triangle indices
}
```

Fig. 6. The storage on each voxel in a constructed CDFR.

such as a sphere. In total, $(4 \cdot n + 5)$ bytes are needed per non-empty voxel, with n being the number of triangles present on that voxel.

For the two industry parts, "connector" and "brevi", we tried resolutions of 128 and 256 resolutions. A 128 resolution limits the average number resident triangles for "connector". However, for "brevi", a resolution of 256 is necessary to cut down resident triangles count per voxel. In Table 2, the "Resolution" column shows the actual dimension of the CDFR volume. The construction time and final sizes of the CDFRs are shown under "Timing" and "Size". "In/Sur/Out" indicates the distribution of interior, surface and outside voxels in the CDFR. Finally, the average number of triangles on surface and interior voxels are presented in the last two columns. Both storage size and construction time of CDFR increase by a factor ranging from 8 to 10 times, as the volume resolution is doubled.

Table 2. CDFR facts for "connector" and "brevi".

Model	Resolution	Timing (sec)	Size (KB)	In/Sur/Out (K voxels)	Avg Tri/Sur	Avg Tri/In
Connector	128,43,58	8.05	970	26.4/20.8/272.0	1.95	2.91
Connector	256,81,112	82.72	7,548	30.7/91.5/1,924.0	1.43	2.53
Brevi	56,52,128	51.53	3,459	100.4/50.5/221.7	3.01	4.66
Brevi	106,98,256	448.2	25,260	1,075.4/205.2/1,379	1.96	3.69

6.4 ARCDF Subdivision Results

For complicated industrial parts, such as connector and brevi, we select the threshold to be 3 times overall average number of triangles (column 3, Table 3) across the volume. Those voxels make up a very low percentage in the total volume (column 4, Table 3). Before subdivision, those voxels having more resident triangles above the threshold, of course, have a relatively high number of average resident triangles (column 5, Table 3). We subdivide those voxels above the threshold to two different levels, 2 by 2 by 2 subdivision (column 6, Table 3) and 4 by 4 by 4 subdivision (column 7, Table 3). As we have discussed in Section 5, the average number of resident triangles does not decrease significantly. Until we use a high subdivision factor of 4 by 4 by 4, the average triangle count only reduces to about half of the number before subdivision.

Table 3. CDFR Triangle Count after Subdivision

Part Name	Initial Res	Avg Tri Cnt	Percentage over threshold	Avg Tri Cnt above threshold	Avg Tri Cnt above threshold (2-sub)	Avg Tri Cnt above threshold (4-sub)
Connector	128	2.55	1.28%	9.57	7.12	4.17
	256	2.34	0.30%	9.60	7.01	4.73
Brevi	128	4.28	1.48%	15.21	11.93	4.92
	256	3.40	0.58%	10.20	10.23	4.52

6.5 Point-based Iso-Distance Contour Extraction Time

Using a higher resolution CDFR has no effect on the accuracy in the final representation. However, it causes an dramatic cubic increase in storage size and construction time. The main motivation in using higher CDFR resolutions is to have more efficient distance contour extraction. In Table 4, we show timings, in seconds, to extract a contour from both 32-res and 64-res CDFRs of the four simpler models. The "Thickness" column shows the iso-distance value chosen. For each CDFR resolution, we collect timings for 3 levels of accuracy and organize the results in regard to which conventional volume resolution the extracted contours would correspond to in accuracy. We list the three corresponding conventional volume resolutions, 128, 192, 384, under both "32 Res CDFR" and "64 Res CDFR". It is obvious, that finer accuracy results in longer extraction time, while higher CDFR resolution shortens extraction time.

Table 4. Time to extract iso-distance contour from simple models. Two CDFR resolutions have been tested with 3 levels of accuracy.

Model	Thickness	Timing (32-Res CDFR) (sec)			Timing (64-Res CDFR) (sec)		
		128 res	192 res	384 res	128 res	192 res	384 res
Cube	0.6	1.13	3.96	29.66	0.84	1.95	14.48
Tetra	0.1	0.22	0.55	4.02	0.15	0.29	1.61
1-Tooth	0.2	0.73	2.15	14.79	0.43	1.61	7.75
6-Star	0.35	0.54	1.53	12.23	0.22	0.61	4.30

As a similar analysis, on the two industrial parts, we chose a thickness of 0.3 inches for "connector", and 4 inches for "brevi". We used 128-res and 256 res CDFRs and the 3 levels of accuracy correspond to conventional volume resolutions of 512, 768 and 1024. The results in Table 5 confirm our findings from Table 4.

ARCDF is designed as a trade-off among high and low resolutions. In terms of contour extraction time, an ARCDF's performance is always in-between those using CDFRs at resolutions corresponding to the root and leave levels of the ARCDF. However, the exact profile of performance trade-off depends upon how evenly geometric details are distributed in a model. In practice, it is suggested to first obtain a histogram of resident triangle counts of a CDFR and then decide a good threshold

Table 5. "Connector" and "brevi" iso-contour extraction timing (sec) and per-point extraction time (μs/point).

Model	128-Res CDFR			256-Res CDFR		
	512 res	768 res	1024 res	512 res	768 res	1024 res
Con Timing (sec)	2.43	7.12	16.92	1.03	2.69	5.67
Con: time/pnt (μs/pnt)	24.08	30.92	41.45	9.88	11.00	13.38
Brevi Timing (sec)	23.66	74.65	174.79	9.25	28.29	64.46
Brevi: time/pnt (μs/pnt)	32.51	45.27	59.72	12.23	16.63	21.36

value based on this histogram, such that effective acceleration can be achieved with a reasonable amount of additional storage.

6.6 A Cutting Edge Application

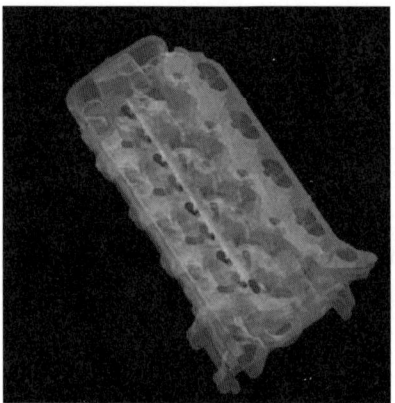

Fig. 7. The thickness contour of a cylinder model, at 8.5mm thickness and 0.137mm error tolerance.

We applied our algorithm to a challenging design part, an engine cylinder head with 135,429 surface triangles. It is modified from a real design model for our research purposes. When built, it weighs 40Kg's, and measures 266 × 480 × 157 cm in size. Typically, the maximal thickness of engine cylinders is only about 9 to 10 mm. In engines blocks, heavy sections are an important source of physical failure. It is highly desired to be able to detect heavy sections at an accuracy higher than 0.15mm. Heavy sections within a part are those regions whose thickness exceeds, i.e. heavier than, a user defined threshold. Unfortunately, there has been no reliable and affordable previous way to perform such detection in the early design stages. Extensive resources have to be spent in the dreadfully long and cyclic design, prototyping and verification process. For conventional volume techniques to handle this task, one needs to build a volume having at least $1774 \times 3200 \times 1046$ voxels in floating-point numbers, amounting to 24GB. Even so, there is still no guarantee of an accurate Euclidean distance field. Moreover, constructing or rendering data sets at such a size is overwhelmingly difficult.

With the CDFR representation, at an CDFR resolution of $142 \times 250 \times 87$, we were able to construct a CDFR of size 37MB in 30 minutes. From this CDFR, we extracted the distance contour as a point-based representation at 0.137 mm accuracy, corresponding to a $1988 \times 3500 \times 1218$ conventional volume resolution. For a thickness contour of 8.5mm, the extraction stage takes about 632 seconds, and the resulting point-based model has 450K points and can be rendered interactively at 2 frames/sec, (Fig. 7). This whole process was done on a 512MB memory, 300MHz processor SGI Octane.

7 Discussions and Future Work

Distance fields are often used in graphics applications for rendering, morphing and modeling. Some applications need to have stringent quantitative measures to control geometric errors, while qualitative error tolerance suffices for others. CAD applications, such as, heavy section detection, thin section detection, tooling feasibility evaluation and die-castability evaluation, require a quantitative control of error. Distance fields constructed using the Nyquist discrete principles cannot support this need.

CDFR has been proposed as an accurate description of distance fields resulting from a surface shape. Unlike Nyquist volumes, resolution does not affect accuracy of a distance field represented by CDFR. ARCDF allows a systematic trade-off to achieve a more efficient iso-contour extraction process, with a controllable amount of additional storage. In this regard, ARCDF can be considered as a hierarchical data structure, just like Octree or ADF. The accuracy of CDFR and ARCDF comes from the fact that no approximation, including interpolation, is used in the computation process. Finally, although the concepts of CDFR may be extended to handle parametric surfaces, the current methods only apply to models represented in polygonal meshes. Further research is necessary to handle parametric surface models.

8 Acknowledgment

This project was funded by Ford Motor Company, Visteon, Inc., and NSF CAREER Award #9876022. A previous version of this paper was published in the Proceedings of IEEE Visualization Conference'2001 [13].

References

1. D. Breen, S. Mauch and R. Whitaker, 3D scan conversion of CSG models into distance volumes , Proc. 1998 IEEE Symposium on Volume Visualization, pp. 7-14, 1998.

2. D.E. Breen, S. Mauch, R.T. Whitaker and J. Mao, 3D Metamorphosis Between Different Types of Geometric Models , Eurographics 2001 Proceedings, September 2001, pp. 36-48.
3. D. Breen, R. Whitaker, A Level Set Approach for the Metamorphosis of Solid Models , IEEE Trans. on Visualization and Computer Graphics, 2001, vol. 7, No. 2, pp. 173-192.
4. H. Cline, W. Lorensen, S. Ludke, C. Crawford, B. Teeter, Two algorithms for the three-dimensional reconstruction of tomograms , Medical Physics, 15(3), May/June, 1988, pp. 320-327.
5. D. Cohen-Or, D. Levin, and A. Solomovici, Three-dimensional distance field metamorphosis , ACM Transactions on Graphics, Vol. 17, No. 2, pp. 116-141, April 1998.
6. H. Pfister, J. Barr, M. Zwicker, M. Gross, Surfel: surface elements as rendering primitives , Proc. of Siggraph 2000, New Orleans, 2000.
7. S. Frisken, R. Perry, A. Rockwood, T. Jones, Adaptively Sampled Distance Fields: A General Representation of Shape for Computer Graphics, Proc. of SIGGRAPH 2000, New Orleans, LA, July, 2000.
8. A. Gueziec, Meshsweeper: dynamic point-to-polygonal mesh distance and applications , IEEE Transactions on visualization and computer graphics, vol. 7, no. 1, pp. 47 - 61, Jan - Mar, 2001.
9. S. Gibson, Using Distance Maps for smooth surface representation in sampled volumes, Proc. 1998 IEEE Volume Visualization Symposium, pp. 23-30, 1998.
10. W. Lorensen, H. Cline, Marching Cubes: a high resolution 3D surface construction algorithm, Computer Graphics (SIGGRAPH 87 Proceedings), 1987, 163-169.
11. J. Huang, R. Yagel, V. Fillipov, Y. Kurzion, An Accurate Method to Voxelize Polygonal Meshes, Proc. of IEEE/ ACM Symposium on Volume Visualization, October, 1998, Chapel Hill, NC.
12. J. Huang, K. Mueller, N. Shareef, R. Crawfis, FastSplats: optimized splatting on rectilinear grids , Proc. of IEEE Conference on Visualization, October, 2000, Salt Lake City, Utah
13. J. Huang, Y. Li, R. Crawfis, S. Lu, S. Liou, A Complete Distance Field Representation , Proc. of IEEE Conference on Visualization, October, 2001, San Diego, CA.
14. D.E. Johnson and E. Cohen, A framework for efficient minimum distance computations , Proc. IEEE Intl. Conf. Robotics & Automation , 1998, pp. 3678–3684.
15. A. Kaufman, An algorithm for 3D scan-conversion of parametric curves, surfaces, and volumes , Proc. of SIGGRAPH 87, pp. 171-179, July, 1987.
16. S. Lu, A. Rebello, R. Miller, G. Kinzel, R. Yagel, A simple visualization tool to support concurrent engineering design , Journal of Computer-Aided Design, Vol. 29, No. 10, pp. 727-735.
17. B. Payne, A. Toga, Distance Field Manipulation of Surface Models , IEEE Computer Graphics and Applications, vol. 12, No. 1, 1992, pp. 65–71.
18. H. Samet, The Design and Analysis of Spatial Data Structures, Addison-Wesley, Reading, MA, 1990.
19. R. Yagel, S. Lu, A. Rubello, R. Miller, Volume-based reasoning and visualization of dicastability In Proc. IEEE Visualization 95, pp. 359-362, 1995.

Fully Dynamic Constrained Delaunay Triangulations

Marcelo Kallmann[1], Hanspeter Bieri[2], and Daniel Thalmann[3]

[1] USC Robotics Research Lab, Computer Science Department, University of Southern California[†] *kallmann@usc.edu*
[2] Institute of Computer Science and Applied Mathematics, University of Bern *bieri@iam.unibe.ch*
[3] Virtual Reality Lab – VRlab, Swiss Federal Institute of Technology – EPFL *daniel.thalmann@epfl.ch*

Summary. We present algorithms for the efficient insertion and removal of constraints in Delaunay Triangulations. Constraints are considered to be points or any kind of polygonal lines. Degenerations such as edge overlapping, self-intersections or duplicated points are allowed and are automatically detected and fixed on line. As a result, a fully Dynamic Constrained Delaunay Triangulation is achieved, able to efficiently maintain a consistent triangulated representation of dynamic polygonal domains. Several applications in the fields of data visualization, reconstruction, geographic information systems and collision-free path planning are discussed.

Key words: Constrained Delaunay Triangulation, Dynamic Constraints.

1 Introduction and Related Work

Delaunay Triangulations and Constrained Delaunay Triangulations are popular tools used for the representation of planar domains. Applications include, for instance, data visualization [21], reconstruction [3], mesh generation [20], and geographic information systems [23].

The Constrained Delaunay Triangulation (CDT) is an extension of the Delaunay Triangulation (DT) to handle constraints. A CDT can be seen as the triangulation closest to the DT that respects given constraints.

The computation of the DT of a set of points in $\mathbb{R}^2$ is a classical problem in computational geometry [15]. Asymptotically optimal algorithms are known, as the $O(n\ log\ n)$ divide-and-conquer [12] and sweepline [9] algorithms. For CDTs, an optimal $O(n\ log\ n)$ divide-and-conquer algorithm is also known [5].

However the most popular implementations for both DT and CDT are those which are incremental and do not require the use of complicated data structures in addition to the triangulation itself. In general, such incremental algorithms take worst-case time of $O(n^2)$ mainly due to point location, which

[†] Work done while at VRlab-EPFL

usually takes $O(n)$ to locate each point to be inserted. However the *jump-and-walk* method of point location [14] allows the incremental computation of DTs within an expected time of only $O(n^{4/3})$ for randomly distributed points. Simple incremental algorithms can thus be competitive with optimal (but more complex) *O(n log n)* approaches. Note that, if dedicated data structures are maintained to optimize point location, the incremental algorithm reaches the expected time of *O(n log n)* [6].

Along this paper we mainly discuss complexity analysis results from existing incremental DT algorithms, expecting that similar asymptotic times are obtained for CDTs without "special cases". Therefore, constraints should have few intersections and not make the CDT become too different from the DT of the vertices of the constraints. Note that constraints can be defined to make a CDT match any given triangulation, thus invalidating time complexity analysis relying on the Delaunay property, as it is the case with the jump-and-walk method [14].

This work presents the implementation of incremental algorithms permitting to efficiently update CDTs in case of online insertions and removals of constraints. We consider constraints to be defined by any set of points and line segments. They may describe open polygonal lines, and simple or non-simple polygons. Edge overlapping, self-intersections and duplicated points are allowed, and are automatically detected and handled online.

Constraints are identified by ids at insertion time so that they can be removed later on. Note that, as edge overlapping and intersections are allowed, a constrained edge in the CDT can represent several input constraint segments. Therefore, we need to coherently keep track of ids during insertion and removal of constraints in order to keep updated which triangulation edges represent which input constraints.

Our implementation for constructing CDTs is strongly related to the approaches taken in previous works [17] [10] [1] [19]. Our main contributions in this paper are extensions to allow keeping track of overlapping and intersecting constraints during insertion and removal operations.

In order to achieve a fast and robust implementation, we follow a *topology-oriented* implementation approach [18], making use of floating-point arithmetic and *epsilons* to better treat degenerate input data. A similar approach has also been successfully used for the implementation of the Voronoi diagram of points and segments [13].

Starting from an initial (trivial) CDT, updates are done respecting all topological properties the CDT has to fulfill. Thus, topological consistency is guaranteed independently of errors in numerical computations. Our aim is that the algorithm never fails, giving always a topologically-valid output.

Our implementation has been successfully tested with several applications requiring representations of dynamic planar domains. Our first application which motivated this work was the navigation of characters in virtual environments. For this case we use the CDT as a cellular decomposition of the free space, so that a simple search over the free triangles is sufficient to determine a free passage joining two input points, if one exists. Our approach becomes particularly interesting because we are able to efficiently update the environment description when obstacles (or constraints) move.

The ability to deal with degenerate constraints and to dynamically insert and remove them in CDTs, leads to a number of new possibilities in certain applications. We present and discuss several examples in different fields: data visualization, reconstruction, geographic information systems and collision-free path planning.

2 Background and Problem Definition

Let V be a finite set of vertices in $\mathbb{R}^2$. Let E be a set of edges where each edge has as its endpoints vertices in V. Edges are closed line segments. A triangulation $T = (V, E)$ is a planar graph such that: no edge contains a vertex other than its endpoints, no two edges cross, and all faces are triangles with their union being the convex hull of V.

Let $T = (V, E)$ be a triangulation. An edge $e \in E$, with endpoints $a, b \in V$, is a Delaunay edge if there exists a circle through a and b so that no points of V lie inside this circle. If T has only Delaunay edges, T is called a *Delaunay Triangulation* (DT) of V. Note that a Delaunay triangulation is not unique in case V has four or more co-circular points.

Let $S = \{C_1, C_2, ..., C_m\}$ be a set of m constraints, such that for each $i \in \{1, 2, \ldots, m\}$, constraint C_i is defined as a finite sequence of points in $\mathbb{R}^2$. C_i may contain a single point, otherwise it represents a polygonal line. Polygonal lines are allowed to be of any form: open or closed, with self-intersections, with overlapped portions or with duplicated points. For simplicity of notation, we call the segments or points of all the constraints in S by simply the segments or points in S.

Let S be a set of constraints as defined above. $T(S) = (V, E)$ is a *Constrained Delaunay Triangulation* (CDT) of S if:

- For each segment $s \in S$, there is a number of edges in E such that their union is equal to s. Such edges are called constrained.
- For each point $p \in S$, there is a vertex $v \in V$, such that $v = p$. Vertex v is said to be constrained, and if it has no adjacent constrained edges, we say that v is *isolated*. Isolated constrained vertices appear due to constraints defined as a single point.
- For every non-constrained edge $e \in E$, e is a Delaunay edge with respect only to the vertices connected with edges to the endpoints of e.

An edge $e \in E$ is constrained if it is equal to, or is a sub-segment of, some segment $s \in S$. All the other edges are non-constrained. Note that, as we allow segments in S to intersect, finding $CDT(S)$ implicitly means to determine all the intersection points, including them as additional vertices in V.

Segments having intersections are split at the intersection points, generating sub-segments which are inserted as edges of E (see Figure 1). Overlapped segments share a same constrained edge in $CDT(S)$.

The number and form of the constraints defined in S determines "how far" $CDT(S)$ is from $DT(V)$. Delaunay triangulations have the nice property of maximizing the minimum internal angle of triangles, and this property is kept in $CDT(S)$ as far as possible. Note that it is possible to define a set of constraints that forces the CDT to become any desired triangulation.

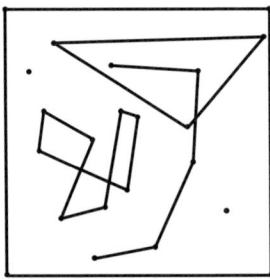 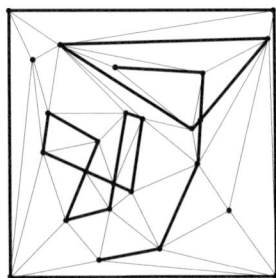

Fig. 1. Left: a set of constraints S composed of two points, a square, a triangle, a non-simple polygon, and an open polygonal line. Right: $CDT(S)$.

Problem definition. Let $S = \{C_1, C_2, ..., C_m\}$ be a set of constraints. The problem we want to solve is to construct and maintain a Dynamic Constrained Delaunay Triangulation of S, more precisely:

- Construct an initial $CDT(S) = (V, E)$, taking into account all possible degenerated cases in S.
- Insert constraints incrementally, i.e., given a new constraint C and a triangulation $CDT(S)$, we want to efficiently compute $CDT(S \cup \{C\})$.
- Remove constraints incrementally, i.e., given a constraint $C \in S$, and a triangulation $CDT(S)$, we want to efficiently compute $CDT(S \setminus \{C\})$.

3 Method Overview

Let $S = \{C_1, C_2, ..., C_m\}$ be a set of constraints. Our algorithm to construct $CDT(S)$ is incremental, and a box B strictly containing all constraints in S must be determined in advance. This is a common initialization requirement that allows considering only the case of inserting constraints in the interior of an existing CDT. B can be any large box that contains all possible constraints to be inserted, as for instance the enlarged (by any factor) bounding box of S. The triangulated B becomes the initial CDT. B constitutes the boundary of $CDT(S)$.

Once the CDT has been initialized we can insert, in any order, any constraint C_i, $i \in \{1, 2, \ldots, m\}$. In order to cope with overlapping segments and to identify constraints for later possible removals, we associate with each edge of the CDT a list containing the indices of the constraints it represents (an empty list means that the edge is not constrained):

```
struct edge_information { list<int> crep; };
```

For example, consider that segment $s_1 \in C_j$ has endpoints $\{(2,0), (3,0)\}$, and segment $s_2 \in C_k$ has endpoints $\{(1,0), (4,0)\}$. During the process of inserting segments s_1 and s_2 in the CDT, the overlapping is detected and a single edge e with vertices coordinates $\{(2,0), (3,0)\}$ is created. Edge e will therefore keep indices $\{j, k\}$ in its list crep. Note that segment s_2 will be represented in the CDT as the union of the three edges $\{(1,0), (2,0)\}, \{(2,0), (3,0)\}$ and $\{(3,0), (4,0)\}$. If s_1 is later on removed, s_2 becomes represented by the edge $\{(1,0), (4,0)\}$.

Vertices in the CDT must also keep track of the constraints it represents, however it is sufficient to count them. Therefore, each vertex in the CDT keeps track of a reference counter, and can be removed from the CDT only if this reference counter is 1. As this is a trivial procedure, we will not make further considerations concerning the management of such reference counters in the remainder of this paper.

Epsilon determination. Another initialization requirement is to determine the value of the epsilon to be considered in numerical computations. As we rely on standard floating-point precision arithmetic, epsilons are used to detect data inconsistencies. The epsilon mainly determines the threshold defining how far distinct points shall be considered the same, and its value reflects the precision of the measurement method. We assume that the user is aware of the nature of the input data and is able to determine the best epsilon to be used.

Held [13] proposes a more sophisticated approach to implement epsilon-based computations. A value is also required as input from the user, but it is considered as an upper bound. The lower bound is automatically determined by the floating-point precision of the used machine. In this way, geometric algorithms start by using the lower bound epsilon, and in case of problems, its value is incrementally grown until these problems are solved. In case the upper bound is reached before, data cleaning is performed and the computation restarts.

Data structure. The algorithms we will present require the triangulation to be represented by an efficient data structure, i.e. capable to give all adjacency relations in constant time. We implemented a data structure following the adjacency strategy of the *quad-edge* structure [12] and integrating element lists and operators as in the *half-edge* structure [27].

Our basic element is a structure representing an oriented edge. Each oriented edge is associated with only one vertex, one edge and one face. We call this basic element a SymEdge. This name is due to the fact that for each

SymEdge, there is another symmetrical one, which is associated with the other vertex and face sharing the same edge.

Each SymEdge structure keeps a pointer to the next SymEdge in the same face (nxt pointer), and a pointer to the next SymEdge rotating around the same vertex (rot pointer). A counter-clockwise orientation is used. The symmetrical element of a SymEdge is obtained by composing the nxt and rot pointers. In addition, three pointers are also stored in each SymEdge in order to retrieve the associated vertex, edge and face elements. The vertex, edge and face elements are organized in lists, and are used to store any application-specific data (as crep lists).

Construction and traverse operators are defined as a safe interface to manipulate the structure. We have implemented the same traverse operators as described in a previous paper [26]. In that work, a benchmark is presented where the mentioned "simplified quad-edge structure" is equivalent to the SymEdge structure described here. The benchmark indicates that the SymEdge structure is one of the fastest to describe general meshes. Note however that specific structures for describing triangulations may have better performance, as discussed in the work of Shewchuk [17].

4 Constraint Insertion

Given a constraint to be inserted in a CDT, we incrementally insert all its points and then all its segments, as shown in the following routine:

```
insert_constraint ( C_i, i )
   for all points p in C_i
      LocateResult lr = locate_point ( p );
      if ( lr is an existing vertex )
         v = lr.located_vertex;
      else if ( lr is on an existing edge )
         v = insert_point_in_edge ( lr.located_edge, p );
      else
         v = insert_point_in_face ( lr.located_face, p );
      add v to vertex_list;
   for all vertices v in vertex_list
      if ( v is not the last vertex in vertex_list )
         v_s = successor of v in vertex_list;
         insert_segment ( v, v_s, i );
```

Point location. The routine insert_constraint requires to locate points. For each point p to be inserted in the CDT, the locate_point routine searches where in the triangulation p is. Point location is an important issue for any incremental algorithm, and several approaches have been proposed. Most efficient methods rely on dedicated data structures, reaching the expected time of $O(\log n)$ to locate one point [6] [4] [11]. Alternatively, *bucketing* has also been used [16] with good performances for well-distributed points. We follow the simpler *jump-and-walk* approach [14], which takes expected $O(n^{1/3})$ time (in DTs). It has the advantage that no additional data structures are needed, which is an important issue in our case: as constraints can be dynamically updated, the use of any additional data structure would also imply additional updates after each operation.

The jump-and-walk method of Mücke, Saias and Zhu [14] first defines a random sample of $O(n^{1/3})$ vertices from the triangulation (n being the current number of vertices), and then determines which one of these is closest to p. Finally, an oriented walk is performed, starting with one triangle t adjacent to the chosen vertex.

The oriented walk [24] [12] method selects one edge e of t, which separates the centroid of t and p in two distinct semi planes. Then, e is used to switch t to the other triangle adjacent to e. This simple process continues until t contains p. However it is only guaranteed to work for DTs [22] [8], and we have included an additional test to ensure its convergence in our CDT search. First, each visited triangle is marked. Then, whenever two edges exist separating the centroid of the current triangle and p in distinct semi planes, the one leading to a non-marked triangle is chosen. Marking triangles does not imply any overhead: we simply keep an integer for each triangle, and for each new search an incremented integer flag is used as a mark.

The geometric test required to determine if two points lie in the same side of a segment is implemented with a standard CCW (counter-clockwise) orientation test. However, as this test is not always robust, we switch to an epsilon-based walk mode when a loop is detected, i.e., when there are no unmarked triangles to switch to during the walk. The epsilon-based walk includes geometrical tests to explicitly check for each triangle t during the walk, if p is equal to some vertex of t, or if p lies on some edge of t (within the epsilon distance).

Point insertion. The point location routine determines (within the epsilon distance), if point p is already in the CDT, if it lies on an edge, or if it lies inside a face of the CDT. If p is already there, it is simply not inserted; otherwise a new vertex v is created in the located edge or face. If p is located in an edge it is first projected to that edge before insertion.

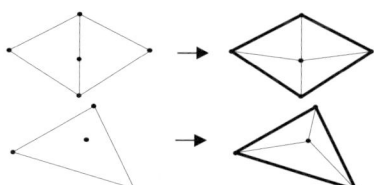

Fig. 2. Point insertion in an edge (top) and in a face (bottom). The edges outlined in bold (right side) constitute the initial edge set $F(p)$.

When p is inserted, non-Delaunay edges introduced in the triangulation need to be corrected. We follow the popular approach of flipping non-Delaunay edges, until all edges become Delaunay [12].

After inserting p, the union of all triangles incident to p forms a star-shaped polygon P. The edges on the border of P constitute the current edges being considered for flipping, and are noted here by the set $F(p)$ (Figure 2). Note that just after p insertion, P is either a triangle or a quadrilateral, and during the whole flipping process $F(p)$ continues to be a star-shaped polygon.

The insertion of one point may require $O(n)$ edge flips, however for DTs with a random input it is known that the expected number of edge flips is constant, no matter how they are distributed [11].

Note that if p is inserted in a constrained edge e, the crep list of indices of e is copied into the crep lists of the two new sub-edges created after the division of e at point p. The final pseudo codes of the insertion routines are as follows:

```
vertex* insert_point_in_edge ( p, e )
    if p is not exactly in e, project p to edge e;
    orig_crep = e->crep;
    v = new vertex created by inserting p in e according to Figure 2;
    set the crep list of the two created sub edges of e to be orig_crep;
    push the four edges of F(p) on stack;
    flip_edges ( p, stack );
    return v;
vertex* insert_point_in_face ( p, f )
    v = new vertex created by inserting p in f according to Figure 2;
    push the three edges of F(p) on stack;
    flip_edges ( p, stack );
    return v;
flip_edges ( p, stack )
    while stack not empty
        edge* e = stack.pop();
        if ( e is not constrained AND e is not Delaunay )
            f = face incident to e, which does not contain p;
            push on stack the two edges of f that are different from e;
            flip ( e );
```

The flip_edges routine tries to flip each edge in the stack, while the stack is not empty. The decision of actually flipping an edge e relies on two tests. The first one simply checks if e is constrained. This test is not subjected to numerical errors. The second test checks whether e is Delaunay or not. This is equivalent to determine if the circle passing through p and the endpoints of e does not contain the opposite point of p in relation to e.

Point-in-circle tests computed with a 4x4 determinant are subject to numerical errors. Different approaches have been used to achieve a robust behavior, such as arbitrary-precision arithmetic [19], or exact computation based on floating-point numbers [17]. Our choice follows our main approach of using epsilons and floating-point arithmetic: we consider p inside the circle c only if: $distance(center(c), p) < radius(c) - epsilon$. It is worth to mention that we have faced never-ending loops during the flipping process when using only a simple determinant evaluation.

Segment insertion. The final task of the insert_constraint routine is to insert segment s defined by a pair of already inserted vertices. Our approach takes three steps: all edges of the CDT which are crossed by s are deleted, s is inserted in the CDT creating a new edge e, and finally the two introduced non-triangular faces on each side of e are retriangulated (see Figure 3).

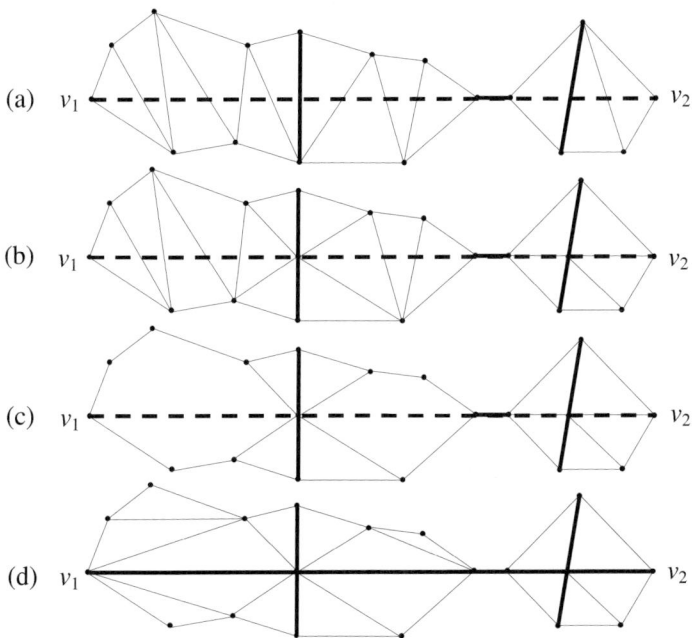

Fig. 3. a) Example of the insertion of a segment $s = \{v1, v2\}$. b) Crossing constrained edges (in bold) are subdivided and the intersection points are inserted. c) Remaining crossing edges are removed. d) Finally, s sub-segments are inserted, and the non-triangular faces are retriangulated.

Our implementation follows the approach taken by Shewchuk [17] and Anglada [1]. However, instead of deleting edges and retriangulating faces, the alternate approach presented by Bernal [2] would also be applicable: only edge flips are performed in order to make the missing segments appear. This seems to be an interesting approach and we intend, in a future work, to compare implementations of both approaches.

Let v_1 and v_2 be the two vertices which are the endpoints of the segment s to insert. The pseudo code of routine insert_segment can be summarized as follows :

```
insert_segment ( v₁, v₂, i )
    // step 1
    edge_list = all constrained edges crossed by segment {v₁,v₂};
    for all edges e in edge_list
        p = intersection point between e and segment {v₁,v₂};
        insert_point_in_edge ( p, e );
    // step 2
    edge_list = all edges crossed by segment {v₁,v₂};
    for all edges e in edge_list
        remove edge e from the CDT;
    // step 3
    vertex_list = all vertices crossed by segment {v₁,v₂};
    for all vertices v in vertex_list
        if ( v is not the last element in vertex_list )
            vₛ = successor vertex of v in vertex_list;
            if ( if v and vₛ are connected by an edge )
                e = edge connecting v and vₛ;
                add index i to e->crep;
            else
                e = add new edge in the CDT connecting v and vₛ;
                add index i to e->crep;
                retriangulate the two faces adjacent to e;
```

To decide if a vertex is crossed by s we take into account the epsilon distance. The determination of lists of elements (edges or vertices) crossed by s can be efficiently done due to the adjacency information stored in the **SymEdge** data structure. In the actual implementation, a list containing all crossed elements is determined once, and used during the entire routine **insert_segment**.

The process of retriangulating the two faces on each side of e is based on incrementally inserting interior edges that are Delaunay. We refer the reader to the work of Anglada [1], where a detailed description is given.

5 Constraint Removal

The removal process of a constraint i is based on two main steps. The first step searches for all edges representing constraint i, and setting that they no longer represent the constraint i. At this point edges may become unconstrained. The second step consists in removing those vertices that are possibly no longer used by any constraint, and removing intersection vertices that are no longer representing intersections (see Figure 4).

One vertex v of the constraint i is required in order to start the (local) search for the edges representing constraint i. Vertex v is saved when the constraint is inserted, and therefore a list is maintained associating one vertex with each constraint index. Note that v needs to be a "corner vertex" of C and not a vertex between two collinear edges representing C. This is to guarantee that v is not removed from the CDT due to the removal of another constraint sharing v. The pseudo-code of the removal process is given below:

```
remove_constraint ( i )
   // step 1
   v = one vertex of the constraint i;
   put on stack all incident edges to v representing the constraint i;
   mark all edges in stack;
   while ( stack is not empty )
      edge* e = stack.pop();
      add e to edge_list;
      push to stack all incident edges to e representing constraint i
      and which are not marked; ensure all edges in stack are marked;
   for all edges e in edge_list
      remove index i from e->crep;
   // step 2
   vertex_list = all vertices which are endpoints of edges in edge_list;
   for all vertices v in vertex_list
      ref = number of different indices referenced by the
      remaining constrained edges adjacent to v;
      n = number of remaining constrained edges adjacent to v;
      if ( n==0 )
         remove_vertex ( v );
      else if ( n==2 )
         let e₁ and e₂ be the remaining constrained edges adjacent to v;
         if ( e₁->crep == e₂->crep AND e₁ is collinear to e₂ )
            let v₁ and v₂ be the two vertices incident to e₁ and e₂, and
            which are different than v;
            crep = e₁->crep;
            remove_vertex ( v );
            edge* e = insert_segment ( v₁, v₂, -1 );
            e->crep = crep;
```

Routine remove_vertex removes a vertex v from the CDT as well as all edges (constrained or not) which are incident to v. After removal, a non-triangular face f appears, which needs to be retriangulated. Currently we follow the approach presented by Anglada [1]. However, the alternate approach based on an *ear algorithm* [7] could also be used with possibly best performance.

When a vertex is removed to simplify two collinear edges, a call to insert_segment is done right after a call to remove_vertex. Certainly more optimized processes can be devised, however they may not be worth to implement as collinear edges are not likely to happen very often. Figure 5 illustrates the importance of removing redundant vertices in collinear edges.

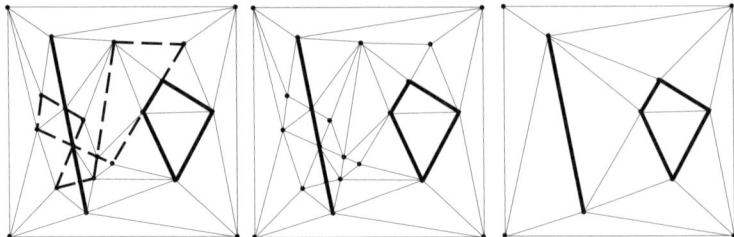

Fig. 4. The constraint C to be removed is drawn as a dashed polygonal line, and the other two intersecting constraints are drawn in bold (left). As soon as all edges of C are identified, they are set to no longer represent constraint C (middle). Finally, points which are no longer required are removed (right).

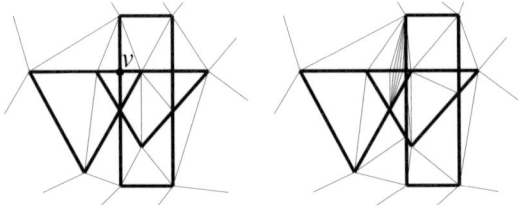

Fig. 5. The left image shows three constraints intersecting at vertex v: two triangles with one collinear edge and a rectangle R. If redundant intersection vertices were not removed, several new intersection vertices would be accumulated during a displacement of R. Such an undesirable situation (right image) is detected and fixed in step 2 of the routine `remove_constraint`.

6 Results

Using the insertion and removal routines it is possible to reinsert constraints to new positions, allowing them to move while the CDT is dynamically updated. Depending on the size of the CDT and on the size and number of constraints being manipulated, constraints can be interactively displaced in applications. This is the case with the examples showed in Figure 6.

Performance. So far, interactive graphical applications maintaining dynamic CDTs with over 10K triangles in a Pentium III 600Mhz computer have been implemented. The speed of these applications greatly depends on the size of constraints being manipulated and on the overhead to update display lists for rendering. Table 1 gives some performance data for the current version of our code.

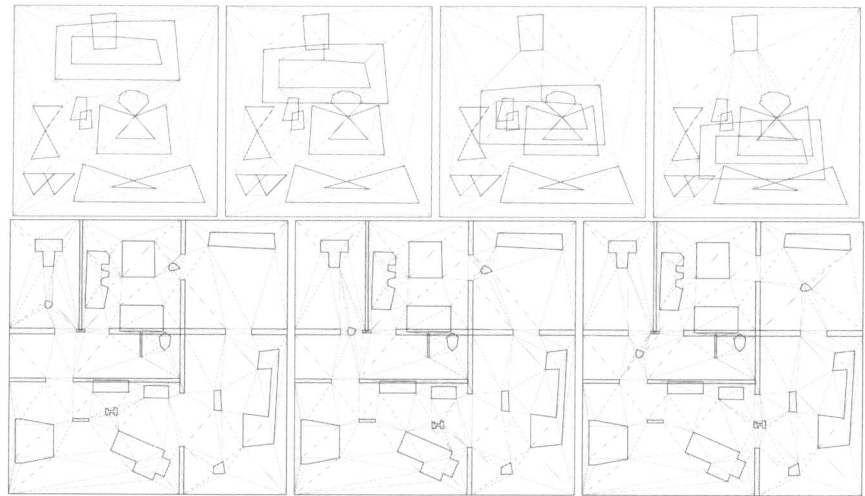

Fig. 6. The top row shows one constraint being displaced in a CDT constructed from a data set containing several intersection and overlapping cases. The bottom row shows the dynamic update of small polygons representing autonomous agents navigating in a planar environment.

Table 1. Performance values. C is a constraint selected to be displaced in each data set. For instance, in the world map data set, C is the Sicilia island. The units are: v=vertices, s=seconds, ms=milliseconds.

data set	|CDT size| |CDT construction|	C size	C removal	C insertion
head (fig.C.12b)	1433 v	2.1 s	8 v	3 ms	4 ms
hexagons (fig.C.12a)	8332 v	3.7 s	6 v	3 ms	4 ms
world map (fig.C.12d)	80652 v	38.1 s	60 v	50 ms	20 ms

7 Applications

We have tested the usefulness of our triangulator with applications in several fields. Having at our disposal a CDT handling overlapping and dynamic constraints, new possibilities are open to many applications.

Visualization. In planar data visualization, data acquisition often provides a set of values associated with unorganized points in the plane. The triangulation of the set of points is often used as a way to create connections between these points [29] [21]. Then, given any point p in the plane, its value can be computed by interpolating the values associated with the vertices of the triangle containing p. In this way, values for all points in the plane can be determined, and a smooth colored visualization of the data can be obtained. This is the case, for instance, when analyzing a temperature or elevation data distribution in a planar domain.

In many cases, to enforce correct interpolation between certain values, it is important to insert constraints representing specific features on the terrain being represented. We have identified three classes of problems in planar visualization that can benefit from the capability to dynamically update constraints:

- Dynamic changes in the data set. Such cases appear whenever features being represented are subject to dynamically change. Examples are the accurate representation of borders of lakes and rivers, continent boundaries, or for data coming from floating maritime devices equipped with GPS.
- Data analysis along paths. For the design and analysis of locations where railway paths, oil tube paths, ski runs, etc, will be installed. Designed paths can be placed as constraints over the triangulation of data points in the target terrain, obtaining data interpolation exactly along the paths. The possibility to update paths dynamically allows an animated visualization of the associated data, opening new methodologies for planning such paths.
- Classification of statistical data according to several binary criteria and visualization of the resulting classes [28]. Let each criterion be defined by a piecewise linear separator which can be modified interactively. A concrete example are habitants of a city represented as points in the plane with "age" and "education level" as their coordinates. One separator could be "qualification for a certain task", another "interest in cultural activities". In order to visualize this data, we construct the corresponding CDT with the two separators as constraints. The basic color of each point is e.g. a monotonous function of the sum of its coordinates. Then the whole triangulation is colored by interpolation. Now, we associate with each separator two additional colors, indicating "criterion fulfilled/not fulfilled", and modify the basic color of each point in the triangulation by the correct one. The result is a visualization of the valid state of the dynamic classification.

Geometric modeling. Our dynamic CDT can be efficiently used to compute Boolean operation on polygons. When inserted in a CDT, all polygons (intersecting or not) are correctly triangulated. Therefore, their union is determined by the union of the triangles which are inside any of the polygons. And their intersection is determined by the union of the triangles which are contained in all polygons. Note that the boundary of such unions can be efficiently determined by simply traversing adjacent constrained edges in the triangulation.

The ability to dynamically update the position of inserted polygons allows the interactive design of planar models using the Constructive Solid Geometry (CSG) approach. The triangulation itself can be hidden from the designer, who works mainly manipulating a tree of Boolean operations among polygons.

Reconstruction. In the field of shape reconstruction from contours (see Figure C.12b,c), a common approach is to consider adjacent contours as constraints in a CDT in order to specify the best strategy to link vertices [3].

The dynamic update of constraints allows, for instance, to reconstruct deformable models at the same time the data set is being captured by some measurement equipment. This is an important kind of application in the medical domain, in the implementation of training and computer-assisted surgery systems.

GIS. Constrained triangulations are important tools in the domain of geographic information systems [23]. They are used for the representation of different kind of data, which can be spatially retrieved using different kind of queries.

The possibility to dynamically update constraints is certainly an important characteristic of systems targeting the design and update of geographical databases. Figure C.12d shows the CDT of a world map. In this example, our algorithm greatly simplifies the construction of the CDT as no requirements concerning the constraints are needed. For instance, we could easily handle overlapping borders between countries, intersections of rivers or lakes traversing country borders, etc.

CDTs like the one shown in Figure C.12d can also serve as a base mesh for other applications, such as the analysis of ocean circulation models [20].

Path planning. The primary application that motivated this work is the determination of collision-free paths among obstacles in the plane. Practical implementations for the determination of shortest paths usually rely on visibility graphs and take quadratic time and space on the number n of obstacle vertices [25]. Using CDTs, approximate shortest paths can be derived in $O(n \log n)$ time (after the CDT construction). This time results from a graph search over $O(n)$ triangles.

Let p_1 and p_2 be two given points in the triangulated domain. First, point location is performed in order to determine the triangles t_1 and t_2 containing points p_1 and p_2, respectively. Then, a graph search is performed over the triangulation adjacency graph, in order to determine the shortest sequence of adjacent triangles (a *channel*) connecting t_1 and t_2, and not traversing any constrained edge. Channels are determined by an A* search rooted at t_1, and switching through adjacent triangles without crossing constrained edges. This process continues until t_2 is reached. If no triangles are available to search before reaching t_2, a collision-free path joining p_1 and p_2 does not exist.

Let P be the boundary of a channel successfully determined with the procedure described above. The problem is now reduced to find the shortest path between p_1 and p_2 inside the simple polygon P. This can be done with the known *funnel algorithm* [25], which runs in linear time.

Our approach becomes particularly interesting because we are able to efficiently update the environment description when constraints move. Moreover, allowing intersecting obstacles greatly simplifies the definition of grown obstacles being used to take into account paths generated for discs and not only for points. Figures C.12e and C.12f illustrate some applications of our collision-free path determination approach.

8 Final Remarks

We have given extensions to known CDT algorithms in order to cope with dynamic and overlapping constraints. Constraints are identified by ids for later removal, and the required management of ids is presented for the insertion and removal procedures.

We exemplified the use of our CDT algorithms by means of several applications, and the performed tests indicate that the algorithm is very robust[1].

As future work, we intend to study the inclusion of an optimized routine to move a constraint, without the overhead of two consecutive remove and insert calls. Extensions to 3D are feasible, but hard to be robustly implemented.

Acknowledgments. The authors thank the anonymous reviewers for their helpful comments.

References

1. Anglada, M.V. (1997): An Improved Incremental Algorithm for Constructing Restricted Delaunay Triangulations. Computer & Graphics, **21(2)**, 215–223
2. Bernal, J. (1995): Inserting Line Segments into Triangulations and Tetrahedralizations. Actas de los VI Encuentros de Geometria Computacional, Universitat Politecnica de Catalunya, Barcelona, Spain
3. Boissonnat, J.-D. (1988): Shape Reconstruction from Planar Cross Sections. Computer Vision, Graphics, and Image Processing, **44**, 1–29
4. Boissonat, J.-D., Teillaud, M. (1993): On the Randomized Construction of the Delaunay Tree. Theoretical Computer Science, **112**, 339–354
5. Chew, L.P. (1987): Constrained Delaunay Triangulations. Proceedings of the Annual Symposium on Computational Geometry ACM, 215–222
6. Devillers, O. (1998): Improved incremental randomized Delaunay triangulation. Proc. of the 14th ACM Symposium on Computational Geometry, 106–115
7. Devillers, O. (1999): On Deletion in Delaunay Triangulations. Proceedings of the 15th Annual ACM Symposium on Computational Geometry, June
8. Devillers, O., Pion, S., Teillaud, M. (2001): Walking in a Triangulation. ACM Symposium on Computational Geometry
9. Fortune, S. (1987): A Sweepline Algorithm for Voronoi Diagrams. Algorithmica, **2**, 153–174
10. De Floriani, L., Puppo, A. (1992): An On-Line Algorithm for Constrained Delaunay Triangulation. Computer Vision, Graphics and Image Processing, **54**, 290–300
11. Guibas, L.J., Knuth, D.E., Sharir, M. (1992): Randomized Incremental Construction of Delaunay and Voronoi Diagrams. Algorithmica, **7**, 381–413
12. Guibas, L., Stolfi, J. (1985): Primitives for the Manipulation of General Subdivisions and the Computation of Voronoi Diagrams. ACM Transactions on Graphics, **4(2)**, 75–123

---

[1] For information concerning the source code please contact the first author

13. Held, M. (2001): VRONI: An Engineering Approach to the Reliable and Efficient Computation of Voronoi Diagrams of Points and Line Segments. Computational Geometry Theory and Applications, **18**, 95–123
14. Mücke, E.P., Saias, I., Zhu, B. (1996): Fast Randomized Point Location Without Preprocessing in Two and Three-dimensional Delaunay Triangulations. Proc. of the Twelfth ACM Symposium on Computational Geometry, May.
15. Preparata, F.P., Shamos, M.I. (1985): Computational Geometry. Springer-Verlag, New York
16. Su, P., Drysdale, R.L.S. (1995): A Comparison of Sequential Delaunay Triangulation Algorithms. Proceedings of the ACM Eleventh Annual Symposium on Computational Geometry, June, 61–70
17. Shewchuk, J.R. (1996): Triangle: Engineering a 2D Quality Mesh Generator and Delaunay Triangulator. First ACM Workshop on Applied Computational Geometry, Philadelphia, Pennsylvania, May, 124–133
18. Sugihara, K., Iri, M., Inagaki, H., Imai, T. (2000): Topology-Oriented Implementation - An Approach to Robust Geometric Algorithms. Algorithmica **27(1)**, 5–20
19. Schirra, S., Veltkamp, R., Yvinec, M. (1999): The CGAL Reference Manual. Release 2.0 (www.cgal.org)
20. Legrand, S., Legat, V., Dellersnijder, E. (2000): Delaunay Mesh Generation for an Unstructured-Grid Ocean General Circulation Model. Ocean Modelling **2**, 17–28
21. Treinish, L. A. (1995): Visualization of Scattered Meteorological Data. IEEE Computer Graphics and Applications, **15(4)**, July, 20–26
22. Weller, F. (1998): On the Total Correctness of Lawson's Oriented Walk Algorithm. The 10th International Canadian Conference on Computational Geometry, Montréal, Québec, Canada, August, 10–12
23. De Floriani, L., Puppo, E., Magillo, P. (1999): Applications of Computational Geometry to Geographic Information Systems, Chapter 7 in Handbook of Computational Geometry, J.R. Sack, J. Urrutia (Eds), Elsevier Science, 333–388
24. Lawson, C. L. (1977): Software for C1 Surface Interpolation. In J. R. Rice (ed), Mathematical Software III, Academic Press, New York, 161–194
25. Mitchell, J.S.B. (1997): Shortest Paths and Networks. Handbook of Discrete and Computational Geometry, Discrete Mathematics & its Applications, Jacob E. Goodman and Joseph O'Rourke, ed., CRC Press, 445–466
26. Kallmann, M., Thalmann, D. (2001): Star Vertices: A Compact Representation for Planar Meshes with Adjacency Information. Journal of Graphics Tools, **6(1)**, 7–18
27. Mäntylä, M. (1988): An Introduction to Solid Modeling, Computer Science Press, Maryland
28. Bennett, K.P., Mangasarian, O.L. (1994): Multicategory discrimination via linear programming. Optimization Methods and Software, **3**, 29–39
29. Nielson, G.M. (1997): Tools for Triangulations and Tetrahedrizations and Constructing Functions Defined over Them. In G.M. Nielson, H. Hagen, H. Mueller (Eds.): Scientific Visualization. IEEE Computer Society Press, 429–525

EVM: A Complete Solid Model for Surface Rendering

Jorge Rodríguez[*,1], Dolors Ayala[1], and Antonio Aguilera[2]

[1] Universitat Politècnica de Catalunya, Diagonal 647, 08028 Barcelona, Spain
{*jrodri, dolorsa*}*@lsi.upc.es*
[2] Universidad de las Américas-Puebla, México, *aguilera@mail.udlap.mx*

Summary. The Extreme Vertices Model (EVM) has been presented as a concise and complete model for orthogonal pseudo-polyhedra (OPP) in the solid modeling field. This model exploits the simplicity of its domain by allowing robust and simple implementations. In this paper we use the EVM to represent and process images and volume data sets. We will prove that the EVM works as an efficient scheme of representation for binary volumes as well as a powerful block-form surface renderer which avoids the redundancy of primitives on the extracted isosurface. In addition, to achieve more realism, the normal vectors at vertices can be added to the model. Furthermore, an efficient tessellator of non-convex orthogonal faces is presented. Useful operating and manipulating tools can be implemented over the EVM, like editing operations via Boolean operators, non voxel-based morphological operations and an improved connected component labeling algorithm. The well-composedness property of the volume can be detected easily.

1 Introduction

A volume data set (or 3D digital image) can be represented as a map $f : Z \otimes Z \otimes Z \to R$, in such a way that every point is assigned a value representing its color. In a binary data the image set is $\{0, 1\}$. A volume can be defined as an union of voxels i.e. upright unit cubes whose vertices have integer coordinates [22]. In the field of digital image processing, the term well-composed has been used to name those digital images with a manifold boundary, i.e, which avoid configurations shown in Fig. 1 modulo reflections and rotations, and have related the concepts of manifoldness and well-composedness [12].

Orthogonal polyhedra (OP) are polyhedra with all their faces oriented in three orthogonal directions. The term orthogonal pseudo-polyhedra (OPP) refers to regular and orthogonal polyhedra with a possible non-manifold boundary. The *Extreme Vertices Model* (*EVM*) is a very concise model in which any OPP can be described using only a subset of its vertices. The EVM is actually a complete (unambiguous) solid model [2]. EVM is also suitable for 2D orthogonal polygons.

---

[*] This author has been supported by a grant from CONICIT (Venezuelan Council of Research in Science and Technology)

From all these concepts, we can say that the continuous analog of a general digital image is an OPP and the continuous analog of a manifold digital image is an OP and both can be represented and operated using the EVM.

In this paper we use the EVM to represent, visualize and operate volumes. Concerning with representation and visualization purposes, we show that the EVM works as an efficient scheme of representation for segmented volumes and that it is a powerful block-form surface renderer which avoids the redundancy of primitives on the generated isosurface. We also present an efficient tessellator of non-convex faces to be used in case that only convex faces would be required for the rendering system. And we offer the possibility to enrich the EVM with the normal vectors at the vertices in order to obtain more realism.

Concerning with operations, we review the EVM useful operating and manipulating tools, which follow a non-voxel-based approach and work in 3D as well as in 2D. We have developed editing operations via Boolean operators with no arithmetic computing. Morphological operations such as erosion and dilation and its derivatives opening and closing have also been studied. Moreover, we have devised an improved connected component labeling algorithm which reduces the number of labels and equivalences produced. This method allows to detect easily the well-composedness property of the volume as well as the critical zones where non-manifold configurations occur.

Finally we will present experimental results obtained with the EVM and as we think that the EVM is closely related to the Semiboundary representation (SB), we will compare results obtained with both methods.

2 Related Work

Concerning volume visualization, it can be achieved following two approaches: indirect volume rendering (IVR), in which an isosurface is extracted from volumetric data and subsequently rendered, and direct volume rendering (DVR), in which the data is directly rendered. See [16] for a little survey on DVR methods. In this paper we are focused on IVR.

There are two main families of models and methods for isosurface extraction. Digital (or block-form) methods represent the surface as a set of voxels or voxel faces [5], [22], [26], [9]. Polygonal (or beveled-form) methods represent the surface as a set of polygons (mostly triangular) [14].

Most of the developed techniques, mainly those of the digital family, are 10th on visualization and are not concerned on other operations and there is a claim for a general model that can carry out the complete set of desirable tasks: visualization, manipulation and editing. Moreover, digital models exhibit formal properties such as closure, orientedness and connectedness whereas polygonal models still lack of a solid modeling foundation [9].

A common drawback of the existing techniques is the redundancy of primitives in the isosurfaces obtained. They consist of lots of little quadrangular (in digital methods) or triangular (in polygonal methods) faces. Several adaptive attempts have been developed to reduce this redundancy [18], [17], [24] but most of them actually make a post-process and produce cracks.

EVM is a digital model, closely related to the Octree and Semiboundary (SB) models and, for this reason, we include a little discussion on these models. Octree [27], and SB [26], [25] models allow to store a volume in a proper data structure in order to carry out operations of visualization and editing efficiently. In [27] the authors point out the large amount of time spent with empty cells and propose the octree-based hierarchy to alleviate this problem which has been also used in other methods [20]. Several other approaches have used this hierarchy to apply multiresolution when dealing with very large data sets [11], [4] and, recently, octrees have been applied to distance fields yielding the so-called adaptive distance fields that allow to represent data and to extract isosurfaces according to a given tolerance [8]. The Octree model is a spatial partitioning which allows to manipulate homogeneous regions of the volume as single entities but which requires volumes with a size of a power of two. It allows Boolean operations to be performed between objects, but it requires a common cubic universe for both operands. Consequently, operations between objects of very different sizes force an important overhead of memory for representing the smaller one. Nevertheless, useful image processing operations such as dilation, erosion and thinning have not been dealt in this scheme.

The SB [26] [9], is a useful model for fast visualization, manipulation and analysis of structures generated from volume data. It only stores the boundary voxels of the data, keeping the information of the interior voxels in an implicit way. SB allows to perform operations such as cut-away, osteotomy, reflections and interactive manipulation, and the authors assure that other data structures such as Octrees and run-length codes will not match the SB efficiency in visualization and manipulation. However, Boolean operations are relatively expensive using this scheme because it is not easy to determine which part of one object lies inside the other one merely by checking the boundary voxels. Furthermore, like the Octree model, SB would require a common cubic universe for both operands involved.

With respect morphological operations, several approaches to improve them have been reported. In [28] a method is proposed for successive dilation operations of volumes by using a queue of contour voxels, but it demands an overhead of memory. In [25], the SB is used in a method for fast eroding and dilating. Both proposals traverse only the contour voxels. In [15] an improvement is presented that decomposes the multi-dimensional structuring elements into 1D elements using a pseudo tensor product approach.

Concerning with connected component labeling (CCL), In [25], the SB is used in a CCL algorithm for volumes. A general approach is formalized in [7] for CCL which can deal with d-dimensional images represented in several data structures such as arrays, quadtrees and bintrees and in [19] an algorithm is suggested for volume rendering based on a 2D connected component search. Almost all the approaches for morphological operations and CCL are voxel-based and therefore their performance decreases as the number of voxels of the image or volume increases.

In most of the consulted bibliography, the authors tend to characterize and avoid non-well-composedness. In [13] the authors say that it is more important to report the non-manifold zones as zones corresponding with determined features than to deal with it. In [23] a method is presented that obtains a well-composed 2D image by doing simple deformations.

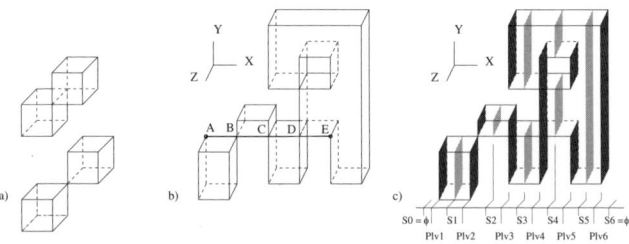

Fig. 1. a) Configurations not allowed in a well-composed volume. b) An OPP with a marked brink from vertex A to vertex E. c) Its planes of vertices and sections perpendicular to the X axis shown in dark and light grey respectively.

3 Review of EVM

Let P be an OPP. A *brink* is the maximal uninterrupted segment built out of a sequence of collinear and contiguous two-manifold edges of P. The ending vertices of a brink are called *extreme vertices* (EV). Fig. 1 b) shows an OPP with a brink from vertex A to vertex E (both EV). Vertices B, C and D are non-extreme The *EVM* represents OPP by its (and only its) set of EV.

A *plane of vertices* (*plv*) is the set of vertices lying on a plane perpendicular to a main axis of P. A *slice* is the region between two consecutive planes of vertices. A *section* (S) is the resulting polygon from the intersection between P and an orthogonal plane. Each slice has its representing section. See Fig. 1 c). Planes of vertices and sections obtained from a 3D object are 2D orthogonal polygons. From them, we can obtain their 1D lines of vertices and 2D slices with their corresponding 1D sections. Lines of vertices and 1D sections are 1D objects which are composed of one or several brinks.

In the EVM the set of EV can be ordered in six possible ways depending on the coordinate values: XYZ, XZY, YXZ, YZX, ZXY, and ZYX.

EVM has several properties. The first one is that coordinate values of non-extreme vertices may be obtained from EV coordinates. The second property allows sections to be computed from planes of vertices and vice-versa:

$$\overline{S_i(P)} = \overline{S_{i-1}(P)} \otimes^* \overline{plv_i(P)}, i = 1\ldots n, \qquad S_0(P) = S_n(P) = \emptyset$$

$$\overline{plv_i(P)} = \overline{S_{i-1}(P)} \otimes^* \overline{S_i(P)}, i = 1\ldots n$$

where $\overline{plv_i(P)}$ and $\overline{S_i(P)}$ denote the projections of $plv_i(P)$ and $S_i(P)$ onto a main plane parallel to P, n is the number of planes of vertices and $\otimes^*$ denotes the regularized XOR operation. Note that in order to operate with the projections we need not take into account the coordinate of the extreme vertices that corresponds to the projecting plane.

Applying the definition of the $\otimes$ operation, this last equation can be expressed as ($i = 1\ldots n$):

$$\overline{plv_i(P)} = \overline{S_{i-1}(P)} \otimes^* \overline{S_i(P)} = (\overline{S_{i-1}(P)} -^* \overline{S_i(P)}) \cup (\overline{S_i(P)} -^* \overline{S_{i-1}(P)})$$

and, thus, we can decompose any plane of vertices into two terms that we will call *forward difference* $(\overline{FD_i(P)} = (\overline{S_{i-1}(P)} -^* \overline{S_i(P)})$ and *backward difference* $(\overline{BD_i(P)} = (\overline{S_i(P)} -^* \overline{S_{i-1}(P)}))$.

$FD_i(P)$ is the set of faces on $plv_i(P)$ whose normal vector points to one side of the main axis perpendicular to $plv_i(P)$, while $BD_i(P)$ is the set of faces whose normal vector points to the opposite side. This property, which guarantees correct orientation, together with the first one provide proof that the EVM is a complete B-Rep model.

There is a property concerning the XOR operation: Let P and Q be two d-D (d $\leq$ 3) OPP, having $EVM(P)$ and $EVM(Q)$ as their respective models, then, $EVM(P \otimes^* Q) = EVM(P) \otimes EVM(Q)$ Then, the XOR operation works in 0D because it is applied to the EV. Sections are obtained from planes of vertices and vice-versa by simply applying XOR to the EV.

See [2] for more explanations concerning EVM.

4 Surface Rendering Using the EVM

In this section we present an isosurface extraction method using the EVM. It belongs to the digital category and reduces, without any post-processing, the excessive redundancy of primitives of most isosurface extraction algorithms. It obtains a complete and closed solid model with an oriented boundary. We also improve the block-like appearance of the objects by using an enriched EVM and show a simple form to tessellate the orthogonal faces if needed.

4.1 Isosurface Extraction Algorithm Using EVM

To obtain an isosurface from a volume, for a given value of density, we first obtain the EVM of the volume and afterwards a hierarchical B-Rep is computed from the EVM. In this section we propose a volume data set to EVM conversion algorithm which complements the EVM to B-Rep conversion algorithm reported in [3].

Given a segmented volume data set, the extreme vertices are those belonging to an odd number of black voxels. This property can be proved by exhaustive analysis over the 256 cases in which 8 voxels (black or white) surround a vertex [2]. The algorithm that obtains the EVM from the volume traverses all its vertices and keeps those which fulfill this property.

procedure $VoxelToEVM$(**Input** $vol : tVoxelization,$ **Ouput** $p : tEVM$)
   var $ev : tVertex$ **endvar**
   for every cut k **in** vol **do**
      for every row i **in** cut **do**
         for every vertex j **in** row **do**
            $odd := XOR(Voxel_{j-1,i-1,k-1}, Voxel_{j-1,i-1,k},$
            $Voxel_{j-1,i,k-1}, Voxel_{j-1,i,k}, Voxel_{j,i-1,k-1},$
            $Voxel_{j,i-1,k}, Voxel_{j,i,k-1}, Voxel_{j,i,k});$
            if odd **then** $ev = (j,i,k);$ $AddVertex(ev, p)$ **endif**
         endfor
      endfor
   endfor
endprocedure

Function XOR applies exclusive-or to all the arguments. $Voxel_{j,i,k}$ is the value, 0 or 1, of the (j,i,k) voxel. *AddVertex* adds a new vertex to the EVM.

4.2 Avoiding the Redundancy of Primitives

The hierarchical B-Rep obtained from the EVM is actually a polyhedral representation of the volume of interest in which the possible coplanar little adjacent faces are already merged in a large one (not necessarily convex).

Fig. 2 show the B-Rep obtained from the SB (SB-BRep) and the B-Rep computed from the EVM (EVM-BRep) for two data sets *skull* (96 × 96 × 69) and *ctHead* (256 × 256 × 94), obtained from CT data segmented by threshold. Table 1 compares the number of faces obtained (squares for EVM and orthogonal polygons for EVM) in the corresponding B-Rep for both volumes. It also shows the number of triangular faces produced for the marching cubes algorithm (MC) [14].

data set	MC	SB-BRep	EVM-BRep	T_1	EVM'	tf	T_2
skull	97467	49030	16355	28.79	23300	1866	1.72
ctHead	336283	180288	54570	118.45	74537	5003	5.98

Table 1. MC, SB-BRep, EVM-BRep: Number of faces. T_1: time (secs.) for the EVM to B-Rep conversion. EVM' and tf: number of total faces (rectangles) of the tesselated EVM-BRep and number of faces that have been to be tesselated, T_2: tesselation time (see section 4.5).

4.3 Efficient Storing of the Volume

In this and the following subsection we study the storage requirements of the EVM and compare them with the SB. We consider a volume data set of size $h_1 \times h_2 \times h_3$, assuming that 5% of voxels are boundary voxels and 3% of vertices are extreme (EV), which are reasonable upper bounds according to our experimental results. Also, we use the fact that a volume data set with $h_1 \times h_2 \times h_3$ voxels has $(h_1 + 1) \cdot (h_2 + 1) \cdot (h_3 + 1)$ vertices and we consider that a coordinate can be stored in 10 bits [26].

The SB of a volume is a tuple (D, Δ), where D is the set of boundary voxels and Δ is an *encoding function* which assigns to every voxel in D a 6-bit number, called the *neighbor code*. The SB is implemented as a pair {P,M}, where P is a pointer array with $h_2 \times h_3$ elements and M is a list of triplets $(d_1, \delta, \mathbf{n})$, where d_1 is the first coordinate of a voxel $d \in D$ (10 bits), $\delta = \Delta$ (6 bits) and $\mathbf{n}$ is a unit normal vector coded in 11 bits, [26]. From these definitions, we have estimated the memory requirements of SB as follow: $h_2 \cdot h_3 \cdot size(pointer)$ for P and $(h_1 \cdot h_2 \cdot h_3) \cdot 0.05$ triplets for M. Given that each triplet requires 27 bits, then M requires $(h_1 \cdot h_2 \cdot h_3) \cdot 0.05 \cdot 27$ bits. Therefore, $Mem(SB) = h_2 \cdot h_3 \cdot size(pointer) + (h_1 \cdot h_2 \cdot h_3) \cdot 1.35 \; bits$

Although for most operations EVM is implemented as a set of ordered EV, it also can be implemented in a way similar to the SB, in order to compare

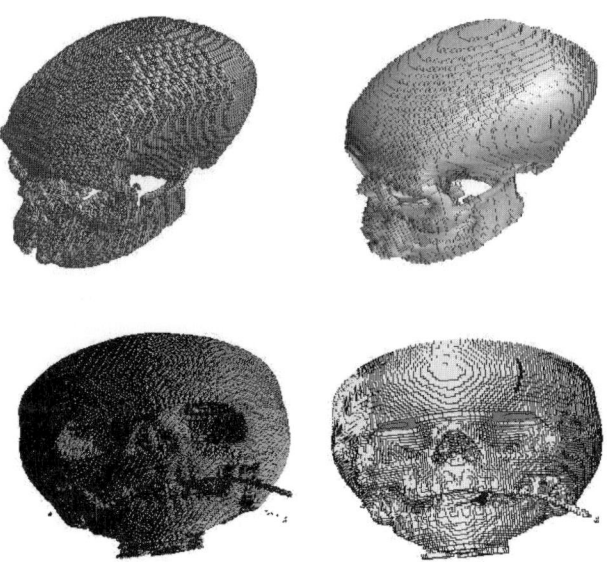

Fig. 2. Top: SB-BRep (left) and EVM-BRep (right) of *skull*. Bottom: SB-BRep (left) and EVM-BRep (right) of *ctHead*

them. Thus, we have a pointer array, P, with $h_2 \times h_3$ elements and a list, M, of brinks (pairs of EV). The memory requirements of this EVM implementation are $h_2 \cdot h_3 \cdot size(pointer)$ for P and $(h_1+1) \cdot (h_2+1) \cdot (h_3+1) \cdot 0.015$ brinks. Note that the number of brinks is estimated as 1.5% of the number of (voxelization) vertices. Given that we require 10 bits per coordinate and that we have two coordinates for each brink (the third one for each EV of the brink), then,

$$Mem(EVM) = h_2 \cdot h_3 \cdot size(pointer) + (h_1+1) \cdot (h_2+1) \cdot (h_3+1) \cdot 0.015 \cdot 20 \; bits$$

Comparing both expressions we get,

$$Mem(SB) > Mem(EVM) \Leftrightarrow (h_1 \cdot h_2 \cdot h_3) \cdot 1.35 > (h_1+1) \cdot (h_2+1) \cdot (h_3+1) \cdot 0.3$$

In order to facilitate the proof we suppose $h_1 = h_2 = h_3 = h$. Then,

$$h^3 \cdot 1.35 > (h+1)^3 \cdot 0.6$$

These cubic functions have an intersection point at $h = 1.54$. As a result, we can conclude that

$$Mem(SB) > Mem(EVM) \Leftrightarrow Min(h_1, h_2, h_3) \geq 2$$

Data set	256^3	512^3	skull			ctHead		
#vox	16777216	134217728	635904			6160384		
	est	est	est	(%)	real	est	(%)	real
#bv	838860	6710886	31795	(5.6%)*	36184	308019	(2.4%)*	153979
#EV	509237	4050170	19758	(3.5%)*	23464	188239	(1.2%)*	80032
M(SB)	2892Kb	22630Kb	122Kb		137Kb	1143Kb		635Kb
M(EVM)	750Kb	5458Kb	42Kb		75Kb	358Kb		323Kb
saving EVM					45%			49%
M(eEVM)					109Kb			440Kb
saving eEVM					20.4%			30.7%

Table 2. Estimated and real memory requirements (EVM and enriched EVM). (%)* is the percentage with respect #vox

Table 2 presents the estimated computed values, based on the previous discussion, for several data sets and real values for two data sets (skull and ctHead). The items on the table are the total number of voxels (#vox), the estimated and real number of boundary voxels (#bv), number of extreme vertices (#EV) and memory requirements for SB and EVM. It also shows the experimental memory saving using EVM instead of SB. Both the estimated values and the experimental ones prove that EVM is a more concise representation scheme.

4.4 Improving Realism

Our surface renderer based on EVM lies in the block-form category and the main drawback of this group is a block-like appearance of the produced surfaces with a weak realism. To improve the appearance, we propose an *enriched EVM* with an associated normal vector to each extreme vertex (EV). This normal vector is estimated by the gradient of grey level directly from the input volume using central differences around the EV. To avoid the overgrowth of memory requirements, we have encoded the normal vector in 12 bits, four per direction, dividing each axis in 16 intervals. Subsequently, when more realism is required, these normal vectors will be used to smooth the shading of the object. Fig. C.13 shows how the enriched EVM improves the realism of the visualization.

The memory occupancy of the enriched EVM grows in 24 bits per brink and therefore

$$Mem(EVM) = h_2 \cdot h_3 \cdot size(pointer) + (h_1+1) \cdot (h_2+1) \cdot (h_3+1) \cdot 0.015 \cdot 44 \, bits$$

Comparing again with the SB, now the cubic functions $1.35h^3$ and $0.66(h+1)^3$ have an intersection point at $h = 3.71$.

As EV are only a subset of the vertices of the object, the remaining non-extreme vertices, which appear when the B-Rep is obtained, receive their normal vectors computed by linear interpolation between the EV of the brinks in which the non-extreme vertex lies. Fig. 3 illustrates such computation. Table 2 allows to compare experimental memory requirements between SB and enriched EVM and Fig. C.14 shows several views of the *ctHead* data set.

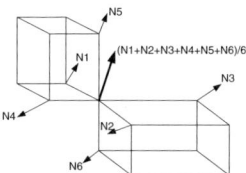

Fig. 3. Normal vector computing on an intermediate vertex

4.5 A Specific EVM-tessellator

The B-Rep obtained from the EVM avoids the redundancy of primitives and obtains large orthogonal faces. In general, these faces are not convex and may contain holes. This feature is not a problem in most visualization systems because they have their own tessellation algorithm for these cases. However, the tessellation strategies used in some of these systems follow general rules and split the face in an arbitrary way with no care about its orthogonal shape. Moreover, there are visualization systems that require convex faces. We present our own tessellation strategy based on the EVM, which profits the knowledge about the geometry of faces to split them in an efficient way.

Let F be a face of an OPP, then F can be expressed as the union of all its slices in a certain orthogonal direction, $F = \cup_{k=1}^{n-1} slice_k(F)$. Each slice define one or more rectangles. The representative 1D sections of each slice are composed by one or more brinks (each brink corresponds to a rectangle) and can be computed doing an XOR operation between the extreme vertices of its previous and next plane of vertices. From these 1D sections, the rectangles of the slice are directly obtained. See table 1 for results concerning tesselation. Furthermore, if triangles would be required for the visualization system, each rectangle could be split along its diagonal (see Fig. 4).

In contrast with general tessellation techniques this method does not check neither collinearity among vertices nor intersection among edges. Moreover, the orthogonal nature of faces allows a very simple test based on a trivial rule to detect non-convexity: an orthogonal polygon is non-convex if its number of vertices is greater than four.

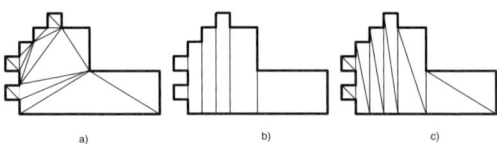

Fig. 4. a) General tessellation (18 triangles). b) EVM-based tessellation (7 rectangles) c) Further triangulation (14 triangles)

5 Volume Operations Using the EVM

In this section we review the EVM developed operations for volumes. All these operations are non-voxel oriented. Editing and morphological operations are based on the sections of the model (see section 3). Connected component labeling is based on a decomposition of the model into big boxes. None of these operations require floating point arithmetic.

5.1 Editing Operations via EVM-Boolean Operators

Editing operations, like cut and paste, over volumes can be achieved via Boolean operators. General Boolean operations between OPP can be carried out by applying recursively the same Boolean operation over the corresponding sections. At the end, the operation is actually performed over 1D sections, which consist of one or several collinear brinks. So, the method is recursive over the dimension. The algorithm has been presented in [1] for solid objects and consists in a geometric merge between the EVM of both operands. In this work we have experimented this algorithm over volumes, obtaining the same good results. Fig. 5 illustrates a simulation of an interactive osteotomy performed via Boolean operations.

5.2 Morphological Operations

The approach for morphological operations, based on the EVM [21] is also recursive over the dimension and, therefore, is suitable for volumes as well as

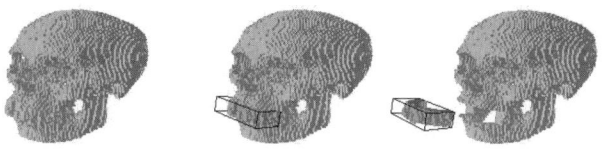

Fig. 5. Osteotomy of an human skull simulated via Boolean operations

for 2D images. The algorithm identifies the inner 2D sections of the volume, which are themselves considered 2D images and, then, decomposes these 2D images into 1D sections. Then it shrinks or elongates these 1D sections. Afterwards, 1D planes of vertices are computed from these 1D sections, obtaining the new 2D sections and, from them, the transformed object, applying recursively the same process.

We can express the erosion of an EVM represented OPP P, $Shrink(P)$, as:

$$\begin{cases} Shrink(S_{i-1}) \otimes^* Shrink(S_i), \ i = 1 \ldots n \ dim > 1 \\ Shrink1D(P) \hspace{4cm} dim = 1 \end{cases}$$

S_i is the *ith* section of P, $i = 0, 1, \ldots n$, dim is the dimension of P (initially, $dim = 3$). *Shrink1D (P)* consists in shrinking each brink of the 1D resulting object. An analogous recursive process can be defined for *Elongate(P)*.

The cost of these operations does not depend directly of the amount of voxels of the input data set, instead, it is associated with the complexity of the orthogonal shape of it, so we are able to operate large sets of adjacent voxels in a single step. Furthermore, our approach admits the use of box-shaped structuring elements of arbitrary size. Fig. 6 shows a 2D example (*Magallanes*) and Fig. 7 a 3D example of a human skull.

Fig. 6. 2D example. Left: Original image. Right: eroded once.

The algorithms have been implemented in C++ on a Sun Microsystems Ultra SPARC 60 machine. Table 3 presents, for these data sets, its resolution, the number of boundary voxels and brinks and the processing time for erosion and dilation operations which increases as the number of brinks grows. The right hand side of the table shows the results obtained when the size of

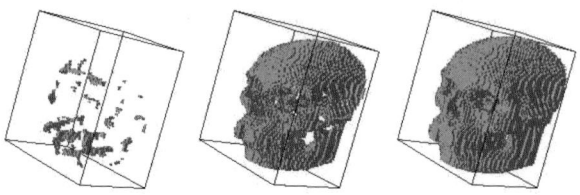

Fig. 7. Left to right: eroded once (*pieces*), original skull and dilated once

the former data sets was scaled 1:2 maintaining the same shape. Note that the number of boundary voxels has increased accordingly but the number of brinks and the processing time remains almost the same.

Data sets	Magall. (400x159)	skull (96^2x69)	Double1 (800x318)	Double2 (192^2x138)
#bv	9692	36184	37124	187592
#brinks	1173	11732	1173	11732
Erosion	0.19s	4.22s	0.28s	4.7s
Dilation	0.19s	6.38s	0.28s	6.78s

Table 3. Morphological processing time (secs.)

5.3 Connected Component Labeling

The method for connected component labeling using EVM [6] is based on a particular decomposition of the model and follows the approach presented in [25]. It works in identical way for 2D images and volumes, is size-independent, improves the performance respect to the voxel-based approaches (reducing the number of labels and equivalences generated in the algorithm) and deals indistinctly with manifold and non-manifold data.

Decomposing EVM into an OUDB

We first obtain a particular partitioning straightforward from the EVM which consists in an ordered union of disjoint boxes (OUDB). This partitioning can be thought as a special kind of cell decomposition which is axis aligned as octrees and bintrees but the partition is done along the object geometry like BSP.

We obtain the OUDB from EVM splitting first the data at every plane of vertices perpendicular to a main axis, say X for instance, obtaining a set of slices. Then, we split each slice at every plane of vertices perpendicular to another main axis, say Y, obtaining a set of boxes for each slice. All the obtained boxes constitute the OUDB partitioning. See Fig. 8. This method is the 3D extension of the method used in section 4.5 to tessellate non-convex orthogonal polygons.

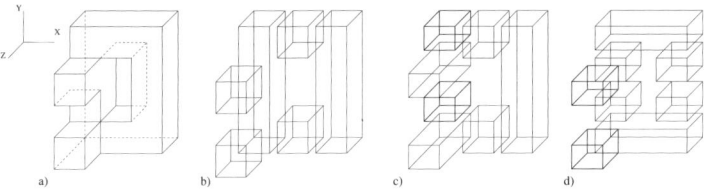

Fig. 8. a) An OPP. b) XZ decomposition (6 boxes). c) XY decomposition (7 boxes). d) YZ decomposition (8 boxes).

Connected component labeling method

Once the OUDB partition has been achieved the connected component labeling process takes place. Previous algorithms use a sequential two-pass strategy which works at voxel-level [10], [25]. In the first pass (labeling) all voxels are labeled but there remain some unsolved label equivalences which are sorted out in a second pass (renumbering). Thurfjell et al. [25] improved this technique by using the semiboundary representation for the volume in order to visit only boundary voxels. Our method follows the classical two-pass approach, but it traverses the set of disjoint boxes(OUDB) instead of a set of boundary voxels, further diminishing the number of elements to label and the number of labels and equivalences generated.

Fig. 9 shows the three steps carried out to in a 2D example. First, the EVM is extracted from the input data and its OUDB partition is obtained (each grey level represent a box), then the labeling of boxes is made and finally the equivalences are solved and the connected components stored as separated objects.

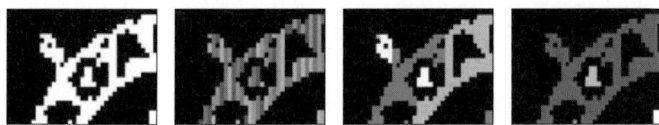

Fig. 9. Left to right: Original image, OUDB decomposition, OUDB labeled before renumbering pass and four connected components as separated objects.

Now we present some results for a 2D example (*magallanes*, shown in Fig. C.15) and for a 3D example (*pieces*, shown in the left hand side of Fig. 7). Table 4 presents several results for each data set. It shows at the left hand side, the number of elements, boundary voxels (#bv) and disjoint boxes (#box). In the middle there is a comparison between the execution time (in seconds) using a voxel-based algorithm and with our approach. At the right hand side there is the number of connected components.

The performance of our algorithm is related with the number of boxes and this number is normally smaller than the number of boundary voxels.

Data sets	elements		voxel-based			EVM-CCL				#cc	
	#bv	#box	Label.	Renum.	Total	OUDB	Label.	Renum.	Total	#M	#B
Magall	4662	1113	0.06	0.03	0.1	0.05	0.02	0.0001	0.07	18	29
Pieces	1160	524	1.45	0.37	1.96	0.08	0.03	0.0001	0.11	56	78

Table 4. CCL results.

This implies a significant reduction on the labels and equivalences generated in the labeling pass and consequently a reduction of both execution time and memory requirements.

We say that our algorithm is image-size independent because if we scale a data set, the number of boundary voxels increases significantly while the number of boxes remains the same. However this is only true for images with the same shape. In general, increasing the size is associated with increasing the shape complexity and, then, the number of boxes also increases.

Two-manifold property

This algorithm allows to detect the well-composedness property of the volume as well as the critical zones where non-manifold configurations occur. This can be done in the labeling step when comparing boxes. If two boxes share a face or part of a face, then they will have the same label. If they are disjoint, then the new box will have a new label. But if they share an edge, a part of an edge or a vertex, then they can have the same label or not. In the former case, the algorithm will maintain the non-manifold zones. In the latter case, it will break the object at these zones and the obtained connected components will be 2-manifold. Fig. 10 and Fig. C.15 show this method applied to a simple 3D example and to the *magallanes* example, respectively. Table 4 shows at the right hand side the number of connected components obtained for the two mentioned examples, maintaining (#M) or breaking (#B) non-manifold zones.

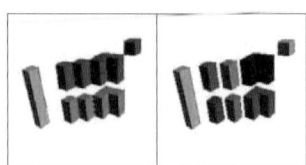

Fig. 10. EVM-CCL algorithm applied to a 3D data set without non-manifold detection(left) and with non-manifold detection(right)

6 Conclusions and Future Work

The aim of this work has been to use the EVM, to represent and process 2D images and volumes. EVM can represent binary data by storing implicitly the boundary of its continuous analog (an orthogonal polyhedron). Although EVM is very concise, it is a complete solid model and enjoys of formal properties like closure, orientedness, and connectedness.

We have studied and experimented the behavior of the EVM representing several data sets. We also have developed a method to obtain an isosurface from the EVM that avoids the excessive redundancy of primitives of most methods. Although our claim is that EVM behaves better for manipulation than for visualization, we have proposed an enriched EVM with normal vectors that give more realism to the rendered images.

We also have reported several operations developed for the EVM. Boolean operations, developed and experimented initially with solid models, behave also well with volume data sets. Non-voxel based approaches have been developed for morphological operations and connected component labeling. We have presented results obtained with the EVM and we have also compared the EVM with the SB representation.

As a future work, we want to study the suitability of EVM to represent and operate multi-modal volumes and large time-varying data sets.

Acknowledgements

This work has been supported by a CICYT grant TIC99-1230-C02-02.

References

1. A. Aguilera and D. Ayala. Orthogonal Polyhedra as Geometric Bounds in Constructive Solid Geometry. *ACM Solid Modeling'97*, pages 56 – 67.
2. A. Aguilera and D. Ayala. Domain extension for the extreme vertices model (EVM) and set-membership classification. In *CSG'98. Ammerdown (UK)*, pages 33 – 47. Information Geometers Ltd., 1998.
3. A. Aguilera and D. Ayala. Converting Orthogonal Polyhedra from Extreme Vertices Model to B-Rep and to Alternative Sum of Volumes. *Computing Suppl. Springer-Verlag*, 14:1 – 28, 2001.
4. C. Andújar, P. Brunet, and D. Ayala. Topology-Reducing Surface Simplification Using a Discrete Solid Representation. *ACM Transactions on Graphics*, 21(2):88 – 105, 2002.
5. E. Artzy, G. Frieder, and H. Gabor. The theory, design, implementation and evaluation of a three-dimensional surface detection algorithm. *Computer Graphics and Image Processing*, 15:1 – 24, 1981.
6. D. Ayala, J. Rodríguez, and A. Aguilera. Connected component labeling based on the EVM model. In *Spring Conf. Computer Graphics 2002*, pages 55–63. Comenius University, 2002.

7. M.B. Dillencourt, H. Samet, and M. Tamminen. A general approach to connected-component labeling for arbitrary image representations. *Journal of the ACM*, 39(2):253 – 280, 1992.
8. S. Frisken, R. Perry, A. Rockwood, and T. Jones. Adaptively sampled distance fields: A general representation of shape for computer graphics. In *Proceedings SIGGRAPH 2000*, pages 249 – 254, 2000.
9. G. J.Grevera, J. K. Udupa, and D. Odhner. An Order of Magnitude Faster Isosurface Rendering in Software on a PC than Using Dedicated, General Purpose Rendering Hardware. *IEEE Trans. Vis. and Comp. Graph.*, 6(4):335 – 345, 2000.
10. T. Kong and A. Rosenfeld. Digital topology: Introduction and survey. *Computer Vision, Graphics and Image Processing*, 48:357 – 393, 1989.
11. E. LaMar, B. Hamann, and K. Joy. Multiresolution techniques for interactive texture-based volume visualization. *IEEE Proc. Vol. Vis. 1999*, p. 355 – 362.
12. L. Latecki. 3D Well-Composed Pictures. *Graphical Models and Image Processing*, 59(3):164 – 172, 1997.
13. C. Lee, T. Poston, and A. Rosenfeld. Holes and Genus of 2D and 3D Digital Images. *CVGIP: Graphical Models and Image Processing*, 55(1):20 – 47, 1993.
14. W. Lorensen and H. Cline. Marching cubes: A high resolution 3D surfaces construction algorithm. *Computer Graphics*, 21(4):163 – 169, 1987.
15. C. Lürig and T. Ertl. Hierarchical volume analysis and visualization based on morphological operators. In *IEEE Visualization'98*, pages 335 – 341, 1998.
16. M. Meiβner et al. A practical evaluation of popular volume rendering algorithms. In *ACM Proceedings Visualization 2000*, pages 81 – 89, 2000.
17. C. Montani, R. Scateni, and R. Scopigno. Discretized marching cubes. In *Proceedings of visualization*, pages 281 – 287, 1994.
18. H. Muller and M. Stark. Adaptive generation of surfaces in volume data. *Visual computer*, 9:182 – 199, 1995.
19. J. Oikarinen, L. Jyrkinen, and R. Hietala. Volume rendering using seed filling acceleration: Supporting cut planes by fast re-seeding. *Computer-aided Surgery*, 4(4), 1999.
20. S. Parker et al. Interactive ray tracing for volume visualization. *IEEE Transactions on Visualization and Computer Graphics*, 5(3):238 – 250, 1999.
21. J. Rodríguez and D. Ayala. Erosion and dilation on 2D and 3D digital images: A new size-independent approach. *VMV'2001*, pages 143 – 150. Infix, 2001.
22. A. Rosenfeld, T. Kong, and A. Wu. Digital surfaces. *CVGIP: Graph. Mod. and Image Proc.*, 53(4):305 – 312, 1991.
23. A. Rosenfeld, T. Y. Kong, and A. Nakamura. Topology-preserving deformations of two valued digital pictures. *Graph. Mod. Image Proc.*, 60(1):24 – 34, 1998.
24. R. Shekhar et al. Octree-based decimation of marching cubes. *IEEE Comp. Graph. and Appl.*, pages 335 – 342, 1996.
25. L. Thurfjell, E. Bengtsson, and B. Nordin. A boundary approach to fast neighborhood operations on three-dimensional binary data. *CVGIP: Graphical Models and Image Processing*, 57(1):13 – 19, 1995.
26. J. Udupa and O. Odhner. Fast visualization, manipulation and analysis of binary volumetric objects. *IEEE Comp. Graph. and Appl.*, 4(1):53 – 62, 1991.
27. J. Wilhelms and A. Van Gelder. Octrees for faster isosurface generation. *ACM Transactions on Graphics*, 11(3):201 – 227, 1992.
28. K. J. Zuiderveld. *Visualization of multimodality medical volume data using object oriented methods*. PhD thesis, U. Utrech. Deutch, 1995.

Topology Simplification of Symmetric, Second-Order 2D Tensor Fields

Xavier Tricoche and Gerik Scheuermann

Computer Science Department, University of Kaiserlautern, P.O. Box 3049, D-67653 Kaiserslautern, Germany, {tricoche, scheuer}@informatik.uni-kl.de

Summary. Numerical simulations of tubulent flows produce both vector and tensor fields that exhibit complex structural behavior. The topological study of these datasets dramatically reduces the amount of information required for analysis. However, the presence of many features of small scale creates a cluttered depiction that confuses interpretation. In this paper, we extend previous work dealing with vector fields to symmetric, second-order tensor fields. A simplification method is presented that removes degenerate points from the topology pairwise, driven by arbitrary criteria measuring their importance in the overall structure. It is based on an important property of piecewise linear tensor fields that we prove in the paper. Grid and interpolation scheme are preserved since the method uses small local changes of the given discrete tensor values to achieve simplification. The resulting topology is clarified significantly though structurally consistent with the original one. The basic idea behind this technique leads back to the theory of bifurcations and suggests and interpretation as a continuous simplification process.

1 Introduction

Tensors are essential mathematical objects involved in the description of a wide range of scientific and technical fields. They are used for instance in fluid flow, fluid mechanics, civil engineering and medical imaging. Consequently, scientists and engineers need methods to extract essential information from very large tensor datasets that are provided by modern numerical simulations. This explains the increasing interest in tensor field visualization during the last decade. The first topology-based visualization of symmetric, second-order, planar tensor fields was presented by Delmarcelle [2]. Basically, one focuses on one of the two eigenvector fields corresponding to the minor or major eigenvalue. This permits the computation of so-called tensor lines that extend the traditional notion of stream line. The foundations of this technique have been laid down by the work of Helman and Hesselink on vector fields [6]. The theoretical background is provided by the qualitative theory of dynamical systems [1] and differential geometry [9]. The visualization results in a graph representation, where the edges are special tensor lines called separatrices and the nodes are singularities (called degenerate points) of the tensor field, i.e. locations where both eigenvalues are equal. This technique proved suitable for tensor fields with simple structure because the extracted topology contains few degenerate points and separatrices, leading to a clear structure description. Nevertheless, turbulent flows provided by Computational Fluid Dynamics (CFD) simulations or experimental measurements create cluttered depictions that are of little help for interpretation. Indeed, the topology of such flows is characterized by the presence of a large number of features of very small scale that greatly complicate the global picture of the data. This shortcoming induces a need for simplification methods that prune insignificant features, driven by qualitative and quantitative criteria specific to the considered application. Several techniques have been presented in the past for the simplification of vector fields [5, 8]. The issue of vector field topology simplification was first addressed by de Leeuw and van Liere [7] who proposed a method to prune critical points from the topological graph. However, their approach provides no vector field consistent with the topology after simplification. In previous work [10, 11], we presented a scheme that merges close singular points, resulting in a higher-order singularity that synthesizes the structural impact of several features of small scale in the large. This reduces the number of singularities along with the global complexity of the graph. Nevertheless, this technique has several limitations. First, it implies local grid deformations to simulate the singularities' merging, combined with local modifications of the interpolation scheme. Second, it is unable to remove singularities completely from a given region since a higher-order singularity is always introduced afterward. This is a problem if the goal is to filter out insignificant local features in a given region. Finally, the simplification can only be driven by geometric criteria (the relative distance of neighboring singularities) which prevents to take any additional qualitative aspect into account. The present method extends previous work on vector fields [12] and has been designed to overcome these drawbacks. The basic principle consists in successively removing pairs of degenerate points while preserving the consistency of the field structure. Each of these removals can be interpreted as a forced local deformation that brings a part of the topology to a simpler, equivalent structure. The mathematical background is provided by the theory of bifurcations, originally developed within the qualitative analysis of dynamical systems (see e.g. [4], an application to the tensor case is described in [13]). Practically, the method starts with a planar piecewise linear triangulation. We first compute the topological graph and determine pairs of degenerate points. We retain those that satisfy both a proximity threshold and some relevance criteria specified by the user. The pairs are then sorted with respect to their distance and processed sequentially. For each of them, we determine a cell pad enclosing both degenerate points and slightly modify the tensor values such that both degenerate points disappear. This deformation is controlled by angular constraints on the new eigenvector values while keeping constant those located on the pad boundary. After the processing of all pairs, we redraw the simplified topology.

The paper is structured as follows. We review basic notions of tensor field topology and briefly present the notion of bifurcation in section 2. The special case of piecewise linear tensor fields is considered from the topological viewpoint in section 3. In particular, an angular property of eigenvectors is proven in this context that plays a key role in the following. In section 4, we show how we determine pairs of degenerate points to be removed and sort them in a priority list. Section 5 presents the technique used to locally deform the tensor field in order to remove both singularities of a given pair. Results for a CFD dataset are shown in section 6.

2 Topology of Tensor Fields

The present method deals with a planar triangulation of vertices associated with 2D symmetric second-order tensor values, i.e. symmetric matrices. The interpolation scheme is piecewise linear and provides a matrix valued function defined over the domain. Therefore, we only consider topological features of first order. In this case, topology is defined as the graph built up of all first-order degenerate points and some particular tensor lines connecting them, called separatrices. The required definitions are given next.

2.1 Tensor Lines and Degenerate Points

A real two-dimensional symmetric matrix M has always two (possibly equal) real eigenvalues $\lambda_1 \leq \lambda_2$ with associated orthogonal eigenvectors $\mathbf{e_1}$ and $\mathbf{e_2}$:

$$\forall i \in \{1,2\}, M\mathbf{e_i} = \lambda_i \mathbf{e_i}, \text{ with } \mathbf{e_i} \in tricoche R^2 \text{ and } \mathbf{e_i} \neq \mathbf{0}.$$

Since the multiplication of an eigenvector by any non-zero scalar yields an additional eigenvector, eigenvectors should be considered without norm nor orientation which distinguishes them fundamentally from classical vectors. Moreover, the computation of the eigenvectors of M is not affected by its isotropic part defined as

$$\frac{1}{\operatorname{tr} M} I_2,$$

where tr M is the trace of M (i.e. the sum of its diagonal coefficients) and I_2 stands for the identity matrix in $tricoche R^2 \times tricoche R^2$. Consequently, we restrict our considerations to the so-called *deviator* that corresponds to the trace-free part of M. The matrix valued function that we processed is thus of the form:

$$T : (x,y) \in U \subset tricoche R^2 \mapsto T(x,y) = \begin{pmatrix} \alpha(x,y) & \beta(x,y) \\ \beta(x,y) & -\alpha(x,y) \end{pmatrix}, \quad (1)$$

where α and β are two scalar functions defined over the considered two-dimensional domain. One defines a major (resp. minor) *eigenvector field* at each position of the domain as the eigenvector related to the major (resp. minor) eigenvalue of the tensor field. For visualization purposes, one restricts the analysis to a single eigenvector field (either minor or major), using the orthogonality of the eigenvectors to extrapolate the topological structure of the other. In an eigenvector field, one defines *tensor lines* as curves everywhere tangent to the eigenvectors. It follows from this definition that these curves have no inherent orientation as opposed to stream lines. Moreover tensor lines cannot be computed at locations where both eigenvalues are equal since every non-zero vector is an eigenvector in this case. At a degenerate point, the deviator value is a zero matrix. In linear tensor fields, these singularities exist in two possible types: *Trisector* or *wedge point* (see Fig. 1). Due to ori-

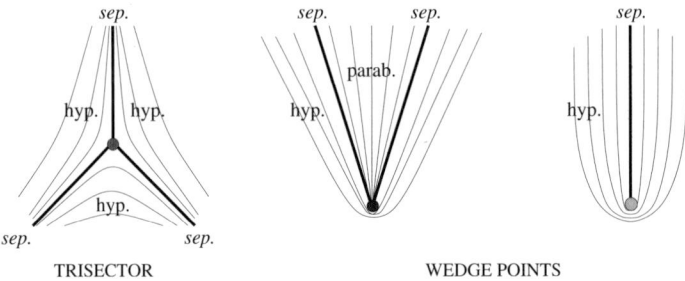

Fig. 1. First Order Degenerate Points

entation indeterminacy of tensor lines, these singularities exhibit structures that would be impossible in the oriented, vector case (consider for example the flow on each side of the single line converging toward the singularity in the second type of wedge points). Remark that more general singularities can be encountered in the piecewise linear case as already mentioned in [11].

In the neighborhood of a degenerate point, the regions where tensor lines pass the singularity by in both directions are called hyperbolic. The regions where they reach the singularity, on the contrary, are called parabolic. The curves that converge toward a degenerate point and bound a hyperbolic region are called separatrices. These special tensor lines constitute the edges of the topological graph. According to this definition, a trisector has three hyperbolic sectors and three associated separatrices while a wedge point has one hyperbolic sector and either one or two separatrices. In the latter case the separatrices bound a parabolic sector. Refer to Fig. 1.

2.2 Tensor Index

A major notion for the structural classification of an tensor field is the so-called tensor index. It is computed along a closed non self-intersecting curve as the number of rotations of the eigenvectors when traveling once along the curve in counterclockwise direction. An illustration is shown in Fig. 2. This extends to tensor fields the essential notion of Poincaré index defined for vector fields. Because of the lack of orientation of eigenvectors, the tensor index is a multiple of $\frac{1}{2}$. The index of a region that contains no degenerate point is zero. If the considered region contains a first-order degenerate point we get an index $-\frac{1}{2}$ for a trisector point while a wedge point has index $+\frac{1}{2}$. The index of a region containing several degenerate points is the sum of

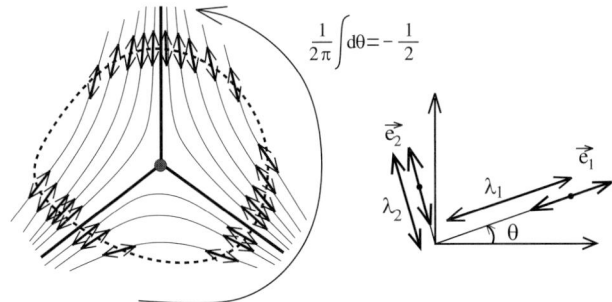

Fig. 2. Tensor index

their individual indices. Remark that in the linear case, since only trisectors and wedges can be encountered, if the index of a closed curve is zero then the enclosed region contains no degenerate point. This property will prove essential in the following.

2.3 Bifurcations

The definitions introduced previously apply to an instantaneous topological state of a tensor field. Now, this stable state may evolve into another one by slight changes of underlying parameters. A typical example is provided by time-dependent tensor fields, the degenerate points of which may move, appear or vanish over time, leading to topological changes. These changes preserve structural consistency and the tensor index acts as a topological invariant. If a topological transition only affects a small region of the field, it is called a *local bifurcation*. If, on the contrary, it leads to a global structural change, it is called a *global bifurcation*. For our purpose we only need to consider a particular kind of local bifurcation: It consists of the pairwise annihilation of a wedge and a trisector point. Since these singularities have global index 0, they are equivalent to a configuration without degenerate point and therefore disappear right after merging. This transition is illustrated in Fig. 3. Additional information on the topic of tensor bifurcations can be found in [13].

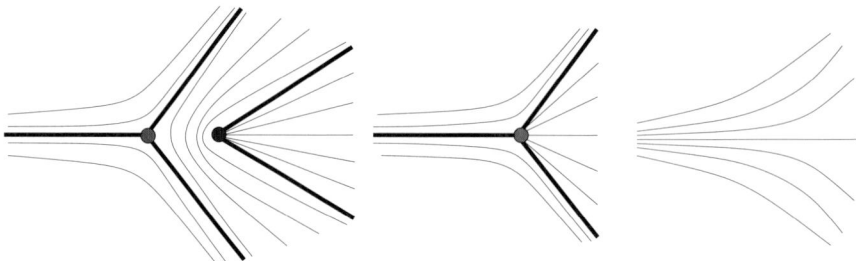

Fig. 3. Pairwise annihilation

Practically, since we want to reduce the number of degenerate points and associated separatrices while being consistent with the original topology, we locally force pairwise annihilations of a wedge and a trisector. This can be done by small local changes in the field values as we show in the following.

3 Linear Tensor Fields

For the simplification method to come, we first need to consider an important property of linear tensor fields from the topological viewpoint.

As discussed previously, we consider deviator tensor fields written in the form of equation 1, where α and β are linear functions of the position (x, y). The eigenvector $\mathbf{e}_\theta = (\cos\theta, \sin\theta)$ identified by its angular coordinate θ satisfies the relation

$$T\mathbf{e}_\theta \times \mathbf{e}_\theta = 0,$$

where $\times$ stands for cross-product. This leads after calculus to

$$\alpha \sin 2\theta - \beta \cos 2\theta = 0,$$

that is

$$\tan 2\theta = \frac{\beta}{\alpha}.$$

Thus, we get the following differential equation

$$d\theta = \frac{1}{2} \frac{\alpha d\beta - \beta d\alpha}{\alpha^2 + \beta^2}. \qquad (2)$$

If we now consider an arbitrary linear interpolated edge $[AB]$ with parametrization $t \in [0, 1]$, we can consider the restriction of T to this edge. We write $\alpha(t) = \alpha_0 + t\alpha_1$ and $\beta(t) = \beta_0 + t\beta_1$. We now compute the angle variation of an eigenvector along $[AB]$ by integrating Equation 2 (remark that this angle variation is the same for both eigenvectors since they are everywhere orthogonal to another):

$$\mathbb{N}_A^B d\theta = \frac{\alpha_0 \beta_1 - \alpha_1 \beta_0}{2} \mathbb{N}_0^1 \frac{dt}{at^2 + bt + c}$$

where a, b and c are functions of $\alpha_{0,1}$ and $\beta_{0,1}$. Furthermore, the discriminant $\Delta = b^2 - 4ac$ is negative. Therefore it follows (after calculus)

$$\mathbb{N}_A^B d\theta = \frac{\text{sign}(\alpha_0 \beta_1 - \alpha_1 \beta_0)}{2} (\text{atan}\mu_1 - \text{atan}\mu_0)$$

where μ_0 and μ_1 are two real scalars that depend on $\alpha_{0,1}$ and $\beta_{0,1}$. Since the function atan maps $tricocheR$ onto the open set $(-\frac{\pi}{2}, \frac{\pi}{2})$, we finally obtain

$$\left| \mathbb{N}_A^B d\theta \right| < \frac{\pi}{2}. \qquad (3)$$

Thus *the angle variation of an eigenvector along a linear interpolated edge is always smaller than $\frac{\pi}{2}$*.

We use this property now to compute the index of a linear tensor field along the edges of a triangle. Since the field is linear, it is determined by the three tensor values at the vertices of the triangle. We denote by θ_0, θ_1, θ_2 the corresponding angle coordinates of one of both eigenvector fields at these positions, enumerated in counterclockwise order. Because eigenvectors have neither norm nor orientation these angle values are defined modulo π (denoted $[\pi]$ in the following). We set by convention $\theta_3 := \theta_0$, so we have

$$\text{index} = \Sigma_{i=0}^{3} \Delta(\theta_i, \theta_{i+1}). \qquad (4)$$

By Equation 3 and using the notation $\delta_i = \theta_{i+1}[\pi] - \theta_i[\pi]$, it comes

$$\Delta(\theta_i, \theta_{i+1}) = \begin{array}{ll} \delta_i & \text{if } |\delta_i| < \frac{\pi}{2} \\ \delta_i + \pi & \text{if } \delta_i < -\frac{\pi}{2} \\ \delta_i - \pi & \text{if } \delta_i > \frac{\pi}{2}. \end{array}$$

4 Selective Pairing of Degenerate Points

As mentioned before, we aim at annihilating pairs of degenerate points of opposite indices. Moreover, the corresponding topology simplification must take geometric and any additional criteria into account to fit the considered interpretation of the tensor field. Our geometric criterion is the proximity of the singularities to be removed pairwise. This choice is motivated by two major reasons. First, close singularities result in small features that clutter the global topology depiction since they can hardly be differentiated and induce many separatrices. Second, piecewise linear interpolation is likely to produce topological artifacts consisting of numerous close first-order singularities, especially if numerical noise is an issue. Therefore, based on a proximity threshold, we determine all possible pairs of wedges and trisectors satisfying the geometric criterion and sort them in increasing distance. Additional criteria may be provided to restrict the range of the considered singularities to those that are little relevant for interpretation. Practically, a quantity is provided that characterizes the relevance of each degenerate point and one retains for simplification only those with a value under a user-prescribed threshold. Thus, if a given singularity is considered important for interpretation, it will be included in no pair and therefore will not be removed from the topology. Remark that compared to the pairing strategy used in previous work for the vector case [12], the connection of both singularities in a pair through a separatrix is not used as criterion. This is because every degenerate point exhibits at least one hyperbolic region which entails that separatrices emanating from a singularity often do not reach any other one.

5 Local Topology Simplification

Once a pair of degenerate points has been identified that fulfills our criteria, it must be removed. To do this, we start a local deformation of the tensor field in a small area around the considered singular points. Practically, we only modify tensor values at the vertices of the triangulation and do not modify the interpolation scheme which obviously ensures continuity over the grid after modification. In the following, we detail first how vertices to be modified are determined and then how new values are set at those vertices to ensure the absence of remaining singularities in their incident cells after processing.

5.1 Cell-wise Connection

The method used here is the same as the one in [12] since the task is the same as in the vector case: Determine well-shaped cell groups that link two singularities over the grid. Consider the situation shown in Fig. 4. We first compute the intersections of the straight line connecting the first degenerate point to the second with the edges of the triangulation. For each intersection point, we insert the grid vertex closest to the second degenerate point (see vertices surrounded by a circle) in a temporary list. After this, we compute the bounding box of all vertices in the list and include all grid vertices contained in this box. Thus, every vertex marked in the former step is included. The use of a bounding box is intended to ensure a well shaped deformation domain, especially useful if many cells separate both singular points. This configuration occurs if the distance threshold has been assigned a large value to obtain a high simplification rate. The vertices concerned with modification are called *internal vertices* and are shown surrounded by squares. Since the

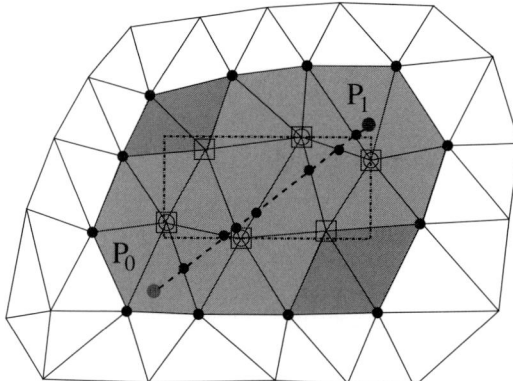

Fig. 4. Cell-wise connection

modification of a vertex tensor value has an incidence on the indices of all triangle cells it belongs to, we include every cell incident to one of the selected vertices in a cell group. These cells are colored in gray. Further processing will have to associate the internal vertices with tensor values that ensure the absence of any singular point in the cell group with respect to the tensor values defined at the *boundary vertices* (marked by black dots in Fig. 4) that will not be changed. The connection may fail if one of the included cells contains a degenerate point that does not belong to the current pair: In this case, the global index of the cell group is no longer zero. If it occurs, we interrupt the processing of this pair. Nevertheless, such cases can be mostly avoided since we simplify pairs of increasing distance.

5.2 Angular Constraints

The basic principle of our simplification technique can be better understood when considering a single internal vertex together with its incident triangles, see Fig. 5. Suppose that every position marked black is associated with a constant tensor value and that the global index of the triangle stencil is zero. The problem consists in determining a new tensor value at the internal vertex (marked white) such that no incident cell contains a degenerate point. This is equivalent to a situation where every incident triangle has index 0 according to what preceedes.

Now, in each triangle the angle coordinates of the eigenvectors defined at the black vertices (say θ_0 and θ_1) induce an angular constraint for the new eigenvector: in equation 3, $\Delta(\theta_0, \theta_1)$ is already set to a value that is strictly smaller than $\frac{\pi}{2}$. The two missing terms must induce a global angle change strictly smaller than π (for the index of a linear degenerate point is a multiple of $\frac{1}{2}$). This condition holds if and only if the new eigenvector value has angle coordinate in $(\theta_1 + \frac{\pi}{2}, \theta_0 + \frac{\pi}{2})$ (modulo π), with $[\theta_0, \theta_1]$ being an interval with

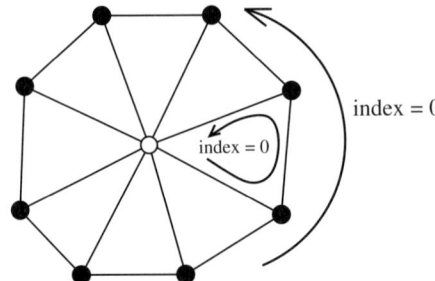

Fig. 5. Configuration with single internal vertex and incident cells

width smaller than $\frac{\pi}{2}$, i.e. the actual angle change along a linear edge from θ_0 to θ_1 (see Fig. 6).

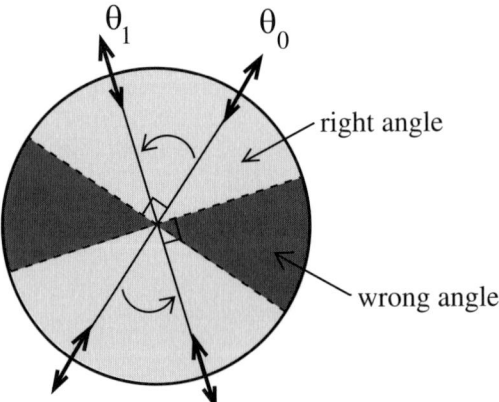

Fig. 6. Angular constraint in a triangle cell

This provides a constraint on the new value for a single triangle. Intersecting the intervals imposed by all incident triangles, one is eventually able to determine an interval that fulfills all the constraints. Note that this interval may be empty. In this case, the simplification is (at least temporarily) impossible. Once a satisfactory angle interval has been found for the new eigenvector, we must provide the vertex with a corresponding tensor value. If θ is an angle in the interval then the following tensor value will be solution:

$$T_{new} = \begin{pmatrix} \cos 2\theta & \sin 2\theta \\ \sin 2\theta & -\cos 2\theta \end{pmatrix}.$$

5.3 Iterative Solution

For each internal vertex (see Fig. 4) we must now find a new tensor value that fulfills all the angle constraints induced by the edges connecting the incident vertices. These incident vertices are of two types: internal or boundary vertices. Edges linking boundary vertices are considered constant and induce fixed angular constraints. Internal vertices still must be provided a final tensor value and introduce flexibility in the simplification scheme. Practically, the problem to solve can be seen as an optimization problem. The quantity to minimize for each internal vertex is the distance of its current angle value to the interval of admissible angles induced by its neighbors. This distance is considered zero if the angle lies within the interval. Initially, the angle values of the internal vertices are undefined. During a first iteration, boundary vertices create angular constraints on the adjacent internal vertices. These constraints are then propagated iteratively to their neighbors in the next steps. If the current angle values of the surrounding vertices correspond to an empty interval, their mean value is used as predictor for the next iteration. Consequently, the whole processing can be interpreted as a local constrained smoothing of the tensor field.

The pseudo-code is as follows.

```
// initialization
for each (internal vertex)
   interval = fixed constraints
   if (interval is empty)
      exit
   end if
   if (no fixed constraints)
      interval = [0, PI[
   end if
end for each

// iterations
nb_iterations = 0
repeat
   succeeded = true
   nb_iterations++
   for each internal vertex
      compute mean_angle of processed incident vertices
      if (interval not empty)
         if (mean_angle in interval)
            current_angle = mean_angle
         else
            current_angle =
               best approximation of mean_angle in interval
         end if
      else
```

```
          succeeded = false
          if (mean_angle in fixed
              constraints)
            current_angle = mean_angle
          else
            current_angle =
              best approximation of mean_angle in interval
        end if
    end for each
until (succeeded or
       nb_iterations > MAX_NB_ITERATIONS)
```

If one of the internal vertices has incompatible fixed constraints, our scheme will fail. Therefore, we interrupt the process during initialization and move to the next pair. If the iterative process failed at determining compatible angular constraints for all internal vertices, we maintain the current pair and move to the next as well.

6 Results

The dataset used to test our method stems from a CFD simulation. This is the symmetric part of the rate of deformation (i.e. first-order derivative) tensor field of a vortex breakdown simulation that was provided by Wolfgang Kollmann from UC Davis. Vortex breakdown is a phenomenon observed in a variety of flows ranging from tornadoes to wing tip vortices, pipe flows, and swirling jets. The latter flows are important to combustion applications where they are able to create recirculation zones with sufficient residence time for the reactions to approach completion. This is a typical case of turbulent global structural behavior. The topology exhibits 67 singularities and 140 separatrices as shown in Fig. 7. The rectilinear grid has 123 x 100 cells. Each rectangular cell is split to result in a triangulation containing about 25000 cells. To simplify this topology we only consider the euclidean distance between degenerate points as a criterion. Remember however that the method does not impose any restriction on the choice of additional qualitative or quantitative criteria characterizing the importance of a singularity or of a given region of the graph. The first simplified topology is obtained with a tiny distance threshold corresponding to 0.2% of the grid diagonal. Every pair consisting of degeneracies that could not be graphically differentiated has disappeared. There are 59 remaining singularities. The modified areas are indicated by rectangular boxes. See Fig. 8.

Increasing the threshold up to a value of 2% of the grid diagonal, one obtains a topology with 35 remaining singularities as shown in Fig. 9. A noticeably clarified graph can be obtained in this case while global strutural properties of tensor field have been preserved. The highest simplification rate is obtained with a threshold of 5% of the grid diagonal. The corresponding

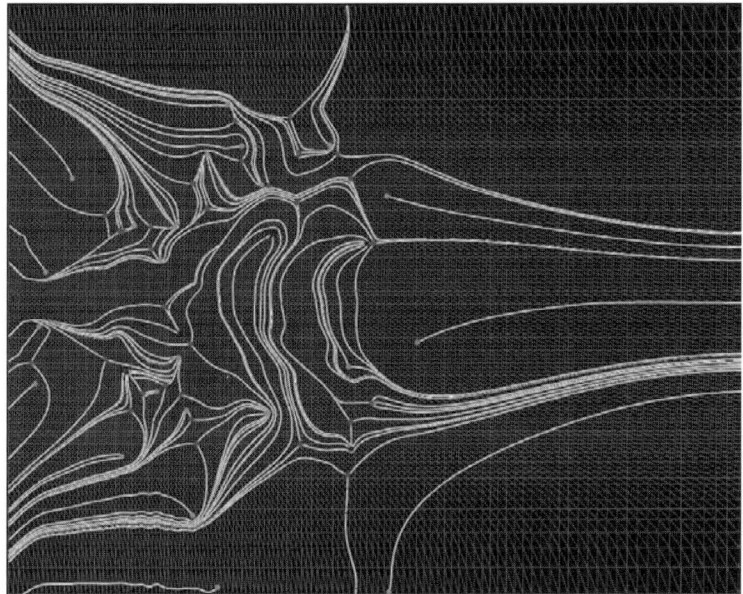

Fig. 7. Initial topology with grid

topology is shown in Fig. 10. The fact that this topology cannot be simplified further (even with a very large geometrical threshold) is explained by the presence of incompatible fixed angle constraints on the boundaries of the cell pads containing the remaining pairs. The local deformation corresponding to the simplified topologies shown so far is illustrated in Fig. 11. The topology is displayed together with the underlying cell structure and the eigenvectors.

7 Conclusion

We have presented a method that simplifies the topology of turbulent planar, symmetric, second-order tensor fields while preserving structural consistency with the original data. The simplification is achieved by means of successive local deformations of the field that entail the pruning of pairs of degenerate points of opposite indices. The pairing strategy can take geometrical as well as any additional criteria into account to fit the domain of application. The theoretical background of this technique is provided by the notion of bifurcation since the disappearance of a pair of singularities corresponds to the pairwise annihilation of a wedge point and a trisector. The method has been tested on a CFD simulation of a vortex breakdown because this kind of datasets exhibit many complex features that clutter the global depiction. The

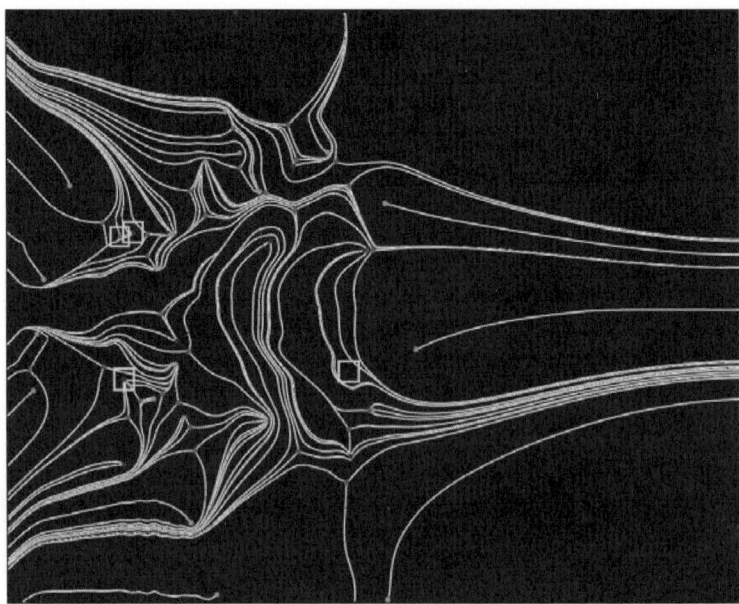

Fig. 8. Simplified topology: distance threshold = 0.2%

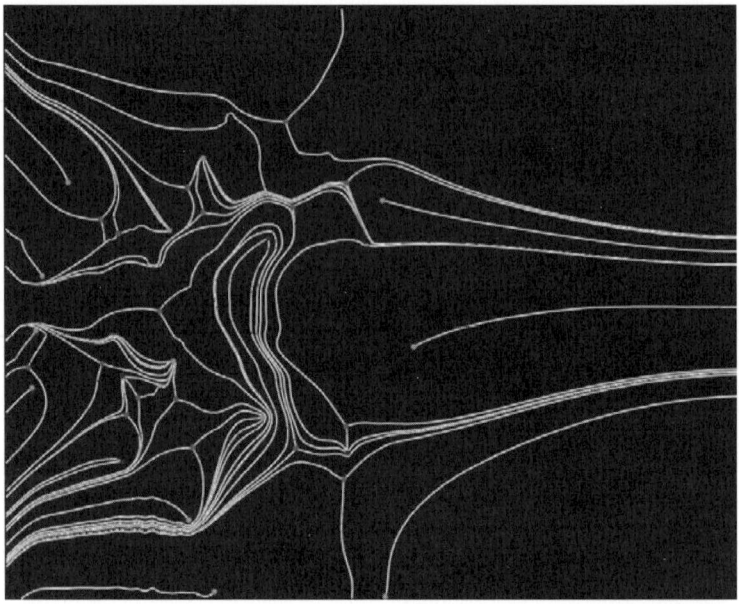

Fig. 9. Simplified topology: distance threshold = 2%

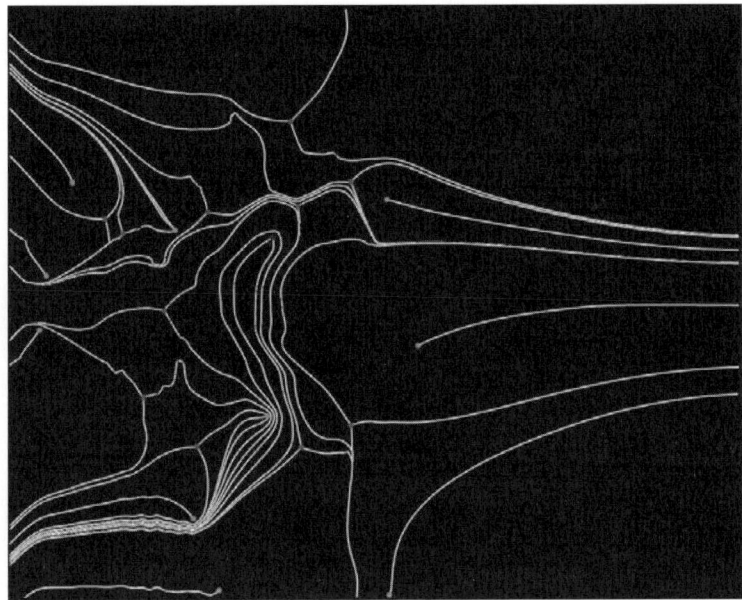

Fig. 10. Simplified topology: distance threshold = 5%

results demonstrate the ability of the method to remove structural features of small scale while letting the rest of the topology unchanged. This clarifies noticeably the depiction and eases interpretation.

Acknowledgment

The authors wish to thank Wolfgang Kollmann, MAE Department of UC Davis, for providing the vortex breakdown tensor dataset. Furthermore, we would like to thank Tom Bobach, David Gruys, Max Langbein and Martin Öhler for their programming efforts.

References

1. Andronov, A. A., Leontovich, E. A., Gordon, I. I., Maier, A. G., *Qualitative Theory of Second-Order Dynamic Systems*. Israel Program for Scientific Translations, Jerusalem, 1973.
2. Delmarcelle, T., Hesselink, L., *The Topology of Symmetric, Second-Order Tensor Fields*. IEEE Visualization '94 Proceedings, IEEE Computer Society Press, Los Alamitos, 1994, pp. 140-147.
3. Delmarcelle, T., *The Visualization of Second-Order Tensor Fields*. PhD Thesis, Stanford University, 1994.

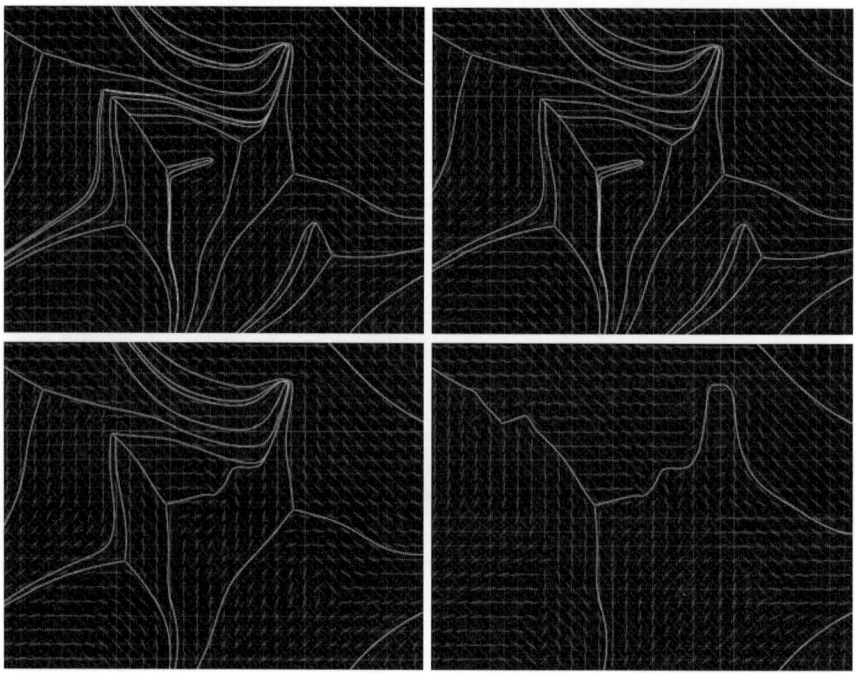

Fig. 11. Local topology simplification: initial graph and simplifications with 0.2%, 2% and 5% as thresholds

4. Guckenheimer, J., Holmes, P., *Nonlinear Oscillations, Dynamical Systems and Linear Algebra*. Springer, New York, 1983.
5. Heckel, B., Weber, G., Hamann, B., Joy, K. I., *Construction of Vector Field Hierarchies*. IEEE Visualization '99 Proceedings, IEEE Computer Society Press,
6. Helman, J. L., Hesselink, L., *Visualizing Vector Field Topology in Fluid Flows*. IEEE Computer Graphics and Applications, 1991. pp.36-46. Los Alimitos, 1999, pp. 19-25.
7. W. de Leeuw, R. van Liere, *Collapsing Flow Topology Using Area Metrics*. IEEE Visualization '99 Proceedings, IEEE Computer Society Press, Los Alamitos, 1999, pp. 349-354.
8. Nielson, G. M., Jung, I.-H., Sung, J., *Wavelets over Curvilinear Grids*. IEEE Visualization '98 Proceedings, IEEE Computer Society Press, Los Alamitos, 1998, pp. 313-317.
9. Spivak, M., *A Comprehensive Introduction to Differential Geometry, Vol. 1-5* Publish or Perish Inc., Berkeley CA, 1979.
10. Tricoche, X., Scheuermann, G., Hagen, H., *A Simplification Method for 2D Vector Fields*. IEEE Visualization '00 Proceedings, IEEE Computer Society Press, Los Alamitos, 2000, pp. 359-366.

11. Tricoche, X., Scheuermann, G., Hagen, H., *Vector and Tensor Field Topology Simplification on Irregular Grids.* Proceedings of the Joint Eurographics-IEEE TCVG Symposium on Visualization in Ascona, Switzerland, D. Ebert, J. M. Favre, R. Peikert (eds.), Springer-Verlag, Wien, 2001, pp. 107-116.
12. Tricoche, X., Scheuermann, G., Hagen, H., *Continuous Topology Simplification of 2D Vector Fields.* IEEE Visualization '01 Proceedings, IEEE Computer Society Press, Los Alamitos, 2001.
13. Tricoche, X., *Vector and Tensor Topology Simplification, Tracking, and Visualization.* PhD thesis, Schriftenreihe / Fachbereich Informatik, Universität Kaiserslautern, 3, 2002.

Automating Transfer Function Design Based on Topology Analysis

Gunther H. Weber[1,2] and Gerik Scheuermann[1]

[1] AG Graphische Datenverarbeitung und Computergeometrie, FB Informatik, University of Kaiserslautern, Germany, {weber,scheuer}@informatik.uni-kl.de
[2] Center for Image Processing and Integrated Computing, Dept. of Computer Science, University of California, Davis, U.S.A., weber@cs.ucdavis.edu

Summary. Direct Volume Rendering (DVR) is commonly used to visualize scalar fields. Quality and significance of rendered images depend on the choice of appropriate transfer functions that assigns optical properties (e.g., color and opacity) to scalar values. We present a method that automatically generates a transfer function based on the topological behavior of a scalar field. Given a scalar field defined by piecewise trilinear interpolation over a rectilinear grid, we find a set of critical isovalues for which the topology of an isosurface, i.e., a surface representing all locations where the scalar field assumes a certain value v, changes. We then generate a transfer function that emphasizes on scalar values around those critical isovalues. Images rendered using the resulting transfer function reveal the fundamental topological structure of a scalar data set.

1 Introduction

Direct Volume Rendering visualizes a three-dimensional (3D) scalar field by using a *transfer function* to map scalar values to optical properties (e.g., color and opacity) and rendering the resulting image. This transfer function presents a user with an additional parameter that influences a resulting visualization. However, quality and significance of the resulting visualization hinge on a sensible choice of the transfer function. Transfer functions are commonly determined manually by trial and error which is time-consuming and prone to errors. Several attempts were made to analyze a data set and generate appropriate transfer functions automatically to aid a user in the visualization process. Pfister et al. [11] give an overview over several techniques and compare results with manually chosen transfer functions.

Apart from DVR, isosurfaces are most commonly used to visualize scalar fields $f(x, y, z)$. An isosurface represents all locations in 3D space, where f assumes a given isovalue v, i.e., where $f = v$ holds. By varying the isovalue v, it is possible to visualize the entire scalar field. Like choosing appropriate transfer functions, determining isovalues where "interesting" isosurface behavior occurs is difficult. Weber et al. [12] have considered the topological properties of scalar fields defined by piecewise trilinear interpolation used on rectilinear grids to determine for which isovalues relevant isosurface behavior occurs. All fundamental changes are tracked: Closed surface components emerge or vanish at local minima or maxima, and the *genus* of an isosurface changes, i.e., holes appear/disappear in a surface component, or disjoint surface components merge at saddles. Values and locations where such changes occur are determined and used to aid a user in data exploration.

Instead of using the resulting set of critical isovalues as indicator for which isovalues expressive isosurfaces result, it is possible to use them to construct a transfer function that highlights topological properties of a scalar data set. DVR commonly uses trilinear interpolation within cells. Thus, critical isovalues extracted by Weber et al. [12] are also meaningful in a volume rendering context. We use the resulting list of critical isovalues to design transfer functions based on the work presented by Fujishiro et al. [5, 7]. We generate transfer functions that assign small opacity to all scalar values except those close to critical isovalues. Colors are assigned using an HLS color model and varying the hue component for different scalar values such that it changes more rapidly close to critical isovalues.

2 Related Work

Few authors utilize topological analysis for scalar field visualization. Bajaj et al. [1] determined a *contour spectrum* for data given on tetrahedral meshes. The contour spectrum specifies contour properties like $2D$ contour length, $3D$ contour area and gradient integral as functions of the isovalue and can aid a user in identifying "interesting" isovalues. Bajaj et al. [3] also developed a technique to visualize topology to enhance visualizations of trivariate scalar fields. Their method employs a C^1-continuous interpolation scheme for rectilinear grids, and detects critical points of a scalar field, i.e., points where the gradient of the scalar field vanishes. Subsequently, integral curves (tangent curves) are traced starting from locations close to saddle points. These integral curves are superimposed onto volume-rendered images to convey structural information of the scalar field.

Fujishiro et al. [5] used a *hyper-Reeb graph* for exploration of scalar fields. A Reeb graph encodes topology of a surface. The hyper-Reeb graph encodes changes of topology in an extracted isosurface. For each isovalue that corresponds to an isosurface topology change, a node exists in the hyper-Reeb graph containing a Reeb graph encoding the topology of that isosurface. Fujishiro et al. [5] constructed a hyper-Reeb graph using "focusing with interval volumes," an iterative approach that finds a subset of all critical isovalues , which has been introduced by Fujishiro and Takeshima [6]. The hyper-Reeb graph can be used, for example, for automatic generation of transfer functions. Fujishiro et al. [7] extended this work and used a hyper-Reeb graph for exploration of volume data. In addition to automatic transfer function design, their extended method allows them to generate translucent isosurfaces between critical isovalues. Considering just the images shown in their paper, it seems that their approach does not detect all critical isovalues of a scalar field.

Critical point behavior is also important in the context of data simplification to preserve important features of a data set. Bajaj and Schikore [2] extended previous methods to develop a compression scheme preserving topological features. Their approach detects critical points of a piecewise linear bivariate scalar field $f(x,y)$. "Critical vertices" are those vertices for which the "normal space" of the surrounding triangle platelet contains the vector $(0,0,1)$. Integral curves are computed by tracing edges of triangles along a "ridge" or "channel." Bajaj and Schikore's method incorporates an error measure and can be used for topology-preserving mesh simplification.

Gerstner and Pajarola [8] defined a bisection scheme that enumerates all grid points of a rectilinear grid in a tetrahedral hierarchy. Using piecewise linear interpolation in tetrahedra, critical points can be detected. Data sets are simplified by specifying a traversal scheme that descends only as deep into the tetrahedral hierarchy as necessary to preserve topology within a certain error bound. This method incorporates heuristics that assign importance values to topological features, enabling a controlled topology simplification.

3 Detecting Critical Isovalues

Our goal is to detect *critical isovalues* of a piecewise trilinear scalar field given on a regular rectilinear grid. Gerstner and Pajarola [8] developed criteria for detecting critical points of piecewise linear scalar fields defined on tetrahedral meshes and used them in mesh simplification. We provide a comprehensive analysis of the topological behavior of piecewise trilinear interpolation and develop criteria to detect critical isovalues for these scalar fields. We further develop methods to use these critical isovalues for volume data exploration.

3.1 Definitions

For a C^2-continuous function f, critical points occur where the gradient ∇f assumes a value of zero, i.e., $\nabla f = 0$. The type of a critical point can be determined by the signs of the eigenvalues of the Hermitian of f. Piecewise trilinear interpolation when applied to rectilinear grids, in general, produces only C^0-continuous functions. Therefore, we must define critical points differently.

Gerstner and Pajarola [8] considered piecewise linear interpolation applied to tetrahedral grids, which also leads to C^0-continuous functions. Considering piecewise linear interpolation, critical points can only occur at mesh vertices. Gerstner and Pajarola's method classifies a mesh vertex depending on its relationship with vertices in a local neighborhood. In the context of a refinement scheme, all tetrahedra sharing an edge that is to be collapsed define a "surrounding polyhedron." Vertices of this surrounding polyhedron constitute the considered neighborhood of a vertex. These vertices are marked with a "+" if their associated function values are greater than the value of the classified vertex; or they are marked with a "-" if their associated function values are less than the value of the classified vertex. Equal values are not considered. Edges of the surrounding polyhedron define an edge graph. In this graph, all edges connecting vertices of different polarities are deleted. A vertex is classified according to the number of connected components in the remaining graph. If this number is one, the classified vertex is a maximum or minimum (depending on the sign of the connected component). If it is two, the classified vertex is a regular point. Otherwise, the vertex is a saddle point. Connected components in an edge graph of a surrounding polyhedron correspond to connected components in a neighborhood of a vertex. This observation leads us to the following definition:

Definition 1 (Regular and Critical Points). Let $F : \mathbb{R}^d \to \mathbb{R}$, $d \geq 2$, be a continuous function. A point $x \in \mathbb{R}^d$ is called a (a) regular point, (b) minimum, (c) maximum, (d) saddle, or (e) flat point of F, if for all $\epsilon > 0$ there exists a neighborhood $U \subset U_\epsilon$ with the following properties: If $\dot{\bigcup}_{i=1}^{n_p} P_i$ is a partition of the preimage of $[F(s), +\infty)$ in $U - \{x\}$ into "positive" connected components and $\dot{\bigcup}_{j=j}^{n_n} N_j$ is a partition of the preimage of $(-\infty, F(s))]$ in $U - \{x\}$ into "negative" connected components, then (a) $n_p = n_n = 1$ and $P_1 \neq N_1$, (b) $n_p = 1$ and $n_n = 0$, (c) $n_n = 1$ and $n_p = 0$, (d) $n_p + n_n > 2$, or (e) $n_p = n_n = 1$ and $P_1 = N_1$.

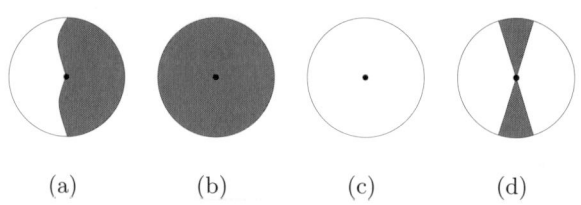

(a) (b) (c) (d)

Fig. 1. (a) Around a regular point $\mathbf{x} \in \mathbb{R}^3$, the isosurface $F^{-1}(F(x))$ divides space into a single connected volume P with $F > 0$ (dark gray) and a single connected volume N with $F < 0$ (white). (b) Around a minimum, all points in U have a larger value than $F(\mathbf{x})$. (c) Around a maximum, every point in U has a smaller value than $F(\mathbf{x})$. (d) In case of a saddle, there are more than one separated regions with values larger or smaller than the value $F(\mathbf{x})$

Remark 1. For (a) – (d), see Fig. 1. Concerning case (e), all points in U have the same value as $F(\mathbf{x})$. It is possible to extend the concept of being critical to entire regions and classify regions rather than specific locations.

Remark 2. The cases $n_p = 2$, $n_n = 0$ and $n_p = 0$, $n_n = 2$ are not possible for $d \geq 2$.

We consider piecewise trilinear interpolation, which reduces to bilinear interpolation on cell faces and to linear interpolation along cell edges. All values that trilinear interpolation assigns to positions in a cell lie between the minimal and maximal function values at the cell's vertices (convex hull property). In fact, maxima and minima can only occur at cell vertices. If two vertices connected by an edge have the same function value, the entire edge can represent an extremum or a saddle. It is even possible that a polyline defined by multiple edges in the grid, or a region consisting of several cells, becomes critical. In these cases, it is no longer possible to determine, locally, whether a function value is a critical isovalue. To avoid these types of problem, we impose the restriction on the data that function values at vertices connected by an edge must differ. Saddles can occur at cell vertices, on cell faces of a cell, and in a cell's interior, but not on cell edges. This fact is due to the restriction that an edge cannot have one constant function value.

Lemma 1 (Regular Edge Points). *All points on edges of a trilinear interpolant with distinct edge-connected values are regular points.*

Proof. By assumption, the two endpoints of the edge have different values. Interpolation along edges is linear, and the derivative differs from zero. The implicit function theorem defines neighborhoods $U_i \times V_i$ and a height function $h_i : U_i \Rightarrow V_i$ in each of the four cubes around the edge, such that the isosurface is a height field in the direction of the edge. Setting U to the smallest interval and determining suitable U_i defines a neighborhood such that the larger and smaller values are above and below a single height field. Therefore, a point on an edge is a regular point, because it is possible to start the construction with an arbitrary small neighborhood around $\mathbf{x}$. □

Thus, in order to detect critical isovalues of a piecewise trilinear scalar field, we only need to detect critical values at vertices of a grid and saddle values within cells and on their boundary faces.

3.2 Critical Values at Vertices

In order to classify a vertex, i.e., to determine whether a vertex is regular or represents an extremum or a saddle, it is sufficient to consider the values at the six edge-connected vertices of a given vertex. We provide a criterion for classification in the following.

Lemma 2 (Local Maximum). *Consider a cell C with vertex numbering as shown in Fig. 2. If $v_0 > \max\{v_1, v_2, v_4\}$, then v_0 is a local maximum in C.*

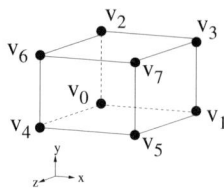

Fig. 2. Vertex numbering

Proof. Choose $m := \max\{v_1 - v_0, v_2 - v_0, v_4 - v_0\} < 0$ and $M := \max\{v_3 - v_0, v_5 - v_0, v_6 - v_0, v_7 - v_0, 1\} \geq 1$. Let $v'_i = v_i - v_0$. If $0 < x, y, z < \epsilon := \frac{|m|}{3|M|}$, then

$$F(x, y, z) - v_0 =$$
$$(1-x)(1-y)(1-z)v'_0 + x(1-y)(1-z)v'_1 +$$
$$(1-x)y(1-z)v'_2 + xy(1-z)v'_3 +$$
$$(1-x)(1-y)zv'_4 + x(1-y)zv'_5 +$$
$$(1-x)yzv'_6 + xyzv'_7$$
$$< x(1-y)(1-z)m + (1-x)y(1-z)M +$$
$$xy(1-z)m + (1-x)(1-y)zM +$$
$$x(1-y)zm + (1-x)yzM + xyzM$$
$$\leq m\epsilon\left[(1-y)(1-z) + (1-x)(1-z) + (1-x)(1-y)\right] +$$
$$M\epsilon^2\left[1 - z + 1 - x + z + 1 - y\right]$$
$$\leq 3m\epsilon(1-\epsilon)^2 + 3M\epsilon^2$$
$$= 3\frac{|m|}{3|M|}\left(\operatorname{sgn}(m)|m|\left(1 - \frac{|m|}{3|M|}\right)^2 + M\frac{|m|}{3|M|}\right)$$
$$= \frac{|m|}{|M|}\left(-|m| + \frac{2}{3}\frac{|m|}{|M|} - \frac{|m|^2}{9|M|^2} + M\frac{|m|}{3|M|}\right)$$
$$= \frac{|m|}{|M|}\left(-\frac{2}{3}|m| + \frac{2}{3}\frac{|m|}{|M|} - \frac{|m|^2}{9|M|^2}\right)$$
$$\leq \frac{|m|}{|M|}\left(-9\frac{|m|^2}{|M|^2}\right) < 0. \qquad \square$$

Lemma 3 (Linear Cell Partition). *Consider a cell C with vertex values v_i and vertex positions $\mathbf{p}_i$ numbered as shown in Fig. 2. If $v := v_0 \neq v_1, v_2 \neq v_4$ holds, then for all $\epsilon > 0$ there exists a $\delta < \epsilon$ such that for the intersection $R = U_\delta(\mathbf{p}_0) \cap C$ the following statements hold: (a) If $v > \max\{v_1, v_2, v_4\}$ then $n_n = 1$ and $N_1 = R$, i.e., all values in the region are less than v. (b) If there exist $i, j, k \in \{1, 2, 4\}$, $i \neq j \neq k$, $i \neq k$, such that $v > \max\{v_i, v_j\}$ and $v < v_k$, then $n_n = n_p = 1$ and R completely contains a surface dividing N_1*

and P_1. Furthermore, all values on the triangle $\mathbf{p_0p_ip_j}$ are less than v. (c) If there exist $i,j,k \in \{1,2,4\}$, $i \neq j \neq k$, $i \neq k$, such that $v < \min\{v_i, v_j\}$ and $v > v_k$, then $n_n = n_p = 1$, and R completely contains a surface dividing N_1 and P_1. Furthermore, all values on the triangle $\mathbf{p_0p_ip_j}$ are less than v. (d) If $v < \max\{v_1, v_2, v_4\}$, then $n_n = 1$ and $N_1 = R$, i.e., all values in the region are greater than v.

Proof. Cases (a) and (d) are symmetrical and follow from Lemma 2. Cases (b) and (c) are symmetrical as well, and it is sufficient to prove one of them. Similarly, the same holds when we choose any other v_i as v and consider its edge-connected neighbor vertices.

Let $\epsilon > 0$. The derivative of F at $\mathbf{p_0}$ is $(v_1-v_0, v_2-v_0, v_4-v_0)$. There exists an $\epsilon > \delta > 0$ such that the derivative has rank 1 in the whole neighborhood $R = U_\delta(\mathbf{p_0}) \cap C$. In this case, the regular value theorem guarantees the existence of an isosurface with function value v_0 dividing $U_\delta(\mathbf{p_0})$ into a single region with larger and a single region with lower function values. If the surface intersects C outside $\mathbf{p_0}$, R is split into exactly two parts. If not, $\mathbf{p_0}$ is a local maximum or minimum. This fact proves the first part of (b) and (c). For small $\epsilon > \delta > 0$, a calculation similar to the proof of Lemma 2 demonstrates that the face with $\mathbf{p_0, p_i, p_j}$ is not intersected inside R by the isosurface in cases (b) and (c). □

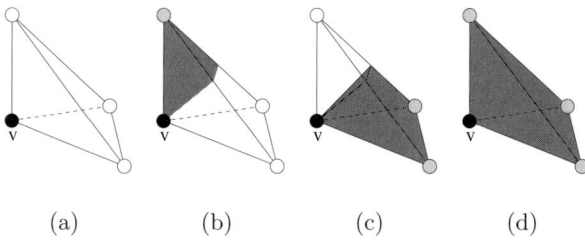

(a) (b) (c) (d)

Fig. 3. When a small neighborhood is considered, a "tetrahedral region" having v as a corner is partitioned in the same way as a linear tetrahedron

Using the L_1-norm[1], the intersection of a neighborhood with a cell corresponds to a tetrahedron. According to Lemma 3, this tetrahedron is partitioned in the same way as a tetrahedron using linear interpolation (even when, as in our case, partitioning surfaces are not necessarily planar), see Fig. 3. A vertex can be classified by considering its edge-connected neighbor vertices. We treat these vertices as part of a local implicit tetrahedrization

---

[1] $\|x\|_1 = \sum_i |x_i|$

surrounding a classified vertex, where the classified vertex and three edge-connected vertices belonging to the same rectilinear cell imply a tetrahedron, see Fig. 4.

Fig. 4. Edge-connected vertices as part of an implicit tetrahedrization

When applying Gerstner and Pajarola's criterion [8] for connected components in an edge graph for the resulting implicit tetrahedrization, we obtain a case table with $2^6 = 64$ entries that maps a configuration of "+" and "-" of edge-connected vertices to a vertex classification. (It can be shown that the connected components in an edge graph correspond to connected components in a neighborhood.) We decided to generate this relatively small case table manually.

3.3 Critical Values on Faces

When linear interpolation is used, critical points can only occur at grid vertices. When piecewise trilinear interpolation is used, critical points can also occur on boundary faces. On a boundary face piecewise trilinear interpolation reduces to bilinear interpolation and the interpolant on a face can have a saddle. This face saddle is not necessarily a saddle of the piecewise trilinear interpolant. The following lemma provides a criterion to whether a face saddle is a saddle of the trilinear interpolant:

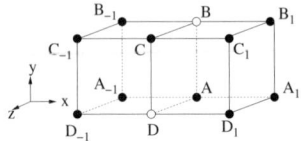

Fig. 5. Vertex numbering scheme used in Lemma 4

Lemma 4 (Face Saddle). *Let* **p** *be a point on the shared face of two cells, where both trilinear interpolants degenerate to the same bilinear interpolant. The point* **p** *is a saddle point when these two statements hold:*

1. *The point* **p** *is a saddle point of the bilinear interpolant defined on the face.*

2. With the notations of Fig. 5, where, without loss of generality, cells are rotated such that A and C are the values on the shared cell face having a value larger than the saddle value, $C(A_1 - A) + A(C_1 - C) - D(B_1 - B) - B(D_1 - D)$ and $C(A_{-1} - A) + A(C_{-1} - C) - D(B_{-1} - B) - B(D_{-1} - D)$ have the same sign.

Otherwise, **p** is a regular point of the trilinear interpolant.

Proof. 1. If **p** is not a saddle of the bilinear interpolant on the face, one partial derivative on the face is different from zero. The regular value theorem implies the existence of a dividing isosurface in both cells in a small neighborhood $U_\delta(\mathbf{p}) \subset U_\epsilon(\mathbf{p})$, leading to a single isosurface in the whole neighborhood that splits into one connected component with values larger than $f(\mathbf{p})$ and one connected component smaller than $f(\mathbf{p})$.
2. Let **p** be a saddle point with respect to the bilinear interpolant on the face. (We adopt an idea from Chernyaev [4].) To simplify notation, we assume that the face is perpendicular to the x–coordinate axis. If we consider any plane x =const, $x \in [0,1]$, parallel to the face the function F becomes $F(y,z) = A_x(1-y)(1-z) + B_x y(1-z) + C_x yz + D_x(1-y)z$ with $A_x = A(1-x) + A_1 x$, $B_x = B(1-x) + B_1 x$, $C_x = C(1-x) + C_1 x$, $D_x = D(1-x) + D_1 x$. As pointed out by Nielson and Hamann [10], the sign of the value at the intersection of the asymptotes $\frac{A_x C_x - B_x D_x}{A_x + C_x - B_x - D_x}$ determines whether the points with value higher than $F(p)$ or lower than $F(p)$ are connected. Since $A_x + C_x - B_x - D_x$ is always positive (by our choice of "cell rotation") for small x, we must consider the sign of $A_x C_x - B_x D_x$. For $x = 0$, this expression is 0 since **p** is a saddle point of the face. Computing the derivative of $A_x C_x - B_x D_x$ with respect to x at **p**, i.e., for $x = 0$, which turns out to be $C(A_1 - A) + A(C_1 - C) - D(B_1 - B) - B(D_1 - D)$, one can determine whether $A_x C_x - B_x D_x$ is positive or negative above **p**. If it is positive, the negative values are connected above **p**. Otherwise, if it is negative, the positive values are connected above **p**. A value of 0 implies bilinear variation in the cube which is not possible, since we have different values along edges. The final criterion results from application of this idea to both cells sharing the face. If the negative or positive values are connected around **p** in both cubes, we have a saddle of the piecewise trilinear interpolant, otherwise we do not have a critical point of the piecewise trilinear interpolant. □

We thus can detect face saddles of piecewise trilinear interpolation effectively by considering all cell faces for a saddle of the bilinear interpolants on faces and checking whether the criterion stated in Lemma 4 holds.

3.4 Critical Values inside a Cell

Saddles of the trilinear interpolant in the interior of a cell are easy to handle as they are always saddles of the piecewise trilinear interpolant as well. Interior saddles are already used by various MC variants to determine isosurface topology within a cell. We compute these saddles by using the equations given by Nielson [9]. Inner saddles of a trilinear interpolant that coincide with a cell's boundary faces or vertices are not necessarily saddles of a piecewise trilinear interpolant. Trilinear interpolation assigns constant values to locations along coordinate-axis-parallel lines passing through the saddle. We currently rule out the possibility of an internal saddle coinciding with a vertex or an edge. Otherwise, our requirement that edge-connected vertices differ in value would be violated. Saddles of trilinear interpolants that coincide with cell faces are also saddles of the bilinear interpolant on the face. As such they are discussed in Sec. 3.3.

4 Automatic Transfer Function Design

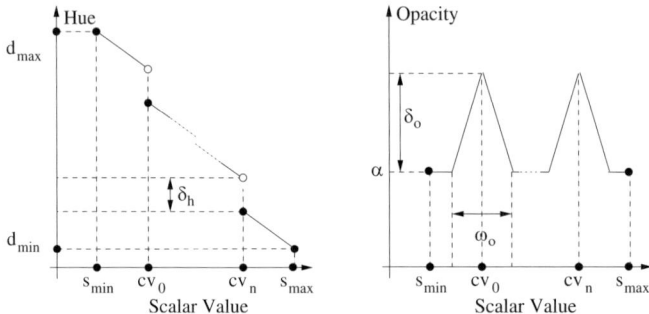

Fig. 6. Transfer function emphasizing topologically equivalent regions

Given a list of critical isovalues we construct a corresponding transfer function based on the methods described by Fujishiro et al. [7]. The domain of the transfer function corresponds to the range of scalar values $[s_{min}, s_{max}]$ occurring in a data set. Outside this range the transfer function is undefined. Given a list of critical isovalues cv_i, we either construct a transfer function emphasizing volumes containing topologically equivalent isosurfaces or a transfer function emphasizing structures close to critical values.

Fig. 6 shows the construction of a transfer function that emphasizes on topologically equivalent regions. The color transfer is chosen such that hue uniformly decreases with the mapped value, except for a constant drop of δ_h at each critical value cv_i. The opacity is constant for all values except for hat-like elevations around each critical value cv_i having a width of ω_o and a height δ_o.

Fig. 7 shows the construction of a transfer function emphasizing details close to critical isovalues. The hue transfer function is constant except for linear descents of a fixed amount δ_h within an interval with a width ω_h

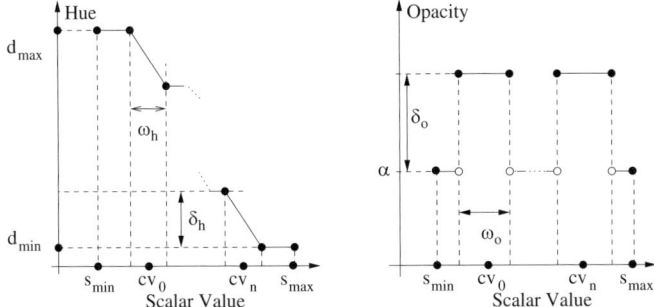

Fig. 7. Transfer function emphasizing details close to critical isovalues

centered around each critical isovalue cv_i. The opacity is constant for all values except in intervals with a width w_o centered around critical isovalues cv_i where the opacity is elevated by δ_o.

When several isovalues are so close together that intervals with a width w_h or w_o would overlap, all isovalues except the first are discarded to avoid high frequencies in the transfer function that could cause aliasing artifacts in the rendered image.

5 Results

Fig. C.16 shows the results of rendering a data set resulting from a simulation of fuel injection into a combustion chamber. (Data set courtesy of SFB 382 of the German Research Council (DFG), see http://www.volvis.org for details.) Fig. 16(a) emphasizes on volumes containing topologically equivalent isosurfaces. Details close to these critical isovalues are more visible in Fig. 16(b).

Fig. C.17 shows the results of rendering a data set resulting from simulating the spatial probability distribution of the electrons in a high potential protein molecule. Fig. 17(a) emphasizes on volumes containing topologically equivalent isosurfaces. Details close to these critical isovalues are better visible in Fig. 17(b).

6 Conclusions and Future Work

We have presented a method for the detection and utilization of critical isovalues for the exploration of trivariate scalar fields defined by piecewise trilinear functions. Improvements to our method are possible. For example, it would be helpful to eliminate the requirement that values at edge-connected vertices of a rectilinear grid must differ. While our approach can be used on data sets that violate this requirement, it fails to detect all critical isovalues for such data. It is necessary to extend our mathematical framework and add the concept of "critical regions" and "polylines." Considering the case of a properly sampled implicitly defined torus, its minimum consists of a closed polyline around which the torus appears. Similar regions of a constant value can exist that are extrema. These extensions will require us to consider values in a larger region; and they cannot be implemented in a purely local approach. Some data sets contain a large number of critical points. Some of these critical points correspond to locations/regions of actual interest, but some are the result of noise or improper sampling. We need to develop methods to eliminate such "false" critical points.

On the other hand it could be useful to consider more noisy data sets and generate a histogram with the number of topology changes for a lot of small isovalues ranges. It should be possible to automatically detect interesting isovalues by looking for values where there are many topological changes. This could be used to detect turbulence in data sets resulting from unsteady flow simulations in which turbulence is usually associated to "topological noise." Histograms could also be used to generate meaningful transfer functions for data sets with a large number of closely spaced critical isovalues.

7 Acknowledgments

We thank the members of the AG Graphische Datenverarbeitung und Computergeometrie at the Department of Computer Science at the University of Kaiserslautern and the Visualization Group at the Center for Image Processing and Integrated Computing (CIPIC) at the University of California, Davis.

References

1. Chandrajit L. Bajaj, Valerio Pascucci, and Daniel R. Schikore. The contour spectrum. In: Roni Yagel and Hans Hagen, editors, *IEEE Visualization '97*, pages 167–173, IEEE, ACM Press, New York, New York, October 19–24 1997.
2. Chandrajit L. Bajaj, Valerio Pascucci, and Daniel R. Schikore. Visualizing scalar topology for structural enhancement. In: David S. Ebert, Holly Rushmeier, and Hans Hagen, editors, *IEEE Visualization '98*, pages 51–58, IEEE, ACM Press, New York, New York, October 18–23 1998.
3. Chandrajit L. Bajaj and Daniel R. Schikore. Topology preserving data simplification with error bounds. *Computers & Graphics*, 22(1):3–12, 1998.
4. Evgeni V. Chernyaev. Marching cubes 33: Construction of topologically correct isosurfaces. Technical Report CN/95-17, CERN, Geneva, Switzerland, 1995. Available as http://wwwinfo.cern.ch/asdoc/psdir/mc.ps.gz.
5. Issei Fujishiro, Taeko Azuma, and Yuriko Takeshima. Automating transfer function design for comprehensible volume rendering based on 3D field topology

analysis. In: David S. Ebert, Markus Gross, and Bernd Hamann, editors, *IEEE Visualization '99*, pages 467–470, IEEE, IEEE Computer Society Press, Los Alamitos, California, October 25–29, 1999.

6. Issei Fujishiro and Yuriko Takeshima. Solid fitting: Field interval analysis for effective volume exploration. In: Hans Hagen, Gregory M. Nielson, and Frits Post, editors, *Scientific Visualization Dagstuhl '97*, pages 65–78, IEEE, IEEE Computer Society Press, Los Alamitos, California, June 1997.

7. Issei Fujishiro, Yuriko Takeshima, Taeko Azuma, and Shigeo Takahashi. Volume data mining using 3D field topology analysis. *IEEE Computer Graphics and Applications*, 20(5):46–51, September/October 2000.

8. Thomas Gerstner and Renato Pajarola. Topology preserving and controlled topology simplifying multiresolution isosurface extraction. In: Thomas Ertl, Bernd Hamann, and Amitabh Varshney, editors, *IEEE Visualization 2000*, pages 259–266, 565, IEEE, IEEE Computer Society Press, Los Alamitos, California, 2000.

9. Gregory M. Nielson. On marching cubes. To appear in IEEE Transactions on Visualization and Computer Graphics, 2003.

10. Gregory M. Nielson and Bernd Hamann. The asymptotic decider: Removing the ambiguity in marching cubes. In: Gregory M. Nielson and Larry J. Rosenblum, editors, *IEEE Visualization '91*, pages 83–91, IEEE, IEEE Computer Society Press, Los Alamitos, California, 1991.

11. Hanspeter Pfister, Bill Lorensen, Chandrajit Bajaj, Gordon Kindlmann, Will Schroeder, Lisa Sobierajski Avila, Ken Martin, Raghu Machiraju, and Jinho Lee. The transfer-function bake-off. *IEEE Computer Graphics and Applications*, 21(3):16–22, May/June 2001.

12. Gunther H. Weber, Gerik Scheuermann, Hans Hagen, and Bernd Hamann. Exploring scalar fields using critical isovalues. In: Robert J. Moorhead, Markus Gross, and Kenneth I. Joy, editors, *IEEE Visualization 2002*, pages 171–178, IEEE, IEEE Computer Society Press, Los Alamitos, California, 2002.

Part V

Multiresolution Data Representation

Simplicial-based Multiresolution Volume Datasets Management: An Overview

Rita Borgo, Paolo Cignoni, and Roberto Scopigno

Istituto di Scienza e Tecnologia dell'Informazione (ISTI), Consiglio Nazionale delle Ricerche, Pisa, Italy - {*borgo, cignoni, scopigno*}*@isti.cnr.it*

Summary. The paper synthetically presents the methodologies proposed for the efficient multiresolution management of large volume datasets. In particular, we review those multiresolution approaches based on simplicial meshes. The methodologies available are classified in two main streams, according to the regular or irregular refinement kernel adopted to build the multiresolution representation. A comparison, highlighting respective strength and weakness of the two classes of methods, is proposed and discussed.

1 Introduction

Many approaches have been proposed to support multiresolution management of volume datasets: naïve sub-sampling, wavelet techniques, hierarchical space subdivisions (e.g. octrees), and simplicial decompositions. The term multiresolution is often used to indicate either discrete or continuous level of detail (LOD) representations. We will cover mostly the second aspect, and therefore we point our attention to those methods that allow to manage selective refinements (or the inverse operation, i.e. selective coarsening) in a dynamic manner, according to run time requirements. Simplicial meshes have been often used in the visualization of volume datasets. The simplicity of the basic cell allows to easily manage isosurface extraction (field is linearly interpolated, no ambiguity) and to implement in an efficient manner direct volume rendering (DVR) solutions [6]. Moreover, tetrahedral-based DVR solution can now be implemented using off-the-shelf graphics hardware, gaining impressive speed-ups with respect to software solutions. Therefore, simplicial decompositions have been often considered in the design of multiresolution methods, not only because they are easy to render but also because they easily adapt to different shapes or to the data field structure/topology. This paper presents an overview and a comparison of the different approaches for simplicial-based multiresolution proposed in the context of volume visualization. We subdivide the existing methods in two main classes, which depend on the refinement kernel used to manage the selective refinement/coarsening: *regular* or *irregular*. Regular techniques starts from a coarse regular base domain and apply recursive regular refinement, resulting in large meshes organized as uniform grid patches. On the other hand, irregular techniques are independent from the topology of the underlying mesh; not being forced to follow a regular subdivision scheme (vertices can be added in any order) irregular techniques result to be more flexible and suitable to resolve complex geometric features and geometry changes.

The paper is organized as follows: Section 2 introduces some basic concepts in multiresolution data management and presents a general representation framework. Then, some representative irregular and regular approaches are presented in a synthetic manner in Sections 3 and 4. A comparative evaluation of the approaches presented is given in Section 5 and concluding remarks in Section 6.

2 Multiresolution Representations for Simplicial Meshes

Current multiresolution models for generic tetrahedral meshes can manage a large set of level of details in a flexible, efficient and compact way. The main idea behind these methods is to exploit, in some way, the information that can be collected during the simplification or the refinement of a volume dataset. The assumption is that we use an iterative simplification or refinement algorithm that progressively reduces/refines the dataset by means of *small* and *local* modifications. This sequence of updates can be organized in structures which encode (either explicitly or implicitly) all the reciprocal temporal dependencies and makes it possible to apply those updates in many different orders. The main goal is to support very fast selective refinement/simplificaction, according to the user dynamic requests on selected portions of the domain.

These data representation approaches are strongly related with the multiresolution and simplification techniques for three-dimensional *surface* meshes [14], that is the domain where those ideas were firstly developed. Many different approaches have been presented to extend the multiresolution data management approach to volume datasets. A general framework which can be used to encompass all of them is the *Multiresolution Simplicial Model* (MSM), introduced by De Floriani, Puppo and Magillo [10] as a multidimensional extension of the 2D 1/2 structure described by Puppo in [24]. We give a brief intuitive description of the MSM scheme in the following subsection and we send back to [10] for a more formal description.

The Multiresolution Simplicial Model (MSM) The main idea of the MSM is that we can infer a multiresolution model by storing as small subsets of tetrahedra Σ_i called *fragments* the sequence of local modifications performed by an iterative simplification algorithm. Obviously these modifications are not independent and the relative dependencies among these fragments can be coded in a partial ordering {Prec: }. In practice this partial ordering represents the fact that a given modification of the mesh cannot be done before another one, usually because they modify a common portion of the mesh and have to be executed in a given order. This partial ordering can be graphically described as a *direct acyclic graph* (DAG), where nodes represent fragments and arcs encode the {Prec: } relation. In Fig. 1 we show an example of a simple two-dimensional MSM encoded as a DAG. Intuitively

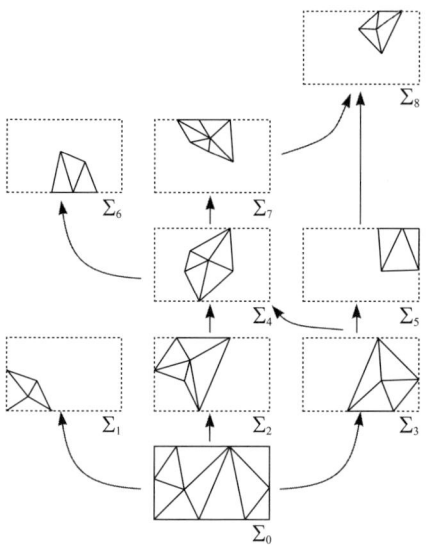

Fig. 1. The DAG describing a simple two-dimensional MSM.

speaking, the fragments describe a portion of the dataset domain at a certain resolution and, by combining them carefully, we can reconstruct many different meshes. For example, let us examine an iterative refinement procedure on a simplicial mesh; the set of simplexes, derived from the substitution of a complex with a more refined one, can be considered as a fragment combined over the existing complex. In this approach the lowest element of the DAG is the coarsest representation of the dataset, and the elements above represent the refinements done onto the mesh (Fig. 1). The combination of fragments can be done through an operator $\oplus$ that replaces the simplexes in the lower fragments with the one in the subsequent fragment, holding the hypothesis that they have some common intersection; the order in which these fragments must be combined is the one induced by the partial ordering. The triangulation resulting from the combination of fragments $\Sigma_0, \Sigma_2, \Sigma_3, \Sigma_4, \Sigma_7$, satisfying a consistent order, is shown in Fig. 4. Note that any other consistent order of combination (e.g. $\Sigma_0, \Sigma_3, \Sigma_2, \Sigma_4, \Sigma_7$) builds the same triangulation. The MSM model is an abstract way of conceiving a generic multiresolution model. In practice, the naive approach to directly map the MSM model into a data structure, representing explicitly the DAG of the set of fragments, does not allow to attain the utmost space and computational efficiency. The importance of the MSM is that it allows to explain and compare all the current *real* multiresolution data structures under a common framework: they can be seen as smart and compact way of coding the elements of a MSM (the DAG and the fragments). For example, multiresolution models based on edge collapses usually represent the MSM fragments in a completely implicit way by storing just the collapse action instead of the explicit list of simplexes modified by the collapse.

In the following two sections we describe the two main classes of approaches: the methods based on *irregular* or *regular* refinement kernels.

3 Irregular Refinement Techniques

Since these multiresolution methods are based on an *irregular* simplification or refinement kernel, we shortly review in the next section the techniques for the simplification or refinement of a tetrahedral complex.

3.1 Simplification of a Simplicial Complex

One of the first approaches for the simplification of tetrahedral datasets was proposed in [4]. It exploits a basic coarse-to-fine refinement strategy, an early technique widely used for approximating natural terrains [9, 13]. An on-line algorithm for Delaunay tetrahedralization is used together with a selection criterion to refine an existing Delaunay mesh by inserting one new vertex at a time. The selection strategy at each iteration is aimed to refine the tetrahedron that causes the maximum error in the current approximation (considering both field value interpolation and mesh warping): the point v_{max} in the initial dataset V holding the maximum error becomes a new vertex of the current simplicial decomposition Σ_i. Adding a point implies to search for the tetrahedron in Σ_i that contains v_{max}, to split it on the new vertex v_{max} (Fig. 2) and to apply a sequence of flipping actions until the resulting mesh satisfies again the Delaunay criterion. This refinement procedure always converges, since the number of points in V is finite. Unfortunately, this approach is limited to datasets whose domain is convex, because the result of a Delaunay tetrahedralization is always convex.

An extension of this approach has been proposed in [6] to deal also with non-convex *curvilinear* datasets: the Delaunay tetrahedralization is computed in the computational domain (i.e. the underlying convex grid), while its image through lifting gives the corresponding mesh in the physical domain. The refinement method described above is difficult to adapt to the case of generic *non-convex irregular datasets*. Major difficulties arise in finding an initial coarse mesh which approximates the original domain Ω of the dataset (Delaunay triangulation is not applicable to non-convex polyhedra) and in the

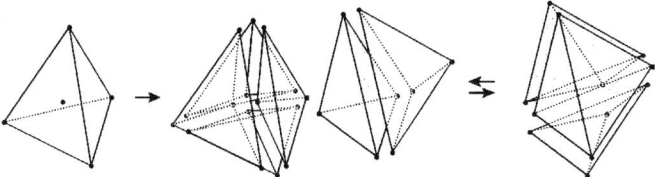

Fig. 2. A tetrahedra *split*, due to the insertion of a new vertex in the mesh, is shown on the left. On the right we show a *flip* action that substitutes two adjacent tetrahedra with three tetrahedra and viceversa.

estimation of the warping of the mesh boundary. Experience in the approximation of non-convex surfaces through 2D triangular meshes suggests that a *simplification* technique might be more appropriate to the case of non-convex irregular volume datasets (see, for example, [15,19]). In the simplification approach we start with an input simplicial mesh decomposition of the dataset V; then, vertices are iteratively discarded as long as the error introduced by removing them does not exceed a given accuracy threshold.

Gross and Staadt [27] present a simplification technique based on collapsing an edge to an arbitrary interior point, and propose various cost functions to drive the collapsing process. Cignoni et al. [6] propose an algorithm based on collapsing an edge to one of its extreme vertices (called half-edge collapse, see Fig. 3), in which the simplification process is driven by a combination of the geometric error introduced in simplifying the shape of the domain and of the error introduced in approximating the scalar field with fewer points. This approach has been extended in [3] by defining a framework for the unified management of the two errors (related to the geometric domain warping and the scalar field interpolation) and by proposing some techniques to forecast or to evaluate efficiently such errors. In fact, selecting a vertex to be removed involves an estimation of how much the error will be increased; the vertex causing the smallest increase is usually selected at each iteration. An exact estimation of error variation can be obtained by simulating the deletion of all vertices in the current mesh. This would be computationally expensive, since, assuming that we are working with a small Euler characteristic and border as in [18] we can state that each vertex has approximately $20 \sim 24$ incident tetrahedra on average. This may involve relocating many points lying inside such tetrahedra. The use of heuristics to estimate apriori how a vertex removal affects error and warping is therefore a wiser choice [3]. Such an estimation is computed for all vertices before decimation starts, and it has to be updated during simplification for all those vertices affected by a change (i.e. when one or more of its incident tetrahedra change).

Trotts et al. [28] perform half-edge collapse as well. They control the quality of the simplified mesh by estimating the deviation of the simplified scalar field from the original one, and by predicting the increase in deviation caused by a

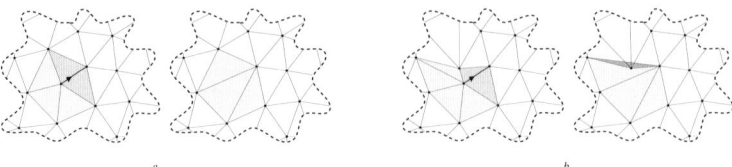

Fig. 3. Half-Edge collapse in 2D: (a) a valid collapse; (b) an inconsistent collapse.

collapse. They also provide a mechanism to bound the deformation of the domain boundary. Another approach for the simplification of generic simplicial complexes, called Progressive Simplicial Complex (PSC), has been proposed by Popovic and Hoppe [23], as an extension of the Progressive Meshes (PM) approach [19]. The PM structure basically encodes the sequence of edge-collapse transformations; the PSC codifies in a similar manner a sequence of more general edge-collapse transformations. It should be noted that while the PSC are quite general, they have been conceived for the management of possibly degenerate 2D surfaces rather than simplicial complexes representing a volume dataset, so the conditions of legality of a sequence of vertex-split operation of a generic complex are not specified, and the problem of evaluating the approximation error introduced in the volume field representation is not considered.

3.2 From Simplification to Multiresolution Models

Each simplification algorithm described in Section 3.1 can be used to build a multiresolution representation. Either a decimation or a refinement algorithm working on a tetrahedral complex produce an "historical" sequence of tetrahedra, namely all tetrahedra that appear in the progressively simplified/refined mesh Σ during its construction. An historical sequence can be also viewed as the sequence of all subdivisions of the whole domain that are obtained through changes, or as an initial subdivision plus a sequence of fragments reflecting the local changes iteratively done to the mesh, which can be partially overlapping and are pasted one above the other to update the existing structure. For example, if we follow the *refinement* heuristic, the initial coarse triangulation is the starting mesh. When we insert a new point v_i in the complex (Fig.2), the new tetrahedra that are built form a new fragment Σ_i; the corresponding fragment replaced by Σ_i is constituted by the tetrahedra that were destroyed by the insertion of v_i. Following the MSM framework, all these fragments (represented by a tetrahedral complex covering a small part of the whole domain Ω) are arranged in a DAG where the order relation between fragments is dependent on their interferences in 3D space. The minimum fragment Σ_0, the coarsest representation of our mesh, has no incoming arcs. Similarly all the triangles on the top of the DAG $\mathcal{S}$, representing the dataset at its full resolution, have no outcoming arcs pointing to new fragments.

A simple data structure to encode a generic MSM was presented in [8]. A much more compact data structure has been proposed for three-dimensional tetrahedral MSM built by a sequence of general edge collapses [5]. This latter structure, customized to the needs of volume visualization, requires three times less storage space with respect to a simple indexed data structure encoding the original mesh at full resolution, and 5.5 times less space than a data structure for the original mesh encoding both connectivity and adjacency information (as required, e.g., by direct volume rendering algorithms based on cell projection [30]). This kind of structure have to maintain, more or less explicitly, the set of all the vertices of $\mathcal{S}$, the set of all tetrahedra $\Sigma_\mathcal{S}$, the set of all fragments in $\mathcal{S}$, the {Prec: } relation among them and for each tetrahedron, the reference to the vertices forming it and the accuracy provided by this cell when we interpolate any point in its interior from the field values hold by the cell vertices.

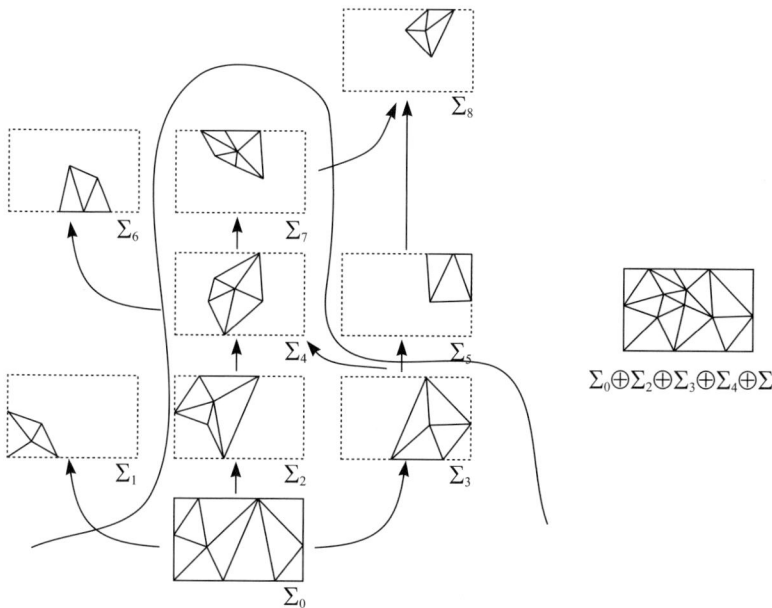

Fig. 4. Extracting a mesh at constant/variable resolution means detecting a proper subset of the fragments, $\mathcal{S}' \subset \mathcal{S}$, corresponding to a cut in the DAG.

3.3 Extracting a Variable Resolution Model

Algorithms for the extraction of a variable resolution model from an MSM have been presented in [5,24]. If we assume that our MSM is by construction monotone (i.e. every refinement action improves the accuracy of the intermediate mesh), the extraction of a *variable resolution model* can be simply performed by defining a boolean acceptance function $c(\sigma)$ which evaluates the accuracy of each tetrahedron σ (or, in reality, of each fragment). The algorithm for the extraction of a variable resolution model builds incrementally the desired solution by adding new fragments to a current solution, performing a breadth-first traversal of the DAG representing the MSM. The traversal starts from the coarsest fragment Σ_0, root of the DAG, and fragments above the current solution are progressively traversed and marked (see Fig. 4). The current solution is maintained as a list of tetrahedra Σ_{Out}. For each fragment Σ that we encounter in the traversal of the DAG, the following two loops are executed:

- we search for fragments before Σ, still not visited and, if found, they are added to the traversal queue Q. All the fragments before Σ can be found by checking, for each tetrahedron $\sigma \in \Sigma$, if the corresponding lower fragment Lower(σ), has been marked.
- for each tetrahedron $\sigma \in \Sigma$, if it satisfies the acceptance function $c(\sigma)$ then σ is added to the current solution, else we add the upper fragment of σ to the traversal queue Q and mark it to be removed from the solution.

The correctness of this algorithm has been proved in [8].

4 Regular Refinement Techniques

A multiresolution data representation can be built by starting from some sort of regularly shaped entities and using a regular subdivision pattern. Simple mathematical rules withhold the basis of a regular subdivision scheme that is usually applied in a recursive manner. The adoption of a predefined rule to perform the subdivision introduces some constrains on the topology of the local region to be refined/simplified, and adapts well mainly on regular dataset.

The simplest approach to perform a regular subdivision on regular voxel grids is the *octree* scheme, but the disadvantages of the octree-based subdivision are also well known (mainly, the discontinuities introduced in the common frontier of cells of different resolution). A common approach to ensure continuity is to adopt methods based on the use of simplicial decompositions, as we have seen in the previous section. Simplicial decompositions can also be used to represent regular datasets and to implement regular subdivision kernels on those datasets. Regular refinement approaches based on the simplicial decomposition have been proposed to manage terrain data represented by regular grids (e.g. [11]), and are nowadays the most common solutions for highly efficient multiresolution terrain rendering. Conversely, just a few approaches have been proposed to manage volume datasets [17, 20–22, 31] and

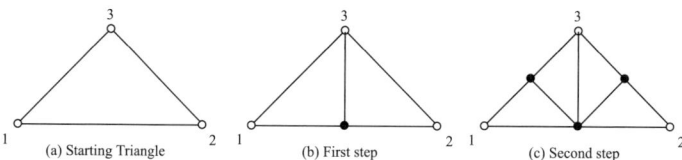

Fig. 5. A 2D example of the Rivara's 4T subdivision approach.

to accomplish a good balance between efficiency of the subdivision technique and easiness of implementation.

Essentially, the refinement heuristic consists of a recursive subdivision of the volume data, driven by a set of mathematical rules which guarantee a progressive improvement of the accuracy of the intermediate representations in an error-controlled manner. The subdivision process in general starts by subdividing the bounding box of the volume (i.e. a single hexahedral cell) in simplicial cells. The subdivision proceeds recursively and can be described either as a per vertex-adding process or, analogously, as a cell subdivision process. Each step picks up a vertex from the original dataset and divides the cell containing it (or a group of adjacent cells) in two (or more) simplicial cells. Because of the regular and hence predictable parametric structure of the refinement process, these techniques always generate meshes with bijective mapping between the coarsest levels and the finest levels. Therefore, going from a coarse to finer level corresponds in executing a regular subdivision of simplices allowing for smooth and continuous changes between different levels of details. The positive advantages of regular refinement techniques are basically their elegant mathematical formulation and the simplicity of the rules for generating different representations (which often allows to avoid explicit representation of the mesh topology). Recursive subdivision schemes usually produce hierarchical multiresolution representations, very efficient to process and store; moreover, regular refinement techniques often guarantee an almost everywhere regular structure of the variable resolution meshes extracted. In the next paragraphs we introduce in detail some of the main contributions in the field of regular refinement techniques.

4.1 4T Algorithm

The 4T algorithm has been introduced first by Rivara [26] and successively extended to n-dimension by Plaza and Carey. This technique can be conceived, in the 2D case, as a triangle refinement based on the longest edge bisection criteria, applied in two steps (see Fig. 5). The refinement process can be applied selectively, but it is important to maintain the refined mesh *conforming*, i.e. the subdivision of a cell on edge e will force the subdivision of all the ones sharing the edge e.

Plaza and Carey extended the 4T algorithm to 3D meshes by working on the "*skeleton*" of the mesh to be refined, i.e. they introduce bisection points on

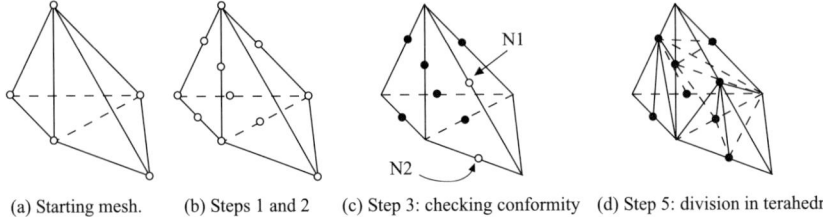

(a) Starting mesh. (b) Steps 1 and 2 (c) Step 3: checking conformity (d) Step 5: division in terahedra

Fig. 6. Application of the 3D subdivision algorithm of Plaza and Carey.

all edges of a cell to be refined. To guarantee the conformity of the mesh, they induce the subdivision of all the simplices incident on a bisected edge. A mesh is considered *conforming* if the intersection of two elements consists of a common face or a common edge or a common vertex or it is empty. The algorithm applies the following actions until no more cells are selected for subdivision:

- Step 1: select the subset of tetrahedra S' which need to be refined, and add a node at the midpoint of each edge of each selected $t \in S'$;
- Step 2: the conformity of each $t \in S-S'$ is checked; for each non-conforming t, a bisection node is added at the midpoint of each non-conforming edge of t and also at the midpoint of the longest edge of t (Fig. 6.c);
- Step 3: each t containing bisection nodes is properly subdivided (Fig. 6.d).

Tetrahedra are selected for refinement following an error indicator (only measures of the shape quality of the mesh are considered in the paper) or to guarantee conformity. The introduction of other bisection vertices in the cell split to guarantee conformity (see nodes $N1$ and $N2$ in Fig. 6.c) is an empirical solution to prevent bad-shaped tetrahedra. Unfortunately, the conformity is obtained at expenses of a significant propagation of each error-driven split; a refinement cannot be very local and tight under the 4T refinement rule, but extends to a much larger area.

4.2 Multiresolution Tetrahedral Framework (MTF)

The Multiresolution Tetrahedral Framework (MTF) [31] also adopts the longest edge bisection criteria in the development of a framework for tetrahedral meshes refinement. The MTF approach starts from the bounding box of the regular volume dataset (seen as a hexahedral cubic cell), which is subdivided into 12 tetrahedra as follows: the center point of the cell is connected to all the 8 vertices forming 6 pyramids, each pyramid is then divided into two tetrahedra by splitting the base along the diagonal. After this initial step the algorithm proceeds as a regular tetrahedra subdivision scheme. Each tetrahedra is recursively subdivided on the midpoint of its longest edge.

The tetrahedra generated by the subdivision can be grouped in three main classes (see Fig. 7.a-b):

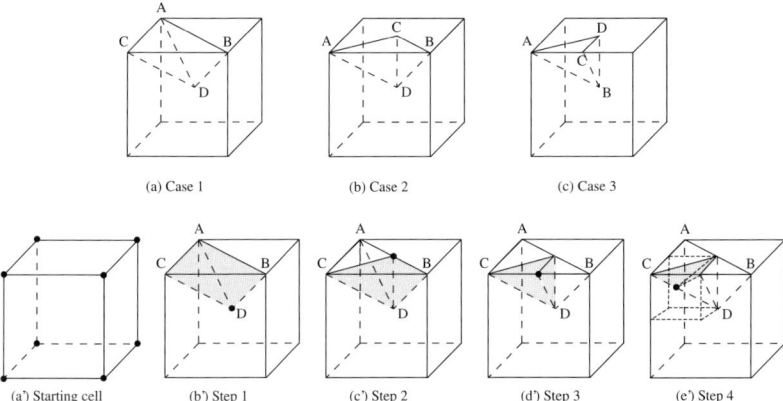

Fig. 7. Zhou et al. approach (MTF): a-b)the three types of tetrahedra generated by the subdivision rule; a'-b')The cyclic cubic cell subdivision.

- *Class 1*: there is only one face parallel to a coordinate plane and there exists only one edge l of that face not parallel to any coordinate axis;

- *Class 2*: there is only one face parallel to a coordinate plane and there exists only one edge l of that face parallel to a coordinate axis;

- *Class 3*: there are two faces parallel to a coordinate plane (the edge that does not belong to any of these two faces is denoted by l).

In all three cases, the edge AB is the edge previously called l, that is the one split along the midpoint in the subdivision process. We show the entire subdivision process in Fig. 7.a'-e'. We start with the initial subdivision of the cubic cell into 12 tetrahedra all belonging to Class 1 (Fig. 7.b'). Those tetrahedra are then subdivided producing tetrahedra belonging to Class 2 (Fig. 7.c'), which are again subdivided producing tetrahedra belonging to Class 3 (Fig. 7.d'). With a further subdivision of a Class 3 tetrahedra we obtain cells of Class 1, and therefore the configuration recursively returns to the initial subdivision step. It is worth noting that each type of tetrahedra belongs to a different step of the subdivision process (modulo 3) and that the overall process has a cyclic behavior.

The MTF approach preserves explicitly the topology of the finest level mesh and this constitutes a problem (excessive storage cost) in the representation of very large datasets .

4.3 Slow Growing Subdivision (SGS)

The Slow Growing Subdivision (SGS), introduced by Pascucci [20], extends the subdivision criteria at the base of the MTF subdivision to generical

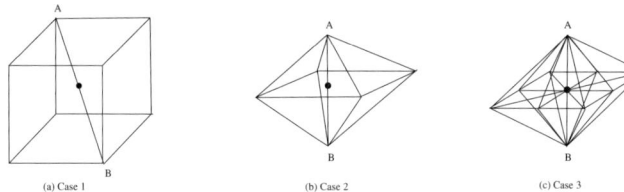

Fig. 8. The three cell types of the Slow Growing Subdivision (SGS) approach.

n-dimensions and, moreover, it is designed to manage huge datasets and presents peculiar features. Analogously to the other approaches, the SGS framework requires a first phase where the decomposition is performed from the coarser cell to the finest ones, with the objective of estimating the degree of accuracy given by any intermediate decomposition, and a run-time phase, where the decomposition is performed according to some application-specific constraints (view-dependent, adaptiveness, error-based criteria).

The starting point of the SGS subdivision is again the volume bounding box, a cubic cell which is divided into 6 pyramids obtained by connecting the boundary faces with the cell center. The SGS approach does not simply divide the cube into tetrahedra, as it happens in MTF. The original idea introduced by the SGS is to setup a cyclic set of *join&split* actions: each of these actions consists in merging some cells produced in the previous subdivision step, forming a new entity; and then splitting this new entity by inserting a new dataset vertex (the vertex introduced is always the center point of the split entity). The SGS method defines three different type of entities (here called *diamonds*, to differentiate them from the tetrahedral or pyramidal cells produced by the split of those entities). Therefore, the second step in the SGS cyclic subdivision process is to build new *diamonds* by joining each of the six pyramids produced in step 1 with the corresponding pyramid generated in an adjacent cube (see Fig. 9.a). This new entity corresponds to an *octahedral-shaped diamond*, whose center corresponds to the center of one of the faces of the cube. This entity is split by inserting the center point and by connecting all its 6 vertices with the center point, thus dividing the octahedral-shaped diamond into eight tetrahedra. This eight tetrahedra correspond each to the eighth part of different *hexadecahedral-shaped diamonds* whose centers correspond to the midpoints of the initial cubes edges (Fig. 9.b). Finally, the sub-cells produced by the subdivision of the hexadecahedral-shaped diamonds will recombine with adjacent ones to form again *cubic-shaped diamonds*, having edge length equal to 1/2 of the initial cubic-shaped cell (Fig. 9.c). This concludes the cycle, which can be repeated until we reach the maximal resolution of the input dataset.

The cells cyclically generated by the subdivision can be grouped in three main classes (Figure 8):

- Class 1 diamond: the cell is *cube-shaped*;

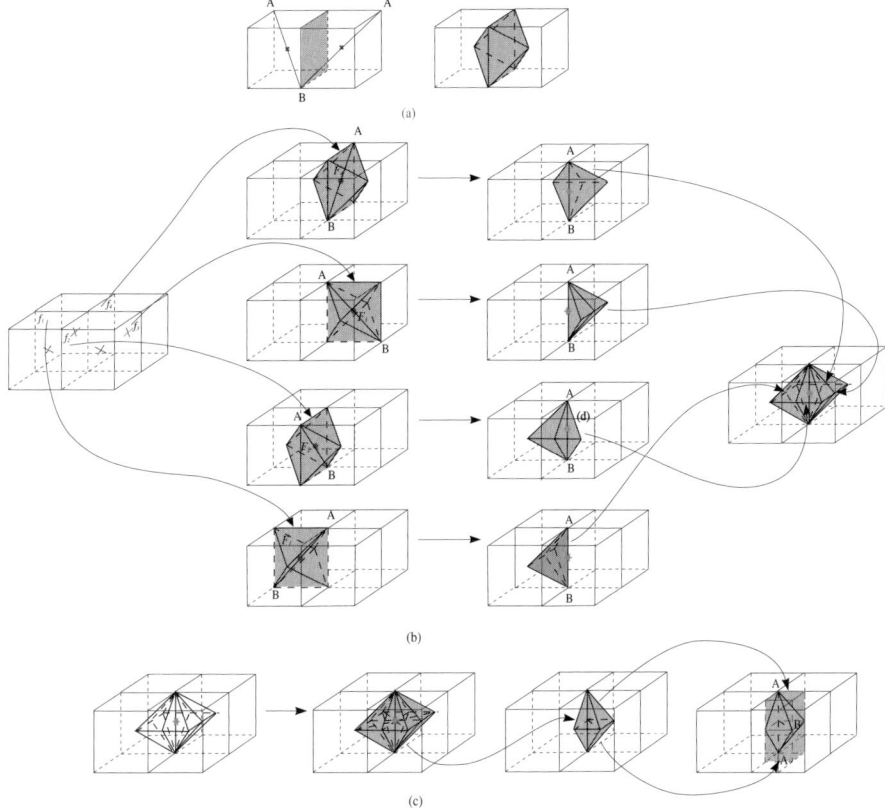

Fig. 9. SGS refinement: a) from the initial cubic cell to a class 2 diamond; b) a class 3 diamond generated for the join of portions of four class 2 diamonds; c) class 1 cells are produced by the subdivision of class 3 diamonds.

- Class 2 diamond: the cell is *hexadecahedral-shaped* and can be oriented along x, y or z axis;
- Class 3 diamond: the cell is *octahedral-shaped* and can be oriented along x, y or z axis.

The entire subdivision process is summarized by the three steps of the cyclic process in Fig. 9.a-c. Each diamonds is characterized by a center that corresponds to the midpoint of edge AB, which is the longest edge of the cell. The midpoint of edge AB is always selected as the dividing point and corresponds respectively to the center of the cube (Fig. 9.a), to the center of the cube faces (Fig. 9.b) and to the midpoint of the cube edges (Fig. 9.c). The subdivision of Class 1 cells (Fig. 8.a) produces cells belonging to Class 2 (Fig. 8.b) that, subdivided again, produce tetrahedra belonging to Class 3 (Fig. 8.c). After the subdivision of Class 3 cells the configuration recursively returns to Step 1. Similarly to the MTF approach, each type of diamond in the SGS subdivision schema belongs to a different step of the subdivision process and the overall process has a cyclic behavior. The difference between the two approaches relies mainly in the cell shape and in the way a cell can be identified. An important advantage of the SGS's approach is the fact that each diamond is completely characterized (i.e. type, orientation and refinement level it belongs to) by the geometrical position of its center. This means that a mesh refinement process can be simply implemented by inserting dataset vertices according to a particular visiting order, and that the topology of the cells has not to be stored explicitly but can be simply reconstructed from the center points coordinates.

A variation to the SGS approach has been proposed by Gregorski et al. [16]; they perform the refinement directly on the tetrahedra that compose the diamonds, and each diamond is seen as an aggregate of tetrahedra and not as the "main" entity. They also exploit the regularity of the scheme to achieve good memory layout. The regularity of the SGS scheme allows to represent the data by storing just the initial volume dataset and a second grid with an error estimate for each voxel node. The subdivision process is regular and the cell produced have not to be represented explicitly, nor their topology. This organization of the data allows efficient storing and LOD's extraction and adaptive traversal. At *rendering time*, for example, for extracting an isosurface at a given view-dependant accuracy, the SGS subdivision process is performed recursively by simply reading voxel nodes from the dataset (according to the peculiar subdivision order) and reconstructing on the fly the new cells introduced. The recursion ends when no more refinement is needed or when we produce class 1 diamonds at the finer resolution level.

4.4 Red/Green Subdivision

The Red/Green Subdivision approach was first introduced by Bank et al. [29] for 2D meshes and successively extended to three dimensional tetrahedral meshes by Bey [1]. Grosso et al. [17] adopted a red/green subdivision schema to formulate a mesh refinement approach based on finite element analysis. Their approach must meet several constraints necessary to perform finite element computations: the mesh should be *conforming* (this condition prevents *hanging nodes*) and *stable* under some measure criteria (degeneracy like zero-angles must be avoided to guarantee stability of the numerical computations); moreover, the subdivision should guarantee *nestedness*, i.e. each element produced during subdivision is obtained by subdividing an element in a coarser level of the refinement sequence.

Since it satisfies the above requirements, the Red/Green refinement algorithm produces a stable and consistent partitioning of the space. The subdivision proceeds following two main rules: a *red*, or regular, subdivision for single elements subdivision, and a *green*, or irregular, subdivision for boundary elements. In the case of 2D meshes, the red refinement rule divides a triangle into four congruent ones by connecting the midpoints of its edges. The green

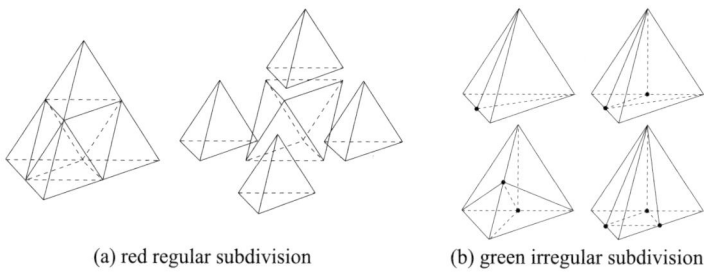

(a) red regular subdivision (b) green irregular subdivision

Fig. 10. Red/green subdivision of a tetrahedron.

refinement consists of simple bisections connecting one edge midpoint with the opposite vertex. In the case of 3D meshes, the regular refinement rule first cuts off four sub-tetrahedra at the vertices of the cell, as shown in Fig. 10.a, leaving an undivided octahedron whose subdivision is not unique and depends on the choice of one of the three possible diagonals. The subdivision of the octahedron is performed following a strategy based on affine transformation to a reference tetrahedron [1] and guarantees the result to be a stable regular refinement. For a tetrahedron there exist $2^6 - 2 = 62$ possible green refinement patterns which can be reduced to 9 using symmetry considerations. Grosso et al. restricts them to four (see Fig. 10.b), performing red refinement on all the remaining patterns. Green refinement rules are normally local, tetrahedra refined through a green rule are inserted only to satisfy border conditions and to avoid hanging nodes. To guarantee stability green refined tetrahedra can be refined only once and, if further subdivision is needed, the originally green refined tetrahedron is re-refined with a red rule.

5 Comparative Evaluation

A comparative evaluation of the two approaches (regular *vs.* irregular) is not straightforward. Both approaches present strength and weakness, different implementations of each method exist with the consequence that the comparison becomes much more complicated. To perform an evaluation we have then identified some issues which depend mainly from the visualization and data management requirements.

Adaptivity. We indicate with the term *adaptivity* how well each approach is able to produce compact and accurate representations of the initial model at different resolutions. The extraction of constant/variable resolution representations can be driven by any of the following objectives: (a) to give a good approximation of the dataset shape (i.e. preserving accurately its external or internal boundaries), or (b) to give an accurate representation of the field values encoded in the input dataset. In general irregular techniques allow for better adaptivity, mainly due to the freedom in the choice of the best-fit refinement pattern which can be selected for each atomic refinement step. Irregular methods based on atomic refinements actions (e.g. edge split) allow to refine in a very selective manner the mesh, either preserving accurately the boundaries or by ensuring accurate representation of the original field values (by the interpolation of the field values on the mesh vertices). Moving from high to low resolution, and vice-versa, the sub-regions, onto which irregular techniques need to operate the update, can be restricted to a quite small number of tetrahedra. Conversely regular techniques need to propagate the change to a wider area, because in most cases those type of techniques allows at most one level of difference in the refinement of adjacent cells. This makes in general the representations, built with an irregular approach, more compact (no. number of cells required) than the one produced with a regular one. For the two-dimensional case this problem has been numerically evaluated and verified in [12], probably similar results could be obtained also for the three-dimensional case. Considering here just the number of cells is justified by the fact that most rendering algorithm have a time complexity that depends on this factor. Please note that the actual storage size does not only depend on the number of components (no. tetrahedral cells), but also by other factors (e.g. space complexity of the auxiliary structures, need to explicitly represent topology).

Flexibility. The term *flexibility* indicates to which extent each method allows to change the parameters used to drive the data refinement/simplification. This is a critical feature for the visualization of datasets with multiple scalar/vectorial field attributes like, for example dataset coming from simulations, where multiple parameters (pressure, temperature, velocity, etc.) should influence an error-based refinement criteria. Changing dynamically the refinement constraints is easier when using regular refinement techniques. The regular subdivision pattern allows, when needed, to execute a fast update of the refinement hierarchy. This is possible because there is no need to change the topology or the temporal ordering of the possible refinement actions (since those actions are defined statically by the selected regular refinement kernel choice). Conversely, the selection of which refinement "pattern" has to be applied and how the compatible sequences of refinement actions should be ordered, are error-driven in irregular methods and depend heavily on the field attribute and error evaluation criterion taken into consideration during the multiresolution representation construction. Therefore, a dynamic change in the parameters used to evaluate the accuracy often requires to rebuild the multiresolution data structure from scratch (and tetrahedral mesh simplification is a costly process on complex datasets).

Efficient Data Access. We indicate with *efficient data access* how well the data can be organized in memory to allow for an efficient data fetching. Irregular techniques usually require random access to the data on disk. On the contrary, regular techniques can be easily organized in memory to guarantee locality in memory access, especially during the refinement actions which are known only at run-time. This issue plays a fundamental rule since the size of the datasets is increasing with a higher rate than the RAM memory available on standard computers. The need to structure data representation in an easy and regular manner is a crucial point to simplify the design of out-of-core solutions [2, 7], which are becoming a must in high-range scientific visualization.

Space Complexity. The adoption of a regular schema allows for an efficient storing of the data, since it is in general possible to avoid to encode explicitly the extent and the topology of the mesh region under refinement and, often, also of the refined section. Conversely, irregular techniques often require the explicit representation of the refined mesh region and of the interdependency relationships that exist between the elements that make up the multiresolution structure. Larger is the amount of information that we can implicitly represent, more compact in space will be the representation schema. The speed of current computers allows us to trade some computation cycles (to reconstruct on the fly either topology or geometry) for saving memory locations and reducing secondary memory fetching overheads.

Implementation Easiness. An initial guess would be that the implementation and the debugging of a regular approach should be simpler. Conversely, our experience in the implementation of both irregular [3, 5, 6] and regular methods [25] leads to a different evaluation. As far as the depth of the refinement chain increases, even on dataset composed by a few million points, the structural complexity of the refinement structure grows exponentially both for the regular and irregular case making quite hard the debugging for both approaches, and very complicated to produce a bug-free and robust code.

Generality of the Method. We indicate with Generality of the method how well a given approach can manage different types of datasets. Irregular techniques show to be more flexible. Regular techniques in fact perform well only on regular datasets and result to be difficult to generalize to irregular ones. Re-sampling an irregular dataset on a regular grid is not convenient for several reason. It is common that the ratio between the sizes of the smallest and the largest cell in those dataset could be easily up to 1:10,000 and large cell may lay in any position across the domain. This would require to use adaptive re-sampling strategies, which produce meshes at different resolutions also not easy to manage with standard regular multiresolution techniques. A second problem arises when we consider that most of the irregular datasets have non-convex domains and the regular grid should properly represent the original data domain. In this case the field has good chances to present sharp discontinuities across the boundary of such domain since it is unknown outside the boundary itself.

6 Conclusions

A synthetic overview of methodologies for the efficient multiresolution management of large volume datasets has been presented. We have introduced the main contributions currently present in literature, subdividing them into regular or irregular refinement techniques, and analyzed strength and weakness of both the approaches. We have seen that irregular techniques, at the expense of a more complex structure, allow a better approximation and tighter refinement of any kind of dataset, while regular techniques mainly allows for efficient refinement of regularly gridded datasets. On their side regular techniques deals better with all the issues arising from the necessity of working out-of-core. The adoption of regular subdivision pattern allows for efficient data organization to improve memory occupancy and efficient access policy to secondary memory, which is becoming a must in the visualization of very large datasets. Visualization results from both techniques are comparable, even if irregular techniques perform better on adaptive refinement while regular techniques deals better with dynamic changes of critical refinement constraint. Both approaches are best suited for specific kinds of problems and it is still not possible to formulate a unique judgement of quality that could prefer one approach instead of another.

References

1. J. Bey, *Tetrahedral grid refinement*, Computing **55** (1995), 355–378.
2. Y. Chiang, C. T. Silva, and W.J. Schroeder, *Interactive out-of-core isosurface extraction*, IEEE Visualization '98 Proceedings, IEEE Press, 1998, pp. 167–175.
3. P. Cignoni, D. Costanza, C. Montani, C. Rocchini, and R. Scopigno, *Simplification of tetrahedral volume with accurate error evaluation*, Proceedings IEEE Visualization'00, IEEE Press, 2000, pp. 85–92.
4. P. Cignoni, L. De Floriani, C. Montani, E. Puppo, and R. Scopigno, *Multiresolution modeling and rendering of volume data based on simplicial complexes*, Proc.of 1994 Symp. on Volume Visualization, October 17-18 1994, pp. 19–26.
5. P. Cignoni, Paola Magillo, Leila De Floriani, Enrico Puppo, and R.Scopigno, *Memory-efficient selective refinement on unstructured tetrahedral meshes for volume visualization*, IEEE TVCG **9** (2003), To appear.
6. P. Cignoni, C. Montani, E. Puppo, and R. Scopigno, *Multiresolution modeling and visualization of volume data*, IEEE TVCG **3** (1997), no. 4, 352–369.
7. P. Cignoni, C. Montani, C. Rocchini, and R. Scopigno, *External memory management and simplification of huge meshes*, IEEE Transactions on Visualization and Computer Graphics **8**.
8. L. De Floriani, P. Magillo, and E. Puppo, *Building and traversing a surface at variable resolution*, Proc. IEEE Visualization 97, October 1997, pp. 103–110.
9. L. De Floriani and E. Puppo, *Hierarchical triangulation for multiresolution surface description*, ACM Transactions on Graphics **14** (1995), no. 4, 363–411.
10. L. De Floriani, E. Puppo, and P. Magillo, *A formal approach to multiresolution modeling*, Theory and Practice of Geometric Modeling, Springer-Velrag, 1997.

11. M. Duchaineau, M. Wolinshy, D.E. Sigeti, M.C. Miller, C.Aldrich, and M. B. Mineev-Weinstein, *ROAMing terrain: Real-time optimally adapting meshes*, Proc. IEEE Visualization 1997, pp. 81–88.
12. William Evans, David Kirkpatrick, and Gregg Townsend, *Right triangulated irregular networks*, Algorithmica **30** (2001), no. 2, 264–286.
13. R.J. Fowler and J.J. Little, *Automatic extraction of irregular network digital terrain models*, Siggraph '79 Proc., no. 3, 199–207.
14. M. Garland, *Multiresolution modeling: Survey & future opportunities*, EUROGRAPHICS'99, State of the Art Report (STAR), 1999.
15. M. Garland and P.S. Heckbert, *Surface simplification using quadric error metrics*, SIGGRAPH 97 Conference Proc., pp. 209–216.
16. B. Gregorski, M. Duchaineau, P. Lindstrom, V. Pascucci, and K.I. Joy, *Interactive view-dependent rendering of large IsoSurfaces*, Proc. IEEE Visualization 2002, pp. 475–484.
17. R. Grosso, C. Lürig, and T. Ertl, *The multilevel finite element method for adaptive mesh optimization and visualization of volume data*, IEEE Visualization 97, pp. 387–394.
18. S. Gumhold, S. Guthe, and W. Straßer, *Tetrahedral mesh compression with the cut-border machine*, Proceedings IEEE Visualization'99, IEEE, 1999, pp. 51–58.
19. H. Hoppe, *Progressive meshes*, Proceedings of SIGGRAPH '96 (1996), 99–108.
20. V. Pascucci, *Slow growing subdivision (sgs) in any dimension: towards removing the curse of dimensionality*, Computer Graphics Forum (Proc. EUROGRAPHICS '02), no. 3, 451 – 460.
21. V. Pascucci and C. L. Bajaj, *Time critical isosurface refinement and smoothing*, Proc. IEEE Symp. on Volume visualization 2000.
22. A. Plaza and G.F. Carey, *About local refinement of tetrahedral grids based on local bisection*, 5th International Meshing Roundtable (1996), 123–136.
23. J. Popovic and H. Hoppe, *Progressive simplicial complexes*, ACM Computer Graphics Proc., Annual Conference Series, (SIGGRAPH 97), pp. 217–224.
24. E. Puppo, *Variable resolution terrain surfaces*, Proceedings Eight Canadian Conference on Computational Geometry 1996, Ottawa, Canada, pp. 202–210.
25. V. Pascucci R. Borgo, *Distributed oriented massive data management: Progressive algorithms and data structures*, CODATA'02, pp. 387–394.
26. María-Cecilia Rivara, *Mesh refinement processes based on the generalized bisection of simplices*, SIAM Journal on Numerical Analysis (1984), no. 3, 604–613.
27. O.G. Staadt and M.H. Gross, *Progressive tetrahedralizations*, Proc. IEEE Visualization '98, IEEE, 1998, pp. 397–402.
28. I.J. Trotts, B. Hamann, and K.I. Joy, *Simplification of tetrahedral meshes with error bounds*, IEEE TVCG **5** (1999), no. 3, 224–237.
29. R.E. Bank A.H. Sherman A. Weiser, *Refinement algorithms and data structures for regular local mesh refinement*, Scientific Computing, IMACS Trans. on Scientific Computation, vol. 1, 1983.
30. P.L. Williams, N.L. Max, and C.M. Stein, *A high accuracy volume renderer for unstructured data*, IEEE TVCG **4** (1998), no. 1.
31. Y. Zhou, B. Chen, and A. Kaufman, *Multiresolution tetrahedral framework for visualizing regular volume data*, Proc. IEEE Visualization '97, pp. 135–142.

Selective Refinement on Nested Tetrahedral Meshes

Leila De Floriani[1] and Michael Lee[2]

[1] Department of Computer and Information Sciences (DISI)
University of Genova
Via Dodecaneso 35, 16146 Genova (Italy)
[2] Department of Computer Science
University of Maryland
College Park, MD - 20742 (USA)

Summary. We consider a multi-resolution representation based on a decomposition of the field domain into nested tetrahedral cells generated by recursive tetrahedron bisection, that we call a *Hierarchy of Tetrahedra (HT)*. We describe our implementation of an HT, and discuss how to extract conforming meshes from an HT so as to avoid discontinuities in the approximation of the associated scalar field. We describe algorithms for selective refinement, which either extract a variable-resolution mesh from scratch through a depth-first, or through a priority-based traversal technique, or which locally refine and coarsen a previously-extracted adaptive mesh through an incremental approach. We show experimental results in connection with a set of basic queries for performing analysis and visualization of a volume data set at different levels of detail.

1 Introduction

We consider the problem of modeling volume data sets, i.e., sets of points spanning a domain in the three-dimensional Euclidean space and describing a scalar field. This problem arises in several applications, including scientific visualization, medical imaging, simulation, and finite element analysis. A *volume data set* S consists of a set V of points in the three-dimensional Euclidean space, and of a field value f associated with such points. A volume data set is often modeled by decomposing its domain through a tetrahedral mesh with vertices at the data points. We consider a regular tetrahedral mesh, i.e., a mesh generated by a recursive decomposition on the points of a regular grid.

In order to analyze volume data sets of large size, a multi-resolution approach can be used. *Multi-resolution models*, also called *Level-OF-Detail (LOD) models*, have been widely used for describing surfaces and two-dimensional height fields (see [3] for a survey). A *virtually continuous* set of simplified meshes at different LODs can be generated from a multi-resolution model. The resolution (i.e., the density of the cells) of an approximating mesh may vary in different parts of the field domain, or in the proximity of interesting field values. The process of extracting meshes at a variable resolution from a multi-resolution model is called *selective refinement*. Selective refinement must be efficiently supported by any multi-resolution data structure.

Here, we consider nested meshes generated by recursive bisection of a tetrahedron along its longest edge, and we will use the term *Hierarchies of Tetrahedra (HTs)* to describe them. Such meshes have been used as domain decompositions for multi-resolution modeling of regularly-spaced volume data sets because of their capability of generating highly adaptive domain decompositions.

In the paper, we discuss our implementation of an HT based on an ordering of the tetrahedra which supports an efficient retrieval not just of the children and the parent of a given tetrahedron, but also of neighboring tetrahedra. We describe different approaches to selective refinement, which either extract a variable-resolution mesh from scratch through a depth-first, or a priority-based search, or which locally refine and coarsen a previously-extracted adaptive mesh through an incremental approach. A major issue here is that the topological consistency of the extracted mesh must be ensured to avoid discontinuities in the approximation of the scalar field. Consistency must be maintained while applying local refinement or coarsening. To this aim, we have developed an efficient technique based on neighbor finding for detecting those tetrahedra which form clusters that must be split, or merged, simultaneously. A worst-case constant time implementation is discussed in [12].

We present results on the performances of the selective refinement algorithms as well as a comparison with a technique based on error saturation. We also show experimental results in connection with a set of basic queries for performing analysis and visualization of a volume data set at different levels of detail, that we call *Level-Of-Detail (LOD)* queries, defined in [1].

The rest of this paper is organized as follows. Section 2 reviews some related work. In Section 3, we review the HT and we describe our implementation. In Section 4, we discuss how to extract conforming meshes from an HT. In Section 5, we describe the depth-first and priority-based approaches to selective refinement, and we present a comparison with a technique based on error saturation. In Section 6, we describe an incremental approach to selective refinement, while in Section 7 we present experimental results in connection with LOD queries. Concluding remarks are drawn in Section 8.

2 Related Work

Nested tetrahedral meshes have been studied in finite element analysis and in computer graphics for describing three-dimensional scalar fields when the field values are given at the vertices of a regular square grid in 3D space. Examples are tetrahedral meshes generated by the so-called *red/green tetrahedron refinement* technique (see, for instance, [10]), or nested meshes formed by tetrahedral and octahedral elements [9].

A common way of generating nested meshes consists of recursively bisecting tetrahedra on their longest edge (see, for instance, [16, 24]). Such meshes have been introduced for domain decomposition in finite element analysis [11, 15, 21], and they have been applied in scientific visualization, for instance, to generate progressive volume models of ultrasound data [22], or for space/time-efficient progressive encoding of isosurfaces at a variable resolution [20]. A generalization and analysis of such meshes in arbitrary dimensions is presented in [18] in connection with adaptive mesh generation. Zhou et al. [24] have first proposed a representation for a nested tetrahedral mesh as a full binary forest, stored as an array. A similar data structure has been used by Gerstner and Rumpf [7] for extracting isosurfaces at different levels of detail. An indexing scheme for out-of-core encoding and traversal has been proposed in [19].

Nested triangle meshes based on recursive triangle bisection have been extensively used for view-dependent terrain rendering (see, for instance, [4,5,13,14,17]). In particular, Evans et al. [5] use a location code to identify a triangle in the nested decomposition, and have proposed algorithms for finding neighbors in time proportional to the depth of the triangle as well as in constant time, by using a relatively small number of arithmetic and bit-wise logical operations. Recently, Lindstrom and Pascucci [14] have designed and implemented a framework for performing out-of-core view-dependent rendering of large terrain surfaces based on nested meshes.

An important issue when using nested tetrahedral meshes is that if the domain is adaptively refined, the field associated with the extracted mesh (and, thus, the resulting isosurfaces) may present discontinuities in areas of transition. One way of ensuring continuity is through error saturation [7,24], thus implicitly forcing all parents to be split before their descendants (see also [14] for an effective saturation technique for terrains). In our approach, the continuity of the field is ensured by efficiently extracting meshes without cracks through a neighbor finding technique. Hebert [11] computes parents, children, and neighbors in a nested tetrahedral mesh in a symbolic way, but finding neighbors still takes time proportional to the depth in the hierarchy. A worst-case constant-time implementation of the neighbor finding technique has been first proposed in [12].

In [8] an algorithm for interactively extracting and rendering isosurfaces of large volume data sets is presented, which extends the ROAMing algorithm introduced in [4]. The authors use a refinement scheme based on tetrahedron bisection, and propose a data structure for representing such meshes that directly encodes clusters of tetrahedra which must be split together to ensure consistency.

3 A Hierarchy of Tetrahedra

The bisection rule for tetrahedra consists of replacing a tetrahedron t with the two tetrahedra obtained by splitting t at the middle point of its longest edge and by the plane passing through such point and the opposite edge in t. This rule is applied recursively to an initial decomposition of the cubic domain into six tetrahedra (see Figure 1a). Splitting a tetrahedron in the initial cube subdivision results in two tetrahedra with a shape identical to that obtained by splitting a pyramid with a square base in half along the diagonal of its base. We call such shape a *1/2 pyramid* (see Figure 1b). Splitting a 1/2 pyramid along its longest edge results in two tetrahedra whose shape we call a *1/4 pyramid* (see Figure 1c). Finally, splitting a 1/4 pyramid along its longest edge results in two tetrahedra whose shape we call a *1/8 pyramid* (see Figure 1d). Each of the six initial tetrahedra also have the shape of a *1/8 pyramid*. These shapes are cyclic in the sense that every three levels of decomposition result in a congruent shape.

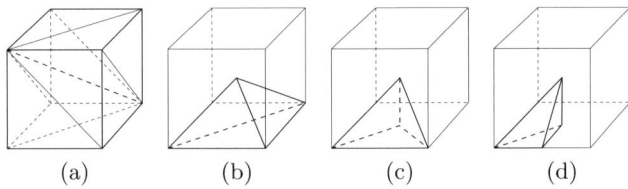

Fig. 1. (a) Subdivision of the initial cubic domain into six tetrahedra. Examples of (b) a 1/2 pyramid, (c) a 1/4 pyramid, and (d) a 1/8 pyramid.

A *location code* for a tetrahedron t in an HT is a pair of numbers, in which the first denotes the level of t in the tree, and the second indicates the path from the root of the tree to t. The location code for a tetrahedron is defined on the basis of a labeling scheme for its children and for the vertices of these children in the hierarchy, as explained below.

For simplicity of computation, we order the vertices of a tetrahedron in such a way that its longest edge is $v_3 v_4$. Since the longest edge in the initial cube is the diagonal, we label the diagonal $v_4 v_3$, where v_4 is the vertex of the cube at the origin of the coordinate system.

Let $t = [v_1, v_2, v_3, v_4]$ be a tetrahedron and v_m is the midpoint of edge $v_3 v_4$ in t. The following labeling conventions are assumed for the tetrahedra in the hierarchy:

- If t is a 1/2 pyramid, then the two resulting 1/4 pyramids are $t_0 = [v_1', v_2', v_3', v_4'] = [v_m, v_1, v_2, v_3]$ and $t_1 = [v_1', v_2', v_3', v_4'] = [v_m, v_1, v_2, v_4]$ (see Figure 2a).

- If t is a 1/4 pyramid and the parent was child 0, then the two resulting 1/8 pyramids are $t_0 = [v_1', v_2', v_3', v_4'] = [v_m, v_1, v_2, v_3]$ and $t_1 = [v_1', v_2', v_3', v_4'] = [v_m, v_1, v_2, v_4]$ (see Figure 2b).
- If t is a 1/4 pyramid and the parent was child 1, then we swap the labels of the children so that $t_0 = [v_1', v_2', v_3', v_4'] = [v_m, v_1, v_2, v_4]$ and $t_1 = [v_1', v_2', v_3', v_4'] = [v_m, v_1, v_2, v_3]$ (see Figure 2b).
- If t is a 1/8 pyramid, then the two resulting 1/2 pyramids are $t_0 = [v_1', v_2', v_3', v_4'] = [v_m, v_1, v_2, v_4]$ and $t_1 = [v_1', v_2', v_3', v_4'] = [v_m, v_2, v_1, v_3]$ (see Figure 2c).

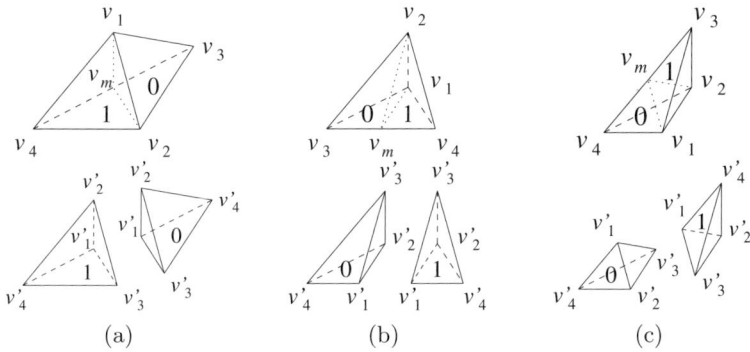

Fig. 2. Labeling of (a) 1/2, (b) 1/4, and (c) 1/8 pyramids.

Our encoding of a hierarchy of tetrahedra makes use of a linear representation instead of a pointer-based one. The data structure consists of:

- a field table containing the field values at the n data points;
- a forest of six almost full binary trees, each of which describes the nested subdivision of one tetrahedron in the initial cube and is encoded as an array containing the errors associated with the tetrahedra corresponding to the internal nodes of the tree.

Note that leaf nodes correspond to the tetrahedra in the mesh at full resolution, and, thus, they have a null error. Also, we do not store location codes, but they are computed on-the-fly and used for indexing the field table and for neighbor finding. To handle large-size data sets, we just store the above data structure on secondary storage, and we perform random accesses to the field and error values.

In our current implementation, we encode with each tetrahedron t the field error associated with each tetrahedron t, computed as the absolute value of the maximum of differences between the actual field value and the interpolated one at each point which falls inside t. An alternative, which is useful in rendering applications, is to store the isosurface error (see [1,8]), which can also be computed from the field error and the gradient.

If n is the number of points in the data set, $12n$ bytes are required for the error values plus $2n$ bytes for the field table, assuming two bytes per error and field value, leading to a total cost of $14n$ bytes. Note that there are $6n$ tetrahedra in the mesh at full resolution, since each cube of the input grid is subdivided into six tetrahedra, and, thus, $6n$ internal tetrahedra. If both the error and the field values are quantized as in [8] to one byte, the storage cost of the data structure reduces to $7n$ bytes. Note that our current HT implementation is a reference implementation to prove the effectiveness of our pointer-based approach, and thus it is not optimized with respect to error and field value quantization or for performing out-of-core queries efficiently.

4 Extracting Conforming Meshes

When we apply tetrahedron bisection, all tetrahedra that share a common edge with the tetrahedron being split must be split at the same time to guarantee consistency. Following the terminology used in multi-resolution meshes [3], we call a *cluster* any set of tetrahedra which share a common edge (a cluster is called a *diamond* in [8, 18]). There are three types of clusters generated by the three congruent tetrahedral shapes, that we call *plane-aligned*, *axis-aligned* and *non-aligned* clusters, respectively (see Figure 3):

- a *plane-aligned* cluster is formed by four 1/2 pyramids, as only 1/2 pyramids share a plane-aligned edge, i.e., on an edge parallel to one of the coordinate planes (see Figure 3a).
- an *axis-aligned* cluster is formed by eight 1/4 pyramids, as only 1/4 pyramids share an axis-aligned edge (see Figure 3b).
- a *non-aligned* cluster is formed by six 1/8 pyramids, as only 1/8 pyramids share an edge which is not aligned to a coordinate axis or plane (see Figure 3c).

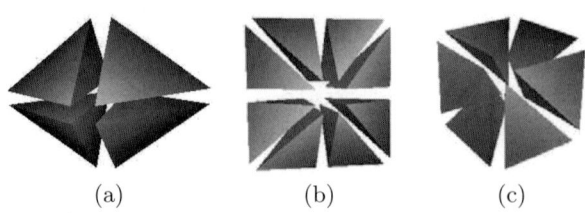

(a) (b) (c)

Fig. 3. (a) A plane-aligned cluster. (b) An axis-aligned cluster. (c) A non-aligned cluster.

We compute clusters on-the-fly when extracting a mesh from an HT by applying a neighbor finding technique. Since all tetrahedra in a cluster share a common edge, the basic operation consists of computing the neighbor of a given tetrahedron along a specified face. We define four neighbor types for a given tetrahedron $t = [v_1, v_2, v_3, v_4]$ depending on the face shared with t: a (v_i, v_j, v_k)-neighbor is the tetrahedron which shares face $v_i v_j v_k$ with t.

The tetrahedra forming a cluster can be computed by starting from an initial tetrahedron as an alternating sequence of (v_1, v_3, v_4)- and (v_2, v_3, v_4)-neighbors (see Table 1 and Figure 4 for the case of a plane-aligned cluster).

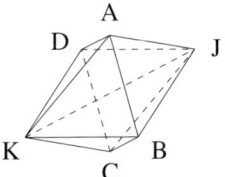

Fig. 4. Computing a plane-aligned cluster.

Count	Neighbor Direction	Tetrahedron
1	Initial tetrahedron	$t_1 = [ABJK]$
2	Find (v_1, v_3, v_4)-neighbor	$t_2 = [ADJK]$
3	Find (v_2, v_3, v_4)-neighbor (if it exists)	$t_3 = [CDJK]$
4	Find (v_1, v_3, v_4)-neighbor	$t_4 = [CBJK]$

Table 1. Steps required to compute a plane-aligned cluster formed by four 1/2 pyramids (symbols refer to Figure 4).

Our neighbor finding algorithm uses an approach similar to the one defined in [23] for region quadtrees. It consists of two steps. Given an input tetrahedron t and a face $v_i v_j v_k$, the first step identifies the nearest common ancestor of t and its neighbor t' along face $v_i v_j v_k$. The second step updates the location code for the neighbor by using the information obtained while finding the nearest common ancestor.

At the first step, the location code of t is simply scanned from right to left until the neighbor direction, defined by face $v_i v_j v_k$, forces us to cross face $v_1 v_2 v_3$ of the ancestor. Since we label the longest edge of the tetrahedron as $v_3 v_4$, the midpoint of this edge together with the remaining two vertices (v_1 and v_2) form the common face. This is face $v_1 v_2 v_3$ in the children, thus any tetrahedra sharing face $v_1 v_2 v_3$ are siblings, and the parent of these sibling ancestors is the nearest common ancestor of tetrahedron t and its neighbor t'.

At the second step, we just invert the one bit corresponding to the child of the nearest common ancestor. This process works regardless of the original neighbor type which we were trying to find. No further work is necessary, as all location codes of the neighbors differ just by this one bit.

The algorithm requires only a slight modification to take into account the first subdivision of the cube into six tetrahedra. In any case, its complexity is proportional to the depth of the input tetrahedron in the hierarchy.

As an example, consider a 1/4 pyramid t of location code 210011 (see Figure 5), and suppose we want to find its (v_1, v_3, v_4)-neighbor. First, we must find the nearest common ancestor using the right to left scan just described. As our neighbor direction forces us to cross face $v_1 v_3 v_4$, we must look at the parent (21001). Keeping the same neighbor direction means that we must now cross face $v_2 v_3 v_4$ of the parent. Again, we must look at the next ancestor (2100). By keeping the same neighbor direction, we now cross face $v_1 v_2 v_3$ of the ancestor. Crossing face $v_1 v_2 v_3$ is our stopping condition, and thus we stop at 2100. The nearest common ancestor is 210 (see Figure 5), but we need to know which child contains t in order to get the appropriate sibling for the neighbor. Since 2100 has a sibling in the desired neighbor direction, we just invert the last bit to point to the new sibling. In this example, the sibling of 2100 is 2101, so the neighbor of 210011 which shares face $v_1 v_3 v_4$ is 210111 (see Figure 5).

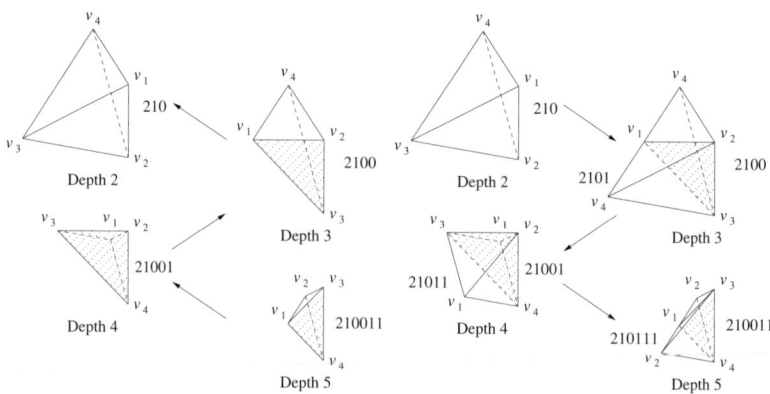

Fig. 5. Finding the (v_1, v_3, v_4)-neighbor of the 1/4 pyramid with location code 210011.

In [12], we have sketched an implementation of the above algorithm which performs neighbor finding in worst-case constant time. The algorithm makes use of the carry property of addition to find a neighbor efficiently without specifically searching for a nearest common ancestor. The iteration in the right to left scan in the neighbor finding algorithm is performed by an arithmetic operation that takes constant time instead of time proportional to depth of the tree. The algorithm makes use of just a few bit manipulation operations which can be implemented in hardware using a few machine language instructions. Of course, the constant-time bound arises because the entire path array for each location code is assumed to fit in one computer word. This, however, allows us to deal with data sets containing up to 10^{10} points.

5 Top-down Algorithms for Selective Refinement

In this section, we describe two algorithms for performing selective refinement on an HT, which are based on a depth-first and on a priority-based approach, respectively. A *selective refinement* operation applied to a multi-resolution mesh M consists of extracting a conforming mesh Σ from M such that Σ covers the domain D of M, the resolution of Σ satisfies some user-defined error requirements, and Σ is the mesh with the smallest number of tetrahedra satisfying the above conditions (see [3]). In [1], we have defined a set of basic queries for analysis and visualization of a volume data set, that we called *Level-Of-Detail (LOD)* queries, and we have shown that all of them are instances of selective refinement.

The depth-first algorithm is a standard refinement algorithm (see, for instance, [15]). It starts from the initial mesh, consisting of the six top-level tetrahedra which subdivide the cube, and initializes the currently extracted mesh (that we call the *current mesh*) with them. For any tetrahedron t, that does not satisfy the error requirements, we split the cluster c of tetrahedra which share their longest edge with t. If a tetrahedron t' in c does not exist in the current mesh, then we split the cluster associated with the parent of t'. This is applied recursively in order to guarantee a conforming mesh. The process continues until all tetrahedra in the current mesh satisfy the error requirements.

A pseudo-code description of the depth-first algorithm is reported below (since we are using it also as part of the incremental algorithm described in the next section). Predicate EXIST(t) returns the value *true* if tetrahedron t is part of the current mesh, and the value *false* otherwise. Function PARENT(t) deletes the last bit from the location code of t, thus returning the parent of t. Functions CHILD_0(t) and CHILD_1(t) add a 0-bit and a 1-bit to the end of the location code of t, respectively thus returning the first and the second child of t, respectively. Function SPLIT(t) replaces t in the current mesh with CHILD_0(t) and CHILD_1(t). Function CLUSTER(t) returns the set of all tetrahedra sharing the splitting edge $v_3 v_4$ with tetrahedron t. This set can be found by using the alternating sequence of face neighbors described in the previous section.

procedure CHECK_SPLIT(t);
/* Check if tetrahedron t satisfies the error requirement and split it if necessary. At start, it is called for each tetrahedron of the initial cube. */
begin

```
      if not(ERROR_REQ(t)) then
        begin
          SPLIT_CLUSTER(t);
          CHECK_SPLIT(CHILD_0(t));
          CHECK_SPLIT(CHILD_1(t));
        end;
    end;

    procedure SPLIT_CLUSTER(t);
    /* Split all tetrahedra within the same cluster as t. */
    begin
      foreach tetrahedron t' in CLUSTER(t) do
        begin
          if not(EXIST(t')) then SPLIT_CLUSTER(PARENT(t'));
          SPLIT(t');
        end;
    end;
```

It can be easily seen that the worst-case time complexity of the algorithm is $O(k)$, where k is the number of tetrahedra in the hierarchy. Since the tetrahedra are only split in SPLIT_CLUSTER if forced by our consistency constraint, and SPLIT_CLUSTER is only called if one of the tetrahedra is beyond the error threshold, the algorithm never splits a tetrahedron unless it is necessary, and thus, it generates the minimum number of tetrahedra which are required to satisfy the error requirements.

The priority-based algorithm applies the splitting process to the tetrahedra according to an error-driven sequence. The tetrahedron with the largest error is split at each iteration. As in the depth-first approach, when a tetrahedron t is split, all the tetrahedra in the same cluster as t must be split as well. If a tetrahedron t' in the cluster is not in the current mesh, the splitting process is recursively applied to the parent of t'. The algorithm makes use of a priority queue of tetrahedra, in which the order is based on the errors associated with the tetrahedra. Tetrahedra are only added to the priority queue if they violate the error requirements. To reduce the number of insertions and the overall size of the priority queue, we simply insert only one of two children of a tetrahedron if they belong to the same cluster, namely the one with the largest error. Only tetrahedra on the queue, or tetrahedra which must be included in the current mesh to maintain consistency, are considered for splitting. It can be easily seen that the algorithm generates the minimum number of tetrahedra necessary to satisfy the error requirements, and that the time complexity of the algorithm is also linear in the number of tetrahedra in the hierarchy.

The priority-based approach produces an interruptible algorithm i.e., it generates a fairly good approximation of the solution, if time has expired, or the number of tetrahedra in the current mesh is above a predefined bound.

On the other hand, using a priority queue increases the storage cost, but we have found experimentally that the size of the queue is on average equal to $10-16\%$ of the size of the output mesh.

Table 2 shows the number of tetrahedron splits and cluster computations per second performed by the in-core and the out-of-core versions of the depth-first algorithm and by the out-of-core version of the priority based algorithm. The results show that the in-core version of the depth-first algorithm is more than 50% faster than the out-of-core version, while the priority based algorithm is about 20% slower than its depth-first counterpart.

Algorithm	Tetrahedron Splits		Cluster Computations	
	Range	Average	Range	Average
Depth-first in-core	560000-695000	641000	115000-122000	118000
Depth-first out-of-core	250000-485000	402000	46000-85000	74000
Priority out-of-core	210000-465000	336000	39000-82000	61000

Table 2. The second column shows the number of tetrahedron splits per second, the third column shows the number of cluster computations per second.

For comparison purposes, we have also implemented a technique for extracting conforming meshes from a hierarchy of tetrahedra based on *error saturation*, as discussed in [16, 24]. First, all tetrahedra belonging to the same cluster are assigned the same error value, which is equal to the maximum of their original error values. Moreover, the approximation error associated with each tetrahedron is saturated to be greater than or equal to the error associated with its children. This implies that, during mesh extraction, if a tetrahedron is refined, then all tetrahedra belonging to the same cluster are refined.

Our experimental comparisons, that we have performed on the basis of LOD queries at a uniform resolution, have shown that the meshes extracted from a saturated HT have, on average, 5% more tetrahedra than those extracted with our method. On the other hand, the computing times of our depth-first algorithm are the same as those of the corresponding algorithm on a saturated HT (which simply performs a top-down traversal of the hierarchy without any neighbor finding computation). The saturation algorithm with the data structure in-core performs between 564000 to 708000 tetrahedron splits per second (645000 on average).

6 An Incremental Algorithm for Selective Refinement

In this section, we describe a selective refinement algorithm based on an *incremental* approach. The algorithm considers the current mesh resulting from a previous execution as the starting mesh and modifies such a mesh according to error requirements. Thus, the current mesh not only may be refined by splitting tetrahedra in a cluster, but also it may be coarsened by merging all tetrahedra incident at a vertex. Such an approach may give sub-optimal solutions if the error does not decrease monotonically, but it is very useful in highly interactive environments, since it minimizes the work performed in updating the mesh. Figure C.18 (see color plate) shows some snapshots of the execution from our implementation of the incremental algorithm: there is a moving box in the domain inside which we enforce the maximum resolution, while we accept an arbitrary resolution outside the box.

A pseudo-code description of the incremental algorithm is reported below (procedure UPDATE_MESH). Function MERGE(t) replaces CHILD_0(t) and CHILD_1(t) in the current mesh with t. The other primitives used in the algorithm description have been introduced in the previous section. Procedure UPDATE_MESH considers each tetrahedron in the input, or in the current mesh. For any tetrahedron, that does not satisfy the error requirements, a downward refinement operation is started. This is performed by CHECK_SPLIT (see description of the depth-first algorithm). For any tetrahedron that is too refined, CHECK_MERGE is called to coarsen the mesh in a bottom-up direction. Note that we never "force" a merge like we do in SPLIT_CLUSTER (where we force other splits). Merging the tetrahedra incident at a vertex can only be performed if all the tetrahedra to be merged are in the current mesh, and all the tetrahedra in the corresponding cluster satisfy the error requirements.

procedure UPDATE_MESH();
/* Update the current mesh so that the resulting tetrahedra satisfy the new error requirement. */
begin
  foreach tetrahedron t in Σ **do**
    if not(ERROR_REQ(t)) **then** CHECK_SPLIT(t)
    else CHECK_MERGE(t);
end;

procedure CHECK_MERGE(t);
/* Check if a merge is possible and then merge as appropriate. */
begin
  foreach tetrahedron t' in CLUSTER(PARENT(t)) **do**
    begin
      if not(EXIST(CHILD_0(t'))) **then** RETURN;
      if not(EXIST(CHILD_1(t'))) **then** RETURN;
      if not(ERROR_REQ(t')) **then** RETURN;
    end;
  foreach tetrahedron t' in CLUSTER(PARENT(t)) **do**
    MERGE(t');

        CHECK_MERGE(PARENT(t));
    end;

Our experiments with the incremental algorithm, when the data structure is maintained out-of-core, show that the approximate number of tetrahedron splits per second is about 280000 on average, which corresponds to approximately 51000 cluster computations per second. The number of tetrahedron merges per second is approximately 300000 on average, which corresponds to approximately 54000 cluster computations per second.

Note that the algorithm described in [8] also performs incremental selective refinement, but it makes use of two priority queues containing candidate tetrahedra to be split or to be merged (see also [4]). The use of priority queues make the algorithm interruptible at the expense of extra storage. In our future work, we plan to implement the priority-based approach in our framework to be able to compare it with our current incremental algorithm.

7 Experimental Results on LOD Queries

In this section, we report performance statistics on LOD queries on the HT implemented with the depth-first algorithm. We have used two volume data sets of different sizes and characteristics:

- **Plasma64** (274,625 vertices, 1,572,864 tetrahedra), a large synthetic data set whose field values represent the 3D Perlin's noise (courtesy of Visual Computing Group, National Research Council, Pisa, Italy).
- **Buckyball** (2,146,689 vertices, 12,582,912 tetrahedra), a very large regular data set (courtesy of AVS Inc).

We have considered just a uniform LOD query where the error is required to be smaller than a given threshold value over the whole domain, and, as an example of a variable LOD query, a query based on a field value (see Figure C.19 (right) in the color plate). In this latter case we require a high accuracy in the specified part of the domain or in the proximity of the specified isosurface, and a lower one elsewhere. Our results are in terms of the number of tetrahedra: this quantity is directly related to the complexity of the queries as the execution time of the selective refinement algorithms depends on such parameter.

Table 3 shows the number of tetrahedra in the extracted mesh for a uniform LOD query and the percentage of tetrahedra with respect to mesh at full resolution for both data sets. The error threshold is expressed as a percentage of the absolute value of the range of the field values in the data sets. Table 4 reports the same statistics for a variable LOD query based on a field value (see Figure C.19) on the Plasma data set. Different values of the error threshold have been selected for the tetrahedra intersecting the isosurface, while an arbitrary large value of the error is allowed on the other tetrahedra.

Uniform LOD Query				
error	**Plasma64**		**Buckyball**	
(% of field range)	tetrahedra	% tetrahedra	tetrahedra	% tetrahedra
0.1	1,493,696	94.9%	9,276,978	73.7%
0.5	620,996	39.4%	2,792,664	22.1%
1.0	337,384	21.4%	1,352,728	10.7%
5.0	16,800	1.1%	247,760	1.9%
10.0	2,818	0.2%	95,400	0.7%

Table 3. Number of tetrahedra in the meshes at *uniform* LOD and percentage with respect to the number of tetrahedra in the mesh at full resolution extracted from the HT representation of the Plasma and Buckyball data sets, respectively.

Variable LOD Based on Field Value		
error	**Plasma64**	
(% of field range)	tetrahedra	% tetrahedra
0.0	19,070	1.2%
0.1	18,266	1.1%
0.5	12,100	0.8%
1.0	9,748	0.6%
5.0	1,912	0.1%
10.0	810	0.1%

Table 4. Number of tetrahedra in the meshes at *variable* LOD and percentage with respect to the number of tetrahedra in the mesh at full resolution extracted from the HT representation of the Plasma data set (see Figure C.19 (right) in the color plate). The error within the proximity of the isosurface is specified in the left column.

8 Concluding Remarks

In this paper, we have considered a multi-resolution representation of a 3D scalar field based on nested tetrahedral meshes generated by tetrahedron bisection, that we call a Hierarchy of Tetrahedra (HT). We have proposed an implementation of an HT, and we have discussed how to extract conforming meshes from an HT by using an efficient neighbor finding algorithm. Our experiments have shown that a simple depth-first selective refinement algorithm based on our neighbor finding strategy extracts a mesh at a uniform LOD as fast as an algorithm which uses error saturation (and, thus, does not require any neighbor finding computation). We have discussed an alternative priority-based implementation of the selective refinement algorithm, which makes it interruptible with a negligible increase in memory (due to the priority queue) and with just 20% increase in computation times.

In [2] we have compared an HT with a multi-resolution model based on unstructured meshes built through edge collapse, called an *Edge-based Multi-Tessellation (Edge-based MT)* on the basis of their selectivity on a set of LOD queries. In our current and future work, we plan to compare our incremental algorithm with a priority-based approach, to investigate more effective solutions to store an HT on secondary storage, and to implement both the construction algorithm and the queries out-of-core. Finally, we are planning to develop data structures for an HT in which not all levels in the tree are necessarily present in order to apply our data structure to modeling implicit shapes defined through adaptive distance fields [6].

Acknowledgments

This work has been partially performed while Leila De Floriani has been a visiting professor at the Department of Computer Science of the University of Maryland, College Park, MD (USA). This work has been partially supported by the collaborative project funded by the Italian Ministry of Education, University and Research (MIUR) "MACROGeo: Algorithmic and Computational Methods for Geometric Object Representation".

References

1. P. Cignoni, L. De Floriani, P. Magillo, E. Puppo, and R. Scopigno. Interactive visualization of large tetrahedral meshes through selective refinement. *IEEE Transactions on Visualization and Computer Graphics*, 2003 (to appear).
2. E. Danovaro, L. De Floriani, M. Lee, H. Samet. Multi-resolution tetrahedral meshes: an analysis and a comparison In *Proceedings Shape Modeling International 2002*, Banff, Canada, May 2002.
3. L. De Floriani and P. Magillo. *Multi-resolution Mesh Representation: Models and Data Structures* . In *Principles of Multi-resolution in Geometric Modeling*,M.Floater, A.Iske, E. Quak (editors), Lecture Notes in Mathematics, Springer Verlag, Berlin (D), 2002.
4. M. Duchaineau, M. Wolinsky, D. E. Sigeti, M. C. Miller, C. Aldrich, and M. B. Mineev-Weinstein. ROAMing terrain: Real-time optimally adapting meshes. In *Proceedings IEEE Visualization'97*, R. Yagel and H. Hagen, eds., pages 81–88, Phoenix, AZ, October 1997.
5. W. Evans, D. Kirkpatrick, and G. Townsend. Right-triangulated irregular networks. *Algorithmica*, 30(2):264–286, March 2001.
6. S.F. Frisken, R.N. Perry, A.P. Rockwood and T.R. Jones. Adaptive sampled distance fields: a general representation of shape for Computer Grpahics. In *Proceedings SIGGRAPH 2000*, pages 249-254, 2000.
7. T. Gerstner and M. Rumpf. Multiresolutional parallel isosurface extraction based on tetrahedral bisection. In *Proceedings 1999 Symposium on Volume Visualization*, ACM Press, 1999.
8. B. Gregorski, M. Duchaineau, P. Lindstrom, V. Pascucci and K. Joy. Interactive View-Dependent Rendering of Large Isosurfaces. In *Proceedings IEEE Visualization 2002*, San Diego, CA, October 2002.
9. G. Greiner and R. Grosso. Hierachical tetrahedral-octahedral subdivision for volume visualization. *The Visual Computer*, 16, pages 357–369, 2000.

10. R. Grosso, C. Luerig, and T. Ertl. The multilevel finite element method for adaptive mesh optimization and visualization of volume data. In *Proceedings IEEE Visualization '97*, pages 387–394, Phoenix, AZ, October 1997.
11. D. J. Hebert. Symbolic local refinement of tetrahedral grids. *Journal of Symbolic Computation*, 17:457–472, 1994.
12. M. Lee, L. De Floriani and H. Samet. Constant-time neighbor finding in hierarchical tetrahedral meshes. In *Proceedings Shape Modeling International 2001*, pages 286–295, Genova, Italy, May 2001.
13. P. Lindstrom, D. Koller, W. Ribarsky, L. F. Hodges, N. Faust, and G. A. Turner. Real-time continuous level of detail rendering of height fields. In *Proceedings of the SIGGRAPH'96 Conference*, pages 109–118, New Orleans, August 1996.
14. P. Lindstrom and V. Pascucci. Visualization of Large Terrains Made Easy. In *Proceedings IEEE Visualization 2001*, pages 363–370, San Diego, CA, October 2001.
15. J. M. Maubach. Local bisection refinement for n-simplicial grids generated by reflection. *SIAM Journal on Scientific Computing*, 16(1):210–227, January 1995.
16. M. Ohlberger and M. Rumpf. Hierarchical and adaptive visualization on nested grids. *Computing*, 56(4):365–385, 1997.
17. R. Pajarola. Large scale terrain visualization using the restricted quadtree triangulation. In *Proceedings IEEE Visualization'98*, D. Ebert, H. Hagen, and H. Rushmeier, eds., pages 19–26, Research Triangle Park, NC, October 1998.
18. V. Pascucci. Slow Growing Subdivision (SGS) in any dimension: towards removing the curse of dimensionality. *Computer Graphics Forum*, Volume 21, Number 3, 2002.
19. V. Pascucci Multi-resolution indexing for hierarchical out-of-core traversal of rectilinear grids. In *Hierarchical and Geometrical Methods for Scientific Visualization*, G.Farin, H.Hagen, B. Hamann (editors), Springer Verlag, Heidelberg, Germany, 2003 (to appear).
20. V. Pascucci and C. L. Bajaj. Time critical isosurface refinement and smoothing. In *Proceedings IEEE Symposium on Volume Visualization*, pages 33–42, Salt Lake City, UT, October 2000.
21. M. Rivara and C. Levin. A 3D refinement algorithm for adaptive and multigrid techniques. *Communications in Applied Numerical Methods*, 8:281–290, 1992.
22. T. Roxborough, G. Nielson Tetrahedron-based, laest-squares, progressive volume models with application to freehand ultrasound data In *Proceedings IEEE Visualization 2000*, pages 93–100, 2000.
23. H. Samet. *Applications of Spatial Data Structures: Computer Graphics, Image Processing, and GIS*. Addison-Wesley, Reading, MA, 1990.
24. Y. Zhou, B. Chen, and A. Kaufman. A multi-resolution tetrahedral framework for visualizing regular volume data. In *Proceedings IEEE Visualization '97*, pages 135–142, Phoenix, AZ, October 1997.

Divisive Parallel Clustering for Multiresolution Analysis

Bjoern Heckel[1] and Bernd Hamann[2]

[1] Plumtree Software, 500 Sansome Street, San Francisco, California 94111, USA, Email: *bjoern_heckel@hotmail.com*
[2] Center for Image Processing and Integrated Computing/Department of Computer Science, University of California, Davis, 2063 Engineering, Davis, California 95616, USA, Email: *hamann@cs.ucdavis.edu*

Summary. Clustering is a classical data analysis technique that is applied to a wide range of applications in the sciences and engineering. For very large data sets, the performance of a clustering algorithm becomes critical. Although clustering has been thoroughly studied over the last decades, little has been done on utilizing modern multi-processor machines to accelerate the analysis process. We propose a scalable clustering technique that benefits from existing parallel computers and networks of workstations. It enables the creation of multiresolution representations for very large geometric data sets. The output of the clustering process can be used for interactive data exploration, supporting techniques like view-dependent rendering, user-guided refinement, or progressive transmission.

1 Introduction

Data sets consisting of giga- and terabytes of information have become increasingly common. The increasing level of detail in engineering applications (e.g., computational fluid dynamics) made possible by advances in computer systems, results in larger volumes of output. Massive data sets also reside in corporate data warehouses, storing information about business and production processes. In digital libraries, millions of documents are accumulated and available to individuals. The complexity of such massive data collections has far surpassed our cognitive abilities to fully understand them as single entities. To gain some form of higher-level insight into massive data it is crucial to develop technology that supports interactive data exploration at different levels of abstraction.

Clustering is a classical data analysis technique that has been studied thoroughly during the last decades [5]. It has also been adopted as a standard technique in the emerging field of data mining [4]. Clustering is used in an enormous variety of applications, ranging from fields like earthquake prediction, whale monitoring, marketing, psychology, biophysics, criminology, information retrieval, image processing, to phonetic taxonomy [2]. Its goal is to establish a set of groups such that objects assigned to the same group have certain similarities while they differ from objects in other groups.

These groups are not known a-priori and must be determined by examining the characteristics of the given objects. Often, one is interested in data partitions providing different levels of granularity - so called multiresolution levels.

Applied to massive data sets, hierarchical clustering can be used for feature extraction, data summary, or creation of categories that allow interactive exploration. For example, applied to a database of customer records, it yields sales patterns that can be used for a focused marketing campaign. Used in context with digital libraries, it produces a hierarchical index that helps in finding related documents and supports browsing by step-wise refinement. For very large data sets however, creating a cluster hierarchy might require significant time. Particularly, with dynamic data sets, i.e., the characteristics of objects change over time or objects are inserted and deleted, the performance of the clustering process is pivotal. In addition, the output of an analysis process depends on various configuration parameters that describe how to interpret the data. When the clustering process is part of the analysis or exploration loop, a fast algorithm is necessary to guarantee an acceptable response time.

In the field of scientific visualization, clustering has been introduced in various form by various researchers, including Heckel et al., for multiresolution analysis solving a variety of problems, ranging from feature extraction [9], and surface reconstruction [10] [11] to vector field compression [12] and mesh simplification [14]. Weber et al. use the multiresolution representation generated by a clustering process for a procedural grid generation method for scattered data approximation and visualization, see [13]. Since massive data sets become increasingly common in scientific visualization, it is important to create an environment that allows processing and exploring such large data sets. Creating multiresolution representations from large data sets $\frac{3}{4}$ as part of the data preprocessing $\frac{3}{4}$ enables interactive data exploration by supporting operations like view-dependent rendering, user-guided refinement, and progressive transmission.

We investigate the design and implementation of a parallel clustering approach (PaC) that is based on a divisive hierarchical paradigm. For a given data set and dissimilarity measure (or distance function) defined on the domain space, PaC creates either a set of multiresolution levels or a multiresolution hierarchy.

We believe that our approach might be helpful in applications such as parallel volume rendering [17] [19] [20] [21], parallelpolygon rendering [22], and remote visualization [18].

In the second section, we discuss sequential clustering approach. The parallelization is described in section three. In the fourth section, we explore the characteristics and performance of PaC. We provide conclusions in section five.

2 Sequential Hierarchical Clustering

The input for our clustering process is a sequence of n-dimensional vectors $\{v_i\}$. Each vector is describing an object o_i. The elements of these vectors a_j are considered attributes characterizing the represented object o_i. For the clustering process, a dissimilarity measure D is used to measure the 'distance' between pairs of objects, and between objects and groups of objects, so-called clusters, and between pairs of clusters. Each cluster is characterized by an attribute vector, called cluster center or cluster centroid, that is most similar to the subset of all objects. The cluster error is the sum of the distance between the cluster center and all objects the cluster represents. The global error is the sum of all cluster errors. The goal of the clustering process is to find a set of partitions of different sizes such that for each partition the global error is minimized. When it is desired to create a multiresolution hierarchy an additional constraint requires the set of partitions to be hierarchically nested.

Our method for creating multiresolution representation is based on an incremental and divisive paradigm. Initially, all data objects are placed in one cluster c_1 that is recursively split until a termination criterion is met. This could be, when a certain number of clusters has been created, a certain global error criterion is met, or no cluster exceeds a certain cluster error bound. In each step of the algorithm, we replace the cluster with the highest internal error by new clusters. For two new clusters and n objects that are assigned to the cluster to be split, there exist $2*n-1$ ways of dividing the cluster into two homogeneous child clusters. Even when the clustering criterion is guaranteed to produce convex clusters only, it is still computationally infeasible to examine all possible divisions. Therefore, a heuristic approach has to be used to calculate the centers of the child clusters. Performing reclassification only locally results in substantial performance gains in comparison to utilizing a global reclassification scheme, which is a NP-hard problem [5]. Effective approaches are, for example, successively removing objects from one cluster to build up the second cluster, selecting the most dissimilar pair of objects in the split cluster as 'seed points' for the child clusters, or - as we do - applying a local k-means algorithm. When it is not desired to create a multiresolution hierarchy, the global error can often be further reduced by reassigning objects in the neighborhood where a cluster has been replaced.

To determine the local neighborhood N_i of a cluster C_i, a neighborhood graph G (e.g., the Delaunay or Gabriel's graph, see [7]) is constructed incrementally using the cluster centers as vertices. A cluster C_1 is considered a neighbor cluster of a given cluster C_2, when the distance in the neighborhood graph does not exceed a neighborhood threshold t_N. The distance $Dis(C_1, C_2)$ in the neighborhood graph corresponds to the minimum number of edges to be traversed to reach C_1 from C_2. Alternatively, the neighborhood N of a cluster can be defined as the k closest neighbors, see [10].

After reclassification, clusters that have been affected by object reassignments are updated (i.e., the cluster center and cluster error are computed).

The neighborhood graph is updated locally by replacing a region R of G that is affected by the reclassification with an updated subgraph R'. Algorithm 1 summarizes our sequential clustering method:

Algorithm 1: Sequential hierarchical clustering.
(1)   While termination criterion is not met {
(2)   Determine cluster C to be split;
(3)   Create w new clusters C_i by splitting cluster C;
(4)   Determine Neighborhood N_i of cluster c_i;
(5)   Reclassify data of clusters in N_i;
(6)   Update changed clusters;
(7)   Update neighborhood N_i;
(8)   Update priority queue;
(9)   }

This approach yields multiple partitions of a data set, called multiresolution levels, for a varying number of clusters. Since generating a multiresolution hierarchy is a special case of computing multiresolution levels (the neighborhood threshold τ_N is zero), we will in the following only consider the more general approach.

Using a priority queue sorted by internal cluster error, the cluster to be split can be determined in constant time. We assume that new clusters can be created and initialized in constant time. The average number of neighbors of a cluster is expected to be constant for a fixed dimensionality of the data domain using, for example, a Delaunay graph or Gabriels's graph to define the local neighborhood. Therefore, the neighborhood of a cluster can be determined in constant time for an incrementally built graph. Since the expected number of neighbor clusters is expected to be constant, the complexity of reclassification as described above is expected to be proportional to the number of objects l to be reclassified . For a reclassification tree of depth d_t, the number of objects l to be reclassified is expected to be proportional to: $l \approx \frac{1}{2^{d_t}}$

Since the number of neighbors is bound by $O(1)$, steps (6) and (7) require constant time. Updating the priority queue is done in $O(log c)$ time, the number of clusters at the current iteration being c. The reclassification is by far the most time-consuming step, especially when clusters are large. Therefore, when designing a parallel clustering algorithm one should focus on exploiting parallelism during the reclassification stage.

3 Parallel Clustering

The parallelization of the algorithm described above is implemented in C++ utilizing MPI [15] as a low-level communication extension. Multiple instances of the parallel program are created on a set of processing nodes. Each instance, subsequently referred to as a task, has a unique identifier in the parallel application, called rank. Each task is assigned to a single processor on a network of workstations and/or a multiprocessor. The association of tasks and processors remains constant during the clustering process. The parallel application uses a master-worker model with a master task controlling the clustering process performed by several worker tasks, see [16].

The parallel clustering process is divided into three phases: In the first phase, the data is distributed over the tasks. In the second phase, all tasks work on computing one iteration of the clustering process (*cooperative parallel clustering*). In the third phase, each task independently computes one iteration (*concurrent parallel clustering*).

3.1 Data Distribution

During the first phase (initialization phase) the data set is read, and each vector is randomly assigned to exactly one of the worker tasks. Two different data distribution mechanisms have been implemented:

1. Data distribution using network transfer:
   The data is read from an input file by the master process and sent to the worker processes. This approach allows using a network of workstations that do not share a file system. To reduce the communication overhead during the data distribution, the master task is caching a certain number of vectors in buckets that are transmitted to a worker task, when an overflow occurs.

2. Data distribution using a shared file system:
   All worker tasks read the input file at the same time. Each read vector is used as a key for a pseudo-random hash function. This function determines which worker task should store a particular datum. Since all tasks use the same hash function and the same key a consistent assignment is guaranteed.

After distributing the data, each vector is stored in the associated task only. This association remains constant during the second phase. In the third phase, vectors might be reassigned and transferred to other tasks. However, each vector is assigned to exactly one task at all times of the clustering process.

The relative performance of the processing nodes can be different, e.g., when using a heterogeneous network of workstations as run-time environment. To minimize idle time of tasks on fast processors at synchronization points during the cooperative clustering phase and to increase overall performance, the number of vectors assigned to a task is chosen to be proportional to the relative performance of the processor a task is running on. The relative performance of a processing node is determined by using the sequential clustering program as a benchmark.

3.2 Cooperative Parallel Clustering

During the cooperative parallel clustering phase all worker tasks are helping to compute one iteration of the clustering process. The objects assigned to each cluster are distributed over all tasks. In each iteration, all worker tasks simultaneously split the same cluster with the highest priority and insert a new cluster. The steps (2) to (4) - that are identical to the sequential clustering - are carried out simultaneously by all tasks. The following local reclassification (step 5) is by far the most time-consuming and is conducted cooperatively. Each task generates a list of vectors in the region that is affected by reclassification. Since the data is randomly distributed over all tasks, each task stores a subset of the region. Then, each task assigns each locally stored vector to its closest cluster. Based on the resulting assignment of the local data, each task is recalculating the cluster centers of the reclassified region. Each task broadcasts a summary of the partial reclassification to the other tasks (step a in Algorithm 2). After receiving this information from all worker tasks (step b in Algorithm 2) each task updates the clusters in the reclassified region (step 6). Finally, all tasks simultaneously update the priority queue (step 7) and the neighborhood graph (step 8).

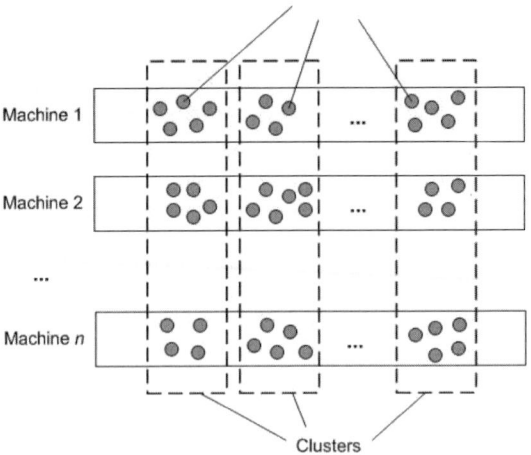

Fig. 1. Data distribution during phase 2, cooperative parallel clustering.

While the classification process progresses, the average cluster size decreases. As a result, the amount of work per iteration decreases, and the ratio of communication overhead and execution time increases. When the average cluster size does no longer exceed some threshold limit τ_S, the classification process enters the third phase, which utilizes a more efficient strategy for large sets of small clusters.

Algorithm 2: Cooperative parallel clustering.
(1)   For all worker tasks: While termination criterion is not met {
(2)   Determine cluster C to be split;
(3)   Create w new clusters C_i by splitting cluster C;
(4)   Determine Neighborhood N_i of cluster C_i;
(5)   Reclassify data of clusters in N_i;
(a)   Broadcast local cluster information;
(b)   Receive cluster information from other tasks;
(6)   Update changed clusters;
(7)   Update neighborhood N_i;
(8)   Update priority queue;
(9)   }

3.3 Concurrent Parallel Clustering

In the third phase, each task is concurrently computing one iteration of the clustering loop. Each cluster with its associated objects is assigned to exactly one task. At the end of the second phase, the domain space is partitioned in different regions, each of them being assigned to a particular worker task. In the third phase, all clusters in one region are initially hosted by the corresponding task. Since at the end of phase 2 clusters are distributed over all tasks, cluster fragments have to be transferred to a single task whenever it must be considered for reclassification by that task.

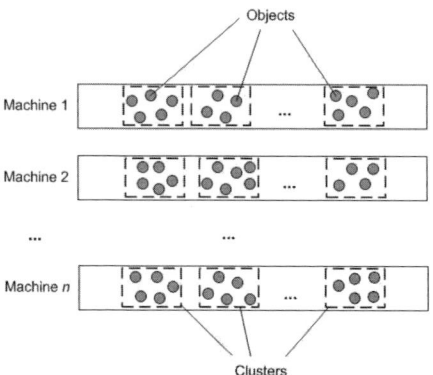

Fig. 2. Data distribution during phase 3, concurrent parallel clustering.

To avoid inconsistencies and ensure a deterministic behavior, a task can only split a cluster provided it has exclusive access to its neighbor clusters. When clusters in the local neighborhood are not present in the worker task, it sends out a request to the master task. The master tasks determines the host of the requested clusters and forwards the requests to the corresponding worker tasks, which transfer the ordered clusters directly to the requesting task. For a sufficiently large value of τ_S, the number of request collisions can be expected to be low. Therefore, the tasks can work relatively independent with very low communication cost. After reclassification, a worker task serves the outstanding cluster requests and transmits these clusters to the requesting task. By checking for requests directly before determining a split candidate, the interference of task is decreased. However, it is possible that a split candidate x is sent from task l to a remote task r when it is member of the split candidate's neighborhood in r. Since x's neighbor's might change after reclassification of r, the list of requested clusters in l has to change as well. Clearly, l cannot proceed unless x has been transferred back to l. When task l receives the previously determined split candidate x, it updates its request list to x's neighbors. For consistency reasons, cluster x is not permitted to be split by task r, or any other task different from l. After it has been identified as a split candidate by i, its 'dirty flag' is set. Dirty clusters are simply ignored during the determination of split candidates. Once x is split by l its dirty flag is cleared, and it might be split by a remote task after it has been transferred.

Algorithm 3: Concurrent parallel clustering.
(1)   For all worker tasks: While termination criterion is not met {
(2)   Determine cluster C to be split;
(3)   Create w new clusters C_i by splitting cluster C;
(4)   Determine Neighborhood N_i of cluster C_i;
(5)   Reclassify data of clusters in N_i;
(c)   Send requests to master task, if not locally present;
(d)   Check for requests and transfer clusters;
(6)   Update changed clusters;
(7)   Update neighborhood N_i;
(8)   Update priority queue;
(9)   }

Algorithm 3: Concurrent parallel clustering algorithm, worker task.

4 Application

To test the performance of our parallel clustering algorithm we have applied it to a set of unorganized points in 3D space. In this case, clustering is used to reconstruct surfaces at different level of resolution, each one described by a triangular mesh, see Figure 3. A detailed discussion of this technique and the application domain is provided in [10].

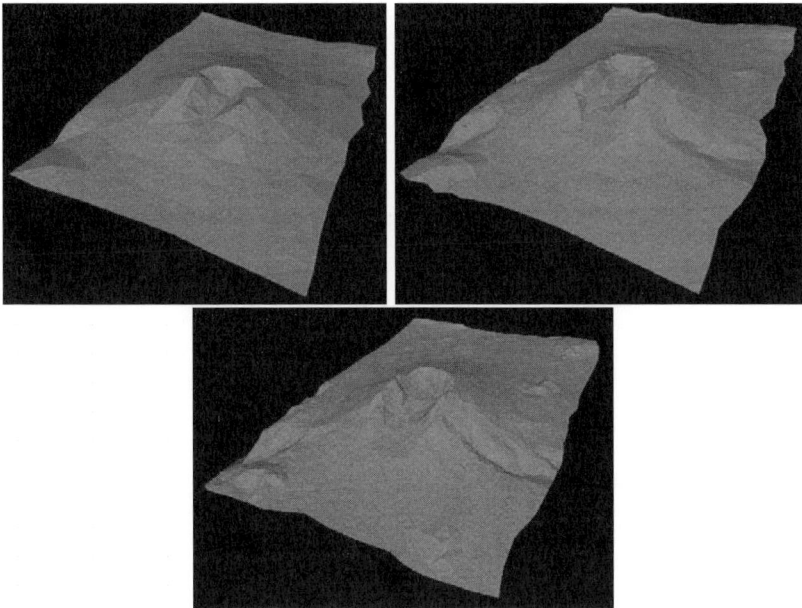

Fig. 3. Three resolution levels of Mt St. Helens data set.

4.1 Data Distribution Costs

Using the network transfer via the master task, the timing for data distribution is linear with respect to the size of the input. For very large data sets, this distribution mechanism requires substantial time and is about twenty times slower than the initialization in the sequential clustering program for our system configuration. This distribution mechanism is therefore a performance bottleneck that highly impacts the performance of the overall parallel clustering application. When all work tasks concurrently read from the input file and assign the data using a pseudo-random hash function as described above, the data distribution shows a performance similar to the sequential clustering algorithm.

4.2 Message Size

Using network transfer via the master task for the data distribution, the message length during the first phase depends on the chosen size of the buckets and the dimensionality of the domain. In the second phase, the message length depends on the dimensionality of the domain and the number of neighbors, which is a function of dimensionality. Our experiments have shown that the message size in phase two varies between about 200 and 400 bytes, with an average of about 260 bytes, for a 3D real space. In the third phase, the message length depends on the size of the cluster transferred between tasks, which is bounded by τ_S and is decreasing while the classification is progressing.

4.3 Performance

We have applied PaC to a topographic data set that consists of roughly 2 million 3D points. Each point describes the height z at a certain location (x,y). Our program is used to reconstruct the relief of the topography at different levels of detail. The size of the input file is 47.3 MB, and 100 clusters are computed. PaC was executed on an SGI Onyx with 4 processors and 512 MB main memory. The clustering time is measured for the parallel application and the sequential version of the clustering program (1*), see

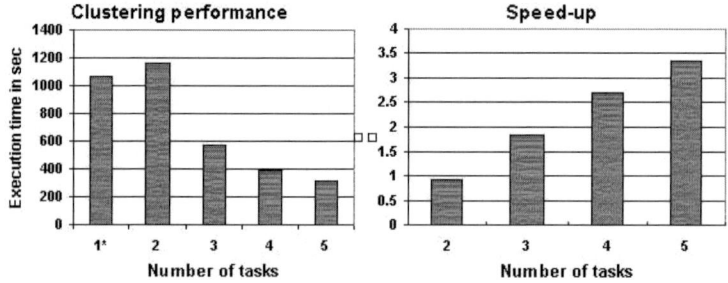

Fig. 4. The speed-up factor is computed by dividing the execution time of the sequential algorithm by the execution time of PaC. With two tasks and only one worker task, PaC exhibits a performance that is 8% slower than sequential program. Utilizing more processors yields an almost linear speed-up.

With four tasks, three processors are fully utilized, while one processor, running the master task, has a low utilization. Adding one task yields a higher performance, but increases the overhead for context switching, since more tasks than processors are used. The state diagram in Figure 5 shows that randomly distributing the data yields a good workload balance for the multiprocessor machine.

4.4 Workload Balancing

When using a network of workstations instead of a single multiprocessor machine, the relative performance of the used machines may vary. In this case,

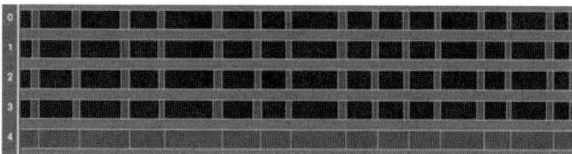

Fig. 5. States of 5 tasks on SGI Onyx 2 - light grey: time spent splitting and reclassifying clusters, dark grey: time spent for synchronization and updating clusters.

using a random distribution may degrade the performance of the existing hardware. In our experiment, we use six machines with varying relative performance, see Figure 6. The relative performance is measured by running the sequential clustering algorithm on the participating machines as a benchmark.

The network of workstations determines 100 clusters for a data set that consists of 100,000 3D points. The execution time is about 36 seconds. The state diagram in Figure 7 reveals that the first three worker tasks (tasks 0 through 3) spend much time on the update phase waiting for task number 4, which runs on the slowest machine (Angela) to reach the synchronization point.

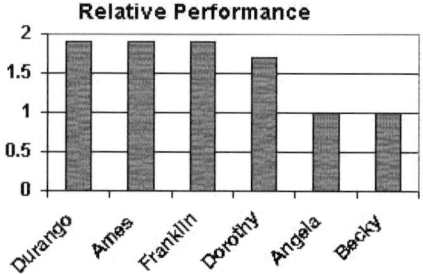

Fig. 6. Relative performance of six machines.

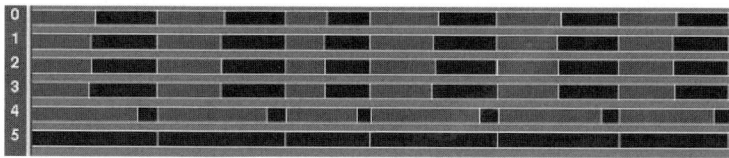

Fig. 7. States of six unbalanced tasks on a network of workstations - light grey: time spent splitting and reclassifying clusters, dark grey: time spent for synchronization and updating clusters.

Weighting the workload by the relative performance reduces the execution time to 26 seconds. The state diagram (Figure 8) shows that considering the relative performance results in a better processor utilization. The clustering time for the sequential program is 110 seconds on Durango, which is the fastest of the used machines. Considering that only five workers are used, this speed-up is almost ideal. The low utilization on the master task's machine also indicates that many more tasks could be used to cluster a sufficiently large data set using the proposed master-worker architecture.

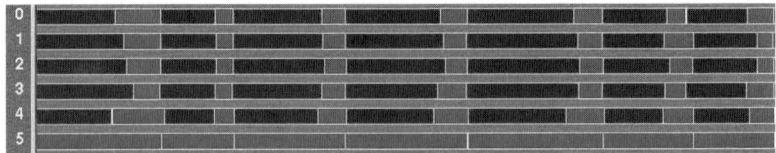

Fig. 8. States of six balanced tasks on a network of workstations - light grey: time spend splitting and reclassifying clusters, dark grey: time spent for synchronization and updating clusters.

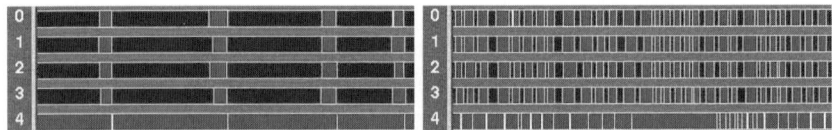

Fig. 9. States of five tasks at the beginning and the end of the clustering process - light grey: time spend splitting and reclassifying clusters, dark grey: time spent for synchronization and updating clusters.

4.5 Granularity

While the clustering in phase two progresses, the average cluster size decreases. This results in an increasing overhead/computation ratio in phase two, see Figure 9. In contrast, during phase three the overhead/computation ratio decreases over time, since the request collision and cluster transfer frequency decrease with an increasing number of clusters. To get the best of both approaches we start with cooperative clustering and switch to concurrent clustering after a sufficiently large number of clusters has been generated. The overhead for transferring the partial cluster fragments is spread out over time, since cluster fragments are only transferred on demand. The optimal value for the parameter τ_S must be determined empirically and is dependent on the system configuration (e.g., number of processing nodes, network topology, type of machines), the characteristics of the data set, and the error metric. For our system configuration, we have determined that the data set and τ_S have to be fairly large to gain considerable speed-ups compared to using phase two only.

5 Conclusion

We have presented a new method that enables to create cluster hierarchies utilizing modern multiprocessors and network of workstations. Our evaluation has demonstrated that our method is scalable and can be used to analyze very large data sets benefiting from existing multiprocessor machines. Our approach should be extremely beneficial for many applications, where the evaluation of multiresolution representations of very large data sets is required.

References

1. Allan D. Gordan, "Hierarchical Classification", in: P. Arabie, L. J. Hubert, G. De Soete, eds., Clustering and Classification, World Scientific Publ., River Edge, 1996.
2. Phipps Arabie, Lawrence J. Hubbert, "An Overview of Combinatorial Data Analysis", in: P. Arabie, L. J. Hubert, G. De Soete, eds., Clustering and Classification, World Scientific Publ., River Edge, 1996.
3. Richard Franke, Gregory M. Nielson, "Scattered Data Interpolation and Applications: A Tutorial and Survey", in: Hagen, H. and Roller D., eds., Geometric Modeling, Springer-Verlag, New York, 1991.
4. Usama Fayyad, Gregory Piatetsky-Shapiro, Padhraic Smyth, "The KDD Process for Extracting Useful Knowledge from Volumes of Data", in: Communications of the ACM, 39(11), pp. 27-34, November 1996.
5. P. Arabie, L. J. Hubert, G. De Soete, "Clustering and Classification", World Scientific Publ., River Edge, 1996.
6. M. de Berg, M. van Kreveld, O. Overmars, O. Schwarzkopf, "Computational Geometry - Algorithms and Applications", Springer-Verlag, Berlin, 1997.
7. A. Okabe, B. Boots, K. Sugihara, "Spatial Tessellations", John Wiley and Sons, Chichester, 1992.
8. R. Cypher, A. Ho, S. Konstantinidou, P. Messina, "A Quantitative Study of Parallel Scientific Applications with Explicit Communications", Journal of Supercomputing, 10(1):5-24, March 1996.
9. B. Heckel, and B. Hamann. Visualization of cluster hierarchies. In Proceedings of Photonics West Electronic Imaging '98, SPIE (The International Society for Optical Engineering), San Jose, California, January 1998.
10. B. Heckel, A. Uva and B. Hamann. Highly effcient generation of hierarchical surface models. In Proceedings of Visualization '98 (Hot Topics), Wittenbrink and Varshney, Eds., IEEE Computer Society Press, Los Alamitos, CA, Oct 1998, pp. 50-55.

11. B. Heckel, A. Uva, B. Hamann and Joy. Surface Reconstruction using adaptive clustering methods. Submitted to IEEE Transactions on Visualization and Computer Graphics.
12. B. Heckel, G. Weber, K. Joy, and B. Hamann. Multiresolution analysis of vector fields. To appear in Proceedings of IEEE Visualization '99, IEEE Computer Society Press, Los Alamitos, CA, Oct 1999.
13. G. Weber, B. Heckel, K. Joy, and B. Hamann. Procedural generation of triangulation-based visualizations. To appear in: Proceedings of Visualization '99 (Hot Topics), A. Varshney, C. M. Wittenbrink, H. Hagen, Eds., IEEE Computer Society Press, Los Alamitos, CA, Oct 1999.
14. B. Heckel. Clustering-based Multiresolution analysis for Scientific Visualization. Ph.D. thesis, Universite of California, Davis, March 2000.
15. W. Gropp, Ewing Lusk, and Anthony Skjellum. Using Mpi : Portable Parallel Programming With the Message-Passing Interface. MIT Press, Scientific and Engineering Computation Series, 1994.
16. Ian T. Foster. Designing and Building Parallel Programs : Concepts and Tools for Parallel Software Engineering. Addison-Wesley, 1994.
17. P. Lacroute. Real-time volume rendering on shared memory multiprocessors using the shear-warp factorization. In: Cox, M., Uselton, S. P. and Wittenbrink, C. M., Eds., Proc. 1995 Parallel Rendering Symposium, Atlanta, GA, October 30-31, 1995, pp. 15–22.
18. P. P. Li, W. H. Duquette, D. W. Curkendall. Remote interactive visualization and analysis (RIVA) using parallel supercomputers. In: M. Cox, S. P. Uselton, and C. M. Wittenbrink, eds., Proc. 1995 Parallel Rendering Symposium, Atlanta, GA, October 30-31, 1995, pp. 71–78.
19. K.-L. Ma. Parallel volume ray-casting for unstructured-grid data on distributed-memory architectures. In: Cox, M., Uselton, S. P. and Wittenbrink, C. M., eds., Proc. 1995 Parallel Rendering Symposium, Atlanta, GA, October 30-31, 1995, pp. 23–30.
20. K.-L. Ma, J. S. Painter, C. D. Hansen, M. F. Krogh. A data distributed, parallel algorithm for ray-traced volume rendering. In: Crockett, T., Hansen, C. and Whitman, S., eds., Proc. 1993 Parallel Rendering Symposium, San Jose, CA, October 25-26, 1993, pp. 15–22.
21. U. Neumann. Parallel volume-rendering algorithm performance on mesh-connected multiprocessors. In: Crockett, T., Hansen, C. and Whitman, S., eds., Proc. 1993 Parallel Rendering Symposium, San Jose, CA, October 25-26, 1993, pp. 97–104.
22. S. Whitman. A load-balanced SIMD polygon renderer. In: M. Cox, S. P. Uselton, and C. M. Wittenbrink, eds., Proc. 1995 Parallel Rendering Symposium, Atlanta, GA, October 30-31, 1995, pp. 63–69.

Hierarchical Large-scale Volume Representation with $\sqrt[3]{2}$ Subdivision and Trivariate B-spline Wavelets

Lars Linsen[1], Jevan T. Gray[1], Valerio Pascucci[2], Mark A. Duchaineau[2], Bernd Hamann[1], and Kenneth I. Joy[1]

[1] Center for Image Processing and Integrated Computing (CIPIC)
  Department of Computer Science
  University of California, Davis
  http://graphics.cs.ucdavis.edu/
  llinsen@ucdavis.edu, {grayj, hamann, joy}@cs.ucdavis.edu
[2] Center for Applied Scientific Computing (CASC)
  Data Science Group
  Lawrence Livermore National Laboratory, Livermore
  http://www.llnl.gov/casc/
  {pascucci1, duchaineau1}@llnl.gov

Summary. Multiresolution methods provide a means for representing data at multiple levels of detail. They are typically based on a hierarchical data organization scheme and update rules needed for data value computation. We use a data organization that is based on what we call $\sqrt[n]{2}$ subdivision, where n is the dimension of the data set. The main advantage of $\sqrt[n]{2}$ subdivision, compared to quadtree ($n = 2$) or octree ($n = 3$) organizations, is that the number of vertices is only doubled in each subdivision step instead of multiplied by a factor of 2^n, i.e., four or eight, respectively. To update data values we use n-variate B-spline wavelets, which yield better approximations for each level of detail. We develop a lifting scheme for $n = 2$ and $n = 3$ based on the $\sqrt[n]{2}$-subdivision scheme. We obtain narrow masks that provide a basis for out-of-core techniques as well as view-dependent visualization and adaptive, localized refinement.

1 Introduction

Multiresolution schemes are used in computer graphics mainly for editing and rendering curves and surfaces at multiple levels of resolution. While most existing schemes could, in principle, be generalized for higher-dimensional data, only a few have been extended to data (or functions) defined over three- or even higher-dimensional domains. The combined subdivision-wavelet scheme we are describing in this paper is driven by the need to represent trivariate data (or functions) at multiple resolution levels for scientific visualization.

Representing volume data hierarchically is important in the context of "volume modeling" and visualizing volume data, e.g., scalar or vector fields defined over volumetric domains. Visualizing inherently trivariate phenomena often requires one to apply rendering operations to volumetric data - examples being volume slicing via a cutting plane, isosurface extraction through marching-cubes-like algorithms, and ray casting. The multiresolution approximation approach we describe in this paper provides an elegant means of hierarchically organizing volume data, and we can use the resulting hierarchy to apply to its various levels volume data visualization methods.

We combine the $\sqrt[n]{2}$-subdivision hierarchy with update rules using n-variate B-spline wavelets to gain an n-dimensional multiresolution data representation. Multiresolution schemes have been studied extensively over the past decade. A survey of the main multiresolution approaches, considering also topological constraints, is given by Kobbelt [11]. These approaches can, for example, be used for a multiresolution representation of isosurfaces. However, when considering (bio-)medical imaging data, for instance, we must be able to switch quickly between isosurfaces corresponding to different isovalues, and when considering, for example, numerically simulated time-dependent hydrodynamics data, we even have to deal with isosurfaces changing over time. It is undesirable to store every single isosurface for all possibly important isovalues at different resolutions and load them during visualization. Instead, we devise a multiresolution volume data representation. We first develop a bivariate B-spline wavelet scheme for $\sqrt{2}$ subdivision and then generalize it to a trivariate B-spline wavelet scheme for $\sqrt[3]{2}$ subdivision. We have applied our techniques to bivariate as well as volumetric data.

For large-scale multiresolution representation, one should use regular rather than irregular data structures, since grid connectivity is implicit and data access simple for regular data. To overcome regular data structures' disadvantage of coarse granularity, we have developed the $\sqrt[n]{2}$-subdivision scheme we discuss in Section 3. Every $\sqrt[n]{2}$-subdivision step only doubles the number of vertices, which is a factor of $\sqrt[n]{2}$ in each of the n dimensions.

When using a wavelet scheme, the data value at a vertex **p** is updated when changing the level of detail, and thus the value varies with varying level of detail. On a coarse level, the value represents the value at **p** itself as well as an average value of a certain region around **p**. This approach leads to better approximations on coarser levels. Wavelets based on the $\sqrt[n]{2}$-subdivision scheme unfortunately have the disadvantage of creating over- and undershoots. For example, for isosurface extraction ($n = 3$) this characteristic can cause creation of isosurfaces (or isosurface components) that are not existent in the full resolution. Therefore, we use n-variate B-spline wavelets, which do not create over- and undershoots, and adjust them to the $\sqrt[n]{2}$-subdivision scheme.

B-spline wavelets have the property that they do not only influence the neighbors of a vertex **p**. Thus, when using out-of-core techniques to operate on or visualize large-scale data, a lot of data must be loaded from external memory with low I/O-performance. Furthermore, the adaptivity for view-dependent refinement techniques is restricted. Lifting schemes with narrow filters can be used to overcome this problem. We review and generalize the lifting scheme from [3] in Section 4. In Sections 5 and 7, we develop a similar lifting scheme for the $\sqrt[n]{2}$-subdivision scheme for $n=2$ and $n=3$. We provide results in Sections 6 and 8.

2 Related Work

Multiresolution volume representation is based on a 3D hierarchical data organization of irregular or regular type. Irregular data structures, see [4,7,8], use non-uniform refinement steps, which makes them highly adaptive. On the other hand, grid information must be stored and data access is not straightforward. Especially for large-scale data, additional memory requirements and memory organization needs are a major disadvantage of irregular structures.

For regular data organizations, octrees, see [15,17,20,24,27,34]; and tetrahedral grids, see [20], are common. For regular structures, grid connectivity is implicit and data is easily and quickly accessed. However, the refinement steps have to conform to the topological constraints, which makes regular structures less adaptive. To overcome this disadvantage, we use the $\sqrt[3]{2}$-subdivision scheme, a regular data organization supporting finer granularity. While, for example, an octree refinement step doubles the number of vertices in every dimension, which leads to a factor of eight, a $\sqrt[3]{2}$-subdivision step only doubles the overall number of vertices. Therefore, $\sqrt[3]{2}$ subdivision will, in general, require less vertices than octrees to satisfy specified image quality error bounds. Since finer granularity leads to higher adaptivity this fact still holds when using adaptive refinement techniques.

The splitting step of the $\sqrt[n]{2}$-subdivision scheme was introduced by Cohen and Daubechies [5] for $n=2$ and Maubach [18] for arbitrary n. It can be described by using triangular as well as quadrilateral meshes ($n=2$) or their counterparts for higher dimensions. In the following, we will consider the quadrilateral case and its generalization.

The refinement step of the approaches described in [6,22,35] is a longest-edge bisection applied to tetrahedral meshes. This step is equivalent to the splitting step of the $\sqrt[3]{2}$-subdivision scheme. However, these approaches do not represent a full subdivision scheme, since the averaging step is missing. Thus, these schemes are restricted to structured-rectilinear grids, where eight cuboids share a common vertex, and the cuboids have the same size. The $\sqrt[3]{2}$-subdivision scheme also applies to structured-curvilinear grids, where hexahedra of arbitrary shape (but with linear edges) are used instead of cuboids. The scheme can even handle extraordinary vertices, see [21].

Recently, Velho and Zorin [32] introduced $\sqrt{2}$-subdivision surfaces ($n=2$) by adding an averaging step. They showed that the produced surfaces are C^4-continuous at regular and C^1-continuous at extraordinary vertices. (For an introduction to subdivision methods, we refer to [33].)

The main advantage of wavelet schemes is the fact that they provide a means to generate good approximations in a multiresolution hierarchy. Stollnitz et al. [28] described how to generate wavelets for subdivision schemes. However, $\sqrt[n]{2}$-subdivision wavelets can lead to over- and undershoots, see Figure 7(b), which are especially disturbing when extracting isosurfaces from different levels of approximation. They can even cause topological changes of isosurfaces when changing the level of resolution. Therefore, we have decided to generate B-spline wavelets for the $\sqrt[n]{2}$-subdivision scheme, which are known to produce good approximations. (For an introduction to B-spline techniques, we refer to [25].)

The computation of wavelet coefficients at a certain vertex for wavelets with good approximation quality like B-spline wavelets is not limited to using only adjacent vertices. Localization, however, is strongly desirable when we want to apply the wavelet scheme to adaptive refinement and to out-of-core visualization techniques. Lifting schemes as introduced by Sweldens [29] decompose wavelet computations into several steps, but they assert narrow filters, see Figure 6. Bertram et al. [2, 3] defined a lifting scheme for 1D and 2D B-spline wavelets using a quadtree organization of the vertices.

Wavelets for general dilation matrices go back to Riemenschneider and Shen [26] who used a box-spline approach. Kovačević and Vetterli [13] and, more recently, Uytterhoeven [31] and Kovačević and Sweldens [12] developed lifting schemes that can be applied to $\sqrt[n]{2}$-subdivision data structures. Uytterhoeven [31] only addressed the 2D case. In [12], the filters that produce good approximations are not narrow enough for our purposes. On the other hand, the update rule for the narrow filters in [12] is the identity, which does not support the creation of good approximations.

Another main difference between our approach and the non-separable filters used in [31] and [12] is the update rule. For example, we update the vertices in a $\sqrt[3]{2}$-subdivision scheme by applying first the 3D, then the 2D, and finally the 1D update rules. This approach automatically includes the boundary cases, which are not sufficiently addressed in [31] and [12]. Moreover, the generalization to arbitrary dimension n is straight-forward.

3 The $\sqrt[n]{2}$-subdivision Scheme

We first describe the case $n = 2$. For a $\sqrt{2}$-subdivision step of a quadrilateral Q, we compute its centroid $\mathbf{c}$, and connect $\mathbf{c}$ to all four vertices of Q. The "old" edges of the mesh are removed (except for the edges determining the mesh/domain boundary). Figure 1 illustrates four $\sqrt{2}$-subdivision steps.

The mask used for the computation of the centroid $\mathbf{c}$ is given in Figure 2(a). Figure 2(b) shows the mask of the averaging step according to [32]. A $\sqrt{2}$-subdivision step is executed by first applying the mask shown in Figure 2(a), which inserts the new vertices, and then (after the topological mesh

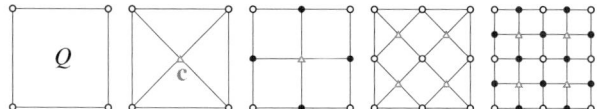

Fig. 1. $\sqrt{2}$ subdivision.

modifications) applying the mask shown in Figure 2(b), which repositions the old vertices.

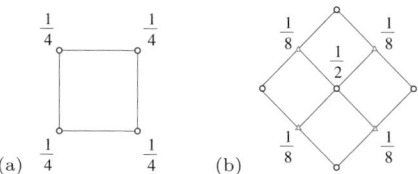

Fig. 2. Masks of $\sqrt{2}$-subdivision step: (a) inserting centroid; (b) repositioning old vertices.

This subdivision scheme for quadrilaterals is analogous to the $\sqrt{3}$-subdivision scheme of Kobbelt [10] for triangles. Therefore, we call it $\sqrt{2}$ subdivision.

We now generalize the subdivision scheme to $\sqrt[n]{2}$ subdivision for arbitrary dimension n. The splitting step is executed by inserting the centroid and adjusting vertex connectivity. The averaging step applies to every old vertex **v** the update rule

$$\mathbf{v} = \alpha \mathbf{v} + (1 - \alpha)\mathbf{w} ,$$

where **w** is the centroid of the adjacent new vertices.

We are especially interested in the case $n = 3$. Little research has been done to date concerning 3D (volumetric) subdivision. One example is the work described in [16]. The $\sqrt[n]{2}$-subdivision scheme in this general setting is discussed in [21]. The literature currently provides no analysis of averaging steps for dimensions larger than two. Thus, at present, we cannot provide a solution for the choice of α used in the update rule.

However, when applying the $\sqrt[3]{2}$-subdivision scheme to large volumetric data sets, we usually deal with structured-rectilinear grids, especially when considering imaging data sets. For structured-rectilinear grids, the update rule does not change the position of the vertices regardless of the specific α value, but it only affects the values at the vertices. In Section 6, we show that the $\sqrt{2}$-subdivision wavelets are not appropriate for our purposes, and we replace them by B-spline wavelets. Thus, we do not need to choose a value for α.

In Figure 3, three $\sqrt[3]{2}$-subdivision steps are shown. In each step, the centroids of the polyhedral shapes are inserted, and the connectivity is adjusted. Three kinds of polyhedral shapes arise. They are shown in Figure 4.

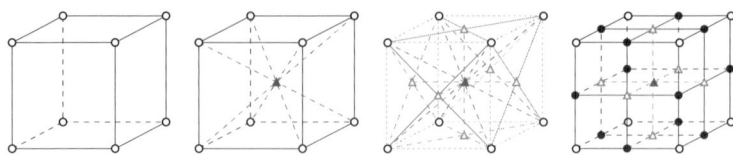

Fig. 3. $\sqrt[3]{2}$ subdivision.

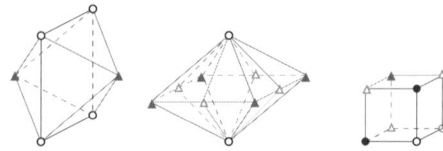

Fig. 4. Polyhedral shapes created by $\sqrt[3]{2}$ subdivision: octahedron, octahedron with split faces, and cuboid.

In the first step, each cuboid (first picture of Figure 3 / third picture of Figure 4) is subdivided by inserting the cuboid's centroid and connecting the centroid to all old vertices (second picture of Figure 3). In the second step, each octahedron (first picture of Figure 4) is subdivided by inserting the octahedron's centroid and connecting the centroid to all old vertices, while all old edges, except the edges inserted in the last subdivision step, are deleted (third picture of Figure 3). In the third step, each octahedron with split faces (second picture of Figure 4) is subdivided by inserting its centroid and connecting the centroid to all old vertices, except the vertices inserted in the next-to-the-last subdivision step (▲), while all old edges, except the edges between the vertices inserted in the next-to-the-last subdivision step (▲) and the vertices inserted in the last step (△), are deleted (fourth picture of Figure 3).

The three subdivision steps can also be described in the following way: The first step inserts the centroid of the cuboid, the second step inserts the centers of the faces of the original cuboid, and the third step inserts the midpoints of the edges of the original cuboid. Three $\sqrt[3]{2}$-subdivision steps produce the same result as one octree refinement step. Hence, for multiresolution purposes, we obtain a much finer granularity through $\sqrt[3]{2}$ subdivision, which reduces the complexity of the scenes to be rendered. If, for example, the resolution in the second picture of Figure 3 suffices to meet a certain screen-space error bound, a $\sqrt[3]{2}$-subdivision hierarchy can provide this resolution, whereas an octree-hierarchy would have to use four times the amount of data (as in the fourth picture of Figure 3). Thus, using the $\sqrt[3]{2}$-subdivision approach, one must render much less data to obtain a desired image quality. The additional effort spent when using the octree approach does not lead to better results, since the improvements projected onto screen are in subpixel range.

4 The B-spline Wavelet Lifting Scheme

In this section, we review and define masks for the 1D lifting scheme of [2] and generalize them to the 2D and 3D cases. In the following sections, we will adjust the 2D lifting scheme to $\sqrt{2}$ subdivision and the 3D lifting scheme to $\sqrt[3]{2}$ subdivision.

The 1D B-spline wavelet lifting scheme makes use of two operations that are defined by the following two masks, called s-lift and w-lift:

$$\text{s-lift}(a, b) : \begin{pmatrix} a & b & a \end{pmatrix}, \tag{1}$$

$$\text{w-lift}(a, b) : \begin{pmatrix} a & b & a \end{pmatrix}. \tag{2}$$

The s-lift mask is applied to the old vertices ○ and their new neighbors ●, whereas the w-lift mask is applied to the new vertices ● and their neighbors ○, see Figure 5(a). For a detailed derivation of the lifting scheme that we use as a basis for this paper, as well as for its analysis (smoothness, stability, approximation order, and zero moments), we refer to [1].

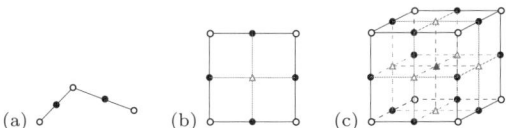

Fig. 5. Refinement step for 1D, 2D, and 3D meshes.

Using the s-lift and w-lift masks, a linear B-spline wavelet encoding step is defined by sequentially executing the two operations

$$\text{w-lift}(-\tfrac{1}{2}, 1) \text{ and}$$
$$\text{s-lift}(\tfrac{1}{4}, 1) .$$

A linear B-spline wavelet decoding step is defined by sequentially executing the two operations

$$\text{s-lift}(-\tfrac{1}{4}, 1) \text{ and}$$
$$\text{w-lift}(\tfrac{1}{2}, 1) .$$

Figure 6 illustrates the 1D lifting scheme.

When applying 2D B-spline wavelets to a quadtree-organized set of vertices, two kinds of new vertices are obtained when executing a refinement step, namely the new vertices inserted at the midpoints ● of old edges and the new vertices inserted at the centers △ of old faces, see Figure 5(b). Therefore, we have two different masks. We derive the needed 2D masks by convolution of the 1D masks in the two coordinate directions. This leads to:

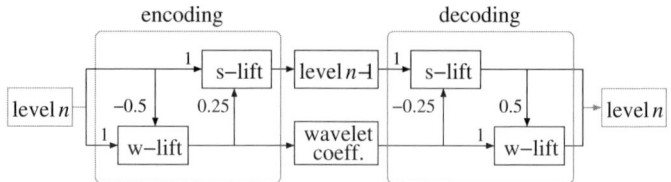

Fig. 6. 1D linear B-spline wavelet lifting scheme.

$$\text{s-lift}(a,b): \begin{pmatrix} a^2 & ab & a^2 \\ ab & b^2 & ab \\ a^2 & ab & a^2 \end{pmatrix}, \quad \begin{pmatrix} a & b & a \end{pmatrix}, \quad (3)$$

$$\text{w-lift}(a,b): \begin{pmatrix} a^2 & ab & a^2 \\ ab & b^2 & ab \\ a^2 & ab & a^2 \end{pmatrix}, \quad \begin{pmatrix} a & b & a \end{pmatrix}. \quad (4)$$

The 1D masks defined by (3) and (4) are applied in both directions. The masks in (3) as well as the masks in (4) are applied simultaneously.

When applying 3D B-spline wavelets to an octree-organized set of vertices, three kinds of new vertices are obtained when executing a refinement step, namely the new vertices inserted at the midpoints ● of old edges, the new vertices inserted at the centers △ of old faces, and the new vertices inserted at the centroids ▲ of old cubes, see Figure 5(c). Therefore, we have three different masks. For 3D masks, we show the structure of the mask and separately define the values for the vertices ○, ●, △, and ▲. We derive the needed 3D masks by convolution of the 1D masks in all three coordinate directions. The s-lift(a,b) masks are defined by this depiction:

$$\begin{matrix} \circ & b^3 \\ \bullet & ab^2 \\ \triangle & a^2b \\ \blacktriangle & a^3 \end{matrix} \quad, \quad \begin{pmatrix} a^2 & ab & a^2 \\ ab & b^2 & ab \\ a^2 & ab & a^2 \end{pmatrix}, \quad \begin{pmatrix} a & b & a \end{pmatrix}. \quad (5)$$

The 1D, 2D, and 3D masks are applied simultaneously to update the vertices ○, ●, and △, respectively. The w-lift(a,b) masks are defined by this depiction:

$$\begin{matrix} \circ & a^3 \\ \bullet & a^2b \\ \triangle & ab^2 \\ \blacktriangle & b^3 \end{matrix} \quad, \quad \begin{pmatrix} a^2 & ab & a^2 \\ ab & b^2 & ab \\ a^2 & ab & a^2 \end{pmatrix}, \quad \begin{pmatrix} a & b & a \end{pmatrix}. \quad (6)$$

5 A Lifting Scheme for $\sqrt{2}$ Subdivision

Using $\sqrt{2}$ subdivision instead of a quadtree-based scheme, we only obtain new vertices at the centers △ of old faces when executing a subdivision step;

at the midpoints ● of old edges, no vertices are inserted, see second picture in Figure 1 and compare to Figure 5(b). Thus, no data is available at the positions of the vertices ●, and we must adjust the 2D masks in (3) and (4).

For encoding with linear B-spline wavelets, the w-lift operation is executed first. Since we have no values at the positions ● required for mask (4), we linearly interpolate the values at the vertices ○. Linear interpolation is appropriate, since we are using linear wavelets. This approach changes mask (4) to

$$\text{w-lift}_{encode}(a,b): \begin{pmatrix} a^2+ab & a^2+ab \\ & b^2 & \\ a^2+ab & a^2+ab \end{pmatrix}. \quad (7)$$

Next, the s-lift operation is executed. Again, we have entries at the positions ● in mask (3). However, the w-lift operation has (theoretically) executed the 1D mask in (4), and we assumed that the values at the vertices ● were linear interpolations of the values at the vertices ○; therefore, the values at the vertices ● have vanished. Mask (3) changes to

$$\text{s-lift}_{encode}(a,b): \begin{pmatrix} a^2 & & a^2 \\ & b^2 & \\ a^2 & & a^2 \end{pmatrix}. \quad (8)$$

For decoding, we first execute the s-lift operation. Prior to executing the s-lift operation of the encoding, the values at the vertices ● have vanished, but the s-lift operation (theoretically) executed the 1D mask in (3). Hence, the values at the vertices ● are now given by linear interpolation of the values at the neighbor vertices △ multiplied by the factor $2a$ of the 1D mask in (3). We rename the factor a to $\bar{a}$ and derive from mask (3) the new mask

$$\text{s-lift}_{decode}(a,b): \begin{pmatrix} a^2+2\bar{a}ab & a^2+2\bar{a}ab \\ & b^2 & \\ a^2+2\bar{a}ab & a^2+2\bar{a}ab \end{pmatrix}. \quad (9)$$

Finally, the w-lift operation is executed again. The s-lift decoding operation has (theoretically) applied the 1D mask in (3). Since the 1D mask in (3) applied by the s-lift decoding operation is the inverse of the 1D mask in (3) applied by the s-lift encoding operation, the values at the vertices ● are the same as before the execution of these two s-lift operations, i.e., they vanish. These considerations define a new mask derived from mask (4), given by

$$\text{w-lift}_{decode}(a,b): \begin{pmatrix} a^2 & & a^2 \\ & b^2 & \\ a^2 & & a^2 \end{pmatrix}. \quad (10)$$

In the 2D case, the masks are as narrow as they can be.

6 Results for the 2D Case

In Figure 7, we provide an example for $\sqrt{2}$ subdivision and 2D wavelets. The original surface shown in Figure 7(a) results from sampling a 2D Gaussian function at 64^2 vertices. The surface is encoded and decoded again. In Figure 7(b), we show a coarse level of detail obtained by $\sqrt{2}$-subdivision wavelets. In Figure 7(c), we show the same level of detail obtained when combining bilinear B-spline wavelets and $\sqrt{2}$ subdivision in the way described in the previous section.

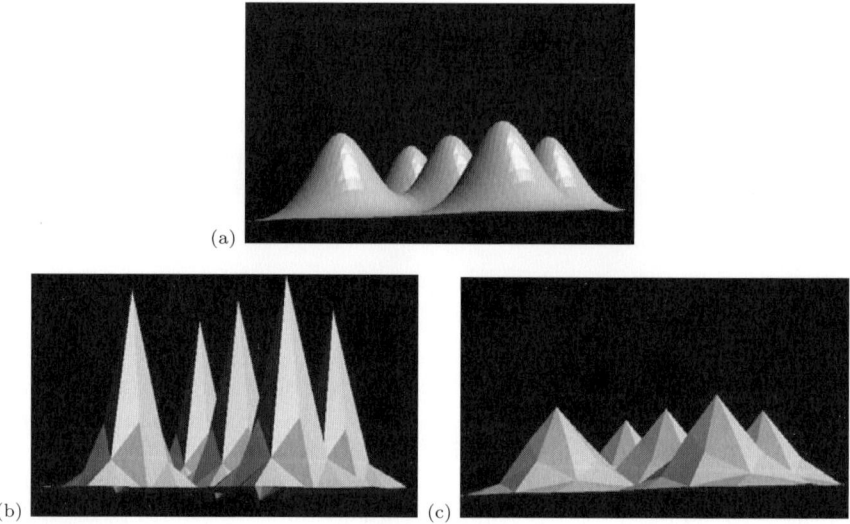

Fig. 7. (a) $\sqrt{2}$-subdivision surfaces; (b) encoded and decoded by $\sqrt{2}$-subdivision wavelets; and (c) bilinear B-spline wavelets.

In Figure 7(b), the over- and undershoots caused by the $\sqrt{2}$-subdivision wavelets can be recognized. No over- and undershoots are visible when combining $\sqrt{2}$ subdivision with linear B-spline wavelets, see Figure 7(c).

We also have developed a lifting scheme for cubic B-spline wavelets, but the masks are not as narrow as in the linear case, three instead of two lifting steps are required, see [2], and, most importantly, over- and undershoots appear again. Since linear B-spline wavelets, contrary to cubic ones, have interpolating scaling functions, interpolating refinement filters are guaranteed, see [12], i.e., no over- and undershoots can appear.

For progressive visualization, e.g. when generating images progressively by loading data from slow external memory or via Internet, the storage of values can be reorganized as shown in Figure 8. Progressive visualization starts by using the upper left block in the right picture, then adding the upper right block, and, finally, adding the lower block. Reordering ensures that data can be read in a continuous stream without reading data multiple times.

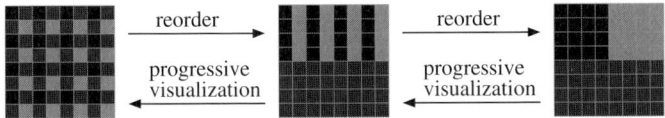

Fig. 8. Reordering data for progressive visualization.

7 A Lifting Scheme for $\sqrt[3]{2}$ Subdivision

In this section, we generalize the ideas of Section 5 to the 3D case. Recalling the steps of a $\sqrt[3]{2}$-subdivision scheme depicted in Figure 3, after the execution of the different steps different kinds of polyhedral shapes arise, see Figure 4. Therefore, we have to distinguish between the different steps. The following description starts with the situation shown in the second picture of Figure 3 (*volume case*), proceeds with the situation shown in the third picture of Figure 3 (*face case*), and finally treats the situation shown in the fourth picture of Figure 3 (*edge case*), which is topologically equivalent to the situation shown in the first picture of Figure 3.

The volume case. To perform linear B-spline wavelet encoding in the situation shown in the second picture of Figure 3, we first execute a w-lift operation. Therefore, we apply masks being similar to masks in (6), subject to the constraint that no values are available at the vertices • and △.

Regarding the structures of masks (6), we assume that the value at a vertex • is defined by linear interpolation of the values at the two vertices ○ (with which the vertex • shares an edge), and that the value at a vertex △ is defined by bilinear interpolation of the values at the four vertices ○ (with which the vertex △ shares a face). One obtains the mask w-lift$_{encode}(a, b)$, depicted as

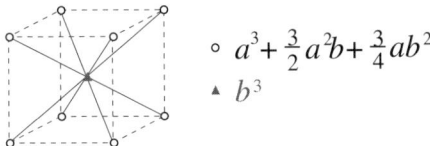

○ $a^3 + \frac{3}{2}a^2b + \frac{3}{4}ab^2$
▲ b^3

The masks being analogous to the 2D and 1D masks in (6) are only "applied theoretically." However, since the values at the vertices • are assumed to be linear interpolations of the values at the vertices ○, and since the values at the vertices △ are assumed to be bilinear interpolations of the values at the vertices ○, the values at the vertices • and △ vanish. Therefore, the mask for the next s-lift operation, which is an analogue of the 3D mask in (5), reduces to the mask s-lift$_{encode}(a, b)$, depicted as

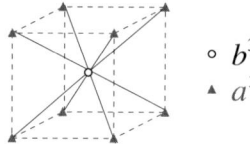

Again, the analogous versions of the 2D and 1D masks in (5) are only applied theoretically.

For the decoding step, we start with the s-lift operation, i.e., we adjust mask (5). Having (theoretically) applied the 2D and 1D masks in (5) with vanishing values at the vertices ● and △, the values at the vertices △ are linear interpolations of the values at the neighbor vertices ▲, multiplied by the factor $2a$, and the values at the vertices ● are bilinear interpolations of the values at the neighbor vertices ▲, multiplied by the factor $4a^2$. By renaming the factor a to $\bar{a}$, we obtain the mask s-lift$_{decode}(a,b)$, depicted as

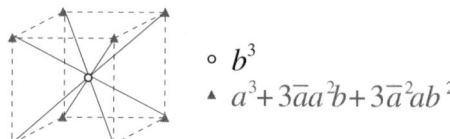

Again, the analogous versions of the 2D and 1D masks in (5) are only applied theoretically. Since the masks (5) of this s-lift operation are the inverse masks of the masks (5) of the encoding s-lift operation, the vertices ● and △ have their former values assigned again, i.e., the values vanish. Hence, the mask for the final w-lift operation, which is the mask being analogous to the 3D mask in (6), reduces to the mask

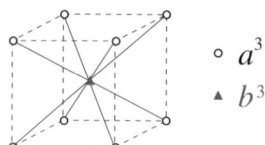

In the 3D case, the masks are as narrow as they can be.

The face case. When applying linear B-spline wavelet encoding to the situation depicted in the third picture of Figure 3, we have to make sure that we do not violate the assumptions made for the volume case. We assume that the values at the vertices △ are bilinear interpolations of the values at the neighbor vertices ○. Thus, when the values at the vertices △ are available, their values should be computed only from the values at the vertices ○. This insight leaves us with the 2D case, and we can apply masks (7) – (10) of Section 5.

The edge case. When applying linear B-spline wavelet encoding to the situation illustrated in the fourth picture of Figure 3, we must not violate the assumption that the values at the vertices ● are linear interpolations of the values at the neighbor vertices ○. When the values at the vertices ● are available, their values should be computed only from the values at the vertices

o. This insight leaves us with the 1D case, and we can apply masks (1) and (2) of Section 4.

It is a significant advantage of our scheme that the face and edge cases cover naturally boundary faces and boundary edges of the domain.

8 Results for the 3D Case

In this section, we compare the results obtained by applying a $\sqrt[3]{2}$-subdivision multiresolution scheme with and without trilinear B-spline wavelet encoding. Since we want to show how our wavelets improve image quality at a low resolution, all examples are provided at a coarse level of detail.

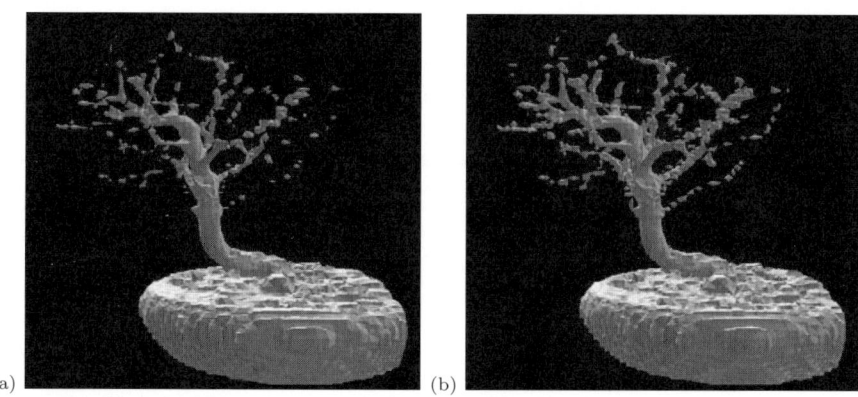

Fig. 9. Comparing $\sqrt[3]{2}$-subdivision hierarchy without (a) and with (b) trilinear B-spline wavelets. Shown is the same isosurface extracted from the level of detail with downsampling ratios 2^6. (Data set courtesy of S. Roettger, Abteilung Visualisierung und Interaktive Systeme, University of Stuttgart, Germany)

The data set used in Figure 9 is a 256^3 uniform rectilinear grid, and at every vertex one scalar value between 0 and 255 is given. The data set represents a "bonsai tree solid." It was obtained by computer tomography. For the visualization of the bonsai tree, we extracted and rendered the isosurface corresponding to the value 80, which was generated by the marching-tetrahedra algorithm described in [9]. All the polyhedral shapes in Figure 4 have a unique subdivision into tetrahedra according to the longest-edge bisection refinement. Thus, all visualization methods based on tetrahedra, including the more sophisticated isosurface-extraction methods in [6, 22], could be applied.

Figure 9(a) shows the isosurface extracted from a coarse level of detail of a $\sqrt[3]{2}$-subdivision hierarchy without using wavelets. Figure 9(b) shows the same isosurface extracted from the same level of detail, where a $\sqrt[3]{2}$-subdivision

Table 1. Root-mean-square errors for three examples at different levels of resolution without (w/o) and with (w/) trilinear B-spline wavelets.

$\mathcal{E}_{RMS}$	Figure 9		Figures 10 and C.20		Figure C.21		Figures 9-C.21
dr	w/o	w/	w/o	w/	w/o	w/	improvement
2^3	1.79%	1.59%	2.84%	2.58%	1.77%	1.57%	11.80%
2^6	3.63%	3.13%	3.84%	3.46%	4.29%	3.90%	12.32%
2^9	6.02%	5.05%	5.07%	4.61%	7.60%	6.98%	12.69%
2^{12}	8.99%	7.51%	6.84%	6.25%	10.94%	9.90%	13.22%
2^{15}	12.26%	9.64%	9.23%	8.72%	14.35%	12.71%	15.31%

hierarchy was combined with the trilinear B-spline wavelet scheme described in the previous section. The resolution is not high enough to represent the finest details, like branches and twigs, but the averaging steps of a wavelet encoding clearly leads to better approximations.

To quantify the improvement in approximation quality, we computed an approximation error for each coarser level of approximation by comparing it to the original, highest resolution level. Given the original function F discretely by sample values at locations $\mathbf{x_i}$, $\mathbf{i} \in [1, n_x][1, n_y][1, n_z]$, we used the root-mean-square error

$$\mathcal{E}_{RMS} = \sqrt{\frac{1}{n_x n_y n_z} \sum_{\mathbf{i}} (F(\mathbf{x_i}) - f(\mathbf{x_i}))^2} \, ,$$

where $f(\mathbf{x_i})$ denotes the approximated function value obtained by trilinear interpolation applied to a "cell" in the coarser level of resolution: If f is defined at corner locations $\mathbf{y_j} = (y_{\mathbf{j},x}, y_{\mathbf{j},y}, y_{\mathbf{j},z})$, and if $\mathbf{x_i}$ is inside the interval $[y_{\mathbf{j},x}, y_{\mathbf{j}+\mathbf{e}_1,x}][y_{\mathbf{j},y}, y_{\mathbf{j}+\mathbf{e}_2,y}][y_{\mathbf{j},z}, y_{\mathbf{j}+\mathbf{e}_3,z})$, the approximated function value $f(\mathbf{x_i})$ results from trilinear interpolation of the eight corner values $f(\mathbf{y_j}), \ldots, f(\mathbf{y_{j+1}})$.

Table 1 lists the root-mean-square errors of the shown examples at various levels of resolution. We scaled the root-mean-square error to the interval $[0, 1]$. The "downsampling ratio" (dr) is defined as the original number of vertices divided by the number of vertices at the used coarser resolution. For all examples and all resolutions, we obtained smaller root-mean-square errors when using trilinear B-spline wavelets. The last row of the table quantifies the "improvement" by listing the average error reduction for each downsampling ratio. We recognize that the improvement increases for coarser resolutions.

Figure 10 shows a biomedical example. The data set represents a human brain. It is given as 753 slices, and each slice has a resolution of 1050 × 970 points, where 24-bit RGB-color information is stored. The original data set was preprocessed with a segmentation algorithm described in [30] to eliminate noise. We applied the wavelet scheme to each color channel independently and, after conversion, used the value V of the HSV color model for isosurface extraction.

Since the data was too large to be stored in main memory, we used out-of-core techniques. Due to the narrow masks of our lifting scheme, at most three slices were used simultaneously.

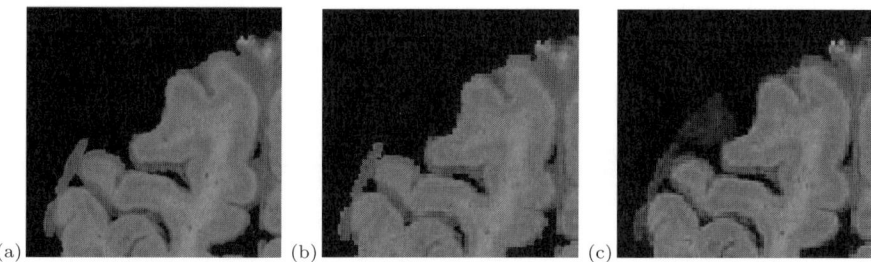

Fig. 10. (a) Slice through 3D brain data set at full resolution; (b) slice at level of detail with downsampling ratio 2^6 without and (c) with B-spline wavelets on a $\sqrt[3]{2}$-subdivision scheme. (Data set courtesy of A. Toga, Ahmanson-Lovelace Brain Mapping Center, University of California, Los Angeles)

For Figure 10, we used an interactive progressive slicing visualization tool, see [23], to generate an arbitrary cutting plane through the brain data set. Figure 10(a) shows the slice at the highest resolution, Figure 10(b) after downsampling with $\sqrt[3]{2}$ subdivision (downsampling ratio 2^6), and Figure 10(c) after downsampling with $\sqrt[3]{2}$ subdivision (downsampling ratio 2^6) and trilinear B-spline wavelets.

Compared to Figure 10(b), the contours of the brain in Figure 10(c) are much smoother. Moreover, the slice in Figure 10(c) does not only contain information of the slice in Figure 10(a) but also of the full-resolution slices next to it. Without the wavelet averaging, some detailed information of the neighbored slices might get lost.

Figure C.20 shows an isosurface for the value 78 extracted from the same data set at the level of detail with downsampling ratio 2^9. For Figure C.20(a), we used a $\sqrt[3]{2}$-subdivision hierarchy without using wavelets, and, for Figure C.20(b), we combined the $\sqrt[3]{2}$-subdivision hierarchy with trilinear B-spline wavelets. Figure C.20(b) exhibits much more detail information than Figure C.20(a).

In Figure C.21, we applied our techniques to numerically simulated hydrodynamics data. The data set is the result of a 3D simulation of the Richtmyer-Meshkov instability and turbulent mixing in a shock tube experiment, see [19]. For each vertex of a 1024^3 structured-rectilinear grid (one time step considered only), an entropy value between 0 and 255 is stored. The figure shows the isosurface corresponding to the value 225 extracted from three different levels of resolution of one time step. Again, we compared the results of the $\sqrt[3]{2}$-subdivision hierarchy without (left column) and with (right column) trilinear B-spline wavelets, partially computed out-of-core.

Considering the example shown in Figure C.21, when using the wavelet approach low-resolution visualizations suffice to understand where the turbulent mixing takes place. For example, Figure C.21(c) shows clearly the big "bubble" rising in the middle of the data set. The bubble can hardly be seen in Figure C.21(a).

9 Conclusions

We have introduced $\sqrt[n]{2}$ subdivision combined with n-variate B-spline wavelets for n-dimensional multiresolution data representation. Visualization of biomedical imaging data and numerically simulated hydrodynamics data, for example, require efficient methods of isosurface extraction. For this purpose, a 3D multiresolution framework is desirable. We first have established a bivariate B-spline wavelet scheme for $\sqrt{2}$ subdivision and have generalized it to a trivariate B-spline wavelet scheme for $\sqrt[3]{2}$ subdivision. The provided examples document the value of our approach for surface and volume modeling and visualization.

By using $\sqrt[n]{2}$ subdivision, instead of using quad- or octrees, a multiresolution hierarchy can be generated that provides much more levels of detail, since, in each subdivision step, the number of vertices is only doubled instead of multiplied by a factor of four or eight, respectively. In the context of view-dependent and adaptive refinement and visualization, this characteristic supports a higher level of adaptivity. Furthermore, $\sqrt[n]{2}$ subdivision does not only work for structured-rectilinear grids, but also for more general structured-curvilinear grids, and even for arbitrary grids, i.e., grids with extraordinary vertices.

By integrating a wavelet scheme into the subdivision approach, we obtain, in general, much better approximations on each level of detail. We have chosen n-variate B-spline wavelets and have developed lifting schemes for $n = 2$ and $n = 3$, which use narrow masks. These narrow masks allow us to utilize the wavelet scheme for view-dependent, adaptive multiresolution visualization of large-scale data.

The wavelet encoding only reorganizes data and does not require additional memory. The $\sqrt[n]{2}$-subdivision scheme also does not require us to store additional connectivity information. Thus, our approach, as a whole, requires no additional storage.

Since the masks of our lifting scheme are of constant size and the number of iterations for our lifting scheme is constant, our algorithms run in linear time with respect to the number of original data. Since the masks are narrow and only two iterations are needed, the run-time constants are small. Considering the examples shown, we conclude that our approach provides a valuable tool for the interactive exploration of volumetric data at multiple level of resolution.

Acknowledgments

This work was supported by the National Science Foundation under contract ACI 9624034 (CAREER Award), through the Large Scientific and Software Data Set Visualization (LSSDSV) program under contract ACI 9982251, and through the National Partnership for Advanced Computational Infrastructure (NPACI); the National Institute of Mental Health and the National Science Foundation under contract NIMH 2 P20 MH60975-06A2; the Army Research Office under contract ARO 36598-MA-RIP; and the Lawrence Livermore National Laboratory under ASCI ASAP Level-2 Memorandum Agreement B347878 and under Memorandum Agreement B503159. We also acknowledge the support of ALSTOM Schilling Robotics and SGI. We thank the members of the Visualization and Graphics Research Group at the Center for Image Processing and Integrated Computing (CIPIC) at the University of California, Davis, and the members of the Data Science Group at the Center for Applied Scientific Computing (CASC) at the Lawrence Livermore National Laboratory. We especially thank Peer-Timo Bremer for supplying us with an implementation of the $\sqrt{2}$-subdivision wavelets.

References

1. Martin Bertram. *Multiresolution Modeling for Scientific Visualization*. PhD thesis, Department of Computer Science, University of California, Davis, California, 2000.
2. Martin Bertram, Mark A. Duchaineau, Bernd Hamann, and Kenneth I. Joy. Bicubic subdivision-surface wavelets for large-scale isosurface representation and visualization. In Thomas Ertl, Bernd Hamann, and Abitabh Varshney, editors, *Proceedings of IEEE Conference on Visualization 2000*, pages 389–396. IEEE, IEEE Computer Society Press, 2000.
3. Martin Bertram, Daniel E. Laney, Mark A. Duchaineau, Charles D. Hansen, Bernd Hamann, and Kenneth I. Joy. Wavelet representation of contour sets. In Thomas Ertl, Kenneth I. Joy, and Amitabh Varshney, editors, *Proceedings of IEEE Conference on Visualization 2001*, pages 303–310. IEEE, IEEE Computer Society Press, 2001.
4. Paolo Cignoni, Claudio Montani, Enrico Puppo, and Roberto Scopigno. Multiresolution modeling and visualization of volume data. *IEEE Transactions on Visualization and Computer Graphics*, 3(4):352–369, 1997.
5. Albert Cohen and Ingrid Daubechies. Nonseparable bidimensional wavelet bases. *Rev. Mat. Iberoamericana*, 9(1):51–137, 1993.
6. Thomas Gerstner and Renato Pajarola. Topology preserving and controlled topology simplifying multiresolution isosurface extraction. In Thomas Ertl, Bernd Hamann, and Abitabh Varshney, editors, *Proceedings of IEEE Conference on Visualization 2000*, pages 259–266. IEEE, IEEE Computer Society Press, 2000.
7. Roberto Grosso and Günther Greiner. Hierarchical meshes for volume data. In *Proceedings of CGI '98, Hanover, Germany*, 1998.

8. Roberto Grosso, Christoph Lürig, and Thomas Ertl. The multilevel finite element method for adaptive mesh optimization and visualization of volume data. In R. Yagel and H. Hagen, editors, *Proceedings of IEEE Conference on Visualization 1997*, pages 135–142. IEEE, IEEE Computer Society Press, 1997.
9. André Guéziec and Robert Hummel. Exploiting triangulated surface extraction using tetrahedral decomposition. *IEEE Transactions on Visualization and Computer Graphics*, 1(4):328–342, 1995.
10. Leif Kobbelt. $\sqrt{3}$-subdivision. In Kurt Akeley, editor, *Proceedings of SIGGRAPH 2000*, Computer Graphics Proceedings, Annual Conference Series, pages 103–112. ACM, ACM Press / ACM SIGGRAPH, 2000.
11. Leif Kobbelt. Multiresolution techniques. In Farin, Hoschek, and Kim, editors, *Handbook of Computer Aided Geometric Design*. Elsevier Science Publishing, Amsterdam, The Netherlands, 2002.
12. Jelena Kovačević and Wim Sweldens. Wavelet families of increasing order in arbitrary dimensions. *IEEE Transactions on Image Processing*, 9(3):480–496, 1999.
13. Jelena Kovačević and Martin Vetterli. Nonseparable multidimensional perfect reconstruction filter banks and wavelet bases for $\mathbf{r}^n$. *IEEE Transactions on Information Theory*, 38(2):533–555, 1992.
14. Lars Linsen, Valerio Pascucci, Mark A. Duchaineau, Bernd Hamann, and Kenneth I. Joy. Hierarchical representation of time-varying volume data with $\sqrt[4]{2}$ subdivision and quadrilinear b-spline wavelets. In Coquillart, Shum, and Hu, editors, *Proceedings of Tenth Pacific Conference on Computer Graphics and Applications – Pacific Graphics 2002*. IEEE, IEEE Computer Society Press, 2002.
15. L. Lippert, M. H. Gross, and C. Kurmann. Compression domain volume rendering for distributed environments. In *Proceedings of the Eurographics '97*, volume 14, pages 95–107. COMPUTER GRAPHICS forum, 1997.
16. Ron A. MacCracken and Kenneth I. Joy. Free-form deformations with lattices of arbitrary topology. In Holly Rushmeier, editor, *Proceedings of SIGGRAPH 1996*, Computer Graphics Proceedings, Annual Conference Series, pages 181–188. ACM, ACM Press / ACM SIGGRAPH, 1996.
17. Donald Maegher. Geometric modeling using octree encoding. *Computer Graphics and Image Processing*, 19:129–147, 1982.
18. Joseph M. Maubach. Local bisection refinement for n-simplicial grids generated by reflection. *SIAM J. Scientific Computing*, 16:210–227, 1995.
19. Arthur A. Mirin, Ron H. Cohen, Bruce C. Curtis, William P. Dannevik, Andris M. Dimits, Mark A. Duchaineau, D. E. Eliason, Daniel R. Schikore, S. E. Anderson, D. H. Porter, and Paul R. Woodward. Very high resolution simulation of compressible turbulence on the ibm-sp system. In Sally Howe, editor, *Proceedings of Supercomputing '99*. IEEE, IEEE Computer Society Press, 1999.
20. Mario Ohlberger and Martin Rumpf. Hierarchical and adaptive visualization on nested grids. *Computing*, 59:365–385, 1997.
21. Valerio Pascucci. Slow growing subdivision (sgs) in any dimension: towards removing the curse of dimensionality. In *to appear in: Proceedings of Eurographics 2002*. COMPUTER GRAPHICS Forum, 2002.
22. Valerio Pascucci and Chandrajit Bajaj. Time critical adaptive refinement and smoothing. In Roger Crawfis and Daniel Cohen-Or, editors, *Proceedings of the ACM/IEEE Volume Visualization and Graphics Symposium 2000, Salt Lake City, Utah*, pages 33–42. ACM/IEEE, 2000.

23. Valerio Pascucci and Randall J. Frank. Global static indexing for real-time exploration of very large regular grids. In *Supercomputing 2001*. ACM, ACM Press, 2001.
24. Dmitriy Pinskiy, Erie Brugger, Henry R. Childs, and Bernd Hamann. An octree-based multiresolution approach supporting interactive rendering of very large volume data sets. In H. Arabnia, R. Erbacher, X. He, C. Knight, B. Kovalerchuk, M. Lee, Y. Mun, M. Sarfraz, J. Schwing, and H. Tabrizi, editors, *Proceedings of the 2001 International Conference on Imaging Science, Systems, and Technology (CISST 2001), Volume 1*, pages 16–22. Computer Science Research, Education, and Applications Press (CSREA), Athens, Georgia, 2001.
25. Hartmut Prautzsch, Wolfgang Boehm, and Marco Paluszny. *Bézier and B-spline Techniques*. Springer-Verlag, Heidelberg, Germany, 2002.
26. Sherman D. Riemenschneider and Zuowei Shen. Wavelets and pre-wavelets in low dimensions. *Journal Approximation Theory*, 71:18–38, 1992.
27. Raj Shekhar, Elias Fayyad, Roni Yagel, and J. Fredrick Cornhill. Octree-based decimation of marching cubes surfaces. In Roni Yagel and Gregory M. Nielson, editors, *Proceedings of IEEE Conference on Visualization 1997*, pages 335–342. IEEE, IEEE Computer Society Press, 1996.
28. Eric J. Stollnitz, Tony D. DeRose, and David H. Salesin. *Wavelets for Computer Graphics: Theory and Applications*. The Morgan Kaufmann Series in Computer Graphics and Geometric Modeling, Brian A. Barsky (series editor), Morgan Kaufmann Publishers, San Francisco, U.S.A., 1996.
29. Wim Sweldens. The lifting scheme: A new philosophy in biorthogonal wavelet constructions. In Andrew F. Laine and Michael Unser, editors, *Wavelet Applications in Signal and Image Processing III*, pages 68–79. Proceedings of SPIE 2569, 1995.
30. Ikuko Takanashi, Eric Lum, Kwan-Liu Ma, Jörg Meyer, Bernd Hamann, and Arthur J. Olson. Segmentation and 3d visualization of high-resolution human brain cryosections. In Robert F. Erbacher, Philip C. Chen, Matti Gröhn, Jonathan C. Roberts, and Craig M. Wittenbrink, editors, *Proceedings of SPIE Visualization and Data Analysis 2002*, pages 55–61. Proceedings of SPIE 4665, 2002.
31. Geert Uytterhoeven. *Wavelets: Software and Applications*. PhD thesis, Katholieke Universiteit Leuven, Belgium, 1999.
32. Luiz Velho and Denis Zorin. 4-8 subdivision. *Computer Aided Geometric Design*, 18(5):397–427, 2001.
33. Joe Warren and Henrik Weimer. *Subdivision Methods for Geometric Design*. Morgan Kaufmann Publishers, San Francisco, U.S.A., 2002.
34. Rüdiger Westermann, Leif Kobbelt, and Thomas Ertl. Real-time exploration of regular volume data by adaptive reconstruction of isosurfaces. *The Visual Computer*, pages 100–111, 1999.
35. Yong Zhou, Baoquan Chen, and Arie E. Kaufman. Multiresolution tetrahedral framework for visualizing regular volume data. In Roni Yagel and Hans Hagen, editors, *Proceedings of IEEE Conference on Visualization 1997*, pages 135–142. IEEE, IEEE Computer Society Press, 1997.

Multiresolution Surface and Volume Representations

Oliver G. Staadt

Computer Science Department
University of California, Davis
One Shields Avenue, Davis, CA 95616, USA
staadt@cs.ucdavis.edu

Summary. We present a wavelet-based geometry compression pipeline in the context of hierarchical surface and volume representations. Due to the increasing complexity of geometric models used in a vast number of different application fields, new methods have to be devised that enable one to store, transmit and manipulate large amounts of data. Based on a multiresolution wavelet representation, we have developed a complete compression pipeline suitable for geometric data on uniform grids in two and three dimensions. Local and global oracles in wavelet space are employed to control the approximation error in lossy compression settings. Two geometry simplification schemes, which are able to build hierarchical mesh representations, are an essential part of the pipeline. In contrast to the two wavelet-based approximation schemes, we have devised the progressive tetrahedralization method, an extension of the popular progressive meshes into volumetric settings. We compare the three approximation schemes with each other using several two- and three-dimensional data sets and provide an extensive error and performance analysis. These results emphasize the individual strengths and weaknesses of each of the discussed methods and concepts.

1 Introduction

During the past decade, the complexity of geometric models in various application fields has increased steadily. With the wide availability of powerful computer systems, many conventional design and production processes are either simulated or carried out entirely on computers. Good examples include the wide field of product design in almost any engineering discipline. The design cycle for many consumer products including cars, toys, appliances, and so on, typically incorporates highly detailed geometric models. Moreover, the simulation of natural phenomena requires the creation of very large and complex geometric models for further computational processing and scientific visualization.

Even though the performance of state-of-the-art computer systems is still increasing in terms of computational power and memory availability, the complexity of geometric models, driven by the demand of applications, is at least increasing at the same rate. Several solutions have been proposed to attack these problems, which can more or less be classified into the following overlapping areas: Geometry simplification, geometry compression and hierarchical mesh representations.

Non-hierarchical geometry simplification schemes have successfully been employed for many years to reduce the number of elements in geometric models. The use of simplified models can speed-up visualization and further numerical processing. Those method allow for the generation of high-quality approximations at a single level of resolution. If the user requests a mesh at a different resolution, however, the whole simplification process has to be repeated from the start.

A completely different approach is to optimize the memory requirements for storing and transmitting large models. Geometry compression methods reduce the memory footprint of a model without compromising its geometrical or topological properties. We distinguish between lossy and lossless compression methods. A certain loss of precision can usually be tolerated for compressing positional information of individual vertices constituting the model. Connectivity information, which specifies the relation between these vertices and defines the model's topology, should be compressed without loss of information.

Although non-hierarchical simplification methods and geometry compression schemes often allow for progressive reconstruction and transmission over networks, they do not inherently define a hierarchical representation of the model. For very large data sets, however, it might be advantageous to inspect or even edit the data at different levels, and hierarchical representations provide build-in support for such applications. Note the difference between "linear" progressive schemes, which also enable the reconstruction of different levels, and fully hierarchical methods: The latter contain additional information about dependencies between successive levels, thus allowing to switch consistently between them. This is especially desirable in multiresolution editing applications, where changes applied at one approximation level are automatically propagated throughout the hierarchy.

The main focus of this paper is a survey and comparison of the aforementioned surface and volume representations, highlighting their advantages and shortcomings.

2 Multiresolution Approximations

In this section we discuss the mathematical fundamentals of the preprocessing we employ for data preconditioning. B-spline wavelets are used as a precoding transform since they combine various advantageous features, such as vanishing moments, continuous approximation, bounded interval definition, linear time algorithms, and localization. For reasons of readability, we first review some basics of cardinal B-spline wavelets. However, our attention is mostly directed to the definition of global oracles, that is, schemes to reject unimportant coefficients. Our global oracle consists of a greedy algorithm resulting from an elaborate analysis of L^2-errors in semiorthogonal settings [10]. Additionally, we will demonstrate how local oracles reject coefficients in unimportant spatial regions and thus enable the construction of *electronic magnifying glasses* for interactive data inspection. For reasons of simplicity, we perform all computations for 1-D functions, but the results extend straightforwardly to higher dimensions.

2.1 Cardinal B–Spline Wavelets

B–spline wavelets were initially introduced by Chui [3], and were extended to bounded intervals by [8] and [22], while nonuniform knot sequences were addressed for instance by [2]. Due to a rich variety of literature in this area, we restrict our introduction to those topics essential for an understanding of our framework.

B–spline wavelets can be constructed from a multiresolution hierarchy of cardinal B–spline scaling functions. Semiorthogonality invokes an additional degree of freedom, however. Thus, approaches as in [8] or [22] end up in slightly different construction schemes. We adapted the methods of Quak et. al [8] to construct B–spline wavelets of arbitrary order bounded to the interval.

Assuming the reader is familiar with some fundamentals of discrete wavelet transforms (DWT), the implementation of the forward transform is carried out by sequences of projection operators $\mathbf{A}^m, \mathbf{B}^m$, where $m = 1 \ldots M$ stands for the decomposition level. An initial function $f(x)$ is mapped from the higher resolution approximation space V^m onto a lower resolution space V^{m+1} and onto its orthogonal complement space W^{m+1}. Given the coefficient vectors $\mathbf{c}^m$ and $\mathbf{d}^m$ for the scaling functions $\varphi_i^m(x)$, and wavelets $\psi_i^m(x)$ in the 1-D setting, with

$$c_i^m = \langle f, \varphi_i^m \rangle \qquad d_i^m = \langle f, \psi_i^m \rangle, \tag{1}$$

$(i : 1 \ldots N/2^m + \text{order} - 1$, order: B–spline order,) the decomposition is carried out by matrix operations

$$\mathbf{c}^{m+1} = \mathbf{A}^m \mathbf{c}^m \qquad \mathbf{d}^{m+1} = \mathbf{B}^m \mathbf{c}^m. \tag{2}$$

Due to the semiorthogonality, we require the inverse operators $\mathbf{P}^m$ and $\mathbf{Q}^m$ to compute the reconstruction with

$$\mathbf{c}^m = \mathbf{P}^m \mathbf{c}^{m+1} + \mathbf{Q}^m \mathbf{d}^{m+1}. \tag{3}$$

It can be easily proven [8] that the operators relate to each other by

$$\begin{bmatrix} \mathbf{A}^m \\ \mathbf{B}^m \end{bmatrix} = [\mathbf{P}^m | \mathbf{Q}^m]^{-1}. \tag{4}$$

In the case of cardinal B–spline wavelets, sparse operators $\mathbf{P}^m$ and $\mathbf{Q}^m$ come along with dense matrices $\mathbf{A}^m$ and $\mathbf{B}^m$. In order to construct linear time algorithms for both decomposition and reconstruction, it is sufficient to know the sequences $\mathbf{P}^m$ and $\mathbf{Q}^m$ to perform an additional base transform of the coefficients into their duals $\tilde{\mathbf{c}}^m$ and $\tilde{\mathbf{d}}^m$ using the inner product matrices $\mathbf{\Phi}^m$ and $\mathbf{\Psi}^m$.

Note that the decomposition involves solutions of the sparse linear systems of type $\mathbf{\Psi}^m \cdot \mathbf{d}^m = \tilde{\mathbf{d}}^m$ for each iteration and $\mathbf{\Phi}^M \cdot \mathbf{c}^M = \tilde{\mathbf{c}}^M$ for the last iteration step. Fortunately, this can be accomplished in linear time as well. For brevity we abandon all mathematical details associated with the construction of these transforms and refer the reader to [8].

2.2 Oracles

One of the most important applications of wavelets is data compression. Instead of using all wavelets ψ_{mp} for reconstruction, it is possible to truncate the approximation. An oracle is used to determine the set of wavelets that are omitted in the truncated approximation.

Global Oracles. A global oracle rejects unimportant wavelet coefficients from the approximation while minimizing a given error norm.

First, let us recall finite dimensional orthogonal approximations for $L^2(\mathbb{R})$. The basis $\{\varphi_i(x)\}_{i=1...N}$ expands a function $f(x)$ as

$$f(x) = \sum_{i=1}^{N} c_i \varphi_i(x), \quad (5)$$

with coefficients $c_i \in V_{m_0}$. Given an integer $1 \leq K < N$, we try to find the best approximation of f with the truncated approximation f'. The L^2 error of $f'(x)$ is determined by the linear combination of the rejected basis functions

$$\begin{aligned}
\|f(x) - f'(x)\|_{L^2}^2 &= \left\| \sum_{i=K+1}^{N} c_i \varphi_i(x) \right\|_{L^2}^2 \\
&= \left\langle \sum_{i=K+1}^{N} c_i \varphi_i(x), \sum_{i=K+1}^{N} c_i \varphi_i(x) \right\rangle \\
&= \sum_{i=K+1}^{N} |c_i|^2 \quad (6)
\end{aligned}$$

Note that the orthogonality of $\langle \varphi_i, \varphi_j \rangle = \delta_{ij}$ cancels out all intermediate terms in Eq. (6), which simplifies the relation. In other words, the magnitude of the coefficient c_i corresponds exactly to the fraction of energy provided by the corresponding basis φ_i. Hence, the L^2-energy of a function $f(x)$ is computed by

$$\|f(x)\|_{L^2}^2 = \langle f(x), f(x) \rangle = \mathbb{N}_{-\infty}^{\infty} |f(x)|^2 \, dx. \tag{7}$$

A globally optimal compression can thus be achieved by sorting the coefficients by their magnitude and by rejecting the K smallest ones. Obviously, this strategy yields an L^2-optimal oracle for orthogonal bases and can be computed in $O(N \log N)$ using a fast sorting algorithm. Unfortunately, in the semi-orthogonal case of B–spline wavelets, the intermediate terms in Eq. (6) are not canceled out, which complicates the construction of an oracle. Maximum distance norm oracles have been proposed by Stollnitz and others [22] for biorthogonal wavelets. Gross [10] and Staadt et al. [21] propose a global oracle for the semi-orthogonal setting. For more details, we refer the reader to these references.

Local Oracles. A local oracle allows one to control the approximation locally in interesting regions. Here, the spatial localization of the basis functions enables one to accentuate particular wavelets while suppressing the influence of others. In this understanding, a straightforward local oracle consists of a weighting function which operates on the coefficients of the transform. A first approach to this is given in Gross et. al [12] who employed a Gaussian weighting. The basic idea is to assume some ellipsoidal weighting area as a local region of interest in the spatial domain. Localization of the wavelet transform allows for the projection of scaled and translated versions of it into wavelet space, where individual coefficients are influenced. The initial version presented in [12], however, did not consider the support regions of individual basis functions, and can lead to some artifacts by rejecting wavelets ranging into the region of interest. Therefore, the method has been extended in [21] by computing the support of the basis functions and by using endpoint-interpolating B-wavelets.

3 Data Compression

We designed a compression pipeline by employing a wavelet-based sub-band codec. The forward compression proceeds as follows: The data set is normalized, transformed into wavelet space and both local and global approximation errors are controlled by the oracles introduced in Section 2.2. Sorting of the individual channels of the wavelet transform (WT) converts the multi-dimensional array into a 1-D data vector, which is quantized and encoded subsequently. Conversely, the decompression pipeline inverts the procedure and prepares the data for subsequent geometric reconstruction. For reasons of brevity, we will only present a short overview of the compression pipeline. For more in-depth coverage we refer to [19, 21].

The main steps of the compression pipeline comprise the following:

1. *Normalization:* As opposed to applications such as image compression, the order of magnitude for geometric data is usually not known in advance, but may vary between data sets and application areas. For that reason, the data are normalized, that is the original values c_{orig} are mapped to the interval $[0,\ldots,1]$ using a simple mapping function $c_{norm} = \frac{c_{orig}-c_{max}}{c_{max}-c_{min}}$
2. *Wavelet Transform:* The subband codec used in this approach employs cardinal B–spline wavelets as introduced in Section 2.1. These basis functions are very well suited for the application of geometry compression, since they share some very important properties: (1) Endpoint-interpolation, (2) a high number of vanishing moments, (3) compact support, and (4) and efficient implementation.
3. *Oracle:* Local and global oracles, which have already been introduced in Section 2.2, are used to derive a truncated series of basis functions. In spite of the fact that semi-orthogonal B–spline wavelets are used for this codec, orthogonal oracles based on the L^2-norm are employed. They can only approximate semi-orthogonal L^2-oracles. Gross [10], however, presents a comparison between orthogonal and semi-orthogonal L^2-oracles, which provides for a justification of this approximation.
4. *Linearization:* To prepare the data for bandwise progressive transmission, the multidimensional coefficient array is mapped into an 1-D vector. Here, the array is traversed from the most significant scaling function coefficients to the high frequency bands representing fine grained detail. This linearization step is carried out similar to the "zig-zag"-traversal of the DCT coefficients in the JPEG image compression standard [15].
5. *Quantization:* The quantization step comprises a multiplication of the initial floating point coefficients with a factor of 2^{n-1}, where n represents the number of bits to be assigned for each coefficient. Subsequent rounding operations transform the floating point value into signed integer formats of size n. Let c_{float} be a coefficient. Its quantized version c_{quant} is obtained by applying a scalar midtreat quantizer of the form $c_{quant} = \text{round}\left(2^{n-1} \cdot c_{float}\right)$
6. *Encoding:* An important task in the proposed compression pipeline is to convert the quantized integer vector into a bitstream of data. Therefore, an entropy coding scheme in the spirit of JPEG [15] is employed. Assuming that many of the coefficients will be equal to zero, encoding is carried out as follows: All non-zero coefficients are represented by two-tuples, where the first element represents the number of bits required to encode the second one. The second element contains the data value itself. All negative numbers are thus replaced by their absolute values, where in the case of a positive number the first bit is cleared. This enables one to encode of the sign elegantly.

4 Quadtree-Based Vertex Removal

The quadtree-based vertex removal scheme is a very fast and efficient method for further reducing the complexity of the data. This bottom-up strategy starts with a dense set of vertices and subsequently removes nodes according to a user-specified error threshold. This process eventually constructs a quadtree data structure which is triangulated by employing an efficient table look-up. The importance of a vertex is determined by analyzing detail coefficients at dyadic points reconstructed from wavelet space. For that reason, a single-step inverse wavelet transform has to be designed, which enables one to reconstruct the different sub-bands in wavelet space individually. Forward transform and compression are carried out in the same fashion as proposed in Section 3. A detailed description of the single-step inverse wavelet transform can be found in [12].

4.1 Vertex Removal in Regular Meshes

So far, some mathematical criteria have been elaborated for approximating a surface data set, sampled on a regular grid, using a multiresolution hierarchy. In order to build an adaptive surface triangulation, however, it is necessary to remove unimportant mesh vertices and to find a triangulation of the remaining ones. The basic criterion by which a mesh vertex is labeled as unimportant is given by the mathematical framework of the wavelet transform. In contrast to existing methods [14] which base on linear spline wavelets, this scheme can be generalized to any type of wavelet. Therefore, the criteria require more elaboration. Keeping in mind that any triangulation of the surface provides a planar approximation, the error between the original surface function $f(x,y)$ and the linear interpolant provided by a triangle has to be bound. Supposing furthermore that the initial data is expanded by wavelet bases, the detail signal in iteration m is used to decide on whether or not each $2^m + 1$th mesh vertex is necessary for the triangle approximation. First, every second vertex is visited and the value of the detail signal Δf^1 of iteration $m = 1$ is analyzed. If the detail signal in some neighborhood of vertex n is sufficiently small, then the vertex is considered not to be important and the approximation can be accomplished by a linear interpolation between vertex $n-1$ and $n+1$. This scheme can now be applied recursively by subsequent computation of the detail signals Δf^m, $m = 1, \ldots, M$ and by visiting all dyadic vertices at positions $n = 2^m k + 1$. Once the detail signal is sufficiently small and the adjacent vertices in step $m-1$ are already removed, the current vertex is labeled as well.

As a consequence, this procedure results in recursively building a quadtree representation of the initial mesh by removing dyadic vertices. Figure 1 illustrates the thinning method which finally figures out a quadtree representation of the mesh vertices. The nodes of the quadtree contain either pointers to some child-nodes, or in case of leaves, point to the entries of a vertex list.

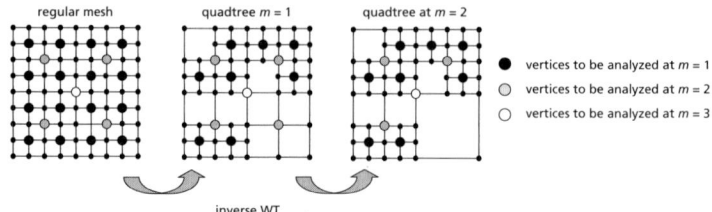

Fig. 1. Bottom-up construction of a quadtree from a regular mesh by analyzing the detail signals of the wavelet transform at each dyadic vertex.

4.2 Vertex Removal Criteria

To finally decide on whether a mesh vertex can be removed or not, we propose the following three criteria, which additionally help to preserve the topology of the tree. A vertex will only be removed from the mesh if all criteria are met:

1. *Wavelet–criterion:* A vertex at iteration level m can be removed if the sum of the squares of its difference signal and those within a 4-neighborhood at resolution m are less than an upper bound ϵ.
2. *Resolution–criterion:* A vertex at iteration level m can be removed if the four surrounding vertices at level $m-1$ have been removed in a previous step.
3. *m to $m-2$ criterion:* A vertex can be removed if the resulting cell is not adjacent to any cell of a level higher than $m-2$. Thus, growth is restricted to cell transitions from m to $m-2$, which simplifies the triangulation algorithm.

Once the tree is built from the above procedure, the quadtree cells have to be triangulated. A generic problem arising from meshing hierarchies of rectangular surface patches is the occurrence of cracks. A crack occurs if adjacency of quadtree cells at different depth and, hence, different resolution has not been considered. The surface may break up, holes may appear and any consistency required for normal interpolation gets lost. Non-manifold surfaces will be the result. We use a scheme based on a two-level look-up table to triangulate the quadtree cells, which has been introduced in [11, 12].

The quadtree-based vertex removal method introduced in this section has some limitations regarding progressivity, adaptivity, and easy extension into higher dimensions. The bottom-up approach requires successful transmission of all first-level detail coefficients before any vertices can be analyzed and eventually be removed. For a 2-D data set, the amount of data to be transmitted is equal to three quarters of the whole data. Furthermore, since only transitions between two consecutive levels are permitted, the adaptivity of the approximated mesh is limited. Extension into three dimensions is possible, but topological complexity and size of the look-up-tables would be increased

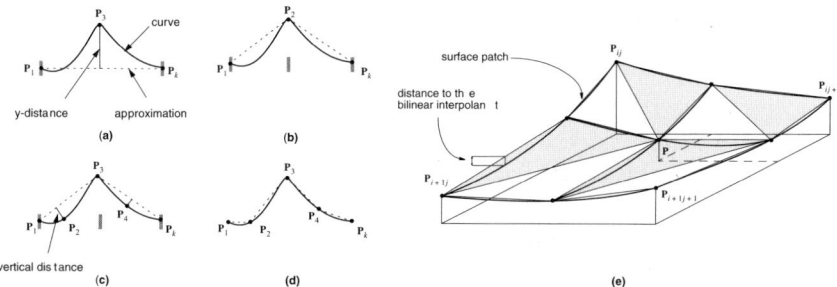

Fig. 2. Recursive algorithm assuming a smooth representation of the underlying curve: (a) $\mathbf{P}_2$ has largest vertical distance; (b) new approximation after insertion of $\mathbf{P}_2$; (c) example for vertical distance metric; (d) final result; (e) extension towards multiple dimensions exemplified for non-parametric data: 2-D version. The underlying B–spline patch is outlined in bold. A new vertex is inserted at position $\mathbf{P}$ and the distance is computed with respect to the bilinear interpolant of $\mathbf{P}_{ij}, \mathbf{P}_{ij+1}, \mathbf{P}_{i+1j}, \mathbf{P}_{i+1j+1}$.

substantially. In addition, the simplicity and elegance of the algorithm for two-dimensional data would not be maintained.

5 Delaunay-Based Vertex Insertion

5.1 Vertex Insertion in 1-D

In order to construct a vertex insertion strategy, we first consider the 1-D setting. Here, the problem reduces to finding a strategy for the reduction of line segments in piecewise linear approximations. Inspired by the algorithm of [7] we extended these ideas in [21] and modified the method to a recursive and progressive algorithm, illustrated in Figure 2. It starts by connecting the first point of a curve, $\mathbf{P}_0$, with the last point $\mathbf{P}_k$. All intermediate points representing the curve are compared against the line segment $\overline{\mathbf{P}_0 \mathbf{P}_k}$ and the point with the largest distance, for instance $\mathbf{P}_2$, is identified. If its distance exceeds a predefined threshold ϵ_0, the vertex is considered *important* and labeled. We split the initial line segment in two halves, on each of which the algorithm can be applied recursively. Obviously, the quality of the removal can be controlled by the distance threshold. The advantage of this extension to the original method lies in the tree type refinement of the vertex analysis coming along with the recurrence relations.

The distance can be computed in different ways, where, however, the computation of the vertical distance, such as depicted in Figure 2c, is computationally much more expensive for general multidimensional settings. Therefore, we recommend computing the y–distance (see Figure 2a) to approximate nonparametric data.

5.2 Generalization to Higher Dimensions

Generalizations of the method towards multidimensional nonparametric data is straightforward. Starting from an initial grid, as in Figure 2e, the algorithm seeks the vertex **P** with the maximum distance and subdivides the field into four (in 2-D) or eight (in 3-D) subcells on which the method is applied recursively. In these cases the distances to the bilinear and trilinear interpolants of the cell vertices are computed, respectively.

Recalling the multiresolution B–spline approximation of the data motivates the extension of the algorithm towards a channelwise progressive vertex insertion. Therefore, the algorithm analyzes mesh vertices progressively and labels unimportant points as new data comes in. In 2-D, for instance, the basic idea is to start from an initial vertex field of resolution 2^{m-M} in each direction, where M represents the maximum iteration. The vertices are provided by the scaling function approximation $f^M(x,y)$ and are processed further by our algorithm. To define a distance metric, we assume a bilinear interpolant between the vertices which approximates the B–spline scaling function representation. If the difference signal $\Delta f^m(x,y)$ is received, the resolution is refined by two and all newly inserted vertices are checked conforming to our distance metric. If required, they will be inserted. In order to compute the intermediate vertices for each iteration, an inverse wavelet transform has to be applied to all coefficients at iteration level m as soon as they are received and decompressed.

For subsequent triangulation in 2-D and 3-D we employ Delaunay triangulation libraries such as [1, 18].

6 Progressive Tetrahedralizations

The wavelet-based vertex removal and insertion schemes introduced in Sections 4 and 5 are very well suited for efficient representations of various kinds of uniform data in multiple dimensions. An increasing demand for purely irregular and unstructured mesh representations, however, shows the requirement for the development of new methods providing even greater flexibility.

Progressive meshes [13] and its generalizations to higher dimensions [16] proofed to be an extremely powerful notion for the efficient representation of triangulated geometric objects at different levels-of-detail. Although a general formulation for arbitrary triangulations has already been given in [16], the special case of progressive tetrahedralizations (PT) is of enormous practical importance, since it can be used as a sophisticated representation for a large variety of computations [4, 20, 23].

This section presents a method for the hierarchical and progressive representation of tetrahedral meshes with arbitrary connectivity and we elaborate on some pitfalls and fallacies people might get caught in when trying to implement the method of edge collapsing for tetrahedral meshes. Specifically, we address the issue of defining appropriate cost functions. Unlike the elegant geometric approach presented in [13], we must account for volume and application specific properties, such as volume preservation, gradient estimation of the underlying data or aspect ratio of the simplex. In addition, we devised a sequence of tests to ensure a robust and consistent progressive tetrahedralization, which have first been introduced in [20]. Dey and colleagues [6] have independently presented an in-depth analysis of topology preserving edge collapses. For reasons of brevity, we refer the reader to [13, 16] for a general introduction to progressive meshes and its higher-dimensional variants.

In the tetrahedral setting, degeneration of tetrahedra into lower dimensional simplices is prohibited. The set of cells sharing an edge e_i will be called $\{icells_i\}$. Thus, an edge split adds the tetrahedra in $\{icells_i\}$ to the list of active elements. Conversely, the set of non-vanishing cells affected by the associated edge collapse is called $\{ncells_i\}$. In order to compute a sequence of robust, non-degenerate and consistent meshes, the following aspects have to be considered: (1) *Cost functions* which determine the order of edge collapse (*ecol*) operations depending on desired mesh optimization criteria. (2) *Feature edges* which should be preserved can be tested during preprocessing. (3) *Intersections* and *inversions* of tetrahedra inside and outside of $\{icells_i\} \cup \{ncells_i\}$ have to be processed at run time.

6.1 Cost Functions

Various elegant algorithms [9, 13] based on the edge-collapse/vertex-split paradigm use cost functions optimized for triangular surfaces, often accounting for distance measures, triangle shape, and others. In tetrahedral meshes, however, we have to redefine the terms of the cost function considering other features, like volume preservation or gradients. Trotts et al., e.g., describe a nice approach in [23]. Although many different measures are conceivable to control the simplification process, the following ones yield a good balance between required degrees of freedom and the difficulty of parameter optimization. Thus, in our setting, for each edge $e_i = (\mathbf{v}_a, \mathbf{v}_b)$, the associated edge collapse operation $ecol_i(a, b) : M^i \leftarrow M^{i+1}$ is assigned the following cost:

$$\Delta E(e_i) = \Delta E_{grad}(e_i) + \Delta E_{vol}(e_i) + \Delta E_{equi}(e_i) + \Delta E_{normal}(e_i). \quad (8)$$

The first term $\Delta E_{grad}(e_i) = |s_a - s_b|$ forms a simplified measure for the difference of underlying scalar volume function along the edge e_i. Hence, edges with considerably differing scalar attributes are assigned high costs. This term of the cost function applies only to volumetric data sets with scalar attributes. The second term

$$\Delta E_{vol}(e_i) = \left(\sum_{T_j \in \{ncells_i\}} vol(T_j) - vol(\bar{T}_j) \right) + \sum_{T_j \in \{icells_j\}} vol(T_j)$$

penalizes volume changes and thus avoids that the mesh is shrinking. $\bar{T}_j$ denotes a tetrahedron after the collapse and $vol(T_j)$ its volume.

In many applications it is required that tetrahedra sustain equilateral shape. $\Delta E_{equi}(e_i)$ can be employed to balance the edge length of individual tetrahedra:

$$\Delta E_{equi}(e_i) = \sum_{T_j \in \{ncells_i\}} \left(\sum_{\{a,b\} \in T_j} (l_{a,b} - m_j)^2 - \sum_{\{a,b\} \in \bar{T}_j} (l_{a,b} - \bar{m}_j)^2 \right)$$

with $l_{a,b} = |\mathbf{v}_a - \mathbf{v}_b|$ and $m_j = 1/|T_j| \sum_{\{a,b\} \in T_j} l_{a,b}$.

Note that the initial mesh M^n will usually be generated from some triangulation scheme. Depending on the application context and the desired mesh features it can be advantageous to include $\Delta E_{edgelen}(e_i) = |\mathbf{v}_a - \mathbf{v}_b|$ into the cost function thereby enforcing short edges to be collapsed earlier.

For triangular surface meshes, the difference between the vertex normals of $\mathbf{v}_a$ and $\mathbf{v}_b$ are a good estimate of the local curvature of the mesh. If e_i lies within a region of high curvature, an edge collapse would flatten that region. Hence, this edge should be associated with an appropriate cost value defined as $\Delta E_{normal}(e_i) = (\mathbf{n}_a \cdot \mathbf{n}_b)|l_{a,b}|^2$, where $\mathbf{n}_i$ denotes the normal vector in vertex $\mathbf{v}_i$.

6.2 Consistency Tests

Unfortunately, brute force selection of edges according to the cost function from above can introduce mesh inconsistencies like degeneration, folding, intersection, or loss of individual features. It is possible, however, to avoid such inconsistencies by carrying out a combination of tests that can either be carried out during preprocessing or dynamically at run-time.

In a preprocessing step, all vertices and edges on the mesh boundary are marked. Table 1 lists the five possible combinations of boundary edge and vertices. Only cases one and five pass the consistency test for legal edge collapses. Allowing for edge collapses in cases two and three may result in "dents" on the mesh boundary, thus, they are optional. Interior edges (case one) have to be further checked dynamically for possible intersections and cell orientation changes.

A normal flipping heuristic [17] can be generalized to circumvent folding or self-intersection of cells. This can easily be implemented by testing for sign changes of surface normal vectors and by analyzing the volume of all tetrahedra $T_j \in \{ncells_i\}$ before and after the collapse. For more details see [19].

Starting from the observation that edge collapses can cause global intersections, we employ the method first introduced in [20] for dynamically testing for cell intersections. We assume that the input mesh is intersection-free and connected. We make no assumption about the genus of the mesh,

Table 1. The five possible combinations of boundary edges and vertices.

case	e	$\mathbf{v}_a$	$\mathbf{v}_b$	valid
1				yes
2		boundary		optional
3			boundary	optional
4		boundary	boundary	no
5	boundary	boundary	boundary	yes

though. The problem can then be divided into checking interior edge and mesh boundary edges.

After passing the static tests described above, we can ensure that any non-boundary edge e_i cannot have a boundary vertex as endpoint. If the set $\{icells_i\} \cup \{ncells_i\}$ contains no boundary edge, its boundary forms a polytope entirely wrapping the edge. A collapse of the edge, however, does not affect the boundary of the polytope, whose disjoint triangulation is given by the tetrahedra $\bar{T}_j \in \{ncells_i\}$. Thus, intersections can only occur with boundary cells and intersection tests can be restricted to the mesh boundary.

In the general case of boundary edges, we have to perform triangle–triangle intersection tests, which can be carried out as follows: First, we define the set of triangles $\{f_k\}$ containing all boundary faces of tetrahedra $T_j \in \{ncells_i\}$. These are the faces which can change after $ecol_i$ since they all share the new vertex $\bar{\mathbf{v}}_a$. Thus, they are our prime candidates for intersection with other boundary faces. To avoid testing these faces against all other boundary tetrahedra, we introduced a more efficient iterative method in [20], which will be used for the experiments in Section 7.

7 Experimental Results and Comparison

This section comprises two comparative experiments for the surface and volume approximation methods in the previous sections. We will analyze potential strengths and weaknesses of each of the methods. This comparison will focus on the geometric approximation performance of the three methods. Although the PT method can process a wide range of unstructured surface and volume data sets, we will restrict these experiments to uniform input data, due to the constraints of the tensor product wavelet transform used by the other two methods.

7.1 Surface Approximation

In this section, we will investigate the approximation behavior of the quadtree-based vertex removal, the Delaunay-based vertex insertion and progressive meshes. We have selected a digital terrain data set of the Matterhorn, which has been cropped to 256 by 256 vertices. The original resolution of the data set is 701 by 481 vertices with 25 meter spacing. For the transformation into the progressive mesh representation, the following cost function terms have been used: Normal energy, edge length energy and equilateral energy, weighted with 50, 1 and 5, respectively. The total transformation took 2'58" on an SGI Octane R10000. The reconstruction parameters for all three methods have been adjusted in order to yield the same number of triangles for each of the five reconstruction levels. The geometric approximation errors have been calculated with Metro [5] and the error visualization is depicted in Figure C.23.

We can see from the statistics and the error visualization that both, the progressive mesh and the Delaunay-based vertex insertion methods outperform the quadtree-based vertex removal. Due to the greater adaptivity of the first two methods, they achieve better approximation results than the latter method, which is constraint by the two-level look-up-table triangulation. A closer look at the quadtree-based method, however, reveals that its performance is still satisfactory. The relative error at the coarsest approximation level is below 0.15 percent.

7.2 Volume Approximation

For the final experiment, a medical CT volume data set of a child's skull has been employed. We have cropped the original data to a representative fragment with 32 by 32 by 32 voxels or 178,746 tetrahedra. Figure C.22 depicts a semi-transparent view of this model.

This data set has been processed with both, the Delaunay-based vertex insertion and the progressive tetrahedralization methods. Performance statistics for this experiment are shown in Figure 3. The timing figures for the PT do not include the initial transformation into the PT representation, which took 10'40" without and 41'3" with full intersection testing. Three tetrahedral intersections were detected during the process and the corresponding collapse operations have been evaded. The original mesh and its approximation have both been sampled at 1,000,000 randomly distributed positions to determine the approximation error.

It is interesting to see that the lines of the distortion graph for both methods take an almost identical course. The performance of the PT methods, however, drops off for the last sample at 10,000 tetrahedra. This behavior can be explained by the fact that the order of priority of the edge collapses is determined by a greedy algorithm. After large number of successful edge collapse operations, it becomes more an more difficult to select edges with low cost without compromising the correct topology of the mesh. This fact can usually be neglected for two-manifold meshes where the topology is much simpler, but it has a considerable influence on the volumetric setting. A global optimization scheme would certainly improve the mesh quality at low approximation levels, but this is probably not feasible for large applications.

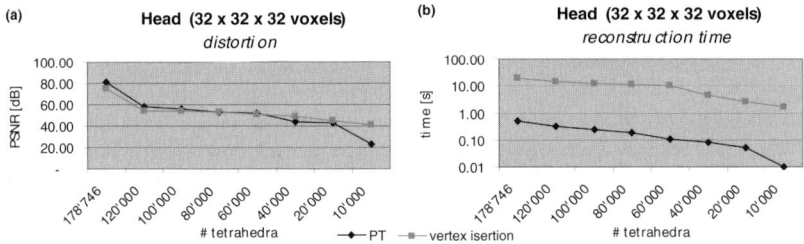

Fig. 3. Performance statistics for the volume comparison experiment; (a) PSNR of both methods for decreasing number of tetrahedra; (b) reconstruction times (note the logarithmic scale of the time axis).

Without taking preprocessing into account, the PT method clearly outperforms the Delaunay-based vertex insertion scheme in terms of reconstruction time by almost two orders of magnitude. The bottleneck of the latter scheme, however, is the Delaunay tetrahedralization which only performs in $O(N \log N)$, where N is the number of voxels. A faster tetrahedralization scheme could significantly improve the overall performance.

8 Conclusions

In this paper we have presented a comparison of different geometry compression and reconstruction methods for uniform data in two- and three-dimensional setting. Results from these methods have been compared to the popular progressive mesh algorithm and our extension to unstructured tetrahedral meshes.

Bottom-up Vertex Removal. The main advantage of the approach described in Section 4 is its fast and elegant implementation, which does not require any geometric calculations upon triangulation. Due to the constraints of the look-up-table, however, the semi-structured approximation limits the adaptivity of the resulting mesh. This is reflected in the numerical results presented in Section 7.1, which show that we cannot achieve the same performance as with the other two methods presented in this paper. Furthermore, the bottom-up nature of the algorithm prevents us from allowing for progressive transmission and reconstruction without precomputing the hierarchy. Before we can start with the construction of the quadtree, we have to receive all detail information at the finest level, which comprises two thirds of the original data (uncompressed).

Top-down Vertex Insertion. Using the top-down method from Section 5 instead of a bottom-up approach opens the possibility for channelwise progressive reconstruction. By transmitting coarse wavelet channels first, which only comprise a small subset of the total number of coefficients, we can refine the geometric representation adaptively. Since we are not constrained by a look-up-table, but employ Delaunay triangulation to determine the mesh connectivity, we can generate approximations which adapt better to the original data. In terms of the geometric approximation error, this method clearly outperforms the quadtree-based scheme. On the other hand, the algorithmic complexity of the vertex insertion method, which is largely determined by the cost of the Delaunay triangulation, is considerably higher. Hence, for very large data sets and applications where fast triangulation on the fly is more important than very low approximation error, the quadtree-based method is still attractive.

Progressive Tetrahedralizations. The accuracy of the progressive mesh and the progressive tetrahedralization methods is comparable to that of our Delaunay-based vertex insertion scheme. Direct comparison for surface triangulations has shown that we could even achieve better results with progressive meshes. This does not come as surprise, because we were able to employ data-specific cost functions which govern the transformation from the input mesh into the progressive mesh representation. Similar results could be obtained for tetrahedral meshes. For very coarse approximations, however, we have realized that the greedy method that determines the ordering of the edge collapse operations, is additionally constrained by the topological considerations described earlier. For that reason, we could not obtain the same accuracy as with our vertex insertion scheme under those circumstances. Another advantage of progressive tetrahedralizations, besides its fine grain progressivity, is its very short reconstruction time – even for larger models. This clearly compensates for the time consuming transformation comprising cost function evaluations, consistency checks and edge collapses, which can be carried out offline as a preprocessing step, though.

Acknowledgments

We would like to thank Markus Gross and the members of the Computer Graphics Lab at ETH Zürich for their support of this work. We would further like to thank the Swiss Federal Office of Topography for the DHM-25 model of the Matterhorn and Advanced Visual Systems Inc. for the CT volume data set.

References

1. C. Barber, D. Dobkin, and H. Huhdanpaa. The quickhull algorithm for convex hulls. *ACM Transactions on Mathematical Software*, January 9 1995.

2. M. D. Buhmann and C. A. Micchelli. Spline prewavelets for non-uniform knots. *Numerische Mathematik*, 61(4):455–474, May 1992.
3. C. K. Chui and J. Z. Wang. A cardinal spline approach to wavelets. *Proc. Amer. Math. Soc.*, 113:785–793, 1991.
4. P. Cignoni, D. Costanza, C. Montani, C. Rocchini, and R. Scopigno. Simplification of tetrahedral meshes with accurate error evaluation. In *IEEE Visualization 2000*, pages 85–92, Oct. 2000.
5. P. Cignoni, C. Rocchini, and R. Scopigno. Metro: Measuring error on simplified surfaces. *Computer Graphics Forum*, 17(2):167–174, June 1998.
6. T. Dey, H. Edelsbrunner, S. Guha, and D. Nekhayev. Topology preserving edge contraction. Technical report, Raindrop Geomagic Inc., Research Triangle Park, NC, 1998.
7. D. Douglas and T. Peucker. Algorithms for the reduction of the number of points required to present a digitized line or its caricature. *The Canadian Cartographer*, 10(2):112–122, December 1973.
8. E. Quak and N. Weyrich. Decomposition and reconstruction algorithms for spline wavelets on a bounded interval. *Appl. Comput. Harmon. Anal.*, 1(3):217–231, 1994.
9. M. Garland and P. S. Heckbert. Surface simplification using quadric error metrics. In *Computer Graphics (SIGGRAPH '97 Proceedings)*, pages 209–216, Aug. 1997.
10. M. H. Gross. L^2 optimal oracles and compresion strategies for semi-orthogonal wavelets. Technical Report 254, Computer Science Department, ETH Zurich, 1996.
11. M. H. Gross, R. Gatti, and O. Staadt. Fast multiresolution surface meshing. In *Proceedings of IEEE Visualization '95*, pages 135–142. IEEE Computer Society Press, 1995.
12. M. H. Gross, O. G. Staadt, and R. Gatti. Efficient triangular surface approximations using wavelets and quadtree data structures. *IEEE Transactions on Visualization and Computer Graphics*, 2(2), June 1996. ISSN 1077-2626.
13. H. Hoppe. Progressive Meshes. In *Computer Graphics Proceedings, Annual Conference Series, 1996 (ACM SIGGRAPH '96 Proceedings)*, pages 99–108, 1996.
14. J. M. Lounsbery. *Multiresolution Analysis for Surfaces of Arbitrary Topological Type*. PhD thesis, University of Washington, Seattle, 1994.
15. W. B. Pennebaker and J. L. Mitchell. *JPEG Still Image Data Compression Standard*. Van Nostrand Reinhold, New York, 1993.
16. J. Popović and H. Hoppe. Progressive simplicial complexes. In *Proceedings of SIGGRAPH 97*, pages 217–224. ACM SIGGRAPH / Addison Wesley, Aug. 1997. ISBN 0-89791-896-7.
17. R. Ronfard and J. Rossignac. Full-range approximation of triangulated polyhedra. In *Proceedings of EUROGRAPHICS '96*, pages C67–C76, 1996.
18. J. Shewchuk. Triangle: A two-dimensional quality mesh generator and delaunay triangulator. Technical report, Carnegie-Mellon Institute, 1995.
19. O. Staadt. *Multiresolution Representation and Compression of Surfaces and Volumes*, volume 12 of *Selected Readings in Vision and Graphics*. Hartung-Gorre Verlag, Konstanz, Germany, 2001. ISBN 3-89649-707-3.
20. O. G. Staadt and M. H. Gross. Progressive tetrahedralizations. In *Proceedings of IEEE Visualization '98*, pages 397–402, 1998.

21. O. G. Staadt, M. H. Gross, and R. Weber. Multiresolution compression and reconstruction. In *IEEE Visualization '97*, pages 337–346, Nov. 1997.
22. E. J. Stollnitz, T. D. DeRose, and D. H. Salesin. *Wavelets for Computer Graphics: Theory and Applications*. Morgan Kaufmann, San Francisco, CA, 1996.
23. I. J. Trotts, B. Hamann, K. I. Joy, and D. F. Wiley. Simplication of tetrahedral meshes. In *Proceedings of IEEE Visualization '98*, pages 287–296, 1998.

Part VI

Biomedical and Physical Applications

Geometric Methods for Vessel Visualization and Quantification - A Survey

Katja Bühler[1], Petr Felkel[1] and Alexandra La Cruz[2]

[1] VRVis Research Center, Vienna, Austria {buehler,felkel}@vrvis.at
[2] Institute of Computer Graphics and Algorithms, Vienna University of Technology, Vienna, Austria alacruz@cg.tuwien.ac.at

Summary. Visualization and quantitative analysis of vessel data is an important preprocessing step in diagnosis of vascular diseases, monitoring, surgery planning, blood flow simulation, education and training of surgeons. This paper surveys several geometric methods to solve basic visualization and quantification problems like centerline computation, boundary detection, projection techniques, and geometric model generation.

1 Introduction

Atherosclerosis is one of the most observed civilization diseases today. Plaque, a mixture of calcium, cholesterol fibrin and other substances accumulated in the vessel lumen, causes stenosis or occlusion of the vessel. Depending on the location of the atherosclerosis strokes, heart attacks, peripheral artery occlusion disease in legs and aneurysms might be the consequence.

Up to now several methods for data acquisition for diagnosis exist, differing in technique (X-ray, computer tomography, ultrasound, magnetic resonance), acquired data (2D or 3D, enhancing different structures like vessels, blood flow, soft tissue or bones), necessity of contrast agent, efforts and expenses (invasive, non-invasive methods, costs and time). The present standards for vessels investigation are Digital Subtracted Angiography (DSA) and 2D Ultrasound (US). Conventional DSA and ultrasound are 2D imaging techniques and do not allow a 3D reconstruction of the vessels. Parts of the vessel tree might be occluded or the presence of noise makes a diagnosis difficult. Recent research focuses on visualization techniques for vessels on basis of two or multiple projections, or (sliced) 3D-data sets of the region of interest allowing a spatial reconstruction of the vessel tree. Datasets produced by CT, MR or 3D US are huge and cannot be handled manually slice-by-slice in an efficient way. In all three cases post-processing (geometry extraction and/or rendering) the data is necessary to extract relevant information.

Geometric processing of vessel data becomes more and more important for visualization, diagnosis and quantification of diseases, for monitoring the disease progress, for surgery planning, and training of surgeons and invasive radiologists using VR techniques. Finite element meshes provide a basis for blood flow simulation needed e.g. for planning of bypass surgery and stent replacement. The generation of geometric vessel models allows a repeatable diagnosis, and fast, interactive visualization. On the other hand, the method is highly sensitive to object selection criteria, and wrong settings might lead to miss important information.

In our paper we will survey and discuss different geometric methods applied to vessel visualization and geometric model generation. In section 2 different kinds of skeletonization methods and applications are discussed. Direct centerline tracking algorithms will be summarized in section 3. Deformable models are presented in section 4. Algorithms to generate vessel models from given contours are discussed in section 5. Section 6 deals with direct isosurface extraction, connected problems, and proposed solutions. Curved planar reformation as a special projection technique for CT angiograms is discussed in section 7. A short summary is given in section 8.

2 Skeletonization of Vascular Structures

The term *skeleton* and its generating *medial axis transform* has been first introduced by Blum [9] in the context of shape recognition in computer vision to describe and characterize geometries of biological shapes. Skeletons are a kind of "stick-figure" representation of an object: the shape is reduced to the set of its medial points which is *the locus of the centers of all maximal inscribable discs / spheres within the boundary of the object*. In the 3D case a skeleton consists of branched 2D manifolds that degenerate to space curves for tube-like structures. The connection of each skeleton point with the radius of its associated disk or sphere allows an error free reconstruction of the shape.

The tubular shape of vessels is particularly suitable for skeletonization: the skeleton of a vessel tree is in the ideal case a tree of connected space curves representing the centerlines of branches of the vessel tree. Skeletonization algorithms applied on segmented angiograms, maximum intensity projections or volume representations of vessels are a powerful and widely used tool for centerline detection [5, 73, 78], path planning for virtual endoscopy [15, 103] and graph based classifications of vessel trees [91].

There exists a wide variety of sequential and parallel skeletonization algorithms for continuous and discrete 2D and 3D data. Algorithms are classified by method (topological thinning, distance map based, or based on Voronoi diagrams) or by input data (continuous, polygonal, discrete). As we focus on techniques for vessel visualization, only skeletonization algorithms for discrete data will be discussed here. A good bibliography on the topic can be found in [66].

In general, a direct application of the definition presented in the last section on discrete data leads to problems: noisy data produces skeletons with too many branches, quality and shape of the approximated skeleton depends highly on the chosen discrete metric. *Branch clipping* algorithms or hierarchical models [70] try to reduce the complexity of the computed skeleton and the influence of noise: heuristics are used to detect those branches representing less important features of the shape. Another problem that arises processing discrete data is based on the fact, that due to the finite resolution of the underlying grid, the skeleton is not uniquely defined and special care has to be taken to preserve connectivity. Nyträm [69] and Chen et al. [15] list some criteria for a good skeleton approximation in discrete data: It preserves the topology of the original shape, and approximates the central axis, it is thin, smooth, and continuous. It should allow full object recovery.

Topological thinning Topological or morphological thinning algorithms are based on the *grass-fire definition* of skeletons which is an illustrative physical interpretation of Blum's skeleton definition: The boundary of a binary object is iteratively peeled off deleting in each iteration step points that fulfill certain geometric and topologic constraints.

Thinning algorithms theoretically preserve topology, but symmetric thinning and connectivity preservation is a difficult task operating on discrete data. Digital topology [23, 73] provides a theoretical basis to overcome this problems: Points are classified according to their neighborhood into different kinds of border, background or foreground, simple, and end points. This allows a derivation of grade of influence on the topology of the object. Another theory applied in thinning algorithms is mathematical morphology [24, 57]. Here the thinning process is defined using so called structuring elements and morphological operators like dilatation, erosion, opening and closing.

Topological thinning has been applied to skeleton extraction of vessel trees by several authors. Palagyi [73] presents a sequential thinning algorithm for segmented data and applies it for center path computations of aortic aneurysms. Dokládal [23] proposes a two step method of skeletonization of 3D grey scale objects using luminosity-driven homotopic erosion. Maglaveras [57] and Eiho [24] extract skeletons of coronary artery trees. Selle et. al. [91] use a thinning algorithm to construct a characteristic graph of the vessel tree to allow a graph based analysis of the structure.

Distance transform Skeletonization algorithms based on distance transformations are a direct application of Blums skeleton definition on discrete binary data. The distance transformation of binary data is the process of labeling each voxel with the (approximated) Euclidean distance to its closest background voxel. The type of discrete approximation of the distance depends on the application. A good survey on Euclidean distance approximations and the influence on the shape of the skeleton can be found in [69, 78]. Puig Puig [78] introduces and discusses the *discrete medial axes transform* (MATD) for segmented 2D and 3D data: The skeleton of the object is defined as set of centers of maximal balls constructed with the labeled distance as radius. Puig Puig also shows examples for skeletonization of (synthetic) vascular structures. Subsets of skeletons necessary e.g. for path planning, can be determined computing minimum-cost spanning trees in the distance map using Dijkstra's shortest path algorithm [5, 15, 22, 97, 103]. For a detailed discussion of tracking algorithms, the reader is referred to section 3.

Hybrid approaches Hybrid algorithms combine topological thinning with distance maps to handle difficulties introduced by anisotropic voxels to ensure the symmetry of the thinning process. Nyström and Smedby [69] adapted an algorithm of Borgefors [11] for skeletonization of volumetric vascular structures. The same hybrid approach has been also used by Selle and Peitgen [90] as basis for graph analysis of vessels.

Voronoi Skeletons Theory and application of Voronoi skeletons have been discussed by Ogniewicz and Ilg [70]. These skeletons allow to overcome the difficulties of topological thinning and distance transform based algorithms replacing the discrete distance map with the Voronoi diagram of boundary points. Due to the discretization of the shape by polygonization, Voronoi skeletons only approximate the true medial axis, but converge against it with increasing sampling rate. Any vertex of the boundary introduces an additional skeleton branch. Branch pruning and multi-resolution representations of the skeleton based on heuristics have been introduced to reduce the skeleton to topological important parts.

A good overview on discrete Voronoi skeletonization algorithms in 2D and 3D is given in [68]. The same paper applies a Voronoi skeletonization algorithm to compute skeletons of organs given as segmented MR datasets. Attali [3] applied Voronoi skeletonization on the triangulated iso-surfaces of heart muscles, using a filtering technique to prune small branches.

3 Direct Tracking of Centerlines

For a rough approximation of the skeleton, and for an extraction of vessel topology, direct tracking of vessels of interest in the raw or binary segmented data is a broadly used technique [28, 42, 80, 101, 106, 109]. The approaches differ in the definition of the starting conditions and in the tracking principle applied. We can classify them into three categories:

Tracking of wave from a seed point is a region-growing approach in the binary segmented data volume, which is enriched by vessel bifurcation detection and topological graph generation [89, 109]. The *wave* is propagated from the seed point in the root of the tree. The bifurcations are detected, when the wave-front splits in separate regions. A similar, 2D approach of wave propagation can be found in [80].

Path tracking from a seed point in given direction is an interactive single vessel centerline detection method [106] in the raw dataset. By setting two start-points, the user defines a possible direction of the vessel centerline.

The method estimates the next candidate point in this direction and then computes its precise position as the point with the highest "likelihood-of-being-center" in the plane perpendicular to this direction (see section 4.3 for details). The method has to be user-supervised and restarted if it leaves the vessel of interest. It also estimates the vessel diameter.

Path tracking from a seed point to given end-point(s) [42] uses the principle of the Dijkstra's shortest (minimal cost) path search in the graph [21,22]. The 3D volume is taken as a graph with *nodes* in places of voxels and with *edges* connecting the neighboring ones. The graph links are assigned a cost, e.g., the absolute difference of the node values, and a monotone increasing function for computation of the cost along the path is defined. The shortest path between two given voxels is found and centered to get the vessel centerline.

Modification of this approach can be used for interactive vessel selection. After pointing to a start-point, all the paths from the start-point to the whole dataset are pre-computed in the preprocessing step [35]. The shortest-path from cursor to the start-point is then interactively displayed.

4 Deformable Models

Segmentation and the generation of geometric models of organic structures is an essential pre-processing step for accurate and repeatable quantitative analysis of medical image data [62]. The application to vessel trees are blood flow simulation [95, 105], vascular surgery planning [95], data reduction to reach real-time frame rates in VR applications [27], tracking [2,17], registering [86], and quantifying [29,47] vascular structures over short and long time periods.

Although extremely time-consuming, today manual segmentation performed slice-by-slice is clinical routine. An automation of this process is a difficult task due to noise, shape complexity, and variability of the human body. Todays techniques are far from full automation. In many cases segmentation is impossible without user interaction.

Deformable models have been identified early as a powerful tool to combine the a-priori knowledge of anatomical structures of a physician with automatic image analysis techniques. First, an initial estimating geometric model is placed by the user, close to the object of interest. Physical, optical and/or statistical forces deform the model automatically in a way that it approximates the true shape of the object.

McInerney and Terzopoulos [62] classify deformable models in three categories whose underlying models will be discussed in the following sections:

- Static energy minimizing (classical snakes, balloons, topological snakes),
- Dynamic (level sets or implicit snakes),
- Probabilistic or shape-based (snakes with probabilistic energy functions, ray propagation algorithms).

When applied to vessel data, deformable models can be used to detect vessel contours in 2D images, to reconstruct the 3D location of the vessel tree from bi- or multiplane angiograms using space curves, and to generate a geometric model of the vessel tree using deformable surfaces or volumes.

4.1 Snakes

Snakes have been introduced by Kass and Terzopoulus as a special case of higher dimensional deformable models [44]. They describe a snake in the following way: *"A snake is an energy minimizing spline guided by external constrained forces and influenced by image forces that pull it towards features such as lines and edges."*

In 2D, snakes are defined as a parametric curve $\mathbf{c}(s) \subset \mathbb{R}^2, s \in [a,b] \subset \mathbb{R}$ representing a controlled continuity spline. The discretization of $\mathbf{c}(s)$ is in general a set of sorted sample points $\mathbf{v}_i \in \mathbb{R}^2, i = 1, ..., n$, also called *snaxels*. The deformation process is driven by minimization of an energy function $\mathcal{E}$ being a combination of internal (shape), external (image) and constrained (user defined) forces [87]:

$$\mathcal{E} = \mathbb{N}_a^b (\mathcal{E}_{internal}(\mathbf{c}(s)) + \mathcal{E}_{image}(\mathbf{c}(s)) + \mathcal{E}_{constr}(\mathbf{c}(s))) ds \qquad (1)$$

The image and constrained forces are combined in some publications to a so-called *external force*: $\mathcal{E}_{external} := \mathcal{E}_{image} + \mathcal{E}_{constr}$. The image energy $\mathcal{E}_{image}$ is often described by an integral of a potential $P(\mathbf{c}(s))$.

The 2D snake concept can be easily transferred to surface or volumetric object representations replacing $\mathbf{c}(s)$ by a parametric surface or volume description and the corresponding discretization by an appropriate mesh. Existing implementations of snakes differ in

- *the underlying geometry*, like snaxels, triangular and quadrilateral meshes, subdivision curves and surfaces, finite element meshs, B-splines, NURBs;
- *the definition of the energy functions*, as the chosen parameters for internal energy, the underlying functions to formulate the image energy (e.g. different kinds of filters, luminance or distances), and possible additional constrained energies like inflation, topological or spring energy;
- *the discretization of the problem*, finite differences, discretization of curvature, finite elements;
- *the chosen optimization strategy*, e.g variational approach, dynamic programming, greedy algorithms, simulated annealing, genetic algorithms.

A full discussion would go beyond the scope of this paper. The reader is referred to [18, 87] where different definitions of energy functions are discussed, as well as discretization methods, and optimization algorithms.

Snakes heavily depend on a proper initialization. The convergence and stability of the deformation process depends on the location of the initial object to avoid that the snake locks at local minima of the energy functional.

The choice of energy functions and related parameters has a significant influence on the quality of the snake: filter-based energy functions can reduce the influence of noise, a local determination of parameters of elasticy and additional inflation or pressure energy can avoid a degeneration (shrinking or flattening) of the snake. Snakes implementing this concept are also called *balloons* [18]: an initial shape inside the object of interest is inflated till it fits the object. A good discussion and some recipes to handle the initialization and degeneration problem can be found in [82]. The inability to adapt to changing topology is another problem of snakes. *Topology adaptive curve and surfaces snakes* [63,64] (T-snakes) have been introduced to overcome this problem. T-snakes also use inflational forces like balloons but are additionally reparametrized during the optimization process. This allows topological changes of the snake like splitting into two separate snakes.

Snakes have been applied successfully to segment vessel data. Klein et al [47] mention that *B-Splines* have characteristics that make them well suited for a segmentation of vessels: they are smooth, continuous, and completely defined by few control points. Furthermore they have an implied internal energy keeping them well shaped - An explicit formulation of $\mathcal{E}_{internal}$ in equation (1) is not necessary, but can be given to extend shape constraints. The piecewise nature and local influence of control points allows to write and compute the curve energy $\mathcal{E}$ as the sum of energy terms for each span.

Planar B-spline snakes have been applied to extract vessel contours in quantitative coronary angiograms [2,47]. (B-Spline) space curve snakes have been chosen as basis for centerline determination in 3D MRA [29], and for reconstruction of catheter paths [67] or the whole vessel topology [12,83] from bi-plane angiograms. Frangi et al [29] also propose surface snakes represented by tensor product B-spline surfaces to model vessel walls. The same model with different energy functions has been used by Huang and Amini [39] for geometric model generation of tubular structures in volumetric 3D image data. A *tubular deformable model* for vessel reconstruction based on triangular meshes similar to the snake model proposed by Frangi [29] has been recently proposed by Yim [107]. He critizises the inflexibility of the mesh due to smoothing constraints and proposes a more generalized deformation process analogous to a mechanical equilibration process. Pujol et al [79] reconstruct vessel walls in intravascular ultrasound images as iso-surfaces of deformable B-spline volumes. Hu [38] proposes a snake with variable stiffness parameters for vessel boundary extraction to allow to adapt to strong and smooth or missing edge features. Different kinds of discrete *statistical snakes* have been applied to segment cross-sections of vessels in intravascular ultrasound images [75,84] and angiographies [98]. McInerney and Terzopolous [64] demonstrated the power of 2D *T-snakes* in application to vessel trees in angiographies. A little seed snake placed within a vessel starts to grow and to segment all vessel branches in a flow-like manner. 3D T-snakes have been applied successfully to compute a triangular mesh representation of the vascular system of the brain based on 3D MRA data [63].

Other interesting methods have not yet been applied to model generation of vessels like surface snakes based on finite element meshes proposed by Cohen [18] and McInerney [61]. Lürig et al. [55] as well as Hug et al. [40] combine the subdivision process of *subdivision surfaces with snake energy functions*. Radeva et al. [81] propose snake based on tensor product B-spline volumes for segmentation and tracking of heart motion of SPAMM MRI data.

4.2 Level Sets

Level sets have been introduced by Osher and Sethian [72] to describe evolving geometries, and to provide an implicit description of boundaries. This method has been applied to track objects in movies, and to recover shapes and structures in medical images, especially vessels.

A level set is formulated as implicit boundary tracking scheme that eliminates many of the difficulties when modeling evolving curves and surfaces using classical snakes: Due to its implicit formulation, level sets are able to handle arbitrary topological changes.

Given an implicit curve g_t propagating in its normal direction with speed v, a level set function $\Phi(x, y;\ t)$ is introduced such that the zero level set $\Phi(x, y;\ t) = 0$ is identified at any time t with the evolving curve g_t. Changes in the geometry and topology of the curve are reflected by changes in the zero-crossing of Φ. The searched boudary curve is equivalent to the zero set of the solution of the following equation:

$$\frac{\partial}{\partial t}\Phi(x, y;\ t) + v\,|\,\nabla_{x,y}\Phi(x, y;\ t)\,| = 0 \quad \text{with} \quad \Phi(x, y;\ 0) = g_0(x, y) \quad (2)$$

The velocity function v determines the evolution of the curve into its normal direction. Implementations of level sets differ in the definition of v that may include geometric and image-dependent constraines like the gradient-based velocity decay or stabilized boundary motion. For a detailed discussion of possible definitions see [105].

The main difference between level sets and snakes is the representation. Snakes explicity store the nodal positions and connectivities. Instead level set methods use an implicit scheme in which neither positions nor connections are directly maintained. Furthermore the method is dimensionality independent, and (2) describes a well-studied type of differential equation with a stable numerical solution [105].

One drawback of this method is the computational cost. Different extensions have been designed to reduce the high computational cost, like the *narrow band* method [1] and *fast marching* [92, 93] methods. The idea of the narrow band method is to consider only pixels which are close to the latest position of the zero level-set contour in both directions (inward and outward). The fast marching method is designed to resolve problems where the speed function never changes sign. Recently a new method has been proposed, called *Hermes algorithm* [74]. This algorithm combines narrow band and fast marching performing selective propagation over a relatively small window. Malladi et al. [58–60] applied narrow band and fast marching methods to recover shapes in medical images. A comparison of level sets with classical snakes and balloons showed that the narrow band method exhibits the best performance when applied to arterial tree recovery in DSA images [60].

The *geodesic active contours model* is a geometric alternative for snakes based on the level set method to handle topological changes for the evolving curves. This method is comparable to classical snakes because it does not depend on the curve parameterization, but, due to the level set implementation, topological changes are easily handled. Lorigo et al. used this method for segmentation of brain vessel using MRA [54], of abdominal aorta using CT images [50], and of cerebral vessel with MRA images [53]. Wang et al. [104, 105] proposed a combination of level sets and thresholding, which has been successfully applied to geometric model generation from MR images for blood flow simulation, as well as for preoperative surgery planning. They also did an analysis of different geometric models for constructing a model of vascular structure from snakes and balloons to level set methods. Level sets turned out to be the best choice with respect to topological adaptability.

Recently Magee [56] combined a 3D deformable model with level sets for segmentation of vascular structure, specifically for the abdominal aorta. The deformable model used is based on a triangulated mesh. The deformation process is knowledge-based, applying the so called Expert Structure Model (ESM). The ESM defines a probability distribution associated to features of interest such a branching vessels. One drawback of this method is the computational cost due to the cost of stochastic process involved.

4.3 Ray Propagation

The ray propagation method consists of drawing rays over the object of interest from inside to outside. The method can be described by a family of curves in 2D $\mathbf{c}(s,t) \subset \mathbb{R}^2$. $s \in I \subset \mathbb{R}$ denotes the curve parameter and $t \in J \subset \mathbb{R}$ the time. The evolution is governed by: $\frac{\partial}{\partial t}\mathbf{c}(s,t) = v(x,y)\,\mathbf{n}(s,t)$ with $\mathbf{c}(s,0) = \mathbf{c}_0(s)$. Where $\mathbf{c}_0(s) \subset \mathbb{R}^2, s \in \mathbb{R}$ denotes the initial curve, and $\mathbf{n}(s,t) \in \mathbb{R}^2$ the normal vector. $v(x,y) \in \mathbb{R}$ denotes the speed of the ray at point (x,y) and determines the deformation process.

Ray propagation has been applied to fast segmentation of vessels and detection of centerlines. In general, intensity gradients are used to describe the velocity function and to stop the propagation of rays. Wink [106] presents a work based on ray propagation to generate the true centerline of a vessel even in presence of calcifications. The algorithm can be described as follows: Given a candidate point in the plane perpendicular to the vessel axis, several rays are casted. The ray stops when the border of the vessel is detected. Gradient information in the image is used to detect the border of the vessel.

The gradient is calculated as a convolution of the original image with a normalized Gaussian derivative. Finally a center likelihood measure assigned to the origin of the rays is defined. The point with the highest center likelihood is selected to be the center of the vessel. Two problems can be solved using this technique: First, the centerline of the vessel is correctly computed even in presence of calcification and ringing artifacts since the gradient is computed in direction of the ray. Second, in general, ray propagation requires an external process to naturally handle the topological changes. In presence of a bifurcation, the presented algorithm allows to select several points with highest likelihood, e.g. by tracking just one of the branches or by user interaction.

A recent work based on ray propagation was presented by Tek [96] proposing the mean shift analysis. This method points towards the maximum increase in density, representing an estimate of the normalized density gradient computed at one point. It is a statistical technique based ray propagation.

Drawbacks of the ray propagation method are, that it requires user interaction, does not handle the topological changes and requires constant parameters for the evolution equation and window size.

5 Model Generation from Given Contours

The generation of a vessel surface model from a set of contours is used in cases, when direct 3D segmentation methods are impossible to use, or in cases, when local corrections of the surface are necessary [31].

Reconstruction of a vessel surface model from a set of cross-sections consists in creation of such a surface, that approximates the "original" vessel as good as possible. The only real knowledge about the vessel is represented by the shape of the cross-sections. Therefore, if we cut the reconstructed surface by the original planes, we must get the same regions in the cross-sections. The contours of these regions are typically approximated by closed simple polygons, which never lie inside others, as the vessels do not contain holes.

5.1 Surface Reconstruction Techniques

Surface reconstruction techniques usually build the surface step-by-step connecting the contour points in adjacent cross-sections by triangular tiles [25, 45, 65, 71]. A method using the information from more neighboring layers has been published by Barequet et al. [4]. Meyers et al. [65] decomposed the surface reconstruction problem into four fundamental subproblems:

The correspondence problem. Which contours in one cross-section should be connected to which contours in other cross-section?

The tiling problem. How should the pairs of given contours be connected? Which vertices and edges should form the triangles?

The branching problem. How to tile the bifurcations, i.e., cross-sections with a different number of contours?

Surface-fitting problem. What does the precise geometry look like? A possible post-processing step smoothing the mesh.

The lack of information about the vessel shape between the cross-sections is solved by local heuristic assumptions. Typical heuristics are: minimizing volume, minimizing surface area, minimizing edge lengths and minimizing angles. Gitlin et al. [32] proved, that there exist polygonal shapes that cannot be tiled without addition of points on the contours or without addition of intermediate layers. Geiger [30] proved, that this problem can be always solved by adding at most two Steiner points onto the contours.

Generalized Cylinders Separate vessel segments and vessel trees can be simply modeled by their centerline and the vessel shape in perpendicular cross-section. With a limitation of the possible cross section shapes (as described below), this is the generalized cylinder model. The mathematical model of the whole vessel-tree surface is defined as a union of *generalized cylinders* [46] that represent each segment of the tree. According to Puig Puig [77] a generalized cylinder of the blood vessel is defined in the following way:

Given a set of cross-sections represented by non-penetrating, simple, closed, convex parametric curves $\mathbf{c}_i(u), u \in [0, 2\pi), i = 1, ..., N$. *Furthermore let* $\mathbf{s}(v), v \in I \subset \mathbb{R}$ *be a continuous and simple parametric curve through the centers (skeletonal points) of all* $\mathbf{c}_i$ *and orthogonal to their supporting planes. The corresponding* Generalized Cylinder $\mathbf{g}$ *is defined as the union of blends between consecutive contour curves*

$$\mathbf{g}(u,v) := \cup_{i=1}^{N-1} \mathbf{b}_i(u,v), \qquad (u,v) \in [0, 2\pi) \times I$$

with $\mathbf{b}_i(u,v) := f_i(\mathbf{c}_i(u), \mathbf{c}_{i+1}(u), v), v \in I_i \subset \mathbb{R}, I_i < I_{i+1}$ *and* $\cup I_i = I$, *and blending functions* f_i *defined in a way that no self-interscetion of the generated surface occurs.*

The skeleton curve and the contour curves can be represented by B-splines or often approximated by polygons [33]. Contour curves can be also simplified to a circle [42]. If the correspondence between the consecutive sections is unique, it simplifies the surface triangulation of the separate segments to a zig-zag triangle pattern. The ends of the generalized cylinders at the branching points can be simply overlapped (but the overlapping cylinders cause flickering during rendering), or they can be smoothly triangulated by any arbitrary contour tiling method with contours in non-parallel cross-sections.

In other approaches, a smooth mesh blend is constructed by means of *convolution surfaces* [7, 8], where the implicit function describing the surface is generated by a convolution of the skeleton curves with the approximation of the Gaussian kernel. The implicit surfaces (class of curved surfaces defined as a solution of some equation $F(x, y, z) = 0$, where the scalar function F assigns a scalar value to each point in the space) are then triangulated by any known method [6, 43]. Another method constructs a smooth surface in two steps [27]: The rough mesh generation is followed by a smoothing by means of the subdivision surfaces [20].

5.2 Volume Reconstruction Techniques

Volume reconstruction techniques [10, 30] construct a tetrahedral mesh between the adjacent cross-sections, with contour points as vertices. New vertices are then added to ensure, that the contours are part of the mesh. Finally, some tetrahedra are eliminated to make the volume consistent with the contours. The reconstructed surface is then the surface of this volume. The volume reconstruction subproblems are [19]:

Meshing Which tetrahedra to use to form the initial mesh?
Conforming How to add points, so that all contour edges appear in the mesh?
Sculpting Which tetrahedra should be removed to get the consistent volume that matches the contours?

5.3 Shape Based Interpolation

The methods described above belong to *direct reconstruction* methods, as the original contour vertices in cross-sections become the vertices of the reconstructed mesh. The use of original vertices limits the shapes of generated triangles, as no criteria can be used to constrain their aspect ratio. Much effort is required for detection and handling of special cases to allow a correct triangulation.

An alternative to the direct reconstruction approach is to use the cross-sections to estimate a 3D function that represents the measure of distance from any point to the surface [99]. Among these method belongs the *shape-based interpolation* proposed by Raya and Udupa [85], which uses the city-block distance to the surface. A more accurate algorithm for computation of the distance values by means of a chamfer distance transformation was used e.g., by Herman et al. [36]. The method works in the following steps: 1. Binary segmentation of the 2D cross-sections. 2. Computation of the distance-field (the distance to the surface) in the cross-section (positive inside, negative values outside the vessel). 3. Interpolation of the distance field between slices (linear, cubic spline, etc...). 4. Iso-surface extraction at zero-level in the interpolated distance field.

The shape-based interpolation methods handle bifurcations and complex shapes, but fails for large shape changes and for significant translations. This problem has been addressed by many authors, trying to align centroids before interpolation, and scale the cross-sections to match bounding rectangles, etc, but still not generate correct surfaces in complex cases. Treece [99] recognized that the problem is in the definition of connectivity, because the correspondence evaluating entire contours is too coarse. He proposed the *disc-guided interpolation*, where the interpolation is guided by using correspondence of *regions* of the cross-sections.

Most of the presented methods handle parallel cross-sections. 3D reconstruction of non-parallel planar cross-sections have been addressed very rarely [19,76], but is at present subject to research, motivated by the evolution of 3D ultrasound techniques [19,99,100].

6 Direct Mesh Generation

One of the simplest method to reconstruct surfaces in scalar or binary volume data is *iso-surface extraction*: A fixed threshold value determines the location of the surface. The standard choice to create triangular meshes based on iso-values is the *marching cubes* algorithm [51]. Based on trilinear interpolation, the algorithm determines step by step the triangulation within each cell belonging to the iso-surface. The original algorithm suffers from ambiguities and creation of non-watertight meshes with a non-optimal triangulation. Many work has been done to overcome these difficulties: Improved case differentiation, crack prevention, triangle reduction, mesh optimization and acceleration [49]. For an overview on existing techniques the reader is referred to [88].

Direct iso-surface extraction works well for objects with clearly determined borders like bones in CT images, but turns out to be problematic for amorphous objects like vessels. Due to image inhomogeneities, noise and other artifacts, iso-surface extraction may be problematic especially for magnetic resonance and ultrasound images. Additional preprocessing steps are necessary to enhance contours or to segment the object.

To cope with inhomogeneities in image intensities, Yim and Summers [108] proposed a local threshold estimation to extract iso-surfaces using a marching cube algorithm. Results are presented for contrast enhanced MRA of thoracic aorta and cerebral ventricles.

Cebral et al. [13] described an iso-surface extraction algorithm to generate CFD meshes for blood flow simulation in arteries (hemodynamics). Several image filtering and segmentation algorithms are applied in a pre-processing step to reduce noise and to enhance image contrast. For the generation of a triangulated iso-surface a simple two step triangulation algorithm is proposed. The triangular mesh is smoothed using the surface fairing algorithm of Taubin [94] and optimized with respect to the number of edges and minimized maximum angles [37]. Based on this initial mesh a finite element mesh is generated. The algorithm has been applied to CTA as well as MRA images. The same authors present a simple CSG algorithm to merge different branches of finite element meshes of vessels to one watertight mesh [14]. Ertl et al. [26] proposed a level-of-detail approach for iso-surface extraction based on multi resolution analysis and wavelets. An adaptively refined tetrahedral mesh is computed which is coarse in homogeneous regions and fine in regions with strong variations. This structure allows a fast and flexible progressive iso-surface extraction. Meshes of iso-surfaces also have been successfully applied to improve phase-contrast flow quantification [110] and hepatic MR angiography [16].

7 Vessel Flattening

As the technique of multi-slice helical CT evolves, it can deliver high resolution datasets covering large anatomic regions. The contrast-enhanced CT of large vessel segments can be scanned in nearly isotropic resolution and can be used for diagnosis with comparable results as invasive DSA. To avoid obscuring of the vessels by other high density structures (mainly bone and inner calcification), a thin slab along the centerline is re-sampled and displayed as a 2D image.

Given a vessel centerline (see Section 3), a line parallel to the horizontal axis of the viewing plane is swept along it, forming a curvilinear surface (a "curved plane"). If we flatten this "curved plane", and display the voxels in the close neighborhood of it, we obtain a 2D image of the vessel. The process of flattening is called a *curved planar reformation (CPR)* [42].

CPR allows one to visualize the vessel lumen together with calcified sediments along the vessel walls in one direction. CPR is highly sensitive to the precise centerline localization—a wrong centerline distorts the vessel lumen and can be misinterpreted as an artificial stenosis.

The simplest "flattening" method is done by projection of the line of samples to the screen—a *(projected) CPR*. Vessel parts parallel to the horizontal plane can be mutually occluded during the projections. Also high intensity structures can occlude the vessel, if parts of the lines intersect bone and these parts are projected to the vessel structures. To overcome the fundamental drawbacks, modifications and enhancements of CPR generation have been proposed (see [41] for details):

Stretched projection - "flattens" the "curved plane" in the viewing direction. Vessel length is displayed completely, avoiding overlap of the swept line for horizontal segments, that causes discontinuities in the projected CPR.
Straightened CPR - sweeps the line perpendicularly to the vessel axis (and parallel to the viewing plane). Therefore, it unfolds the vessel in the "left-right" direction as well, producing a straight line vessel projection.
Rotating CPR - overview of the whole vessel by 180° view animation [41].
Multi-path projection (or *Medial Axis Reformation (MAR)* [35]) simultaneously displays a tree of vessels in one 2D image, by means of overlapping of CPRs of more vessels. For a single vessel it is identical to CPR.
Thick CPR projects a slab of certain thickness. It is therefore less sensitive to the precise center-line detection. The values in the slab can be composed by averaging, maximal (MIP) or minimal intensity projection (MinIP)

8 Summary

Geometric processing of vessel data is a challenging task due to image inhomogeneities, noise, artifacts, and the undetermined human anatomy. Geometric vessel models allow repeatability of quantitative analysis, objective comparison of data, and provide the basis for blood flow simulation. Special projection techniques like CPR enables one to resolve occlusions and to investigate a complete vessel branch. The most important geometric methods for vessel visualization and quantification have been surveyed in this paper. All presented methods are actual topics of research:

Skeletonization and center path tracking are the basis for many applications and an important pre-processing step for other algorithms, like graph-based analysis of vessel trees, geometric model generation from given contours and centerlines, and curved planar reformation.

Deformable models are one of the most powerful and widely used tools for the treatment of vessel images: segmentation and geometric model generation in 2D and 3D are the most important application areas.

Models generated from given contours or centerline and radius information are a simple method to generate approximative meshes.

Iso-surface extraction is the classical method to generate mesh representations for vessels

CPR is a standard technique for analysis of CT or MR angiographies available in most commercial analysis tools on the market.

Due to the huge amount of existing publications it was not possible to mention all applied geometric methods like projection techniques used for 3D reconstruction of vessel trees from bi- and multiplane angiograms/projections [34, 102], and fractal structures for the modeling of vessel trees [48].

References

1. D. Adalsteinsson and A. Sethian. A fast level set method for propagating interface. *J. Comp. Phys.*, pages 269–277, 1995.
2. A. Amini, J. Huang, A. Klein, P. Radeva, and M. Elayyadi. Flexible shapes for segmentation and tracking of cardiovascular data. In *Int. Conf. on Image Processing*, volume 2, pages 5–9. IEEE, 1998.
3. D. Attali and J.-O. Lachaud. Delaunay conforming iso-surface, skeleton extraction and noise removal. *Computational Geometry*, 19:175–189, 2001.
4. G. Barequet, D. Shapiro, and A. Tal. Multi-level sensitive reconstruction of polyhedral surfaces from parallel slices. *Visual Comp.*, 16(2):116–133, 2000.
5. I. Bitter, A. Kaufman, and M. Sato. Penalized-distance volumetric skeleton algorithm. *IEEE Trans. on Visual. and Comp. Graphics*, 7(3):195–206, 2001.
6. J. Bloomenthal. An implicit surface polygonizer. In P. Heckbert, editor, *Graphics Gems IV*, pages 324–349. Academic Press, 1994.
7. J. Bloomenthal. *Skeletal Design of Natural Forms*. PhD th., U. Calgary, 1995.

8. J. Bloomenthal and K. Shoemake. Convolution Surfaces. *ACM Siggraph '91, Computer Graphics*, 25(4):251–256, 1991.
9. H. Blum. A transformation for extracting new descriptors of shapes. In W. Wathen-Dunn, editor, *Models for the perception of speech and visual form*, pages 362–380. MIT Press, Cambridge, 1967.
10. J.-D. Boissonnat and B. Geiger. Three dimensional reconstruction of complex shapes based on the Delaunay triangulation. Research report 1697, INRIA, Sophia Antipolis, April 1992.
11. G. Borgefors, I. Nyström, and G. Santini Di Baja. Computing skeletons in three dimensions. *Pattern Recognition*, 32(7):1225–1236, 1999.
12. C. Canero, P. Radeva, R. Toledo, J.J. Villanueva, and J. Mauri. 3D curve reconstruction by biplane snakes. In *Int. Conf. on Pattern Recognition*, volume 4, pages 563–566. IEEE, 2000.
13. J.R. Cebral and R. Löhner. From medical images to CFD meshes. In *Int. Meshing Roundtable*, 1999.
14. J.R. Cebral, R. Löhner, P.L. Choyke, and P.J. Yim. Merging of intersecting triangulations for finite element modeling. *J.Biomech.*, 36(6):815–819, 2001.
15. D. Chen, B. Li, Z. Liang, M. Wan, A. Kaufman, and M. Wax. A tree-branch searching, multi-resolution approach to skeletonization for virtual endoscopy. In K. Hanson, ed., *Medical Imaging–Image Processing*, vol. 3979, *SPIE*, 2000.
16. P.L. Choyke, P. Yim, H. Marcos, VB. Ho, R. Mullick, and R.M. Summers. Hepatic MR angiography: a multiobserver comparison of visualization methods. *Am. J. Roentgenol.*, 176(2):465–470, 2001.
17. L.D. Cohen. Deformable surfaces and parametric models to fit and track 3D data. In *IEEE Int. Conf. on Systems, Man, and Cybernetics*, volume 4, pages 2451–2456, 1996.
18. L.D. Cohen and I. Cohen. Finite-element methods for active contour models and balloons for 2-D and 3-D images. *IEEE Trans. Pattern Anal. and Mach. Intell.*, 15(11):1131–1147, 1993.
19. C. Dance and R. Prager. Delaunay reconstruction from multiaxial planar cross-sections. TR273, Cambridge University Engineering Department, 1997.
20. T. DeRose, M. Kass, and T. Truong. Subdivision surfaces in character animation. In *SIGGRAPH'98*, pages 85–94. ACM, 1998.
21. R.B. Dial. Algorithm 360: Shortest path forest with topological ordering. *Commun. of the ACM*, 12:632–633, 1969.
22. E. W. Dijkstra. A note on two problems in connection with graphs. *Numerische Mathematik*, 1:269–271, 1959.
23. P. Dokládal, C. Lohou, L. Perroton, and G. Bertrand. A new thinning algorithm and its application to extraction of blood vessels. In *Biomedsim' 99*, pages 32–37. ESIEE, 1999.
24. S. Eiho and Y. Qian. Detection of coronary artery tree using morphological operator. *Computers in Cardiology*, 24:525–528, 1997.
25. A.B. Ekoule, F.C. Peyrin, and C.L. Odet. A triangulation algorithm from arbitrary shaped multiple planar contours. *ACM Trans. Graph.*, 10(2):182–199, 1991.
26. T. Ertl, R. Westermann, and R. Grosso. Multiresolution and hierarchical methods for the visualization of volume data. *Future Generation Comp. Sys.*, 15(1):31–42, 1999.

27. P. Felkel, A. L. Fuhrmann, A. Kanitsar, and R. Wegenkittl. Surface reconstruction of the branching vessels for augmented reality aided surgery. In J. Jan, J. Kozumplík, and I. Provazník, eds. *Biosignal'02*, pages 252–254, 2002.
28. P. Felkel, R. Wegenkittl, and A. Kanitsar. Vessel tracking in peripheral CTA datasets – an overview. In R. Ďurikovič and S. Czanner, editors, *SCCG'01*, pages 232–239. IEEE, 2001.
29. A.F. Frangi, W.J. Niessen, R.M. Hoogeveen, T. van Walsum, and M.A. Viergever. Three-dimensional model-based stenosis quantification of the carotid arteries from contrast-enhanced MR angiography. *IEEE Trans. Med. Imaging*, 18(10):946–956, 1999.
30. B. Geiger. *Three-dimensional modeling of human organs and its application to diagnosis and surgical planning*. PhD thesis, Ecole des Mines de Paris, 1993.
31. A. Giachetti, M. Tuveri, and G. Zanetti. Measurable models of abdominal aortic aneurysms on the web. In *Machine Meets Virtual Reality'01*, pages 158–160, 2001.
32. C. Gitlin, J. O'Rourke, and V. Subramanian. On reconstructing polyhedra from parallel slices. *Int. J. Comp. Geom. Appl.*, 6(1):103–122, 1994.
33. K.K. Hahn, B. Preim, D. Selle, and H.-O. Peitgen. Visualization and interaction techniques for the exploration of vascular structures. In T. Ertl, K. Joy, and A. Varshney, editors, *Visualization'01*, pages 395–402. IEEE, 2001.
34. P. Hall, P. Andreae, and M. Ngan. Reconstruction of blood vessel networks from a few perspective projections. In *Int. Two-Stream Conf. on Artif. Neural Networks and Expert Systems*, pages 325–328. IEEE, 1995.
35. S. He, R. Dai, B. Lu, C. Cao, H. Bai, and B. Jing. Medial axis reformation: A new visualization method for CT angiogr. *Acad Radiol.*, 8(8):726–733, 2001.
36. G.T. Herman, J. Zheng, and C.A. Bucholtz. Shape-based interpolation. *Comp. Graph. Appl.*, 12(3):69–79, 1992.
37. H. Hoppe, T. DeRose, T. Duchamp, J. McDonald, and W. Stuetzle. Mesh optimization. *Computer Graphics*, 27(3):19–26, 1993.
38. Y.-L. Hu, W.J. Rogers, D.A. Coast, C.M. Kramer, and N. Reichek. Vessel boundary extraction based on a global and local deformable physical model with variable stiffness. *Magnetic Resonance Imaging*, 16(8):943–951, 1998.
39. J. Huang and A.A. Amini. Anatomical object volumes from deformable B-spline surface models. In *ICIP'98*, pages 732–736. IEEE, 1998.
40. J. Hug, Ch. Brechbühler, and G. Székely. Tamed snake: A particle system for robust semi-automatic segmentation. In *MICCAI '99*, 1999.
41. A. Kanitsar, D. Fleischmann, R. Wegenkittl, P. Felkel, and E. Gröller. CPR - curved planar reformation. In *Visualization'02*. IEEE, 2002.
42. A. Kanitsar, R. Wegenkittl, P. Felkel, D. Fleischmann, D. Sandner, and E. Gröller. Computed tomography angiography: a case study of peripheral vessel investigation. In *Visualization'01*, pages 477–480. IEEE, 2001.
43. T. Karkanis and A. J. Stewart. Curvature-dependent triangulation of implicit surfaces. *Comp. Graph. Appl.*, 21(2):60–69, 2001.
44. M. Kass, A. Witkin, and D. Terzopoulos. Snakes: Active contour models. *Int. J. Comp. Vision*, pages 321–331, 1988.
45. E. Keppel. Approximating complex surfaces by triangulation of contour lines. *IBM J. of Research and Development*, 19:2–11, 1975.
46. M. Kim, E. Park, and H. Lee. Modelling and animation of generalized cylinders with variable radius offset space curves. *J. Vis. Comp. Animation*, 5(4):189–207, 1994.

47. A.K. Klein, F. Lee, and A.A. Amini. Quantitative coronary angiography with deformable spline models. *IEEE Trans. Med. Imaging*, 16(5):468–482, 1997.
48. G. Kókai, Z. Tóth, and R. Ványi. Modelling blood vessel of the eye with parametric L-systems using evolutionary algorithms. In W. Horn, Y. Shahar, G. Lindberg, S. Andreassen, and J. Wyatt, editors, *Artificial Intelligence in Medicine*, volume 1620 of *LNCS*, pages 433–443. Springer, 1999.
49. Y. Livnat, S. Parker, and C.R. Johnson. Fast isosurface extraction methods for large imaging datasets. In Isaac Bankman, editor, *Handbook of Medical Image Processing*. Academic Press, 2000.
50. S. Loncaric, M. Subasic, and E. Sorantin. 3-D deformable model for abdominal aortic aneurysm segmentation from ct images. In *Int. Worksh. on Image and Signal Processing and Analal*, pages 139–144. IEEE, 2000.
51. W.E. Lorensen and H.E. Cline. Marching cubes: a high resolution 3D surface reconstruction algorithm. *Comp. Graph.*, 21(3), 1987.
52. L.M. Lorigo, O. Faugeras, W.E. Grimson, R. Keriven, and R. Kikinis. Segmentation of bone in clinical knee MRI using texture-based geodesic active contours. *Medical Image Comp. and Comp.-Ass. Intervention*, 1998.
53. L.M. Lorigo, O. Faugeras, W.E. Grimson, R. Keriven, R. Kikinis, A. Nabavi, and C.-F. Westin. Codimension-two geodesic active contours for the segmentation of tubular structure. In *Conf. on Computer Vision and Pattern Recognition*, pages 444–451. IEEE, 2000.
54. L.M. Lorigo, O. Faugeras, W.E. Grimson, R. Keriven, R. Kikinis, and C.-F. Westin. Co-dimension 2 geodesic active contours for MRA segmentation. In *Int. Conf. Information Processing in Medical Imaging*, 1999.
55. Ch. Lürig, L. Kobbelt, and T. Ertl. Hierarchical solutions for the deformable surface problem in visualization". *Graphical Models*, 62(2):2–18, 2000.
56. D. Magee, A. Bulpitt, and E. Berry. 3D automated segmentation and structural analysis of vascular trees using deformable models. In *Ws. on Variational and Level Set Methods in Computer Vision*, pages 119–126. IEEE, 2001.
57. N. Maglaveras, K. Haris, S.N. Efstratiadis, J. Gourassas, and G. Louridas. Artery skeleton extraction using topographic and connected component labeling. In *Computers in Cardiology'01*, pages 265–268. IEEE, 2001.
58. R. Malladi, R. Kimmel, D. Adalsteinsson, G. Sapiro, V. Caselles, and J.A. Sethian. A geometric approach to segmentation and analysis of 3D medical images. In *Worksh. on Mathematical Methods in Biomedical Image Analysis*, pages 244–252. IEEE, 1996.
59. R. Malladi and J.A. Sethian. A real-time algorithm for medical shape recovery. In *Int. Conf. on Computer Vision*, pages 304–310. IEEE, 1998.
60. R. Malladi, J.A. Sethian, and B.C. Vemuri. Shape modeling with front propagation: a level set approach. *IEEE Trans. on Pattern Analysis and Machine Intelligence*, 17(2):158–175, 1995.
61. T. McInerney and D. Terzopoulos. A dynamic finite element surface model for segmentation and tracking in multidimensional medical images with application to cardiac 4D image analysis. *Computerized Medical Imaging and Graphics*, 19(1):69–83, 1995.
62. T. McInerney and D. Terzopoulos. Deformable models in medical image analysis: A survey. *Medical Image Analysis*, 1(2):91–108, 1996.
63. T. McInerney and D. Terzopoulos. Topology adaptive deformable surfaces for medical image volume segmentation. *IEEE Trans. Med. Imaging*, 18(10):840–850, 1999.

64. T. McInerney and D. Terzopoulos. T-Snakes: Topology adaptive snakes. *Medical Image Analysis*, 4(2):73–91, 2000.
65. D. Meyers, S. Skinner, and K. Sloan. Surfaces from contours. *ACM Trans. on Graphics*, 11(3):228–258, 1992.
66. P. Min. Shape analysis bibliography, 2002.
http://www.cs.princeton.edu/~min/mc/bib.cgi?template=complete.
67. C. Molina, G. P. Prause, P. Radeva, and M. Sonka. 3-D catheter path reconstruction from biplane angiography using 3D snakes. In K. Hanson, editor, *Medical Imaging*, volume 3338 of *SPIE Conf. Proceedings*, 1998.
68. M. Näf, O. Kübler, R. Kikinis, M. Shenton, and G. Székely. Characterization and recognition of 3D organ shape in medical image analysis using skeletonization. In *Ws. on Math. Meth. in Biomed. Image Anal.*, pp 139–150. IEEE, 1996.
69. I. Nyström and Ö. Smedby. Skeletonization of volumetric vascular images-distance information utilized for visualization. *J.Comb. Opt.*, 5(1):27–41, 2001.
70. R. Ogniewicz and M. Ilg. Voronoi skeletons: Theory and applications. In *IEEE Conf. of Comp. Vision and Pattern Recog.*, pages 63–69, 15–18 1992.
71. J.-M. Oliva, M. Perrin, and S. Coquillart. 3D reconstruction of complex polyhedral shapes from contours using a simplified generalized voronoi diagram. *Computer Graphics Forum*, 15(3), 1996.
72. S. Osher and J. A. Sethian. Front propagation with curvature-dependent speed: Algorithm based on Hamilton-Jacobi formulations. *J. Comput. Phys.*, pages 12–49, 1988.
73. K. Palágyi, E. Sorantin, E. Balogh, A. Kuba, C. Halmai, B. Erdohelyi, and K. Hausegger. A sequential 3D thinning algorithm and its medical applications. *LNCS*, 2082:409–15, 2001.
74. N.K. Paragios and R. Deriche. Geodesic active contours and level sets for the detection and tracking of moving objects. *IEEE Trans. on Pattern Analysis and Machine Intelligence*, pages 266–280, 2000.
75. X.M. Pardo and P. Radeva. Discriminant snakes for 3D reconstruction in medical images. In *ICPR 2000*, volume 4, pages 336–339. IEEE, 2000.
76. B.A. Payne and A.W. Toga. Surface reconstruction by multiaxial triangulation. *Comp. Graph. Appl.*, 14(6):28–35, 1994.
77. A. Puig Puig. Cerebral blood vessels modelling. Technical Report LSI-98-21-R, Universitat Politècnica de Catalunya, 1998.
78. A. Puig Puig. Discrete medial axis transform for discrete objects. Technical Report LSI-98-22-R, Universitat Politècnica de Catalunya, 1998.
79. O. Pujol, C. Canero, P. Radeva, R. Toledo, J. Saludes, D. Gil, J.J. Villanueva, J. Mauri, B. Garcia, J. Gomez, A. Cequier, and E. Esplugas. Three-dimensional reconstruction of coronary tree using intravascular ultrasound images. In *Computers in Cardiology*, pages 265–268. IEEE, 1999.
80. F.K. Quek and C. Kirbas. Vessel extraction in medical images by wave-propagation and traceback. *IEEE Trans. Med. Imaging*, 20(2):117–131, 2001.
81. P. Radeva, A. Amini, Jiantao Huang, and E. Marti. Deformable B-solids and implicit snakes for localization and tracking of SPAMM MRI-data. In *Worksh. on Mathem. Meth. in Biomed. Image Anal.*, pages 192–201. IEEE, 1996.
82. P. Radeva, J. Serrat, and E. Marti. A snake for model-based segmentation. In *Int. Conf. on Computer Vision*, pages 816–821. IEEE, 1995.
83. P. Radeva, R. Toledo, C. Von Land, and J. Villanueva. 3D vessel reconstruction from biplane angiograms using snakes. In *Computers in Cardiology*, pages 773–776. IEEE, 1998.

84. P. Radeva and J. Vitria. Region-based approach for discriminant snakes. In *Int. Conf. on Image Processing*, pages 801–804. IEEE, 2001.
85. S. P. Raya and J. K. Udupa. Shape-based interpolation of multidimensional objects. *IEEE Transaction Med. Imaging*, 9(1):32–42, 1990.
86. D. Rotger, P. Radeva, C. Canero, J. Villanueva, J. Mauri, E. Fernandez, A. Tovar, and V. Valle. Corresponding IVUS and angiogram image data. In *Computers in Cardiology*, pages 273–276. IEEE, 2001.
87. J.A. Schnabel. *Multi-Scale Active Shape Description in Medical Imaging*. PhD thesis, University of London, 1997.
88. H. Schumann and W. Müller. *Visualisierung. Grundlagen und allgemeine Methoden*. Springer Verlag, 2000.
89. D. Selle. *Analyse von Gefäßstrukturen in medizinischen Schichtdatensätzen für die computergestützte Operationsplannung*. PhD thesis, Uni Bremen, 1999.
90. D. Selle and H.-O. Peitgen. Analysis of the morphology and structure of vessel systems using skeletonization. In M. Sonka and K.M. Hanson, editors, *Medical Imaging*, volume 4322 of *Proceedings of SPIE*, 2001.
91. D. Selle, W. Spindler, B. Preim, and H.-O. Peitgen. Mathematical methods in medical image processing: Analysis of vascular structures for preoperative planning in liver surgery. In Enquist and Schmid, editors, *Mathematics Unlimited - 2001 and Beyond*, pages 1039–1059. Springer, 2000.
92. J.A. Sethian. A fast marching level set method for monotically advancing fronts. *Proc. Nat. Acad. Sci*, pages 1591–1595, 1996.
93. J.A. Sethian. *Level Set Methods and Fast Marching Methods*. Cambridge University Press, 2 edition, 1999.
94. G. Taubin. A signal processing approach to fair surface design. In R. Cook, editor, *SIGGRAPH 95*, pages 351–358. Addison Wesley, 1995. held in Los Angeles, California, 06-11 August 1995.
95. C.A. Taylor. Finite element modeling of blood flow: Relevance to atherosclerosis. In P. Verdonck and K. Perktold, editors, *Fluid Structure Interaction, Intra and Extracorporeal Cardiovascular Fluid Dynamics.*, volume 2 of *Computational Fluid-Structure Interaction in the Cardiovascular System, Advances in Fluid Mechanics*, pages 249–289. WIT Press, 2000.
96. H. Tek, D. Comaniciu, and J. Williams. Vessel detection by mean-shift based ray propagation. In *Works. on Math. Meth. in Biomed. Image Anal.*, 2001.
97. A. Telea and J. van Wijk. An augmented fast marching method for computing skeletons and centerlines. In D. Ebert, P. Brunet, and I. Navazo, editors, *IEEE TCVG Symposium on Visualisation*, pages 251–259, 2002.
98. R. Toledo, X. Orriols, P. Radeva, X. Binefa, J. Vitria, C. Canero, and J.J. Villanueva. Eigensnakes for vessel segmentation in angiography. In *Int. Conf. on Pattern Recognition*, volume 4, pages 340–343. IEEE, 2000.
99. G. Treece. *Volume Measurement and Surface Visualisation in Sequential Freehand 3D Ultrasound*. PhD thesis, University of Cambridge, 2000.
100. G. Treece, R.W. Prager, A.H. Gee, and L. Berman. Surface interpolation from sparse cross-sections using region correspondence. *IEEE Trans. Med. Imaging*, 19(11):1106–1114, 2000.
101. B. Verdonck, I. Bloch, H. Maître, D. Vandermeulen, P. Suetens, and G. Marchal. Accurate segmentation of blood vessels from 3D medical images. In *Int. Conf. on Image Processing*, volume III, pages 311–314, 1996.
102. A. Wahle, H. Oswald, and E. Fleck. 3D heart-vessel reconstruction from biplane angiograms. *Comp. Graph. Appl.*, 16(1):65–73, 1996.

103. M. Wan, F. Dachille, and A. Kaufman. Distance-field based skeletons for virtual navigation. In *Visualization'01*, 2001.
104. K.C. Wang, R.W. Dutton, and C.A Taylor. Improving geometric model construction for blood flow modeling. *IEEE Engineering in Medicine and Biology Magazine*, 18(6):58–69, 1999.
105. Kenneth C. Wang. *Level Set Methods for Computational Prototyping with Application to Hemodynamic Modeling*. PhD thesis, Department of Electrical Engineering, Stanford University, USA, 2001.
106. O. Wink, W.J. Niessen, and M.A. Viergever. Fast delineation and visualization in 3-D angiographic images. *IEEE Trans. Med. Imaging*, 19(4):337–346, 2000.
107. P.J. Yim, J.J. Cebral, R. Mullick, H.B. Marcos, and P.L. Choyke. Vessel surface reconstruction with a tubular deformable model. *IEEE Trans. Med. Imaging*, 20(12):1411–1421, 2001.
108. P.J. Yim and R.M. Summers. Analytic surface reconstruction by local treshhold estimation in the case of simple intensity contrast. In *Medical Imaging*, volume 3660 of *Proceedings of SPIE*, pages 288–300, 1999.
109. C. Zahlten. *Beiträge zur mathematischen Analyse medizinischer Bild- und Volumendatensätze*. PhD thesis, Universität Bremen, 1995.
110. M. Zhao, F. T. Charbela, N. Alperinb, F. Lothc, and M.E. Clark. Improved phase-contrast flow quantification by three-dimensional vessel localization. *Magnetic Resonance Imaging*, 18(6):697–706, 2000.

An Application for Dealing with Missing Data in Medical Images, with Application to Left Ventricle SPECT Data

Oscar Civit Flores[1], Isabel Navazo[2], and Àlvar Vinacua[2]*

[1] Departament de Tecnología
  Universitat Pompeu Fabra
  Estació de França
  Barcelona, Spain
  oscar.civit@tecn.upf.es
[2] Institut de Robòtica i Informàtica Industrial
  Universitat Politècnica de Catalunya
  Parc Tecnològic de Pedralbes, edifici U
  c. Llorens i Artigas 4–6, 2na planta
  E-08028 Barcelona, Spain
  {isabel,alvar}@lsi.upc.es

Summary. When using data-capture methods designed to retrieve functional information, we may have incomplete morphological information in the presence of desease. This is specially so for SPECT data of an infarcted heart.

In this paper we present two techniques to deal with such missing data, appropriate for different scenarios. The first one reconstructs small infarcts that show as holes through the ventricle's wall. This method uses two discrete elastic membranes that are deformed to wrap the myocardium without touching each other.

The second method handles cases where large portions of the heart are missing, by computing approximating ellipsoids to the walls of the myocardium, and filling the gaps.

1 Introduction

Medical imaging is an area that is offering a number of interesting problems to the scientific visualization community, and is correspondingly exhibiting intense activity. In it we need to extract and present the maximum amount of precise information that can be obtained from the data captured through specialized and sophisticated devices, that nonetheless are affected by diverse insufficiencies. Obvious limitations bound our capacity to improve the quality of the data gathered arbitrarily (as the implications for the patient impose severe restrictions). Therefore it becomes necessary to develop instead procedures that help us utilize to its full whatever data we do have.

---

* This work has been supported by the Spanish Government through grant TIC-2000-1009

A distinction must be made between two cases [8]: the medical imaging equipment sometimes gathers data related to the patient's anatomy (for example an x-ray shows the patient's bones, or an MRI—Magnetic Resonance Imaging— shows his soft tissues); other devices gather information about the body's function, like PET—Positron Emission Tomography- to display brain activity, or SPECT—Single Photon Emission Computed Tomography— data to measure the blood perfusion in the heart muscle (myocardium).

In either case one needs to reconstruct the anatomical data from the data gathered, in order to evaluate the results or to isolate relevant from non relevant information. However, the methods yielding functional data may often yield incomplete anatomies: the portion of a heart that has suffered an infarction is not irrigated and therefore presents near zero perfusion, yielding data undistinguishable from the background.

In this paper we present an application that helps reconstruct a three dimensional model of an anatomical structure from incomplete sample volume data. The application is motivated by and focused in the problem of reconstructing a patient's left ventricle from gated SPECT data.

To gather these images, physicians inject a radioactive marker in the patient's bloodstream. This marker attaches to muscle tissue and remains active for a short span. During this time, the images are computed based on radiation counts from the patient in different directions. In the case studied here, the images capture the portions of the myocardium that are well irrigated. Areas that are ischemic (lacking proper irrigation) appear instead black. Figure 1 shows the data obtained from a phantom healthy heart ([18], [4]) by

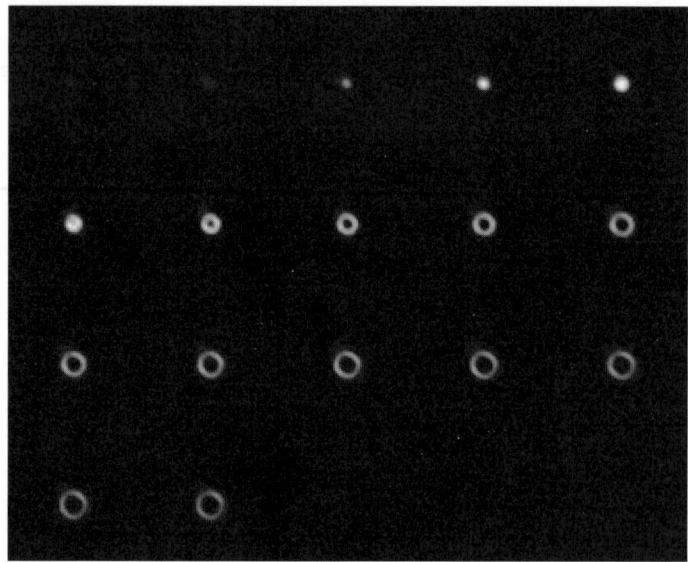

Fig. 1. Sample SPECT images of the short axis sections of a phantom heart

this method. Observe that these sections correspond to a shape resembling a thick mug with a narrow base (the apex). The phantom is a plastic container that physicians can fill up with a liquid suspension of the radioactive tracer used with patients, and then take the SPECT data from it as if it were a patient. In this case the pictures obtained are specially clean, as there is no noise from nearby muscles and bowels that usually appear in real data. Also the phantom is perfectly still during the data gathering, but of course the patient's heart is beating and data-gathering must be synchronized with the patient's heart by analyzing his electrocardiogram, a procedure that adds still more noise to the data. Furthermore, these data are intrinsically noisy (even in the case of phantoms) because of the random nature of the radioactive decay that produces the data. Notice the blurry aspect of the pictures because of this. Also their resolution is low compared to other medical imaging techniques. This region of interest in these pictures, for instance, has a resolution of roughly 35 by 35 pixels on each slice.

Nonetheless this test is widely used [3] in clinical practice to evaluate the amount of damage sustained by an infarcted heart. These studies focus on the left ventricle, as it is the largest functional part of the heart and the one charged with the hardest work. The availability of tools to accurately detect and measure this ill-irrigated portions is therefore important.

In the case of an infarcted heart, obviously, the direct application of surface extraction techniques to this kind of data would yield wrong results, as they would fail to detect the non-functional part of the heart, and thus give an incorrect morphology for the heart. These errors bar the calculation of important indicators of clinical relevance (see Fig. 7).

We use two techniques to complete the volume of the heart's muscle, depending on the size of the missing data. The first one is based on the deformation of a discrete elastic membrane (see Sect. 3.1 and [7], [6]) gives good results when the missing data (damaged tissue) appears localized. A second technique is needed for hearts with extensive damages or damages in the heart's apex. In this cases the previous technique yield large underestimations of both the infarcted area and the internal volume of the ventricle. In this case we automatically compute three dimensional ellipsoids that approximate the internal and external surfaces of the heart's ventricle. Restrictions are used so that these correspond closely to the expected shape of the heart even when only small portions of the heart's wall are available.

In Sect 4 we present results with both methods and provide volume visualizations of the results. We also contrast their validity by computing relevant indicators (like the fraction of infarcted tissue) with respect to known phantom models. We conclude that the methods proposed are efficient and allow a robust computation of these clinical indicators. We expect these techniques to be adaptable to other applications with missing or ill-segmented volume information.

2 Previous Work

From the slices provided by the data-capture devices, it is immediate to construct a voxel model of the region of interest by stacking the different slices at adequate (known) distances. Standard surface reconstruction techniques from such voxel data are based on the Marching Cubes algorithm ([14], [17]) or on deformable models ([16], [21]). In both cases they deal poorly or plainly break if there are missing data. Deformable models can be equipped with constraints (amounting to a certain stiffness) so that they will stay on the surface and avoid wandering inside the cavity of the heart through a hole produced by missing data, but that has a large computational cost. Deformable models have received extensive attention in the literature. Many improvements and specializations have been developped, including systems that learn to recognize certain features (see [12]). Others deal with the problem (very relevant to our case) of simultaneously segmenting two almost parallel surfaces (see [19,22]). However these schemes are not designed to handle missing data, and start from much higher resolution data gathered with MRI. It is not clear that these methods would be able to yield correct results when the missing data allow for the two surfaces to touch.

To overcome this problem, Germano proposed ([10], [4]) algorithms to extrapolate the heart's surface on each section. Their algorithm approximates the midmyocardial surface, and extraplates the endocardium and epicardium as offsets of this one, as opposed to our methods that try to infer both surfaces independently. Their system is aprioristic, but standardized in its application. We are preparing a comparison of the results yielded by both techniques on a large database of clinical cases, but this is of interest mainly to the medical audience.

Some authors (see [13]) have proposed superquadrics to approximate the left ventricle's surface. These papers deal with the problem of representing a known surface with few parameters, rather than extracting the surface from volume data, and definitely do not deal with missing data. However, since the superquadric representation of the surface contains few parameters, this can be exploited for surface extraction through optimization (see Sect. 3.2)

3 Our Approach

Our approach consists in first detecting the regions where data are missing (i.e. voxels belonging *anatomically* to the left ventricle yet without perfusion, presumably because of an infarction). These voxels are tagged as infarcted, but are given adequate values for the surface reconstruction algorithms to be able to compute an adequate result in those regions.

We use two different techniques to do this, whose adequateness depends on the size and location of the missing data. The first one is an adaptation of a general technique (see [7], [6]) for extracting geometry from non-uniformly sampled sets of points from the surface enclosing a solid, and is discussed in the following section. It is especially crafted to yield reasonable reconstractions when the data that are missing form a hole in the ventricle's wall. The next section then deals with our second method of handling missing data, which is best suited for extensive damages or those including the ventricle's apex. This second method, however, is specific to the case at hand, as it is based on the known shape of the left ventricle.

Once the data have been amended in this way, one can apply standard algorithms, like discrete marching cubes or surface deformation, to extract both the internal (endocardium) and external (epicardium) surfaces of the ventricle.

3.1 Elastic Membrane

This method for completing missing data can be seen as a voxel-based variant of active contours ([16], [2], [15]). The algorithm starts with a six-connected collection of voxels that one should think of as describing a discrete closed elastic surface. In its more general incarnation cited above, this elastic membrane is pushed inwards with cubes of decreasing size (the first one having the size of the whole voxel set) until the membrane reaches a data voxel. This is reminiscent of alpha-shapes, on which there is extensive literature (see for example [5], [11] and [1]). When the cubes being used are sufficiently small, they can get inside the volume of the solid through gaps in the data set. To avoid this, the algorithm keeps a history of its evolution, and analyzes the environment on the inside of the membrane. Reaching a voxel that contains data points but is already on the other side of the boundary means it has somehow gotten inside the solid, so the algorithm backtracks and adds new pseudo-data voxels to complete the solid's surface where the small cube started pushing. If instead it finds a different portion of the elastic membrane, it considers this a passing hole, and modifies the topology of the membrane to adapt itself to solids whose topology is not that of a sphere.

In our case we need a different behavior. To begin with, the topology of the left ventricle is fixed and known in advance, so the adaptive topology features of the algorithm are disabled. Also, we use the fact that we know the inside and outside of the ventricle to start up with two such elastic membranes. One is formed by all of the voxels on the boundary of the model (the voxels with extreme indices), and the other by a small collection of voxels inside the void within the ventricle.

The algorithm then pushes these flexible surfaces towards the model as in the generic variant. The small ball within the ventricle is inflated, pushing the surface outward, whereas the surface at the boundary of the voxel universe is pushed inwards. This is adequate to reconstruct the epicardium, whereas the first (growing from the inside of the ventricle) is used to detect the endocardium. But instead of the heuristic used for general shapes, here the algorithm backtracks when the two elastic surfaces touch.

When the algorithm stops (it cannot push the outer membrane inwards further, nor can it inflate any more the inner membrane), most of the membranes are inside boundary voxels of the left ventricle. Those which are not correspond to areas of the corresponding surface (epicardium or endocardium) that the algorithm judges missing. this produces the additional voxels needed to allow the surface extraction techniques to operate correctly. Furthermore, the voxels trapped between data voxels and this inner and outer estimated surfaces are counted to estimate the volume of the infarcted muscle.

Figure 2 shows a schematic of this algorithm. The figure represents a portion of a voxel model (here represented as two dimensional) where red squares

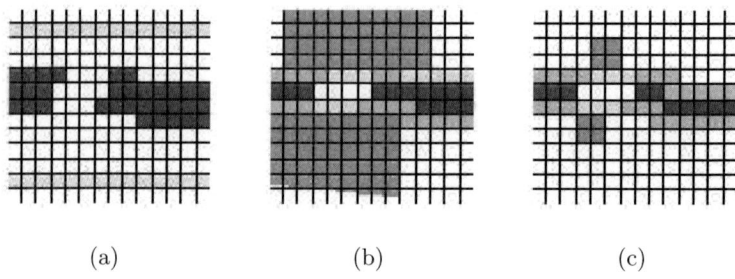

(a)   (b)   (c)

Fig. 2. (A version in color is at the end of this volume) Evolution of the discrete membranes (yellow). Red squares represent perfusion data.(a) is the initial stage. The green squares in (b) represent the blocks that push the membranes towards the data. Pink squares represent portions where the membrane has solidified against the perfusion voxels. The remaining portions of flexible membrane in (c) will be solidified when a through hole is detected by a sufficiently small block depicted in the same green.

correspond to voxels with perfusion. In Fig. 2(a), the yellow squares represent the inner and outer elastic membranes, that have been pushed until that position by some very large cubes (not shown). Further along the execution, fairly large cubes (shown in cyan in Fig. 2(b)) eventually push the elastic membranes into cubes with perfusion data, and there it stops (these final positions are depicted in pink). Yellow cubes are candidates to be further pushed by yet smaller cyan cubes. However, when the algorithm reaches the situation shown in Fig. 2(c), the cubes pushing each elastic membrane hit the opposite membrane, detecting a portion of missing data. At this point the pushing cubes backtrack to the position where they started at that resolution level, and mark the offending portion of the elastic membrane as no longer candidate to be pushed (which is the case for all the yellow squares in Fig. 2(c)). These are precisely the portions that constitute reconstructed portions of the boundary of the myocardium.

3.2 Reconstruction of Large Regions of Missing Data

This method tries to approximate the geometry of the left ventricle with an ellipsoid and then generates the missing data using that ellipsoid as a guide, but not exactly on its surface, as we also consider the width of the heart tissue in order to extrapolate the voxels stabbed by the endocardium and the epicardium. The process is divided in two parts which are described in the remainder of this section.

Ellipsoid Fitting The first part of the process calculates an ellipsoid that fits the existing heart voxels by minimizing the *distance* $d(q, P_i)$ between each data voxel and the surface of the ellipsoid, where q is the vector parameters that define the ellipsoid and P_i is the center position of the i-th voxel.

For efficiency and simplicity the distance minimized is not the *geometric distance* (minimum Euclidean distance between point and surface) but the *algebraic distance* (see [9] for an exhaustive discussion). The algebraic distance is given by the following equation:

$$d(x,y,z) = \frac{(x-x_0)^2}{a^2} + \frac{(y-y_0)^2}{b^2} + \frac{(z-z_0)^2}{c^2} - 1$$

This distance is zero for points on the surface, negative for points inside the surface and positive for the ones outside, so the problem of finding the ellipsoid that fits the tissue voxels can be formulated as the solution to the system of non-linear equations:

$$d(q, P_1) = 0$$
$$d(q, P_2) = 0$$
$$\vdots$$
$$d(q, P_N) = 0$$

where $\{P_1, \ldots, P_N\}$ are the voxel centers and $q = [x_0, y_0, z_0, a, b, c]^T$ are the parameters that define the position and sizes of the ellipsoid, for which we solve the equation. This equation can be written in a more compact form as $F(q) = 0$, where $F(q)$ is a vectorial function from $\mathbb{R}^6$ to $\mathbb{R}^N$.

We do not solve for the ellipsoid's orientation because the initial model is conveniently oriented so that the heart's main axis is parallel to the voxel world z axis. This fact simplifies the formulation and resolution of the problem, as taking orientation into account would not only increase the number of free parameters, but also introduce numerical instabilities.

The solution is found using an iterative method (Newton-Raphson). This allows us to monitor convergence and to stop the process when the desired precision is reached or when the maximum computing time has been used, which is necessary to ensure interactive-rate visual feedback of the fitting process.

The main calculation required by this method is computing the solution to the system of linear equations $J\Delta q = -F(q)$, where $J = \frac{\partial F}{\partial q}$. The least-squares solution to this system is computed with a slightly modified implementation of SVD (Singular Value Decomposition) from the one found in *Numerical Recipes* [20].

The process starts from an initial vector of default parameters $q_0 = [0, 0, 0, \frac{1}{4}Z_0, \frac{1}{4}Z_0, \frac{1}{4}Z_0]^T$. The initial position is the origin and the initial sizes are $\frac{1}{4}$ of the Z length of the voxel world. These values have been set empirically and generally produce good approximations fast.

Improving the quality of the ellipsoids. The described method successfully calculates least-squares ellipsoidal approximations to the tissue voxels, but the resulting ellipsoids are not guaranteed to be good and can sometimes be far from optimal.

We increase the quality of the computed ellipsoids in several ways. One improvement is to use a weighted version of the algebraic distance that approximates the result which the geometric distance would produce. The problem of using the algebraic distance is that it tends to neglect points far from the center of the ellipsoid as shown in [9]. The implemented weighting scheme is inspired on the one proposed in [9] and its expression is:

$$w_i = radius(q, P_i)$$

where $radius(q, P_i)$ is the distance between the z axis of the ellipsoid and the intersection point between the surface of the ellipsoid and a segment from its center to P_i. This weight is multiplied by the left side of each equation in the system $F(q) = 0$.

A second way in which we improve the quality of the ellipsoids is by using the empirical knowledge that we have of the heart's geometry and of the standard alignment of the model to define additional constraints on the ellipsoid parameters:

- Eccentricity minimization: We know that real hearts have a limited eccentricity, and we can use this fact to add the following equation to the system:

$$g_1(q) = \rho_1\left((a-b)^2 + (a - K1*c)^2 + (b - K1*c)^2\right) = 0$$

where ρ_1 is a user-tunable parameter (initially set to 0.1) that controls the weight of this equation in the system and $K1$ is a scaling factor for the size of the ellipsoid in the z axis, which is set by default to 0.5.

This equation primes solutions where the size in z (c parameter) is twice that in the orthogonal direction (a and b parameters), as our models are oriented so that the z axis corresponds to the main axis of the heart.

- Distance from origin minimization: The input model is also centered, so we use this fact to add the following constraint on the solution that minimizes the distance from the center of the ellipsoid to the origin of the voxel world coordinate system:

$$g_2(q) = \rho_2 \left(x_0^2 + y_0^2 + K2^2 * z_0^2 \right) = 0$$

where ρ_2 is a user-tunable parameter (initially set to 0.01) that controls the weight of this equation in the system and $K2$ is a factor that allows to tune the weight of the z component of the position of the ellipsoid in the constraint, as it should be less constrained than x and y due to the orientation of the input models. We nitially set $K2 = 0.25$.

Our equations require that the distance between data voxels and the surface of the ellipsoid be small, but not the converse. The solution will thus usually overshoot on the side opposite the apex (see Fig. 3). Therefore we

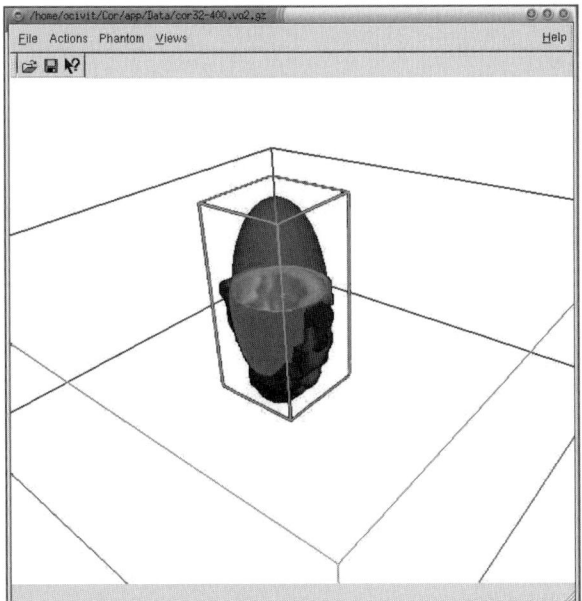

Fig. 3. The image shows an ellipsoid already fitted to the heart voxel model represented by its isosurface. The red part of the ellipsoid is the part that will be used in the next phase of the reconstruction process.

trim the excess off with a maximum height plane (that despite its name may be slightly inclined). This plane is automatically computed as close as possible to the top data voxels. It can also be manipulated at will by the user, as we discuss next.

Interface. The fitting process is mostly automatic, and in general produces ellipsoids that successfully approximate the heart geometry. However, it is always desirable for the user to have full control on the ellipsoid that will be used to repair the heart model. For this purpose an intuitive, direct manipulation, interface has been implemented and allows the user to:

- Set the position and sizes of the ellipsoid.
- Set the plane that defines the maximum height of the ellipsoid considered in the reconstruction.
- Run the automatic fitting process using the current ellipsoid parameters as an initial approximation.
- Set the parameters of the automatic fitting process (ρ_1, ρ_2)

The sequence of actions to obtain the desired fitting ellipsoid will generally consist of one or a few executions of the fitting command until the desired precision is reached, but sometimes the user may define an initial ellipsoid manually and run the automatic process to improve the fitting and iterate this actions to have a closer control on the resulting ellipsoid.

Ellipsoid-guided Voxel Generation Using the approximating ellipsoid just computed, we proceed to tag the voxels that correspond to heart tissue. The voxel generation is done independently for each slice of the heart from the bottom to the maximum height plane. These slices are ellipse-shaped.

For each ellipse we calculate the arc that doesn't intersect any tissue voxel. This arc is defined by the angular range $[u_{min}, u_{max}]$ on the parametric definition of the ellipse (see Fig. 4).

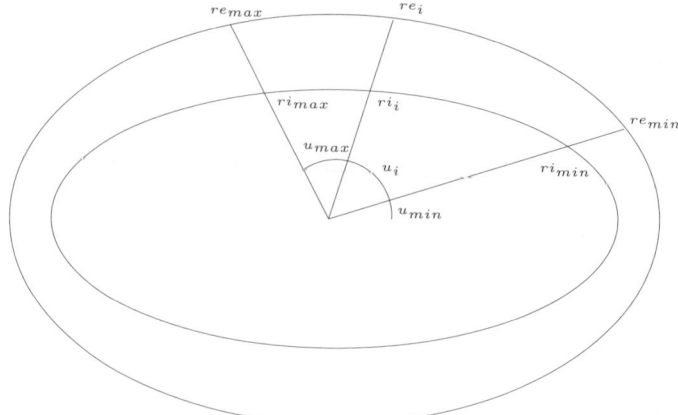

Fig. 4. The diagram shows graphically the quantities involved in width interpolation and angular sampling for the tissue voxels generation

As we want to reconstruct the heart taking into account the tissue width, we calculate the internal and external radii $ri_{min}, re_{min}, ri_{max}, re_{max}$ at the intersection points between the arc and the heart tissue (at u_{min} and u_{max}).

For a number of sample angles $u_0, \ldots, u_N$ in the arc $[u_{min}, u_{max}]$ we calculate the estimated inside and outside radii ri_i, re_i by interpolation (using u_i as the parameter) between the radii at the ends of the arc. Then we finally generate tissue voxels in a number of sample points in the parametric interval $[ri_i, re_i]$ of the ray that goes out from the center of the slice in the u_i direction.

4 Results, Discussion and Future Work

The three dimensional reconstruction of the data in Fig. 1 is shown in Fig. 5. The surface displayed has been constructed using marching cubes in all the examples in this section. An interesting feature of phantoms is that physicians

Fig. 5. Reconstruction of the healthy phantom

can simulate different injuries by introducing obstacles not permeable to the liquid containing the marker isotope, thus producing an infinite variety of calibrated test cases.

For example, figure 6 shows the same phantom with a small infarction on a wall, and the result of completing the myocardial volume using the method in Sect. 3.1. The pseudo-data voxels added by the algorithm are shown semi-transparent in green.

For an example of a reconstruction using the method described in Sect. 3.2, figure 7 shows again the same phantom with a large infarction of about 32% in volume, and the result of the reconstruction. Notice how the algorithm extrapolates the shape of the ventricle.

Table 1 shows the differences between the percentual damage introduced in the phantom and the percentual damage computed by our program. We have chosen to list here the results attained through automatic approximation of the infarcted region, although in all cases an experimented user can

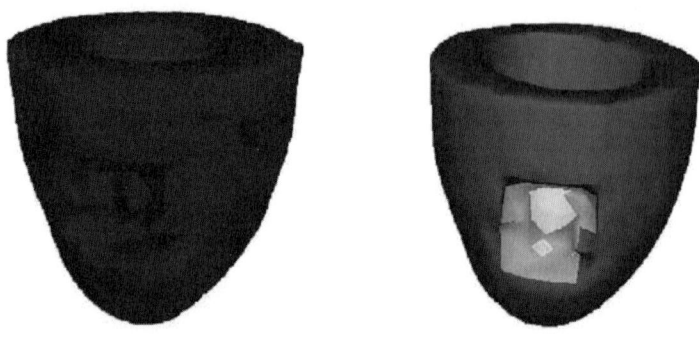

Fig. 6. Localized missing data, reconstructed with the discrete bands (see color plate at the end° of this volume)

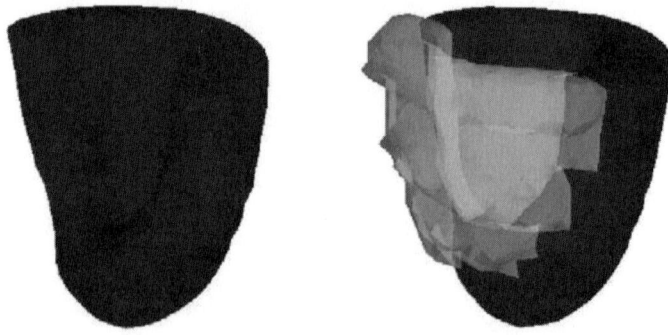

Fig. 7. Extended missing data, reconstructed by the second method (see color plate at the end° of this volume)

quickly obtain much better results using the interactive interface. This is specially so in the case of the 10% leisure, where the approximating ellipsoid is almost completely contained in data voxels, hence not triggering the width estimation almost anywhere. A skilled operator can detect this deviation on-screen and compensate it, but our algorithm is not yet capable of doing so automatically.

Readers should also note that the volumes listed here are computed as counts of stabbed voxels, not taking into account the actual counts in data voxels, so a certain amount of aliasing adds noise to our figures.

Table 1. Comparison of the estimated leisure size on different phantom models

Real size (%)	Estimated size (%)
10%	5.05%
32%	30.56%
53%	54.82%

The application presented here appears to achieve reasonably good approximations. We have so far tested it in a small number of real datasets as well as some synthetic datasets as those presented in this paper. These results have been validated by practicing specialists.

Furthermore the application runs on simple hardware (standard PC's with either Linux or Windows) and is fast enough to achieve interactive operation under a user-friendly environment. The user may manually override the program's decisions, although experts place importance in the automatic operation as it provides repeatability of results, something they do not have with current technology.

The application uses two complementary techniques depending on the size and localization of the damage. The first of them is easily adaptable to other kinds of data, whereas the second one exploits the known shape of the target organ, and is therefore specific to the application discussed here.

We are presently working on extending the capabilities of this application in several directions. Among them are more accurate volume computations to improve the figures presented here. Also we are looking into more flexible interpolants for the shape of the ventricle to properly handle cases with malformations. Superquadrics are the most likely candidate as they represent a much more reacher family of surfaces with only a modest increase in the number of parameters to optimize (see [13]).

5 Acknowledgements

We are grateful to Jordi Esteve for facilitating his original implementation of the elastic membrane code on which our current software is based for what is discussed in Sec. 3.1. We also want to thank Oriol Esteve and Dani Marin for their help with the implementation. We are deeply indebted to Dr. Jaume Candell and Dr. Santiago Aguadé for providing the data and for their help. We also want to thank the anonymous reviewers for their help in improving this paper.

References

1. Fausto Bernardini and Chandrajit L. Bajaj. Sampling and reconstructing manifolds using alpha-shapes. In *Proc. 9th Canadian Conf. Computational Geom-*

etry, pages 193–198, August 1997.
2. Andrew Blake and Michael Isard. *Active Contours*. Springer Verlag, 1998.
3. Candell-Riera, Castell-Conesa, and Aguadé-Bruix. Myocardium at risk and viable myocardium. *Developments in Cardiovascular Medicine*, 234, 2001.
4. J. Candell-Riera, J. Castell-Conesa, and S. Aguadé-Bruix, editors. *Myocardium at risk and viable myocardium. Evaluation by SPET*. Developments in cardiovascular medicine. Kluwer Academic Publishers, 2001.
5. Herbert Edelsbrunner and Ernst P. Mücke. Three-dimensional alpha shapes. *ACM Transactions on Graphics*, 13(1):43–72, January 1994.
6. J. Esteve, P. Brunet, and A. Vinacua. Aproximación de un conjunto de puntos disperso de densidad variable mediante una erosión gradual. In R. Joan-Arinyo, I. Navazo, and R. Quirós, editors, *CEIG 2000, X Congreso Español de Informática Gráfica*, pages 235–248. Universitat Jaume-I, 2000.
7. J. Esteve, P. Brunet, and A. Vinacua. Scattered data approximation through the erosion of discrete bands. In *Research Report*. Software Department, Universitat Politècnica de Catalunya, 2002.
8. N. Ezquerra, I. Navazo, T. Infantes, and E. Moncl·s. Graphics, vision and visualization in medical imaging. pages 21–80, 1999.
9. Walter Gander, Gene H. Golub, and Rolf Strebel. Least-squares fitting of circles and ellipses. *BIT*, 34(4):558–578, December 1994.
10. G. Germano, H. Kiat, P.B. Kavanagh, M. Moriel, M. Mazzanti, H. Su, and D.S. Van Train, K.and Berman. Automatic quantification of ejection fraction from gated myocardial perfusion spect. *The Journal of Nuclear Medicine*, 36(11):2138–2147, 2001.
11. Baining Guo, Jai Menon, and Brian Willette. Surface reconstruction using alpha shapes. In David Duke, Sabine Coquillart, and Toby Howard, editors, *Computer Graphics Forum*, volume 16(4), pages 177–190. Eurographics Association, 1997.
12. M. Leventon, W. Grimson, and O. Faugeras. Statistical shape influence in geodesic active contours. In *Proceedings of the IEEE Conference on Computer Vision and Pattern Recognition (CVPR-00)*, pages 316–323, Los Alamitos, June 13–15 2000. IEEE.
13. P. Linares, L. Rodríguez, and G. Montilla. Genetic algorithm fitting of deformable superquadric applied to left ventricle visualization. In *Computers in Cardiology*, volume 25, pages 657–660. IEEE, 1998.
14. W. Lorenson and H. Cline. Marching cubes: a high resolution 3d surface reconstrution algorithm. *ACM Computer Graphics*, 21:163–169, 1987.
15. T McInerney and D. Terzopoulos. Deformable models in medical image analisys: a survey. *Medical Image Analysis*, 1(2):91–108, 1996.
16. T McInerney and D. Terzopoulos. Topology adaptative deformable surfaces for medical image volume segmentation. *IEEE Transaction on Medical Imaging*, 18(10):840–850, 1999.
17. C. Montani, R. Scateni, and R. Scopigno. Discretized marching cubes. In *IEEE Visualization'94*, pages 281–287. IEEE, IEEE Computer Society, 1994.
18. M. K. O'Connor, R. J. Gibbons, J. E. Juni, J. O'Keefe, and A. Ali. Quantitative myocardial spect for infarct sizing: feasibility of a multicenter trial evaluated using a cardiac phantom. *The Journal of Nuclear Medicine*, 36(6):1130–1136, 1995.

19. Nikos Paragios. A variational approach for the segmentation of the left ventricle in mr cardiac images. In *Proceedings of the IEEE Workshop on Variational and Level Set Methods in Computer Vision, 2001*, pages 153–160, 2001.
20. William H. Press. *Numerical recipes in C: the art of scientific computing.* Cambridge University Press, 1992.
21. A. Susn, I. Navazo, A. Vinacua, and P. Brunet. Dymanic recognition and reconstruction of the human heart. In A. Sanfeliu, JJ. Villanueva, M. Vanrell, R. Alquézar, J. Crowley, and Y. Shirai, editors, *Proceedings of 15th International Conference of Pattern Recognition*, pages 88–93. IEEE Computer Society, 2001.
22. X. Zeng, L.H. Staib, R.T. Schultz, and J. S. Duncan. Volumetric layer segmentation using coupled surfaces propagation. In *Proceedings of the IEEE Conference on Computer Vision and Pattern Recognition (CVPR-98)*, pages 708–715, June 23–25 1998.

Constraint-Based Astronometric Modeling Tools

Andrew J. Hanson[1], Chi-Wing Fu[1], and Priscilla C. Frisch[2]

[1] Computer Science Department
  Indiana University
  Bloomington, IN 47405 USA
  hansona@indiana.edu, cwfu@indiana.edu
[2] Department of Astronomy and Astrophysics
  University of Chicago
  Chicago, IL 60637 USA
  frisch@oddjob.uchicago.edu

Summary. Self-consistent estimation of three-dimensional structures is a challenging task in the analysis of astronomical data, which is predominantly two-dimensional. Obtaining explicit and scientifically justifiable spatial models is also an essential task for visualization scientists working with astronomers to develop comprehensive simulations of complex astrophysical structures and phenomena.

We address the fundamental problem of facilitating the interactive utilization of two-dimensional stellar spectral data to impose consistent constraints on three-dimensional interstellar cloud structures. This enables the construction and verification of models for spectrally absorbing clouds of gas and dust lying between the Earth and nearby stars. Our data construction tool, while specific to the astronomy application, is a prototype for the design of any expertise-driven system for resolving ambiguous initial data into a plausible complete model. In this case, we use spectral information and distances derived from stellar measurements such as the Hipparcos satellite data; the domain expert exploits the initial estimate of dust cloud thickness estimated from these data to adjust the 3D shape of the cloud. Various additional sources of information may be combined to decide on interactive changes in the parameters, yielding a qualitatively self-consistent shape model.

The examples tested highlight the importance of allowing the scientist to interact with and tune a multitude of variables and parameters in the course of the data analysis, and suggest that fully automatic tools cannot be expected to function acceptably in the construction of 3D models of interstellar media for visualization purposes.

1 Introduction

Motivation. Reliable modeling of three-dimensional (3D) structures is one of the more challenging tasks in the analysis of astronomical data, and is required for understanding the relationships between multispectral observations of the *interstellar matter* (ISM) in the local galactic neighborhood of the Sun.

Such three-dimensional reconstructions are valuable because they yield information on interstellar cloud distributions, properties, and the energy transfer between stars and the ISM. For instance, the physical properties of warm, cold, hot, neutral, and ionized interstellar clouds depend on the relative distribution of stars and clouds. Cloud boundary properties, equilibrium, and stability also cannot be understood without accurate three-dimensional knowledge.

Visualization scientists working with astronomers can then exploit these 3D models to produce simulations and virtual explorations of complex and poorly understood astronomical environments.

Task. We report on a method for combining the shadowing properties of the ISM with astrometric data to visualize and derive the 3D distribution of interstellar material within a few hundred parsecs of our Sun. As an example of an important application of this methodology, we study the Local Bubble, which is a large-scale artifact of star-formation activity in the solar neighborhood.

We describe the scientific context, design, and features of our interactive visualization tool that utilizes measurements of spectral data (emission and absorption of light) to impose consistent distance constraints on the structure of interstellar matter. Because a major impact of the tool is to allow a skilled astronomer to infer and modify the radial location of cloud surfaces based on spectral measurements, we refer to the software system as the *Distance Editing Tool* (DET for short). The system enables the construction and verification of models for spectrally absorbing and emitting gas clouds lying between the Earth and nearby stars. The most extensive data set of interstellar absorption information comes from photometric properties of the 10^5 stars whose distances have been measured directly by the Hipparcos satellite [6]. Alternative data sets include spectral data on interstellar absorption lines at specific atomic transitions towards nearby stars (e.g., neutral sodium lines). The default application is to use stellar photometric data from the Hipparcos database, and to supplement this data as needed with data sources sampling specific absorption or emission lines, possibly combined with theoretical models. The three dimensional objects determined from this analysis can in turn place constraints on related features seen in emission data sets such as the HI hydrogen hyperfine 21 cm all-sky emission maps [11, 13], and features identified in the soft X-ray background, as seen, e.g., by [19].

After selecting and loading a data set, which typically provides only cloud thickness estimates, but no distance information, the scientist uses the graphics interface to adjust cloud boundary locations for mutual consistency, possibly introducing supplementary data as available. The examples provided in the results section highlight the importance of allowing an expert scientist to interactively tune distances and local variables such as cloud volume densities for each line of sight.

Graphics and Visualization Issues. This paper focuses on a specific domain application in astronomy. However, we believe that this example sheds light on significant generic issues, as well as the challenging problem of "connecting" between visualization scientists and domain scientists to produce a productive collaboration. In particular, we present the following observations:

- **Interdisciplinary Communication.** Developing this system took years. Translating a scientist's idea into a software system was a long-term, evolving interactive process. Typically, the scientist's idea of what was wanted was (a) difficult to define, (b) had parts that were impossibly difficult that sounded trivial to them, and (c) had parts that were completely trivial to implement that sounded impossible to them. As a consequence, there was often something easy that was not asked for until long after it was needed, and complicated things that took months to implement that were found wanting and later abandoned. Much domain knowledge had to be assimilated by the visualization team, and much knowledge of the limitations as well as the power of interactive graphics had to be assimilated by the scientist. The reader of this paper can get an accurate impression of the remarkable detail of the interplay between the disciplines that occurred during this work, and perhaps have a more realistic expectation of how to approach interdisciplinary problems similar to ours.
- **Free Model Editing vs Domain-Constrained Editing.** Graphics practitioners routinely construct complex dynamic geometric models for use in Hollywood animations, game environments, and simulations. The model creation and editing environment is "free" in that the modeler can add, move, distort, and modify elements to obtain a variety of alternate artistic effects. In this paper, we deal with a very important sub-domain, the requirement of consistency with the measurements of scientific instruments, that required a very specific type of editor, one that *freed* the scientist from worrying about consistency with global experimental constraints by allowing only *constrained* activities with guaranteed consistency. Many problems have similar features; while each is different, our approach provides a prototype for a scientifically-acceptable alternative to free-form modeling for the construction of models that are guaranteed to maintain compatibility with experimental measurements. It is instructive to note the parallels with motion-sensor-based human figure animation, and the imposition of joint constraints in that context (see, e.g., [14]).
- **Exploiting Expertise to Resolve Ambiguities.** Traditionally, huge amounts of measured data are resolved into models of some sort using automatic procedures. The domain we are interested in has unknown features that must be resolved by applying knowledge that is very difficult to express in a computer program. The pattern that we provide attempts to supply automated support for every aspect of the analysis *except* the ambiguous; we claim that, in general, this is the way to design domain-specific modeling systems – develop every possible automatable aspect of

the modeling, leaving the core of the unresolved or unautomatable to the human expert, thereby making the most efficient possible use of the scientist as modeler. This accomplishes a principal goal of visualization: using interactive computing to empower individual humans, thereby extending their productivity.
- **User Interface Features.** We make no attempt to describe the full range of user interface features of our system in this paper; however, once again, their nature can be characterized in generic terms. A key feature is touch-of-the-button background information: every object can be touched to display details from a feature database. Every scenario can have additional "context information" added to boost the confidence of the expert — for example, we provide a large library of multispectral background sky images, each generated by a different type of astrophysical process and giving a different clue to the analyst. (Almost) every imaginable computable quantity can be displayed in some way, and, conversely, can be selected and deselected to eliminate clutter and visual confusion. Our system, and any system with a similar problem scenario, can benefit from literally looking at the possible quantities that can be computed from the instrumental data readings, and providing a suite of display options to access that information.

2 Scientific Background

In this section we provide more detailed information on the motivations for creating this tool, and the astrophysical data analysis techniques and concepts that make it possible.

The need for this tool. An appreciable amount of structure is seen in cold interstellar gas, over scale size ranges ~ 10 astronomical units (an AU is the Earth-Sun distance, about 1.496×10^{11}m) up to over 100 parsecs (a parsec is 3.085×10^{16}m $= 3.262$ light years). The ability to project all-sky texture maps makes it possible to compare parsec-scale structures seen in absorption with two-dimensional cloud structures visible in multispectral emission data. All-sky maps of emission from warm and cold neutral gas (e.g. H^o 21-cm hyperfine transition), warm and hot ionized plasma (e.g., H^+ $H\alpha$ emission, soft X-ray emission at ~ 0.25 keV), and dust (e.g. 100 μ m infrared) emission do not have distances attached to the emitting material. Such distances must be determined by comparison with the properties of interstellar gas towards stars with known distances (e.g., reddening of starlight and astrometric distances from the Hipparcos satellite [6]). The presence of small scale structure in the ISM makes this comparison both cumbersome and error-prone when performed by hand. In contrast, the combination of texture maps based on multispectral emission data and star positions simultaneously displayed to the user of the DistEdTool allows multispectral interstellar catalogs to be readily evaluated. Thus, the Distance Editing Tool provides an easy-to-use graphical interface for detecting the alignment of stars with prominent interstellar features traced by emission data.

Observations of Interstellar Matter. Interstellar material within 500 parsecs, with densities 10^{-4} to 10^4 atoms cm^{-3} and temperatures 15 K to 10^6 K, emit or absorbs radiation between energies ~20 MHz and ~1 keV. The dominant component of interstellar gas clouds is hydrogen (90% by number), and H^o has an an average density in our galactic neighborhood of ~ 1 atom/cm^3. However large volumes of interstellar space are filled with low-density, hot (10^6 K) plasma; an example is the "Local Bubble," which fills a large region of space enclosing the Sun itself. These "bubbles" appear to be remnants or fossils of supernova explosions that blast holes in the ISM, leaving behind cavities filled with hot low-density gas and surrounded by shells of cooler, compressed material at the boundaries, The Northern sky happens to be dominated by the largest and most distinct of these remnants, Loop I, which forms one end of the Local Bubble. The Sun itself is embedded in a low-density outflow of ISM from Loop I which may be a fragment of a decelerated superbubble shell generated by an ancient supernova explosion. This diffuse material surrounding the Sun (known as the "Local Fluff,") is sparse and undetectable in reddening data. However, it is clearly distinct from the hot plasma which fills most of the Local Bubble. The juxtaposition of the Local Fluff diffuse clouds and Local Bubble hot plasma provides an example of why multispectral data is required when evaluating the three-dimensional morphology (geometry and topology) of interstellar clouds.

Interstellar cloud positions are given by ISM "shadows" in stars of known distances, while atomic absorption lines in the optical and ultraviolet provide crucial information about the physical conditions of ISM between a star and the Earth. The most useful information is from atomic transitions characterizing elements and molecules in the ISM which create observable absorption lines in the light of background stars. These spectra, particularly optical lines which are readily observable, provide an important database of information about locations and velocities of interstellar clouds.

Another set of tracers for these clouds are interstellar dust grains, which couple tightly to interstellar gas and redden the light of background stars. Optical photometric observations of the stars create a database of photometric information which provide a measure of the amounts of both gas and dust towards the background star. Although absorption lines, such as the yellow D-lines of neutral sodium, can be used in the Distance Editing Tool for determining the morphology of the Local Bubble, in this paper we will focus on the reddening of starlight, which is based on an extensive body of readily available photometric information.

Measurable ISM Components. We next summarize the types of measurements that can in principle be used to assist us in our task of constructing 3D models of interstellar materials. These include (see Appendix A):

- *Star colors*, which are conventionally measured through a set of predefined spectral filters (e.g. the Johnson B and V filters) and combined with the intrinsic star properties to yield a measurement of intervening dust and gas.
- *Interstellar Dust*, which also polarizes starlight and emits infrared emission.
- *Molecular Clouds*, where hydrogen is mainly molecular, and trace amounts of carbon monoxide (CO) and other molecules are found.
- *Interstellar Neutral Gas*, which is traced throughout the galaxy by H^o 21 cm hyperfine transition line.
- *Interstellar Ionized Gas*, ionized and excited hydrogen, emitting in the hydrogen alpha at 6535 Angstroms (distinct from the Lyman alpha atomic levels), and in the radio spectrum at 408MHz.
- *Hot Plasma* , which typically emits X-rays in the > 200 electron Volt (eV) range.

3 Gas Column Density and Length Determination

The basic principle used to estimate cloud sizes from, e.g., spectral measurements of starlight, is that attenuation of the expected intensity gives an estimate of the *column density* N, the total number of atoms contained in a column of material one square centimeter in cross-sectional area and arbitrary height (in a given direction in the sky). The length of the column represents the total cloud length sampled by the observed feature.

In three-dimensional space, the volume density is conventionally written as n, in atoms per cm^3. If this is known or can be estimated, then the length l of a column may be determined from the equation

$$N = n \times l \tag{1}$$

Typically, the DistEdTool is used to construct the 3D spatial distribution of interstellar gas by combining accurate photometric data with accurate stellar distances. The astrometric and spectral data contained in the Hipparcos catalog can be used to evaluate the distribution of interstellar gas as follows: We compare the observed color of the star (Johnson $B-V$, in magnitudes) to the intrinsic color indicated by the spectral type of the star [3] to evaluate the amount that the starlight is reddened by interstellar dust (the "color excess" $E(B-V)$, also in magnitudes). $E(B-V)$ can be converted directly to an amount of interstellar gas using the mean gas-to-dust ratio determined from Copernicus satellite data [2]:

$$\frac{N(\mathrm{H}^o) + 2N(\mathrm{H}_2)}{E(B-V)} = 5.8 \times 10^{21} \mathrm{cm}^{-2}\mathrm{mag}^{-1} , \tag{2}$$

Here, H^o is neutral hydrogen and H$_2$ is molecular hydrogen, while $N(\mathrm{H}^o)+2N(\mathrm{H}_2)$ (cm^{-2}) is the *column density* of interstellar hydrogen in the sight-line to the star. The color excess $E(B-V)$ is calculated from the photometric and spectral data; $E(B-V)$ is then combined with Eq. (2) to create the hydrogen column density, the length of the gas column "core sample" along a star sight-line, as

$$l = \frac{N(\mathrm{H^o}) + 2N(\mathrm{H_2})}{n} \ . \qquad (3)$$

Here n is the local density of atoms in the interstellar material under consideration; n ranges from 10^{-4} to 10^{+4} atoms/cm^3 for the variety of phases of the interstellar medium near the Sun. The tool permits the user to adjust individual stellar volume density parameters to achieve perceived global consistency; ISM is often wispy, with local density variations that do not conform to the overall average.

In practice, this procedure yields the most accurate results when applied to relatively massive stars (spectral types O, B, A, F) distant enough to sample measurable amounts of dust, and relatively bright stars (V<7 magnitudes) where spectral data are most reliable.

The DistEdTool is suitable for determining the distance distribution for any set of interstellar clouds where both interstellar column density and star distance are known.

The near and far boundaries of the volumetric objects for which the column densities provide evidence are signals for surfaces located at volume-density discontinuities. To construct these surfaces, we project the selected stars with reliable evidence to points on the Cartesian latitude-longitude plane and perform a Delaunay triangulation in 2D [16] that is projected radially to 3D. In order to generate a full-sky inner and outer cloud boundary, the standard triangulation must be extended to the sphere; we have implemented our own noise-insensitive version resembling that of Renka [18].

While the main focus of the current effort is on full-sky models, we may, by selecting small portions of the sky and selecting stars within a boundary containing only column densities above a given threshold, model localized material clouds as well.

4 Design

The Distance Editing Tool is an OpenGL application using the public GLUI [17] user interface toolkit, with minor local modifications to provide enhanced functionality and interface consistency. The overall structure, as shown in Figure 1, breaks into several stages:

- **Data Preparation.** Typical data sets contain large amounts of supplementary data that can be used to exercise filtering and selection processes. The filters thus support the straightforward construction of sets of stars based on a variety of geometric and physical characteristics. A

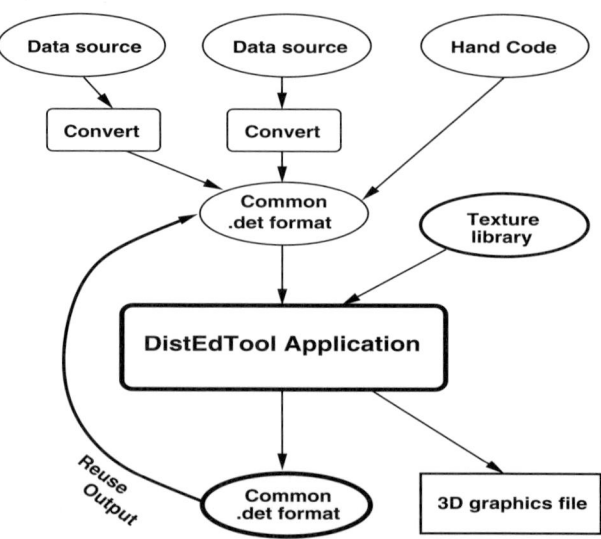

Fig. 1. Structure of the Distance Editing Tool.

family of Perl scripts, `hip2det.pl` ("Hipparcos catalog format to DET"), `absl2det.pl` ("absorption line data to DET"), and `bare2det.pl` ("bare data to DET"), process various forms of inputs from outside sources into a consistent and uniform internal format, the `.det` file. In an Appendix, we give a synopsis of the `hip2det.pl` command line options enabling data set selection from catalogs of stars [6] and their velocities [1]. The options range from simple rectangle selection to choices like `rselect`, which allows one to ignore stars with unreliable spectral types or distant stars with unreliable parallax values by requiring the spectral and parallax-based distances to agree within a given tolerance. Alternative astronomical datasets can be parsed into DET format for use in the tool; a domain expert can edit the data by hand if necessary to handle special circumstances. For example, a pulsar catalog can be parsed into the DET format to compare locations with those of clouds apparent in the multispectral texture maps choosable as backgrounds. The domain expert can work directly with the data in plain text to add information from outside sources and to correct inconsistencies.
– **Textures.** In addition to the `.det` file data providing to the properties of specific stars, it may be useful to refer to specific photographic images of multispectral emission data that are relevant to various types of clouds. We provide this facility in the form of spectrally-identified texture maps of the full sky. Having these images displayed against a particular star collection can provide independent reference material or corroborating evidence for specific clouds. Images to be used in this way can cover any subset of the sky, but must be formatted as a full-sky image in Earth-centered Carte-

sian galactic (l,b) coordinates, with the galactic center at the origin, and should be provided with a spectral label, e.g., the dominant wavelength in meters. The user interface provides for both individual image selection and a continuously varying spectral slider permitting fluid comparison of neighboring wavelengths.
- **Input: Data and Textures.** The program itself takes a .det file and an optional texture specification file as input on the command line. The texture specification file may be customized to provide any desired multi-spectral list of full-sky images as background texture maps.
- **Running the Program.** The elaborate user interface allows dozens of different helpful display options, detailed later in the user interface description. The selected stars can be displayed in flat Cartesian galactic (l,b) coordinates, flat Aitoff (equal area) galactic coordinates, or Sun-centered spherical coordinates, with or without an "exploded" reference sphere to reduce clutter in the display. Each star is displayed as a line segment representing the radial distance from the Sun to the star; optional displays include radial distance error estimates and stellar 3D velocities, with error estimates. Attached to the line segment is a "sliding tube" representing a hypothesized cylindrical gas volume between the star itself and the Earth-based observer. The star's radial line segments from the Earth to each star constrain the motion of this material, which resembles a section of straight macaroni sliding on a wire. Although there are many options and refinements provided by the interface, the scientist's main task is to interactively adjust (or "edit") the radial position ("distance") of this tube to achieve a globally consistent model. The length of any individual material tube can be adjusted by modifying original global estimated volume density (see Eq. (1)).
- **Output and Iteration.** When the sliding tubes representing star-aligned samples of interstellar material density have been adjusted to the satisfaction of the scientist, the current state of the system can be saved to a new .det file. This file can be reused and edited to add refinements based on other data sources, or simply to continue a session later: it may take lengthy periods of intermittent evidence gathering and consideration to develop a model that is acceptable to a particular scientist.
- **3D Model Output and Visualization.** The visualization function of the system is supported by the options in the Surface tool set. Since each tube or column of material has a beginning, the material onset, and an end, where we either run out of material, or perhaps encounter the star itself within the ISM, there are two intrinsic resultant surfaces, basically the front and the back of the modeled volume. These two 2D surface boundaries for the cloud in question are constructed using a Delaunay-triangulation algorithm to turn the collection of 3D sample points into a surface. There are actually two modes: a Cartesian surface in latitude:longitude space, and a closed-sphere adaptation of the Delaunay triangulation for modeling

actual shapes in 3D galactic coordinates. In spherical display mode, there are two sub-modes: one for editing, in which each star is displaced by an arbitrary amount from a common center to facilitate spacing the star columns for editing, and a "true geometry mode" in which all the star rays originate from a single point approximating the position of the Sun.

In the .det file, individual stars can be marked as ignored in the triangulation, but retained to be added back in a later pass. Triangulated cloud surfaces can be viewed with a number of options within the editing tool itself, including wire-frame, flat shaded surface, smooth-shaded surface, and with column-density texture values as shading. Both the inner and outer surfaces of the material model can be output in standard simple 3D triangle formats. At this time OFF files, viewable (possibly with minor format adjustments) using, e.g., Geomview [15], and VRML 2.0 files, viewable using, e.g., CosmoPlayer, are supported. Surface files can be written out at any time for separate examination.

5 Applications and Results

Background. A prime motivation for the construction of the modeling system described in this paper is its application to the reconstruction of the three-dimensional boundaries of the Local Bubble from observational data. The relative geometrical positions of the interstellar matter delineating the boundaries of this void, compared to stars, supernova, and star-forming regions, determines ISM properties in the solar neighborhood. The Sun has been within the Local Bubble void for millions of years (which may not be a coincidence [8]). The current and historical physical properties of the heliosphere and interplanetary environment of the solar system are directly determined by the galactic environment of the Sun. This facilitates comparisons of features such as the solar space motion, interstellar clouds, and the Local Bubble boundaries; such observations are of interest in understanding historical variations in the interplanetary environment of the solar system. The reliable reconstruction of self-consistent three-dimensional structures from data is a challenging task that, when successful, will advance our understanding of our overall astrophysical context.

Visualizing Stellar Properties. As a precursor to analysis of material boundaries such as the Local Bubble, the DistEdTool user interface can be used to support a wide variety of preliminary investigations. One of the most powerful tools is the ability to load multispectral full-sky background maps; in fact, the interface supports loading any number of such maps tagged by their principal wavelength, and interpolating smoothly between them in real time from one end of the electromagnetic spectrum to the other. In Figure C.29, we present a Hydrogen Balmer alpha-emission spectral map [7], based on a combination of WHAM, SHASSA, and VTSS data. This map is rendered on the galactic sphere, together with a selection of type O stars; these stars are so hot and massive that they ionize their local interstellar medium and many can be seen clearly as the *centers* of hot spots in the H-alpha image. the Orion neighborhood and Barnard's loop comprise the lower left bright area, and the Gum nebula the upper right. The column density (N(H)) for each star is displayed as a green "bead" that is interactively adjusted during 3D editing.

Gould's Belt and the Local Neighborhood. Adding a larger number of bright, massive young stars in the type O and B classes, we can see a distinctive structure of the local galactic neighborhood known as Gould's belt in Figure C.30. Gould's Belt is tilted by ~18 degrees with respect to the galactic plane, and the ISM associated with these young stars exhibits a similar asymmetry that is difficult to represent with conventional 2D map projections. The Local Bubble reflects the effect of Gould's belt dominating star formation activities in the solar neighborhood (<500 parsecs).

Local Bubble. We remind ourselves that the "Local Bubble" refers to the absence of dense interstellar clouds within ~100 parsecs of the Sun. (To be precise, this region lacks stars whose light shows evidence of large column densities, $N(H) > 10^{19.5}$ cm^{-2}.) This void in the nearby ISM leaves a characteristic footprint in a wide range of multispectral data sets, including various absorption properties of the ISM such as interstellar absorption lines and the reddening of starlight by interstellar dust grains, HI 21-cm emission, CO molecular clouds, soft X-ray data (<0.5 keV), infrared dust emission, the polarization of starlight, and radio continuum synchrotron emission. The Local Bubble dominates the observational characteristics of the ISM in the solar neighborhood. The DistEdTool has been specifically tuned to yield a comprehensive model of the Local Bubble morphology, including comparisons with multispectral datasets. The scientific results will be presented in a separate publication [10].

In Figure C.31, we show the application of the DET software to developing a detailed model of the Local Bubble. The boundaries of the Local Bubble are found from the reddening of starlight, which can be determined from the photometric data in the Hipparcos catalog. The operative quantity is the "color excess" of the stars, E(B-V), determined from the standard photometric B and V magnitudes (which are indices of the star colors). Local Bubble walls are defined by color excess values E(B-V)>0.08 magnitudes, corresponding to the amount of ISM found when molecular hydrogen exceeds 1% of the gas, or $N(H) = 4.6 \times 10^{20}$ cm−2 (using Copernicus satellite results). For a nominal cloud density n=10 cm^{-3}, this corresponds to a cloud length of 14.2 parsecs. To minimize contributions from inaccurate photometric data, tool options are included to limit selected stars to objects brighter than a selected magnitude V_o (V_o ~7 magnitudes here) showing self-consistent astrometric and spectral distances. Either incorrect spectral classifications or incorrect photometric data will yield incorrect N(H) values, so the consistency check between astrometric and photometric distances removes the most egregious uncertainties. The DET software permits zooming and rotation, and allows the adjustment of Local Bubble wall positions for any given sight-line. The data producing the model in Figure C.31 have had minor smoothing applied to remove data that are inconsistent in small spatial interval samples; in particular, type A star data have been rendered more reliable using the `-reselect 0.1` option (see Appendices), which forces estimated spectral distances and measured parallax distances to agree within 10%. Extensive smoothing is conventional in 2D projections of the Local Bubble walls, but removes information about the small scale structure of the ISM that is exposed by the DET-derived models.

Focusing on Cloud Details. Our final example of the use of the DET is presented in Figure C.32, where the stars of the galactic center, including the constellation of Scorpius, have been selected interactively from a much larger data set using the selection tool (red box on the plane). This reconstruction of the Scorpius-region dust cloud surface is set against a background texture from the Dickey and Lockman full-sky compilation of the 21cm hydrogen hyperfine structure [5]. Stars are denoted by blue cubes, and the distance along a line of site from the Earth at a particular galactic longitude and latitude (l,b) are shown as white lines, measured against the interactively movable yellow yardstick (marked in units of parsecs). Galactic coordinates are marked on the (l,b) plane; column density markers and an optional set of grid lines have been deselected for decluttering. The indicated surface is constructed from a Delaunay triangulation of the star locations in the (l,b) plane, and then edited by an astronomer to conform to the 3D location of the inner low-to-high material density transition characterizing the limit of the Local Bubble.

6 Example Interaction Scenarios.

Like any specialized tool, the DistEdTool has special techniques that one gradually learns for efficient utilization. Among these are:

- **Using the System as a Whole with the Filtering Tools.** An example user scenario that selects a set of stars from the Hipparcos database with maximum magnitude 5 and stellar types "OBA" and invokes the DET interface is presented in an Appendix.
- **Moving Beyond the Limits.** To accommodate the fact that cloud length estimates may be inaccurate due to variable volume densities, the bottom edge of the cloud can be moved some distance below zero, and the upper edge some distance beyond the star itself. This is an indication that the estimated volume density is too low, giving an excessive length prediction

for the gas sample. It is also possible for a star to be immersed in a material cloud, so the columns should be extended some distance beyond the star. After aligning one edge of the cloud, the user would adjust the volume density in the output file to reduce the cloud thickness appropriately.
- **Grouping Velocities.** Radial velocity data are available for many stars in the catalogs [1]. When velocity information is provided, the system can color code the gas column depictions to make it easy to tell whether two gas sections covering nearby stars have the same velocity and therefore are likely to belong to the same surface interface. Supplementary information relevant to separating same-velocity data into distinct sections is providable in background texture maps such as the HI 21cm data (see, e.g, [11]), as well as from direct examination of interstellar absorption line Doppler shifts.

7 Conclusions

In response to a need in astronometric data analysis, we have developed, tested, and refined a tool that combines visualization of astronometric gas cloud data with interactive editing of the gas cloud properties. This "Distance Editing Tool" can easily accommodate many types of initial interstellar data and volume density estimates, and can support estimation and revision of cloud surface coordinates using neighboring consistency, background sky map data in any frequency range, and, most importantly, expert knowledge about anomalies available only to users with highly specialized astrophysical knowledge. Because of the inherent difficulty in the manual procedures needed to analyze these data, the availability of an interactive visualization tool is a significant advantage.

While we have focused on the example of the large-scale Local Bubble structure, there are also other, more subtle structures that could also be studied in the future using the DET: one region of great interest is the so-called "Local Fluff" [9], which is the interstellar material in which the moving Sun is currently immersed. This material flows past the Sun in a direction away from the point at the galactic center. The Local Fluff consists of a group of cloudlets with a complex structure showing an upstream flow velocity that is consistent with an expanding supernova shell from the Loop I region.

We anticipate that future work will extend these techniques to include the treatment of stellar and gas cloud velocities, velocity distributions, and error distributions, including the direct display and investigation of errors in the estimated and measured data and their effects.

Acknowledgments

This research was supported in part by NASA grant NAG5-8163. Special thanks to Dr. Eric Wernert and the Indiana University Advanced Visualization Laboratory.

References

1. M. Barbier-Brossat and P. Figon. Catalogue général de vitesses radiales moyennes pour les étoiles galactiques. Mean radial velocities catalog of galactic stars. *Astron. Astrophys. Supl.*, 142:217–223, March 2000.
2. R. C. Bohlin, B. D. Savage, and J. F. Drake. A survey of interstellar H I from L-alpha absorption measurements. *Astrophys. J.*, 224:132–142, 1978.
3. A. N. Cox. *Allen's Astrophysical Quantities*. AIP Press, 2000.
4. T. M. Dame, D. Hartmann, and P. Thaddeus. The Milky Way in Molecular Clouds: A New Complete CO Survey . *Astrophys. J.*, 547:792–813, February 2001.
5. J. M. Dickey and F. J. Lockman. H I in the Galaxy. *Ann. Rev. Astron. Astrophys.*, 28:215–261, 1990.
6. M. A. C. Perryman et al., editor. *The Hipparcos and Tycho Catalogues*. ESA Publications Division, The Netherlands, 1997.
7. D. P. Finkbeiner. A Full-Sky H-alpha Template for Microwave Foreground Prediction. *ArXiv Astrophysics e-prints*, pages 1558–+, January 2003.
8. P. C. Frisch. G-star astropauses - A test for interstellar pressure. *Astrophysical Journal*, 407:198–206, April 1993.
9. P. C. Frisch. Characteristics of nearby interstellar matter. *Space Sciences Reviews*, 72:499–592, April 1995.
10. P.C. Frisch, A.J. Hanson, and Chi-Wing Fu. The Local Bubble: Constraint-based modeling of 3d astrophysical objects, 2003. In preparation.
11. D. Hartmann and W. B. Burton. *Atlas of Galactic Neutral Hydrogen*. Cambridge University Press, Cambridge, 1997.
12. C. G. T. Haslam, H. Stoffel, C. J. Salter, and W. E. Wilson. A 408 mhz all-sky continuum survey. ii - the atlas of contour maps. *Astron. Astrophys. Supl.*, 47:1ff, January 1982.
13. Carl Heiles, 1999. Unified HI 21 cm data, private communication.
14. L. Herda, R. Urtasun, A. Hanson, and P. Fua. Automatic determination of shoulder joint limits using experimentally determined quaternion field boundaries. *International Journal on Robotics Research*, 2002. In press.
15. Mark Phillips, Silvio Levy, and Tamara Munzner. Geomview: An interactive geometry viewer. *Notices of the Amer. Math. Society*, 40(8):985–988, October 1993. Available by anonymous ftp from SourceForge.net.
16. F.R. Preparata and M.I. Shamos. *Computational Geometry: An Introduction*. Springer-Verlag, New York, 1985.
17. P. Rademacher. Glui user interface library. GLUI is a GLUT-based C++ user interface library which provides controls such as buttons, checkboxes, radio buttons, and spinners to OpenGL applications. Avalailable from the University of North Carolina.
18. Robert Renka. Algorithm 772: Stripack, delaunay triangulation and voronoi diagram on the surface of a sphere. *ACM Transactions on Mathematical Software*, 23(3), 1997.
19. S. L. Snowden, R. Egger, M. J. Freyberg, D. McCammon, P. P. Plucinsky, W. T. Sanders, J. H. M. M. Schmitt, J. Truemper, and W. Voges. Rosat survey diffuse x-ray background maps. ii. *Astrophys. J.*, 485:125ff, August 1997.
20. S. L. Snowden, M. J. Freyberg, K. D. Kuntz, and W. T. Sanders. A Catalog of Soft X-Ray Shadows, and More Contemplation of the 1/4 KEV Background. *Astrophys. J. Supl.*, 128:171–212, 2000.

A Data Features Pertaining to ISM

Multispectral Data Pertaining to Local Bubble	
Interstellar Dust	
A_V, E(B-V), E(b-y)	Reddening of starlight (optical) [Lucke (1978), Perry et al. (1982)]
Polarization of stars	Polarization of starlight by interstellar dust (optical) [Leroy (1999), Heiles (2001)]
Dust diffuse infrared (100 μm) emission	Cleaned all-sky maps compiled from IRAS/ISSA and COBE/DIRBE data [Schlegel et al. (1998)]
Heated dust	Hot stars heating nearby dust [e.g., Gaustad, van Buren (1993)]
Molecular Clouds	
CO/H_2 distribution	Galactic plane, high and low latitudes (1-0 rotational transition, 115 GHz) [Dame (2001), Magnani et al. (2000), Hartmann et al. (1998)]
Interstellar Neutral Gas	
N(H°) to star	UV, optical, and EUV absorption (both lines and EUV continuum) [e.g., Frisch, York (1983), Paresce (1984), Warwick et al. (1993), Diamond et al. (1995), Welsh et al. (1999), Sfeir et al. (1999)]
H° 21-cm emission	All-sky maps with cloud velocity, resolution <0.5° [Colomb et al. (1980), Hartmann, Burton (1997), Heiles(1999)]
X-ray shadows	Shadows in soft X-ray background, caused by ISM with log N(H°)>19.5 cm^{-2} [Snowden et al. (2000)]
Interstellar Ionized Gas	
H^+ Hα emission	All-sky Hα emission (northern hemisphere with velocities) [Haffner et al. (1999), Gaustad (1997), Finkbeiner (2003)]
Radio continuum	All-sky 408 MHz emission, ionized gas, cosmic ray synchrotron emission [Berkhuijsen (1971), Haslam (1982)]
10^6 K Plasma	
X-ray emission	ROSAT 1/4 keV PSPC all-sky survey [Snowden et al (1997)]
Stars	
Hipparcos catalog	Distances (10% accuracy at 100 pc) and photometry for 118,000 stars, including 20,500 O, B, A, and F stars within 500 pc (up to one star per $\sim$0.5 square degrees on the sky) [Perryman (1997)]
Nearby stars (Gliese)	Complete catalog of stars within 25 pc (3800 objects with small astrometric distances) [Gliese (1995)]

B DET Data Format

Several Perl scripts support the preparation of data for the DistEdTool, producing standardized **DET** files. The scripts themselves are self-documenting, listing the extensive options when no arguments are provided. The DET file format, summarized below, follows a community convention for single-line ASCII data that, in this particular case, is also well-adapted to the often necessary hand-editing of stellar features based on expert-acquired ancillary information.

		DET input data fields
1	HD	Unique "Henry Draper" number of the star
2	Tri	Triangulation bit (0/1): is this star selected for triangulation?
3–4	L,B	Galactic coordinate (in degrees)
5	DStar	distance to the star (in parsecs)
6	%Err	percentage error in DStar, [0,100]
7	DCloud	distance to the gas cloud (in parsecs)
8–10	Star_vel	star velocity in galactic coordinates (km/s)
11–13	%Err	percentage error in Star_vel
14	Leng	length of gas column (in parsecs)
15	VolDen	Volume Density (atoms cm^{-3})
16	Err	error of length of gas column (in parsecs)
17	Vmag	magnitude of the star
18	SpType	spectral type of the star
19	v_ks	radial velocity of the absorption line
20	b_ks	Doppler constant of the absorption line
21	ComName	Common Name
22	Ref	Reference material, if any

A supplementary radial velocity material file is used to compute the Star_vel field when combined with the Hipparcos-supplied proper motions (tangential star velocities). Below we give an example that creates a reliability-selected data file of O, B, and A stars with magnitudes 5 or brighter, and a distance within 100 parsecs, followed by an invocation of the Distance Editing Tool with a customized list of texture maps to load.

```
> hip2det.pl -t OBA -maxM 5 -s -i -rmax 100 -rselect 0.1 \
      ../Data/hip_main.dat ../Data/hipOBA5.det

> distEdTool ../Data/hipOBA5.det ../Images/COdata.textures
```

Geometric Modelling for Virtual Colon Unfolding

Anna Vilanova[1] and Eduard Gröller[2]

[1] Department of Biomedical Engineering
   Eindhoven University of Technology
   a.vilanova@tue.nl
[2] Institute of Computer Graphics and Algorithms
   Vienna University of Technology
   meister@cg.tuwien.ac.at

Summary. A virtual endoscopic view is not necessarily the best way to examine a hollow organ, such us, the colon. The inner surface of the colon is where polyps are located, and therefore what is examined by the physicians. A flight through the colon using a common endoscopic view shows a small percentage of the inner surface. Virtually unfolding of the colon can be a more efficient way to look at the inner surface. We propose two methods to unfold the colon: a method that unfolds the colon locally using local projections, and a method that obtains a global unfolding of the colon by achieving a suitable parameterization of its surface.

1 Introduction

Most of the virtual endoscopy applications presented in the last years concentrate on simulating the view of a real endoscope. This is the view that endoscopists are used to, and it is useful for certain applications, like in an intraoperative scenario. However, it is not necessarily the best way to inspect the inner surface of an organ. Actually, a real endoscope and organ are subject to physical limitations that a virtual endoscope and organ do not have. In this chapter, we concentrate on virtual colonoscopy, which focuses on the examination of the colon.

Physicians are mainly interested in visualizing the inner surface of the colon which is where polyps can be detected with endoscopy. It is important that the physician can estimate the size of polyps, since large polyps are more likely to develop into malignancies. The usual endoscopic view visualizes just a small part of the surface. Furthermore, it is difficult to detect polyps that are situated behind the folds of the colon. An efficient way to inspect the inner surface would be to open and unfold the colon, and then examine its internal surface. Unfortunately, this cannot be done in reality, if we want that the patient survives. On the other hand, there is no patient damage if this dissection of the organ can be achieved virtually with the medical data obtained by CT or MRI (i.e., the virtual organ). The resulting unfolded model has to facilitate the physician's inspection and detection of polyps.

In this chapter, we present different approaches to unfold the colon. After an overview of the existing methods, we present in detail two methods: a method that unfolds the colon locally using local projections, and a method that obtains a global unfolding of the colon.

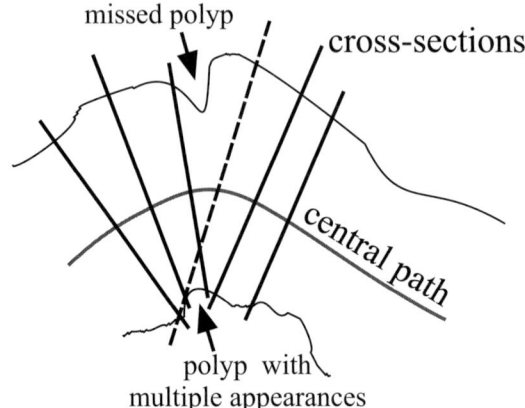

Fig. 1. Illustration of the possible undersampling and multiple appearance of polyps due to intersections of the cross-sections in high curvature areas. The dashed cross-section line produces a multiple appearance of the polyp.

2 Related Work

Wang et al. [20] proposed a technique to straighten and unravel an organ virtually. Their approach starts with defining a path which is placed as close to the center of the object as possible. Then, a sequence of frames is calculated. For each frame, a cross-section orthogonal to the path tangent is calculated. The central path is straightened and the cross-sections are piled to form a stack. As a last step, the straightened colon is unfolded by tracing rays in radial directions to the path. The result is a volumetric model of the unfolded colon. The model is displayed afterwards using standard volume rendering techniques. This method can be seen as a resampling and parameterization operation.

However, one of the main problems of this technique appears in high curvature areas of the central path, i.e., at path locations where the radius of curvature is bigger than the organ diameter. In such cases, orthogonal cross-sections intersect each other in some regions or are far apart in some other regions (see figure 1). As a consequence, a polyp can appear more than once in the unfolded model or it can be missed completely. These problems are the consequence of undersampling and an ambiguous parameterization of the organ surface.

In later works Wang et al. [18, 19] try to overcome these problems. The authors use electrical field lines generated by a locally charged path to govern curved cross-sections instead of planar sections. The cross-sections tend to diverge avoiding conflicts. If the complete path is charged then the curved cross-sections will not intersect. However, for each point of the field lines the contribution of each charge on the path must be calculated. This operation is computationally so expensive that the authors propose to just locally charge the path. A small segment of the path contains the charges for each cross-section. In this way, the method is feasible in practice, but it cannot ensure anymore that the curved cross-sections will not intersect each other. In other words, it cannot ensure that the parametrization of the space will be unambiguous. Furthermore, the undersampling is still a problem.

Other authors propose methods to flatten a polygonal representation of the colon surface. These techniques involve tasks that have already been used in texture-mapping for computer graphics. A major step thereby is to come up with a suitable surface parameterization. For texture mapping, this parameterization is used to assign texture values to surface points. For surface flattening the parameterization allows to display surface values (e.g., color) in the 2D parameter space [13]. Many techniques are dealing with texture distortions which generally cannot be avoided entirely. The distortions depend on the chosen surface-parameterization characteristics (e.g., length and area preserving [2], angle preserving [6] or a combination of both [3, 9]).

Haker et al. [5] use conformal (i.e., angle preserving) texture mapping to map the polygonal colon surface to a plane. One of the main problems of this method is that a highly accurate segmentation is necessary to ensure good results for diagnosis. The entire polygonal surface is flattened. The result is a triangulated plane where the polyps have also been flattened. Shading is applied to the flattened surface using the normals of the original surface and the color-coded mean curvature. This is the only information which helps the physician in identifying polyps in the unfolded plane. Furthermore, the surface needs to be smoothed to achieve a good mean-curvature calculation which can imply smoothing and missing polyps.

In the previous methods, the whole colon surface was unfolded or flattened. To increase the visible surface, Paik et al. [12] propose to use different camera projections. With a normal endoscopic view just 8% of the solid angle of the camera is seen in each frame. Paik et al. project the whole solid angle of the camera by map projection techniques used for geographical charts. They suggest to use the Mercator projection for mapping the solid angle to the final image. This technique samples the solid angle of the camera, then the solid angle is mapped onto a cylinder which is mapped finally to the image. This method generates a video that the physician has to inspect.

Serlie et al. [14] present a method based on image-based rendering. In a pre-processing step, a cubic environment map is calculated at each camera position along the central path. These environment maps can be used to obtain real-time navigation using image-based rendering, as proposed by Wegenkittl et al. [21]. Serlie et al. [14] also propose to display these cubic maps unfolded in order that the physician has a 360 degrees field of view.

All these methods introduce some kind of deformation. Flattening a surface in 3D space onto a 2D plane introduces distortions unless the surface has zero Gaussian curvature. [11]

3 Local Colon Unfolding

In this section, we propose a method to unfold the colon using a new camera projection technique. This method [17] generates a video where each frame is a local unfolding of the organ. It allows to inspect locally unfolded regions such that multiple appearances of polyps do not occur. This method is similar to the one proposed by Serlie et al. [14].

The presented method involves moving a camera along the central path of the colon. Several techniques can be used to generate a smooth central path (see, e.g., Vilanova et al. [15]).

At each camera position along the central path, an orthogonal coordinate system is taken which specifies the location and orientation of a cylinder. One coordinate axis is given by the tangent vector of the central path. The other axes are in the plane orthogonal to the central path at the camera position. The Frenet frame is commonly used to define a coordinate system for a point on a curve. However, it is not a good choice in our case. Firstly, the Frenet frame is not defined in linear portions of the central path. Secondly, by moving along the path, the two vectors orthogonal to the tangent vector can rotate considerably, thus reducing coherence between adjacent frames. Instead of the Frenet frame, we use a rotation-minimizing coordinate frame as presented by Klok [8].

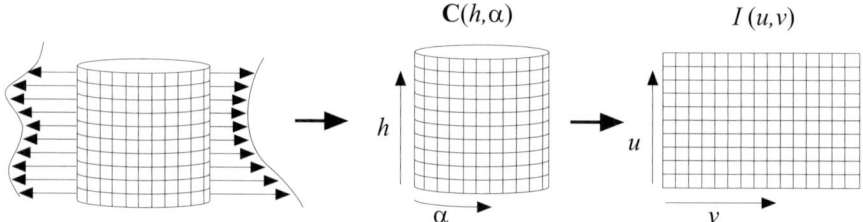

Fig. 2. Illustration of the projection procedure. A region of the organ surface is projected onto the cylinder $\mathbf{C}(h,\alpha)$. Then, the cylinder is mapped to the image $I(u,v)$.

For each camera position, a small cylinder $\mathbf{C}(h,\alpha)$[1] tangent to the path is defined. The point in the middle of the cylinder axis corresponds to the camera position. For each ray, direct volume rendering is used to calculate the color which corresponds to the cylinder point where the ray was projected. Finally, the colored cylinder with the sampled rays is developed into a 2D image $I(u,v)$ by a simple mapping function $\mathbf{f} : (h,\alpha) \rightarrow (u,v)$ (see figure 2). The simplest mapping function is the identity where $\mathbf{f}(h,\alpha) = (h,\alpha)$.

The cylinder axis must be short enough, such that the cylinder does not penetrate the surface of the colon. This can be done by taking into account the distance of the path to the organ surface.

The result is a video where each frame shows the projection of a small part of the inner surface of the organ onto a cylinder. If the camera is moved slowly enough the coherence between frames will be high and the observer will be able to follow the movement of the surface and polyps.

In high curvature areas, the intersection of cross-sections also occurs (see figure 1). However, crossing of rays can happen just between frames, which does not cause a multiple appearance of a polyp within a single image. Moving along the central path in a high curvature area, a polyp might move up and down (due to

---

[1] Throughout this chapter, scalars are given in italics and vectors in bold typeface. Angles are denoted by Greek letters. For example, $\mathbf{C}(h,\alpha)$ is a function which returns a vector (point) and has as parameters two scalars h and α, where α is an angle.

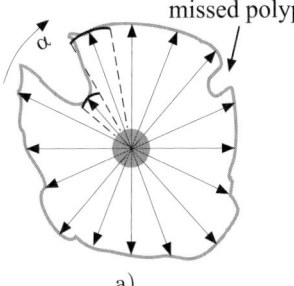 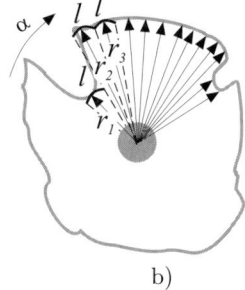

Fig. 3. a) Constant angle sampling: it is shown that different surface lengths are represented by the same length in the cylinder. **b)** Perimeter sampling: same length but different angle.

crossing rays), but it is clearly identified as a single object. The user is able of tracking the polyp movement if the coherence between frames is high enough.

The sampling distance (i.e., the distance between two consecutive rays) in the h-direction is constant, and it must be at most half of the size of a voxel (see figure 2). In this way, correct sampling (with respect to the Nyquist limit) is possible in the h-direction.

In the next sections, two methods are described which project the organ surface onto the cylinder depending on the sampling of angle α, i.e., constant angle sampling and perimeter sampling.

3.1 Constant Angle Sampling

Constant angle sampling means that the angle between consecutive rays in the α-direction is constant for rays with the same h-value. Figure 3a illustrates how this sampling is done. Using this method, the cylinder is sampled uniformly but not the organ surface.

The advantage of this method is that the relationship between both directions is preserved locally. Therefore, the angles are locally preserved too. An image generated by this method can be seen in figure 4a.

On the other hand, the area of the projected region is not preserved (see figure 3a). Therefore, the size of a projected polyp depends on the distance of the cylinder axis to the organ surface cavity. Consequently, the physician cannot trust the sizes of the projected polyps. With constant sampling, polyps can be missed if the angle increment is too large (see figure 3a). If the sampling distance is too small, rays are traced where it would not be necessary. This makes the method inefficient.

3.2 Perimeter Sampling

With perimeter sampling, rays are calculated such that the surface length that they represent is constant. A constant sample length l is defined. l must be at most half the size of a voxel to stay above the Nyquist frequency and therefore not to miss any important feature. l should have the same value as the sampling distance in the h-direction to preserve the ratio, or proportion, in the final mapping.

The algorithm incrementally calculates the ray directions which are in the plane defined by a certain value of h. The angle between the current ray and the next one is computed such that the length of the surface sample that the current ray represents is l in the α-direction (see figure 3b). r_i is defined as the distance from the cylinder axis to the surface point hit by the ith ray. The surface sample length in the α-direction that a ray represents is approximated by the arc with radius r_i. Therefore, the value of the angle increment for the next ray is estimated as $\frac{l}{r_i}$ radians. This projection method projects the organ surface to a generalized cylinder

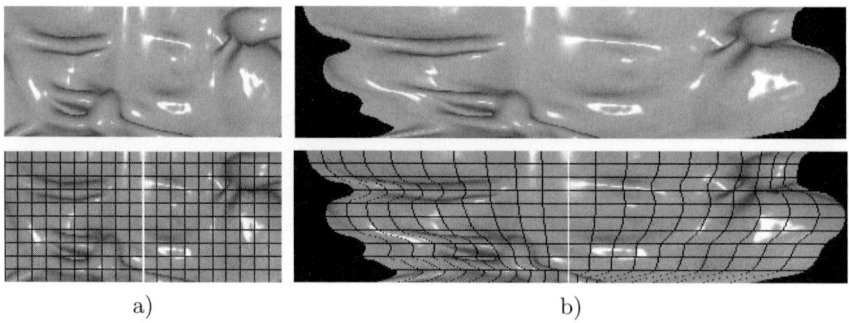

Fig. 4. a) Constant angle sampling of the organ surface. **b)** Same camera position as a) but with perimeter sampling. The bottom images show a grid with constant angle α.

whose radii are not constant within the cylinder. Moving along the central path, varying perimeter lengths are represented by a varying number of rays. Therefore, the generalized cylinder is not mapped to a complete rectangular domain (see figure 4b). The mapping function **f** in this case maps each sampled ray to a pixel in the image (i.e., each pixel corresponds to an area of size $l \times l$ on the surface). The projected point that corresponds to the first ray is positioned on a vertical line in the center of the image. Then, from left to right, the ray values are mapped onto the image until the perimeter length is reached.

This projection is area preserving. The relative sizes of surface elements are preserved in the image plane and do not depend on the distance of the cylinder axis to the surface. On the other hand, a distortion is introduced with respect to the h and α-directions, so the angles are not preserved anymore. At the vertical center line of the image, no distortion occurs, but the distortion increases progressively when we move to the left or right. Figure 4b shows an image generated with perimeter sampling. The superimposed grid corresponds to a regular grid in a constant angle sampling of the cylinder.

4 Nonlinear Colon Unfolding

The method presented in section 3 requires that the physician reviews a video and cannot visualize the complete surface at once. In this section, we describe a method [16] to obtain a complete model of the unfolded colon. This method uses similar ideas as the method by Wang et al. [20]. The technique provides solutions to multiple appearances of polyps, distortion and undersampling. In our approach, the colon unfolding does not produce a surface but a height field (distance of the colon surface to a central path). This avoids that the polyps are flattened as with the methods proposed by Haker et al. [5]. Furthermore, the height field gives a more natural visualization than a color-coded flattened surface.

Unfolding the colon can be divided into three main steps: nonlinear ray casting, which solves the problem of multiple appearances of polyps; nonlinear 2D scaling, which reduces the distortion due to nonuniform sampling; and resampling, which avoids to miss polyps.

As in local unfolding, a central path is calculated. A distance map is generated from the central path [10]. The distance map contains the distance to the nearest point on the central path. A coordinate frame is moved along the path. For each path

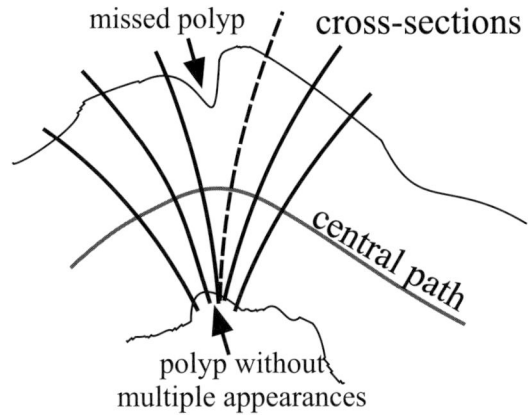

Fig. 5. Elimination of multiple polyp appearances by nonlinear ray casting.

position, rays are initialized in the plane orthogonal to the path, following radial directions (constant angle sampling). To avoid multiple appearances of polyps in high curvature areas of the path, the rays follow the negative gradient direction of the precalculated distance map. The rays are not straight lines anymore. They do not cross each other, but converge at most (see figure 5). Nonlinear ray casting has already been investigated before by several authors (e.g., Gröller [4]). Section 4.1 explains how these rays are traced.

Along each curved ray, direct volume rendering is performed. The ray is terminated when it hits the surface of the colon. The result of the nonlinear ray casting can be interpreted as a 2D cylindrical parameterization of the inner colon surface. One parameter corresponds to the position along the path. The second parameter specifies the ray within the plane orthogonal to the current path position. The distances between ray origins on the central path and intersected points on the colon surface determine a height field. The height field is unfolded, and the result corresponds to a parallel projection of the unfolded height field.

Nonlinear ray casting samples the height field nonuniformly. A straightforward unfolding to a regular grid (i.e., a uniform parameterization) contains severe area distortions and is therefore not optimal. In a second step, an iterative scaling transforms the previously generated 2D parameter grid in order to compensate for these distortions. After the scaling, the ratios between the area that the samples represent and their area in the 2D grid are approximately equal. The second step is based on nonlinear 2D scaling that is used in a similar way for magnification fields in information visualization [7]. In section 4.2, the algorithm is described in detail. Afterwards, the colon surface is resampled with an adequate minimum sampling rate using the transformed 2D grid.

4.1 Nonlinear Ray Casting

The central path of the colon is described by a parametric curve $\mathbf{c}(v)$. We define $dist(\mathbf{p}) : \mathbb{R}^3 \rightarrow \mathbb{R}$ as a function which gives the minimum distance between a point $\mathbf{p}$ and $\mathbf{c}(v)$. $dist(\mathbf{p})$ is sampled in a discrete distance map $Dist(\mathbf{q}) : \mathbb{N}^3 \rightarrow \mathbb{R}$ where $\mathbf{q}$ is a voxel position in the volume. A reconstruction filter is applied to $Dist(\mathbf{q})$ to approximate $dist(\mathbf{p})$ (see Vilanova et al. [16] for details).

$dist(\mathbf{p})$ is continuous in the first derivative nearly everywhere. Exceptions are ridge and valley lines of the distance map $dist(\mathbf{p})$. $dist(\mathbf{p})$ induces a vectorfield that is defined by the gradient, $-\nabla dist(\mathbf{p})$. It is known that trajectories of such vectorfields will not cross each other and are unambiguous (see Abraham et al. [1] for details). These trajectories will correspond to our nonlinear rays. The nonlinear rays are traced from the central path in uphill direction, i.e., along the negative gradient direction $-\nabla dist(\mathbf{p})$. Furthermore, in our situation, trajectories will not produce cycles, since it is impossible to return to the same point by always moving uphill. In the worst case, the nonlinear rays will merge in ridge and valley lines, but they will not cross. With these curved rays the multiple appearance of polyps is avoided and an unambiguous and correct parameterization of the inner colon surface is obtained.

Casting of Nonlinear Rays

The first step to trace the nonlinear rays is to move a coordinate frame along the curve $\mathbf{c}(v)$. We again use the rotation-minimizing coordinate frame of Klok [8]. For each position on the path, a constant number of rays is traced. The initial point of each ray is placed in the plane orthogonal to the path. Note that the gradient is not defined along the path $\mathbf{c}(v)$ since it is a valley line of the distance map $dist(\mathbf{p})$. Therefore, the initial points are placed circularly at a small distance from the path position. Once the initial points have been determined (u parameterization), the rays are traced integrating the negative gradient of the distance map.

The rays have the tendency to be perpendicular to the path $\mathbf{c}(v)$. This is the direction of maximal change of $dist(\mathbf{p})$ in linear segments of the path. The rays become curved in areas where the curvature of the path increases.

Colon Surface Parameterization

In the previous section, nonlinear and non-crossing rays were traced from the central path $\mathbf{c}(v)$ towards the colon surface. While the rays are traced, direct volume rendering is performed. The ray terminates when the colon surface is hit. The result of the nonlinear ray casting is a sampling of the inner surface of the organ.

The tracing of the nonlinear rays defines an unambiguous parameterization of the inner colon surface $\mathbf{s}(u,v)$. Here, v is the parameter along the central path $\mathbf{c}(v)$, and u is the radial angle along which the nonlinear rays are started ($u \in [0, 2\pi]$).

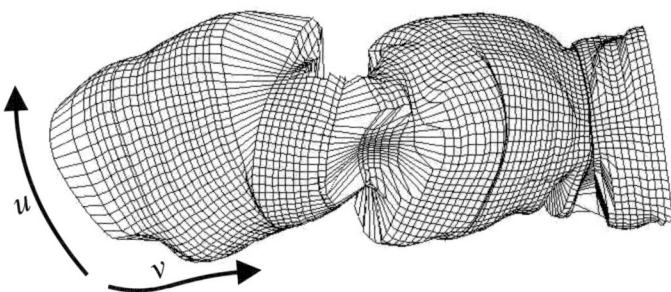

Fig. 6. Surface obtained after nonlinear ray casting. The sampling of the surface is nonuniform.

Figure 6 shows $\mathbf{s}(u,v)$ which results from applying nonlinear ray casting to a colon piece. The lines correspond to the isolines of the parametric surface $\mathbf{s}(u,v)$. The parameter space is sampled uniformly in the u and v direction, but this does not correspond to a uniform sampling of $\mathbf{s}(u,v)$.

Unfolding of the $\mathbf{s}(u,v)$ surface can easily be done by mapping $\mathbf{s}(u,v)$ onto a regular grid in the 2D u,v-parameter space. In figure 7a, parameterization of the colon surface is done with straight rays (ambiguous parameterization and non-uniform sampling). In figure 7b, a parameterization of the colon surface is done with curved rays (unambiguous parameterization, but still non-uniform sampling).

Nonlinear ray casting avoids that features appear more than once, but on the other hand the sampling of the surface is far from being uniform. There are over-sampled areas, which lead to geometric deformations, and also undersampled areas exist. In the latter case, deformations appear but also features of the surface can be missed.

In figure 7a, the solid circles indicate areas where features appear more than once. Using nonlinear ray casting, the polyps do not appear more than once, and instead an enlargement of the feature appears (figure 7b). The areas marked by dashed circles indicate undersampled areas and therefore areas where features are possibly missed. Note that the same undersampled areas are present in both figures.

In the next section, an algorithm is presented to obtain an unfolding of the parametric surface $\mathbf{s}(u,v)$ which avoids geometric deformations and undersampling.

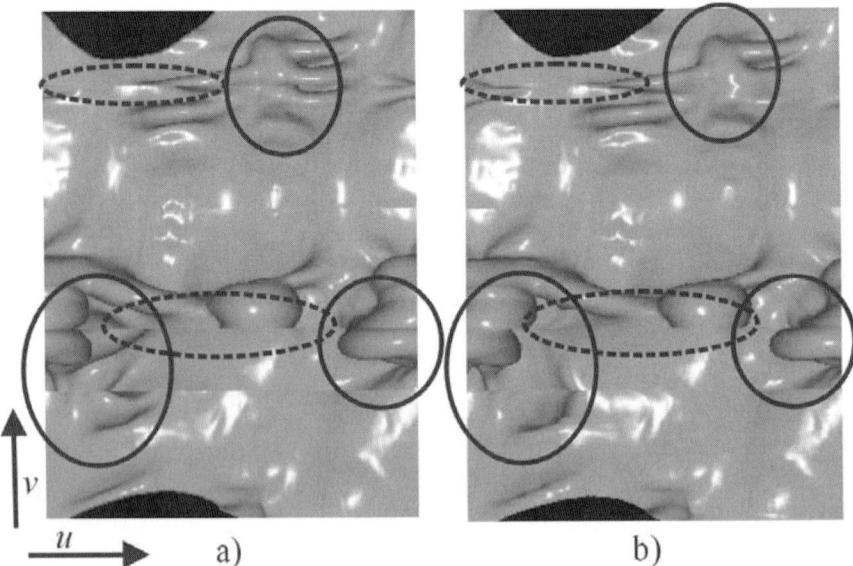

Fig. 7. a) Unfolding of the colon surface of the data set presented in figure 6 using straight rays. Solid circles indicate areas where polyps appear more than once. Dashed circles indicate undersampled areas. b) Parametric surface generated using nonlinear rays as shown in figure 6. Unfolding is done onto a regular grid. Multiple appearance of polyps disappear, but undersampled areas not.

4.2 Nonlinear 2D Scaling

In the previous section, an unambiguous parameterization of the inner colon surface projected to the central path has been introduced. The sampling of the surface $s(u, v)$ defines a valid and non self-intersecting quadrilateral mesh on the colon surface (see figure 6). Furthermore, the distance between the surface point $s(u, v)$ and the corresponding path position $c(v)$ defines a height field $r(u, v)$.

The goal of nonlinear 2D scaling is to achieve a 2D grid (i.e., parameter space) which approximates a parallel projection of the unfolded height field (see 4.1). The unfolded height field shall approximately preserve the length of the edges of $s(u, v)$ in u- and v-direction.

Height-field Unfolding

In the nonlinear ray casting, a 3D quadrilateral mesh is obtained. We know the distances b_i between adjacent quadrilateral vertices (i.e., the length of the edges of the quadrilateral). If we preserve these distances in the 2D grid (i.e., parameter space), the sizes of the quadrilaterals will be preserved (see figure 8b). However, by preserving the 3D edges of the quadrilateral mesh, we flatten the surface and the polyps. This is due to the fact that we do not take the height field $r(u, v)$ into account. We want that the edges of the 3D quadrilateral mesh are preserved in the unfolded height field. This implies that the distance e between edges in the 2D grid should correspond to the length of the projection of the edges onto the grid plane (see figure 8c).

Using these observations, we define e such that the unfolded height field preserves the length of the edges of the 3D quadrilateral mesh in u and v direction (see Vilanova et al. [16] for details). In the next section, an algorithm to obtain such a 2D grid is presented.

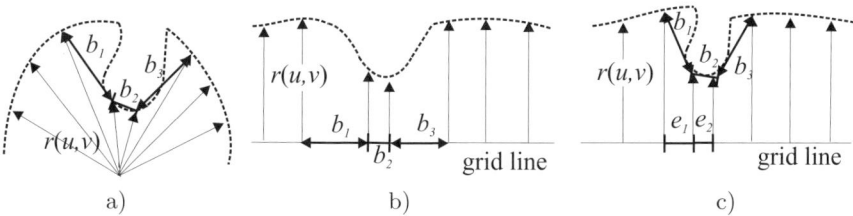

Fig. 8. Illustration of height field unfolding in u direction: a) cross-section of $r(u,v)$ for a fixed value of v, b) unfolding preserving the edge lengths b_i of the 3D quadrilateral mesh in the 2D grid, and c) unfolding preserving the edge lengths b_i of the 3D quadrilateral mesh in the height field.

Nonlinear 2D Scaling

The objective of the nonlinear 2D scaling algorithm is to generate a 2D grid whose edges preserve the length $e(i,j,k,l)$, calculated as explained in the previous section. In $e(i,j,k,l)$, i and k correspond to the ith and kth sampled parameter value in u-direction, and respectively j and l correspond to the jth and kth sampled parameter value in v-direction. Finding an analytical solution to this problem is too complex, so a numerical solution is adopted. We use an approach similar to the one presented by Keahey et al [7]. The main difference is that our algorithm not only preserves areas, but also the edge lengths.

We define a function $\mathbf{T}(i,j) : \mathbb{N}^2 \to \mathbb{R}^2$ as a transformation of a 2D regular grid. $\mathbf{T}(i,j)$ has to be C^0-continuous and it should preserve the order (i.e., no flipping edge or grid node).

We define a 2D scaling field S as a field of scalar values for each edge. Each scalar value indicates the scaling factor that a transformation $\mathbf{T}$ has applied to the edge. The 2D scaling field S for an edge defined between $\mathbf{T}(i,j)$ and $\mathbf{T}(k,l)$ is $S(i,j,k,l) := \|\mathbf{T}(i,j) - \mathbf{T}(k,l)\|$. A 2D scaling field S is defined for any transformation $\mathbf{T}$.

The goal of the nonlinear 2D scaling algorithm is to find a transformation $\mathbf{T_g}$ such that the equation $e(i,j,k,l) = \|\mathbf{T_g}(i,j) - \mathbf{T_g}(k,l)\|$ holds for all values of (i,j), where (k,l) is a 4-connected neighbor of (i,j). In other words, we want to find a transformation $\mathbf{T_g}$ whose 2D scaling field is $S_g(i,j,k,l) = e(i,j,k,l)$ for each edge of the grid.

The major problem is to find the coordinates (x,y) of the transformation $\mathbf{T_g}$, given the scalar values of the 2D scaling field S_g. It is clear that for the same 2D scaling field several transformations are possible. We have used an iterative method which provides a numerical solution. The goal of the algorithm is to find a transformation $\mathbf{T_a}$ that provides a good approximation of $\mathbf{T_g}$.

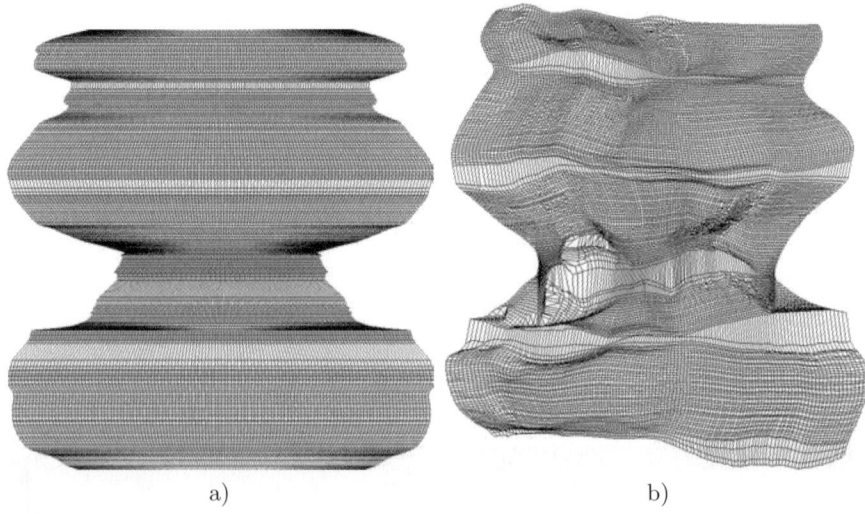

Fig. 9. Illustration of the nonlinear 2D scaling algorithm using the same data set as in figure 6. a) Initial T_a corresponding to a 128x171 grid. b) T_a after 960 iterations of the algorithm.

Given a transformation T_a, the corresponding scaling field S_a can easily be calculated. A scaling field error can then be computed by $S_e = S_g - S_a$. S_e gives the difference between the current scaling field S_a and the desired scaling field S_g.

The iterative algorithm starts with T_a as a regular grid. Then S_a and S_e are calculated. The algorithm iterates over each node of the grid. For each node, the value of S_e at each of the 4-connected neighbors is investigated. If $S_e > 0$ (i.e., the edge is not long enough) then the neighbor is moved away from the node . If $S_e < 0$ (i.e., the edge is too long) then the neighbor is pulled towards the node. The edge is modified by a length of $S_e(i,j,k,l)C_r/2$ where $C_r \in [0,1]$ is a parameter of the algorithm. The division by 2 is necessary because each edge is treated twice, once for each vertex of the edge. Changing an edge is thus done by modifying each of its vertices. An important requirement of the algorithm is to preserve the order. So the neighbors are moved as far as S_e and C_r allow without flipping edges.

The neighboring nodes are changed with coordinate-aligned movements. These movements have the tendency to preserve the rectangular appearance defined by $\{T_a(i,j), T_a(i+1,j), T_a(i+1,j+1), T_a(i,j+1)\}$, which, for example, does not degenerate to a triangle.

Once the iteration has run for all the nodes, the new T_a is generated. Then, a new S_a and a new S_e are calculated from the resulting T_a. S_e is calculated just once per iteration.

The convergence factor is measured using the distance between the approximating scaling field S_a and the desired 2D scaling field S_g. This is expressed as the root mean squared error σ of S_e.

The convergence of the algorithm can be improved by starting with a T_a which is a closer approximation of the desired result than a regular grid. The length of the edges within a horizontal line (i.e., the horizontal edges between nodes with the same j value) are set such that the line length approximates the perimeter of the colon in the corresponding cross-section. The distance between two consecutive horizontal lines is set to the average of the vertical edge lengths in S_g which join the nodes between the two lines (see figure 9a). This initial modified grid leads to a faster convergence by generating basically the same result as a regular grid would give.

The complexity of the nonlinear 2D scaling algorithm is $O(k \cdot n \cdot m)$ where $n \times m$ is the size of the original grid and k is the number of iterations needed for convergence.

Figure 9a shows the initial grid $\mathbf{T_a}$ for the segment of the colon presented in figure 6. The resolution of the grid is 128x171 and the initial value of σ is 0.8008. After 960 iterations $\mathbf{T_a}$ has been evolved into the grid in figure 9b. The value of σ is 0.2808.

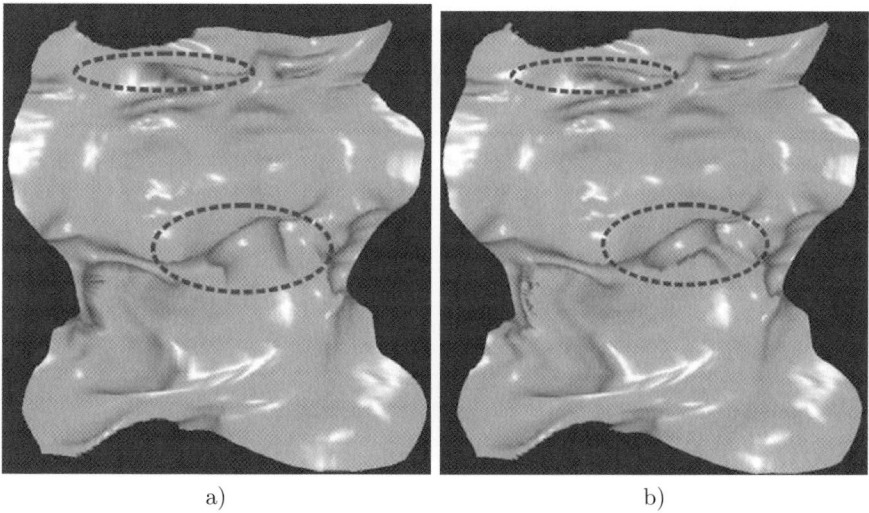

Fig. 10. Resampling after the nonlinear 2D scaling. a) 128x171 shaded grid using bilinear interpolation. b) Shading of the resampled grid.

4.3 Resampling

The nonlinear 2D scaling provides a mapping between the 3D quadrilateral mesh and a 2D grid avoiding area deformations. The color of each ray obtained in the nonlinear ray casting step is assigned to its corresponding node in the 2D grid. Bilinear interpolation is used to fill the quadrilaterals of the grid. An example can be seen in figure 10a. The areas encircled by dashed ellipses are the same as in figure 7. Some features are missing due to undersampling.

The undersampled areas are easily identifiable from the 2D grid. A minimum sample step for the 2D grid is defined. The sample step corresponds directly to a sample step in the 3D space. For each quadrilateral, the subdivision consists of generating a subgrid whose edge lengths are smaller than or equal to the sample step.

Each of the newly generated nodes in the grid can easily be identified with its corresponding point in 3D using linear interpolation. The 3D points do not correspond to surface points, but they are close to the colon surface. A short nonlinear ray segment through the point is investigated to locate the correct surface point. Then the rays are traced forward again to find the correct surface point.

The resulting color values are mapped directly to the corresponding point in the 2D grid. The results of the resampling procedure can be seen in figure 10b. The encircled areas show regions where features that were not present in figure 10a have been identified.

4.4 Results

The CT volume data of an extracted colon with a resolution of 381x120x632 is used in our experiments (see figure C.33 right top). The colon is 50 cm long and contains 13 polyps. The unfolding of this colon can be seen in figure C.33 to the left. All the polyps could be detected easily by inspection. The extracted colon was physically dissected and several pictures of the dissected colon were also taken. These pictures enable a qualitative comparison between the real data and the results of the presented algorithm (see figure C.33).

5 Conclusions

Simulating an endoscopic view is not the most suitable visualization technique in many endoscopy procedures. For example, if the physician is interested in inspecting the inner surface of the organ, the endoscopic view visualizes just a very small percentage of it. This chapter presented various techniques which generated an unfolded model of a colon. This unfolded model allows a more efficient visualization of the inner surface. They are generated by projection, resampling and/or adequate parameterization of the organ.

Section 3 describes a technique that locally unfolds the colon, and generates an animation sequence from consecutive unfolded regions. The images allow the physician to visualize most of the surface, and to easily recognize polyps that would be hidden in an endoscopic view by folds or would be hard to localize. For more examples refer to www.cg.tuwien.ac.at/research/vis/vismed/ColonFlattening/.

With the previous method, the physician has to inspect a video to be able to visualize the whole surface. The colon-unfolding technique in section 4 enables the physicians to get a fast overview of the entire organ surface within a single image. This approach solves the problem that previous techniques had [20]. Compensation of area distortions due to the unfolding is achieved using an iterative method. However angle distortions are not taken into account. For more results refer to www.cg.tuwien.ac.at/research/vis/vismed/ColonUnfolding/.

The methods presented in chapters 3 and 4 have been tested with a data set that enabled a qualitative comparison of the resulting images with images of the corresponding real extracted colon with satisfactory results.

These techniques present a new way to parameterize the organs in order to inspect their inner surface.

Acknowledgements

The work presented in this publication has been funded by the Adapt project. Adapt is supported by *Tiani Medgraph*, Vienna, and the *Forschungsförderungsfonds für die gewerbliche Wirtschaft*, Austria. We thank Dr. Erich Sorantin from the Department of Radiology in Graz for his collaboration and for providing the data sets and the images of the dissected colon.

References

1. R. H. Abraham and C. D. Shaw. *Dynamics: The Geometry of Behavior*. Addison-Wesley, 1992.
2. C. Bennis, J.M. Vézien, and G. Iglésias. Piecewise surface flattening for non-distorted texture mapping. In *SIGGRAPH'91, Conference Proceedings*, pages 237–246, 1991.
3. M.S. Floater. Parametrization and smooth approximation of surface triangulation. *Computer Aided Geometric Design*, 14:231–250, 1997.
4. E. Gröller. Nonlinear ray tracing: Visualizing strange worlds. *The Visual Computer*, 11:263–274, 1995.
5. S. Haker, S. Angenent, A. Tannenbaum, and R. Kikinis. Nondistorting flattening maps and the 3D visualization of colon CT images. *IEEE Transactions on Medical Imaging*, 19(7):665–671, July 2000.
6. S. Haker, S. Angenent, A. Tannenbaum, R. Kikinis, G. Sapiro, and M. Halle. Conformal surface parameterization for texture mapping. *IEEE Transactions on Visualization and Computer Graphics*, 6(2):181–189, April-June 2000.
7. T.A. Keahey and E.L. Robertson. Techniques for non-linear magnification fields. In *IEEE Information Visualization'97, Conference Proceedings*, pages 51–58, 1997.
8. F. Klok. Two moving coordinate frames for sweeping along a 3D trajectory. *Computer Aided Geometry Design*, 3:217–229, 1986.
9. B. Lévy and J.L. Mallet. Non-distorted texture mapping for sheared triangulated meshes. In *SIGGRAPH'98, Conference Proceedings*, pages 343–352, 1998.
10. G. Lohmann. *Volumetric Image Analysis*. Chichester Wiley, 1998.
11. B. O'Neill. *Elementary Differential Geometry*. Academic Press, 1966.
12. D.S. Paik, C.F. Beaulieu, R. B. Jeffrey, Jr. C.A. Karadi, and S. Napel. Visualization modes for CT colonography using cylindrical and planar map projections. *Journal of Computer Tomography*, 24(2):179–188, 2000.
13. M. Samek, C. Slean, and H. Weghorst. Texture mapping and distortion in digital graphics. *The Visual Computer*, 2:313–320, 1986.
14. I.W.O. Serlie, F.M. Vos, R.E. Van Gelder, J. Stoker, R. Truyen, Y. Nio, and F.H. Post. Improved visualization in virtual colonoscopy using image-based rendering. In *VisSym '01 Joint Eurographics - IEEE TCVG Symposium on Visualization, Conference Proceedings*, pages 137–146. Springer, 2001.

15. A. Vilanova Bartrolí, A. König, and E. Gröller. Cylindrical approximation of tubular organs for virtual endoscopy. In *Computer Graphics and Imaging 2000, Conference Proceedings*, pages 283–289. IASTED/ACTA Press, November 2000.
16. A. Vilanova Bartrolí, R. Wegenkittl, A. König, and E. Gröller. Nonlinear virtual colon unfolding. In *IEEE Visualization 2001, Conference Proceedings*, pages 411–418, October 2001.
17. A. Vilanova Bartrolí, R. Wegenkittl, A. König, E. Gröller, and E. Sorantin. Virtual colon flattening. In *VisSym '01 Joint Eurographics - IEEE TCVG Symposium on Visualization, Conference Proceedings*, pages 127–136, May 2001.
18. G. Wang, S.B. Dave, B.P. Brown, Z. Zhang, E.G. McFarland, J.W. Haller, and M.W. Vannier. Colon unraveling based on electrical field: Recent progress and further work. In *SPIE, Conference Proceedings*, volume 3660, pages 125–132, May 1999.
19. G. Wang, E. G. McFarland, B. P. Brown, and M. W. Vannier. GI tract unraveling with curved cross sections. *IEEE Transactions on Medical Imaging*, 17:318–322, 1998.
20. G. Wang and M.W. Vannier. GI tract unraveling by spiral CT. In *SPIE, Conference Proceedings*, volume 2434, pages 307–315, 1995.
21. R. Wegenkittl, A. Vilanova Bartrolí, B. Hegedüs, D. Wagner, M. C. Freund, and E. Gröller. Mastering interactive virtual bronchioscopy on a low-end PC. In *IEEE Visualization 2000, Conference Proceedings*, pages 461–464, October 2000.

Appendix

Color Plates

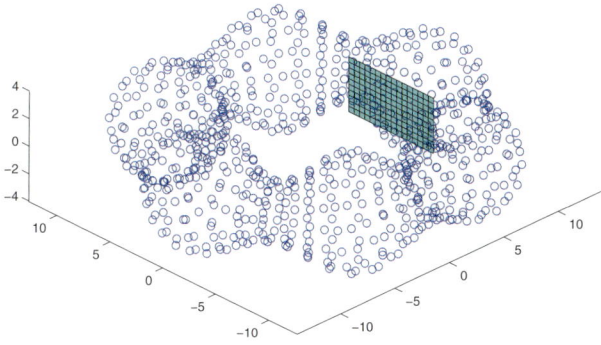

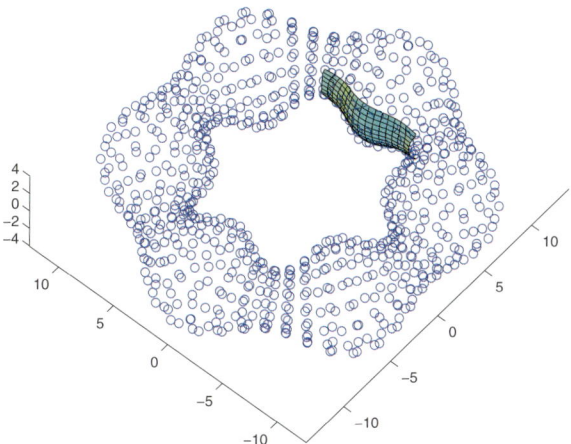

Fig. C.1. Upper part - the data points, and a plane segment L near it. Lower part - the projection $P_2(L)$.
(LEVIN, PP. 36–50)

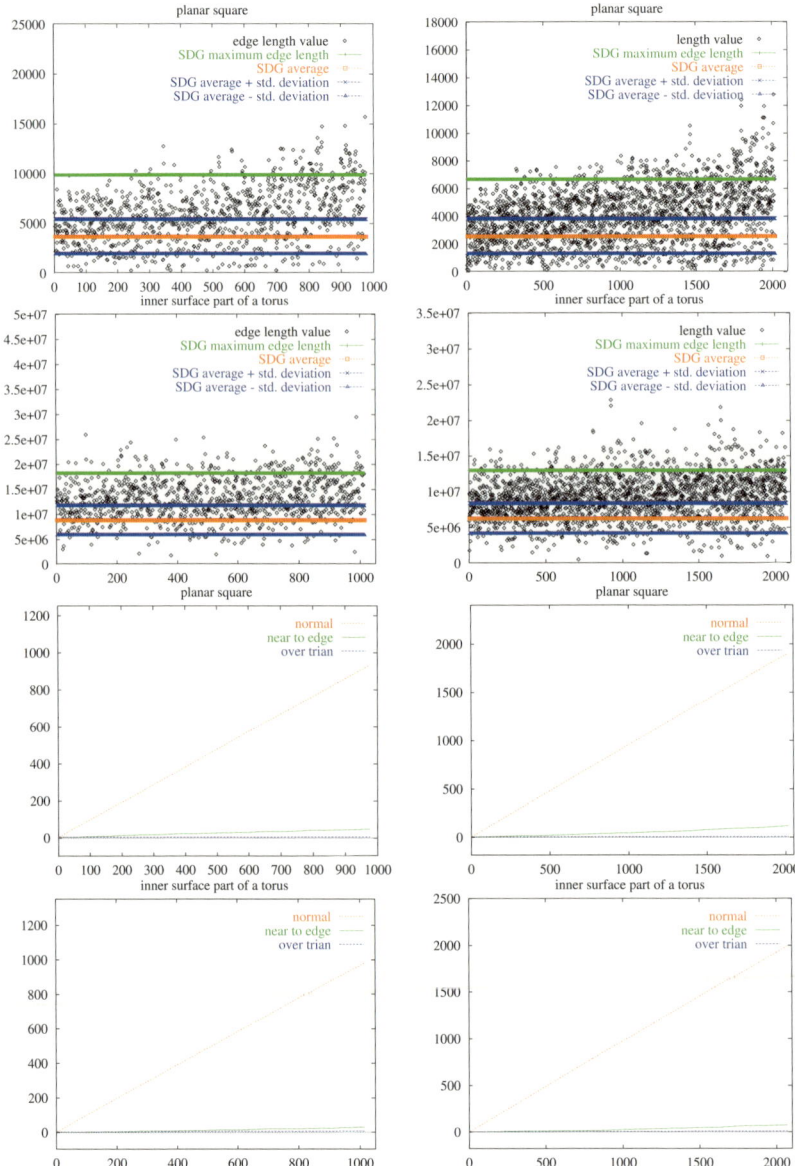

Fig. C.2. First two rows: The lengths of the inserted edges, taken for a sampled square (first row) and the sampled inner region of a torus (second row). For comparison, the average edge length, the standard deviation, and the maximum edge length of the initial SDG are depicted, too. Data sets with about 500 (left column) and about 1000 points (right column) have been used. Last two rows: The number of occurrences of condition 2 ("over trian"), of condition 3 ("near to edge"), and of the case that none of the conditions hold ("normal"). The data are measured for a square (first row) and the inner region of a torus (second row), and about 500 (left) and 1000 (right) data points.
(MENCL AND MÜLLER, pp. 50–66)

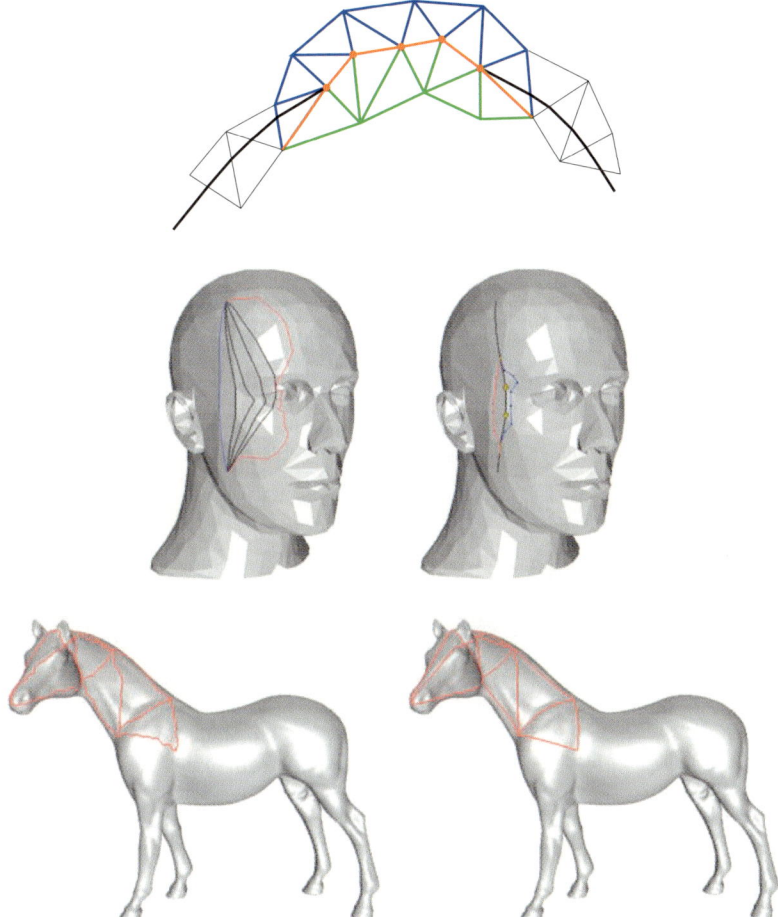

Fig. C.3. Top: Topological step. The red points show the PL vertices that have merged with the mesh vertices. The red edges show the triangle strip edges that are adjacent to these vertices. The green triangles are included in the new triangle strip. They replace all but the first and the last blue triangles. **Middle:** Iterative smoothing process (left). The topological modifications (right). Details are given in section 6. **Buttom:** The top image shows the layout of 19 initial PL connecting 10 user specified triangles. The bottom image shows the final smooth PL using the linear filter. Such a net of smooth PL may be used for surface parameterization. (BONNEAU AND HAHMANN, PP. 69–85)

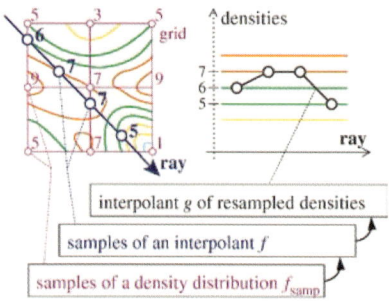 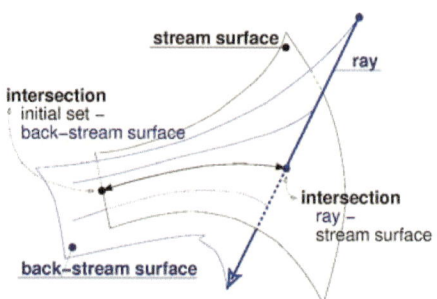

Fig. 2. 2D example of root finding along a ray for iso-surface ray tracing – often two-fold reconstruction is used!

Fig. 3. Lazy evaluation ray casting of stream surfaces by the use of a "back-stream surface".

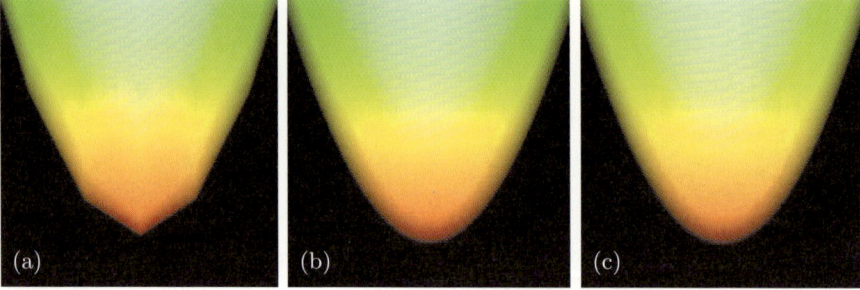

Fig. 10. Gaussian curvature plot using linear (a) and cubic (b) function reconstruction vs. analytic computation (c).

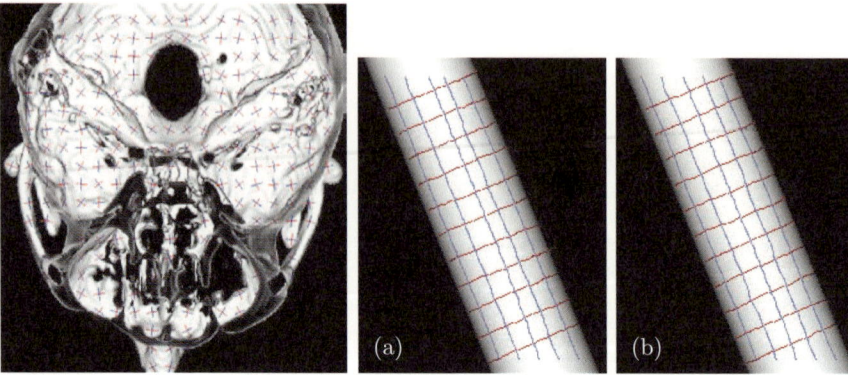

Fig. 7. Iso-surface computed for ten slices, scanned from a human head, with curvature crosses.
(HAUSER ET AL., PP. 106–123)

Fig. 11. Quality of curvature calculations depending on the function reconstruction scheme – linear (a) vs. cubic (b) reconstruction.

Color Plates 475

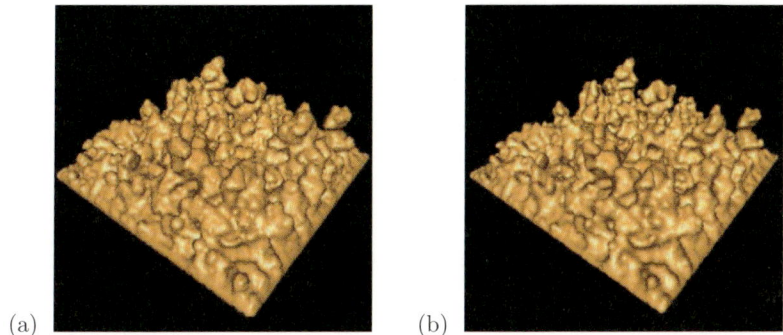

Fig. C.4. DLA simulation 1, $115 \times 115 \times 49$ grid: (a) initial surface with $\epsilon = 3$, (b) 30 iterations

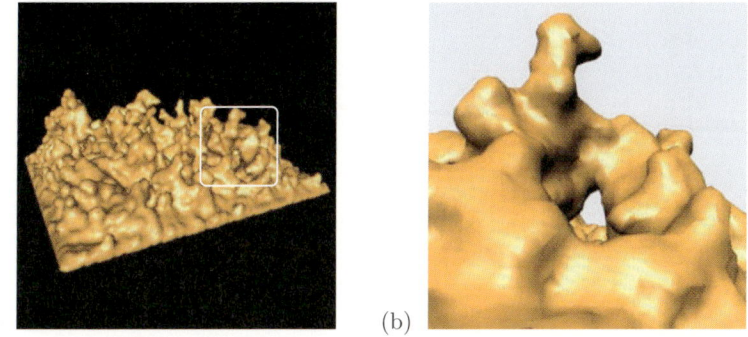

Fig. C.5. Arches from DLA simulations 1: (a) arch structure in the boxed region, (b) zoom-in view of the region

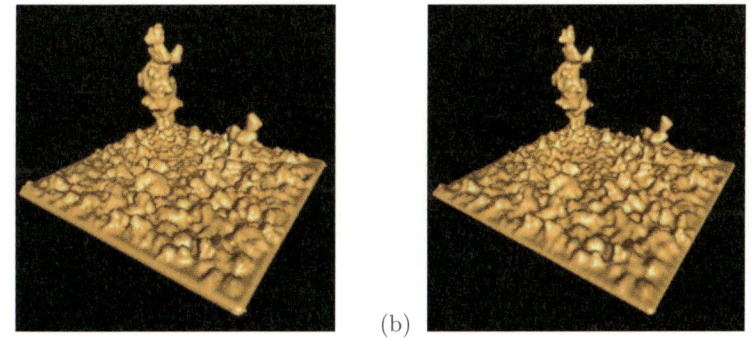

Fig. C.6. DLA simulation 2, $115 \times 115 \times 75$: (a) initial surface with $\epsilon = 3$, (b) 30 iterations
(KIM ET AL., PP. 123–138)

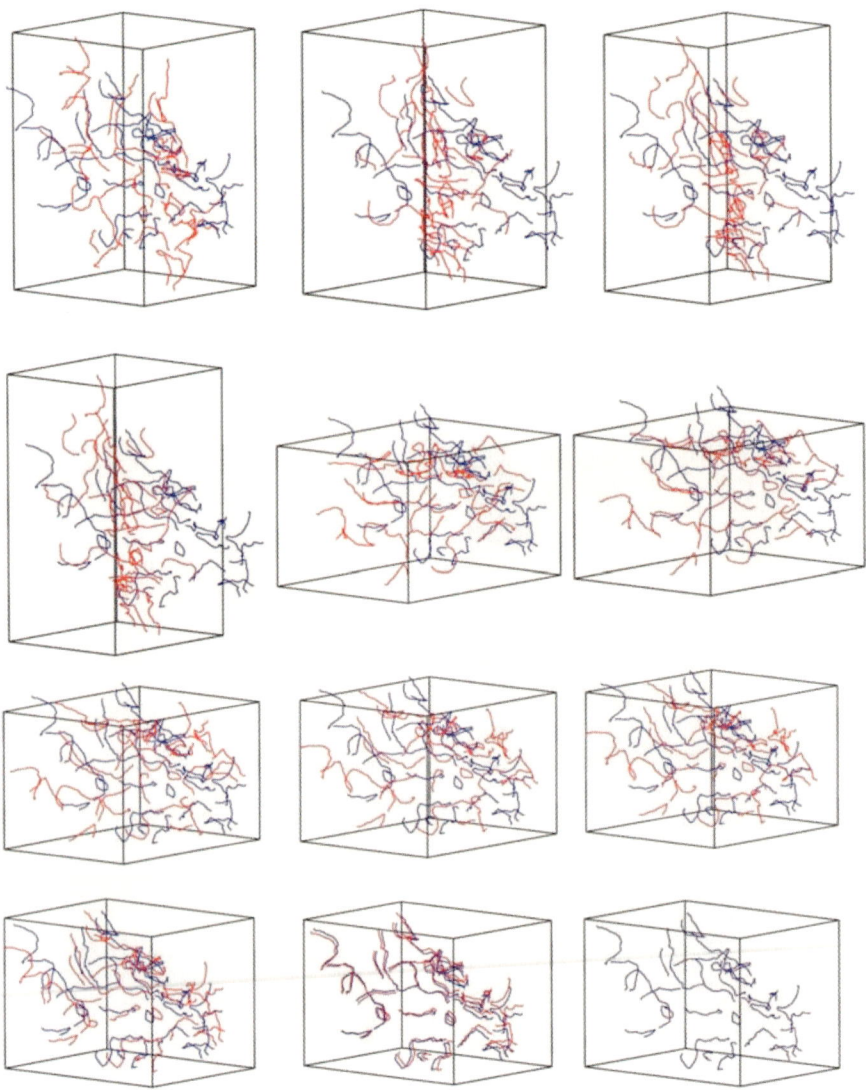

Fig. C.7. Rigid Registration using RICC with target surface's rotation angle (0,0,45). The figures show the feature curves of source (red) and target (blue) with the bounding box of source.
(STYLIANOU, PP. 138–150)

Self-Optimized Texture Maps

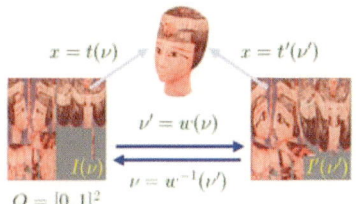

$x = t(\nu)$ $x = t'(\nu')$
$\nu' = w(\nu)$
$\nu = w^{-1}(\nu')$
$I(\nu)$ $I'(\nu')$
$Q = [0,1]^2$

Problem and Approach

- How to optimize the texture image in the rate-distortion sense?
- Image down-sampling may introduce large distortions
- Texture image can be warped and texture coordinates recomputed to compensate, before down-sampling, to assign more pixels to detail areas
- Texture mapping (software/hardware) performs inverse warping

1) Construct the Stretch Function

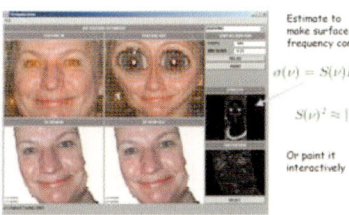

Estimate to make surface frequency constant

$\sigma(\nu) = S(\nu)F_I(\nu)$

$S(\nu)^2 \approx \|\frac{\partial x}{\partial \nu}\|^2$

Or paint it interactively

2) Compute Warping Function

- Relax a regular grid so that nodes move to high areas high stretch stretch value

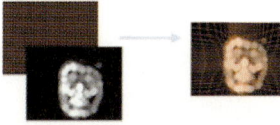

- By Minimizing Stretch Energy

$$\sum_{e=(i,j)} \sigma(\nu_i)\sigma(\nu_j)\|\nu_i - \nu_j\|^2$$

3) Warp the Texture Image

- Apply inverse warping to each target pixel location to determine color from source image
- Down sample

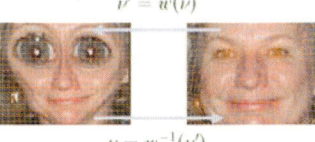

$\nu' = w(\nu)$
$\nu = w^{-1}(\nu')$

4) Warp the Texture Coordinates

- For each vertex need to locate cell in warped grid containing texture coordinates
- Efficient implementation with 2D spatial data structure

$\nu' = w(\nu)$

Fig. C.8. Space-Optimized Texture Maps. (TAUBIN, PP. 207–223)

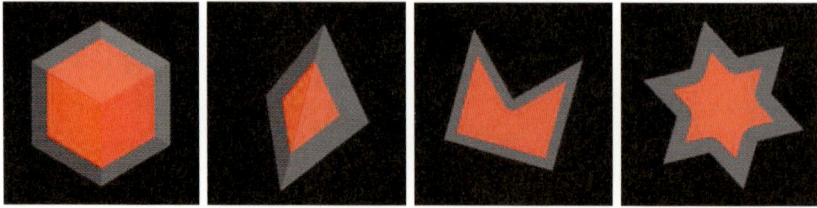

Fig. C.9. A cube and tetrahedron, with the surface mesh shown in semi-transparency. The distance contours (shown in red, perpoint shaded) are of thickness (a) 0.6 inch and (b) 0.1 inch. Two concave examples, a 6-pointed star and a one-ended tooth. (c) For small thickness values (0.2 inch), the distance field retains most corners and edges, with little smoothing. (d) As the thickness increases (0.35 inch), the distance field evolves into the model, showing more smoothing.

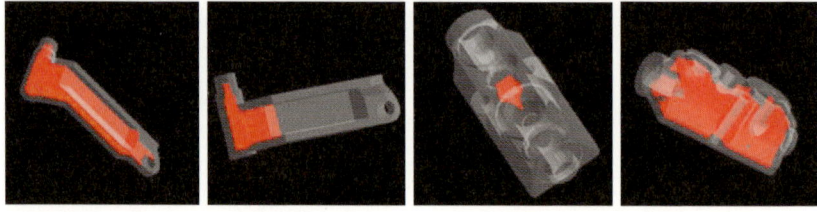

Fig. C.10. (1)Results of "connector' . The surface mesh is shown in semi-transparency, and the per-point shaded distance contours are at thicknesses (a) 0.2 inch and (b) 0.35 inch. (2) Sample images of "brevi". With the per-point shaded contours at thicknesses (a) 10 inches and (b) 4 inches.

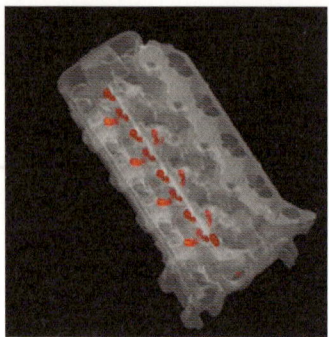

Fig. C.11. The thickness contour of a cylinder model, at 8.5mm thickness and 0.137mm error tolerance.
(HUANG AND CRAWFIS, PP. 225–241)

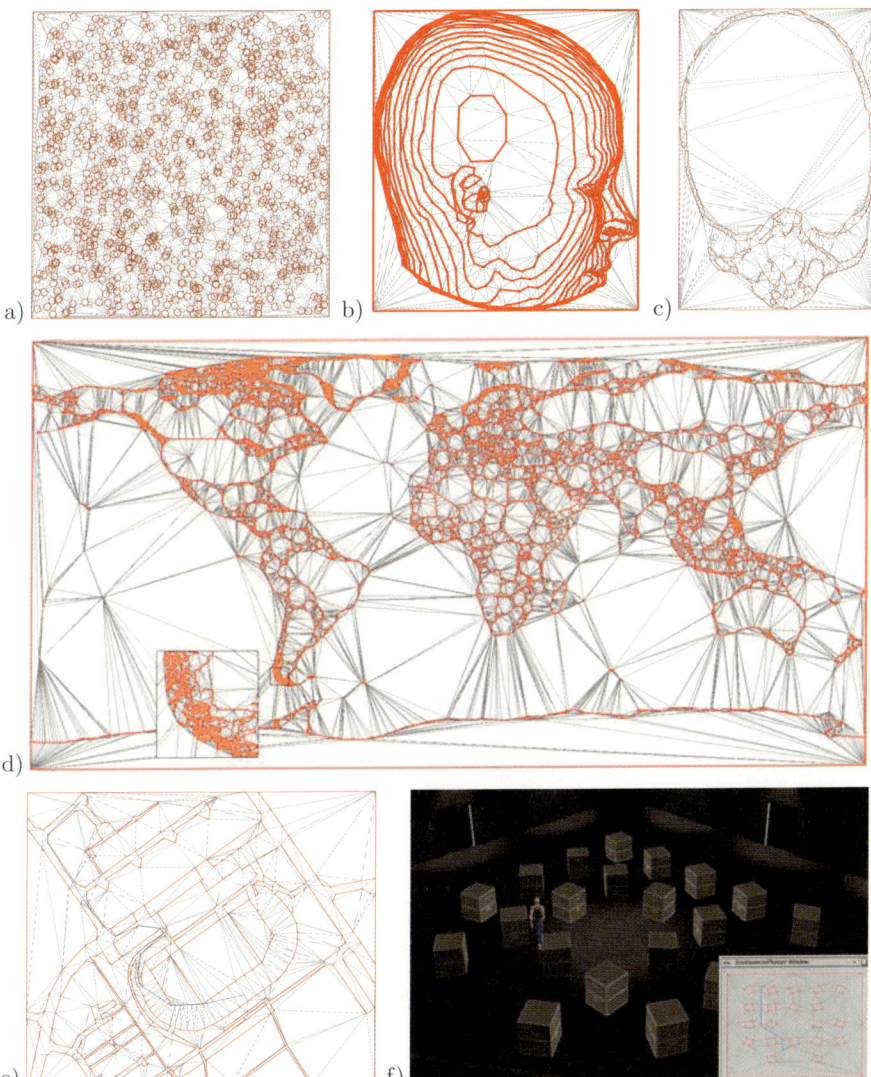

Fig. C.12. a) 960 hexagonal constraints inserted randomly in a square. b) Contours scanned from a 5-month old boy's skull with premature ossification. c) One slice of the boy's skull. d) A world map with 2800 inserted constraints delimiting countries, islands and lakes. e) A path joining two points in the reconstructed town of Grangemouth, Scotland. f) Screenshot of an interactive application where a virtual human is able to walk to any selected location without colliding with boxes inside the room. Boxes are also subject to change position. Data sources: Gill Barequet web page at Tel Aviv University (b,c), DEWA/GRID center for data and information management of the United Nations Environment Programme - UNEP (d), and CROSSES IST European Project (e).
(KALLMANN ET AL., PP. 241–258)

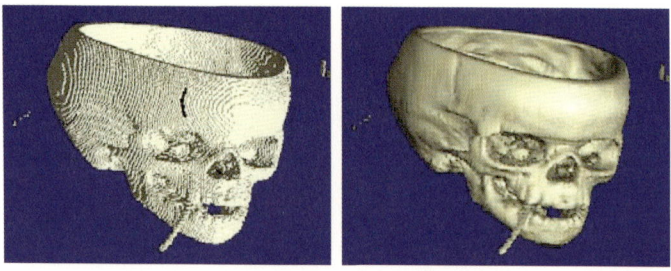

Fig. C.13. Left: rendering from EVM. Right: rendering from Enriched EVM

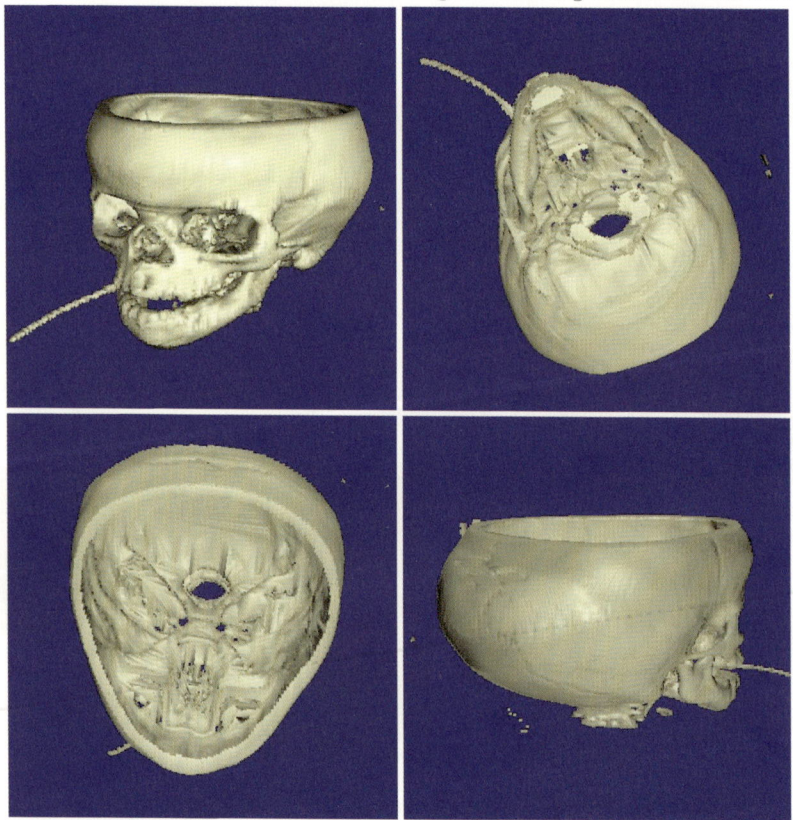

Fig. C.14. Several views of *ctHead*

Fig. C.15. CCL maintaining (right) or breaking (left) non-manifold zones. (RODRIGUEZ ET AL., PP. 258–275)

Color Plates 481

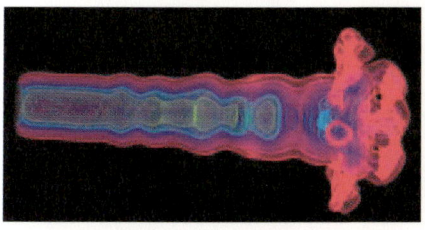 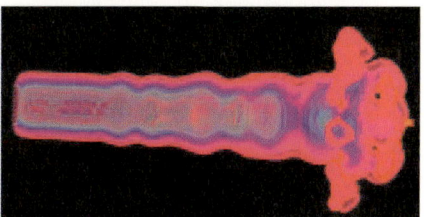

(a) Transfer function emphasizing topologically equivalent regions.

(b) Transfer function emphasizing structures close to critical isovalues.

Fig. C.16. "Nucleon" data set. Data set courtesy of SFB 382 of the German Research Council (DFG), see http://www.volvis.org

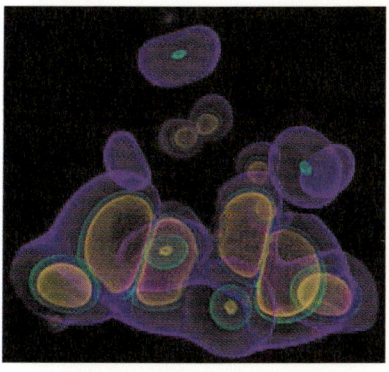 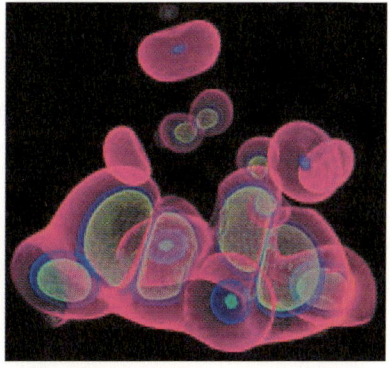

(a) Transfer function emphasizing topologically equivalent regions.

(b) Transfer function emphasizing structures close to critical isovalues.

Fig. C.17. "Neghip" data set. Data set courtesy of VolVis Distribution of SUNY Stony Brook, NY, USA, see http://www.volvis.org
(WEBER AND SCHEUERMANN, PP. 292–306)

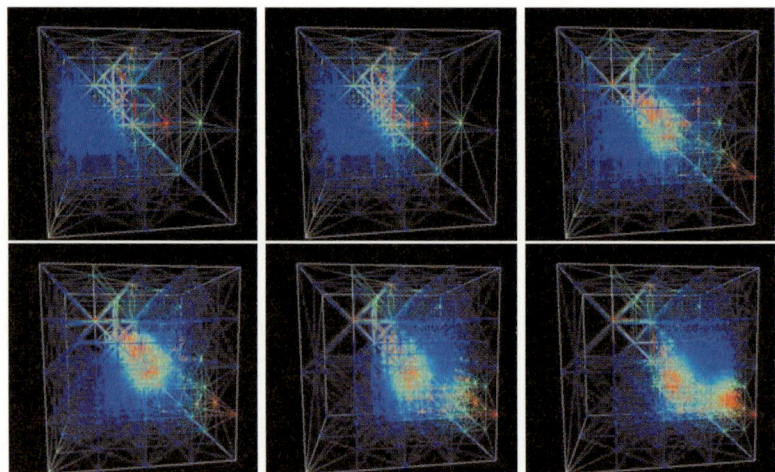

Fig. C.18. Snapshots of the incremental algorithm.

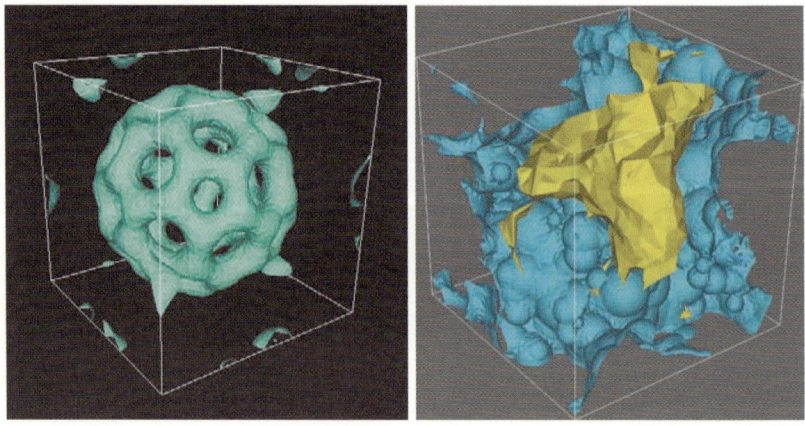

Fig. C.19. Uniform LOD extraction (left): error threshold equal to 5.0% of the field range of the whole domain. The isosurface for a field value equal to 100.0 is shown. Variable LOD extraction based on a field value (right): error threshold equal to 0.1% of the field range enforced near isosurface of value 1.27 (blue). The isosurfaces for field values equal to 1.27 and 1.45 are shown.
(DE FLORIANI AND LEE, PP. 328–345)

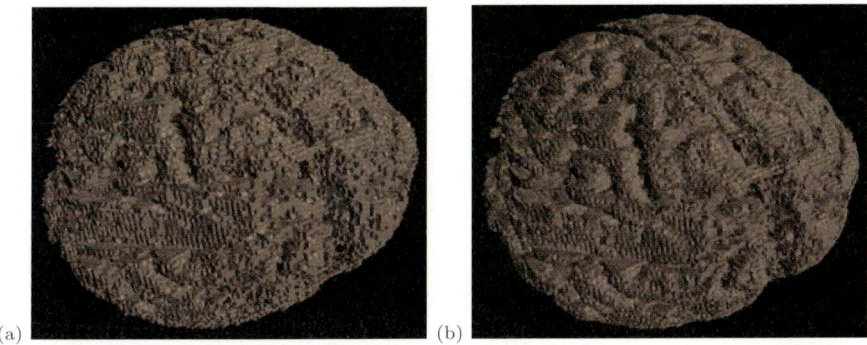

Fig. C.20. Hierarchical visualization of brain data set, (a) based on $\sqrt[3]{2}$-subdivision without and (b) with B-spline wavelets. (Data set courtesy of A. Toga, Ahmanson-Lovelace Brain Mapping Center, University of California, Los Angeles)

Fig. C.21. Entropy in a 3D simulation of Richtmyer-Meshkov instability, visualized by isosurface extraction from a $\sqrt[3]{2}$-subdivision hierarchy without (left column) and with (right column) B-spline wavelets (downsampling ratios 2^{15} and 2^9). (LINSEN ET AL., PP. 359–378)

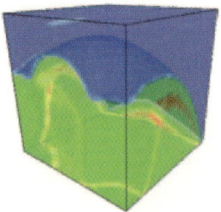

Fig. C.22. Fragment of the child's skull data set which is employed for the comparison of the Delaunay-based vertex insertion and the progressive tetrahedralization methods. (Data set courtesy of Advanced Visual Systems Inc.)

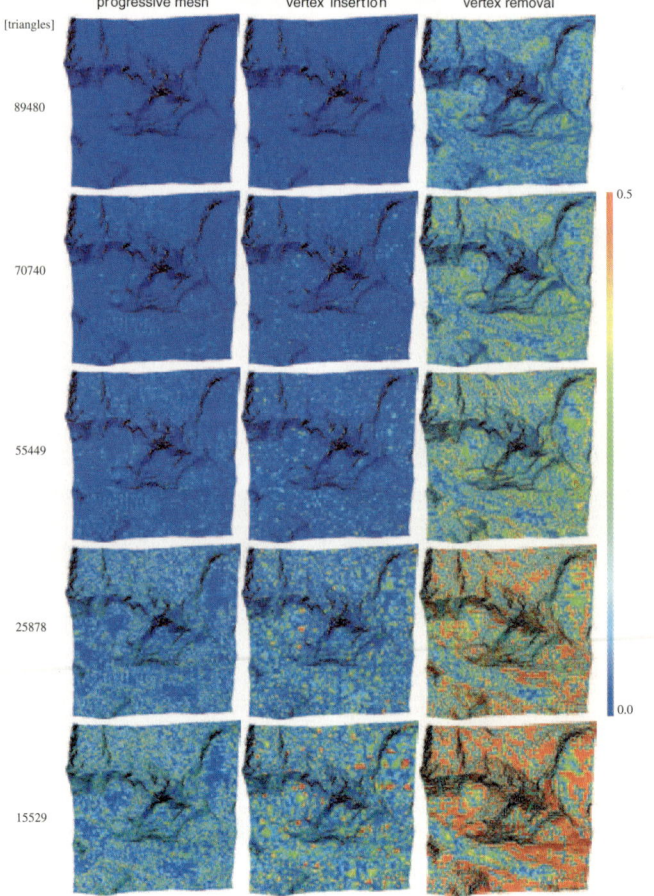

Fig. C.23. Error visualization of the Matterhorn model for the three different mesh approximation methods at various reconstruction levels. The magnitude of the absolute mean-square geometric error corresponds to the color bar on the right.
(STAADT, PP. 378–397)

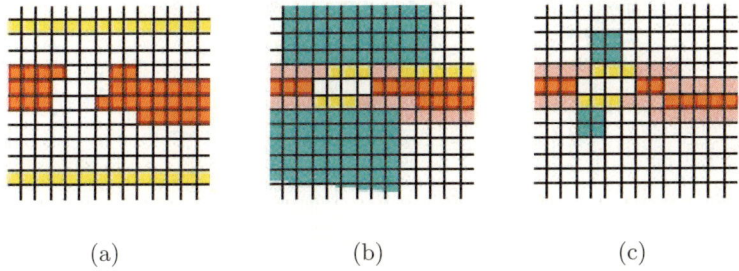

(a) (b) (c)

Fig. C.24. Evolution of the discrete membranes (yellow). Red squares represent perfusion data.(a) is the initial stage. The green squares in (b) represent the blocks that push the membranes towards the data. Pink squares represent portions where the membrane has solidified against the perfusion voxels. The remaining portions of flexible membrane in (c) will be solidified when a through hole is detected by a sufficiently small block depicted in the same green.

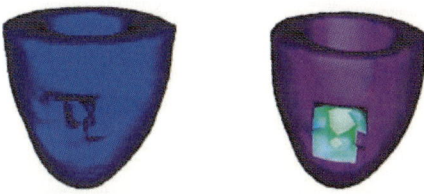

Fig. C.25. Localized missing data, reconstructed with the discrete bands

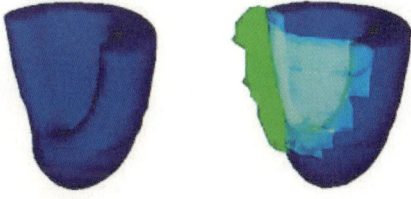

Fig. C.26. Extended missing data, reconstructed by the second method (CIVIT FLORES ET AL., PP. 420–436)

486   Appendix

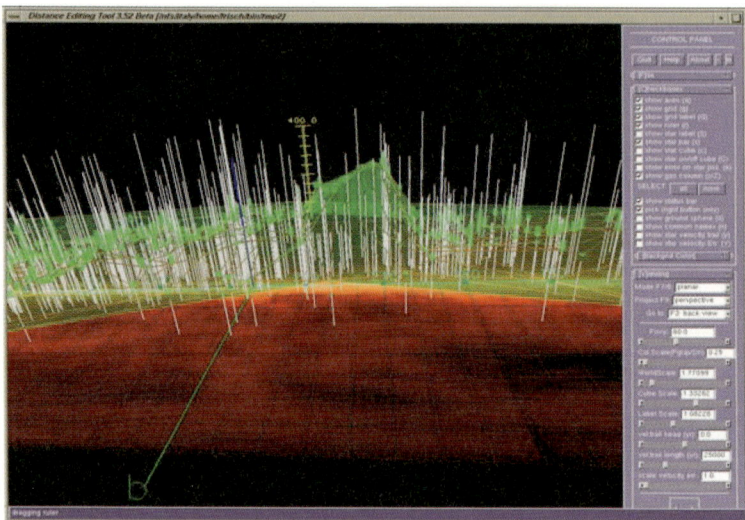

Fig. C.27. Cartesian Tool View. Here the selected stars are displayed against a flat Cartesian grid representing galactic longitude, from 0 to 360 degrees horizontally, and galactic latitude, from -90 to +90 degrees vertically. The horizontal display center can be adjusted interactively, e.g., to place galactic longitude zero at the center. The vertical scale ruler can be adjusted interactively, and a selection rectangle can be defined for detailed editing of partial data groups without reloading or changing the data set.

Fig. C.28. Spherical Tool View. Here the selected stars are displayed radially against a constant-radius reference sphere. The background is Rosat X-ray data [19]. Examples of the control interface options appear in the right-hand control panel.

(HANSON ET AL., PP. 436–453)

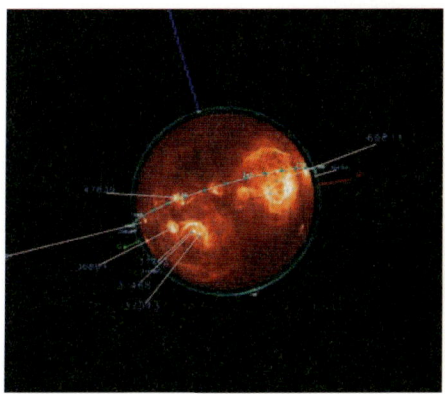

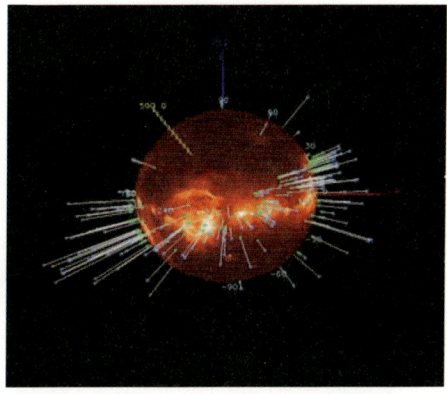

Fig. C.29. Hot massive type O stars ionize the surrounding interstellar clouds producing Hydrogen Balmer H-alpha emission. Blue dots are stars; line length gives relative distances from the Sun. The texture map is from [7].

Fig. C.30. Gould's belt, a tilted band of nearby stars, is defined by stars of types O and B. Galactic longitude 270 degrees is at image center, galactic center to the right. Orion at the lower left and Scorpius at the upper right characterize the geometry of Gould's belt.

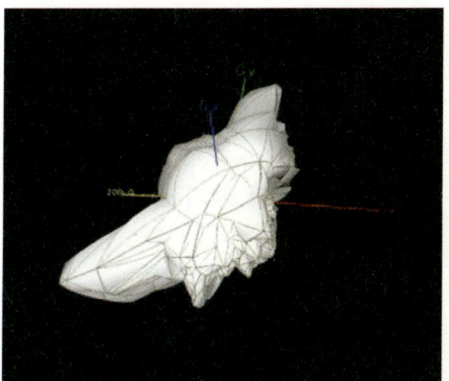

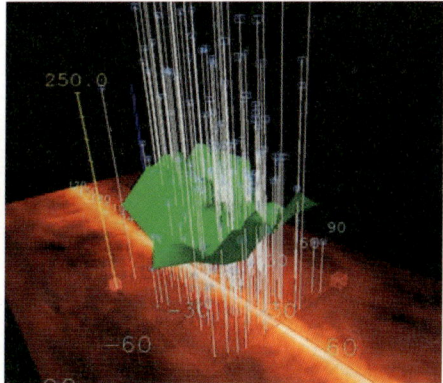

Fig. C.31. View of Local Bubble from a viewpoint above the galactic plane, with galactic center to the right. The radial surface of this void around the Sun varies from 80pc to 300pc, possibly reflecting the geometry of the Milky Way's local spiral arm. The geometry is reconstructed using hand-edited selections from the E(B-V) reddening data for 700 stars in the Hipparcos catalog.
(HANSON ET AL., PP. 436 453)

Fig. C.32. Reconstruction of the dust cloud surface around the direction of the galactic center, including the constellation of Scorpius. The background texture is the Dickey and Lockman 21cm hydrogen hyperfine compilation [5] for the full sky.

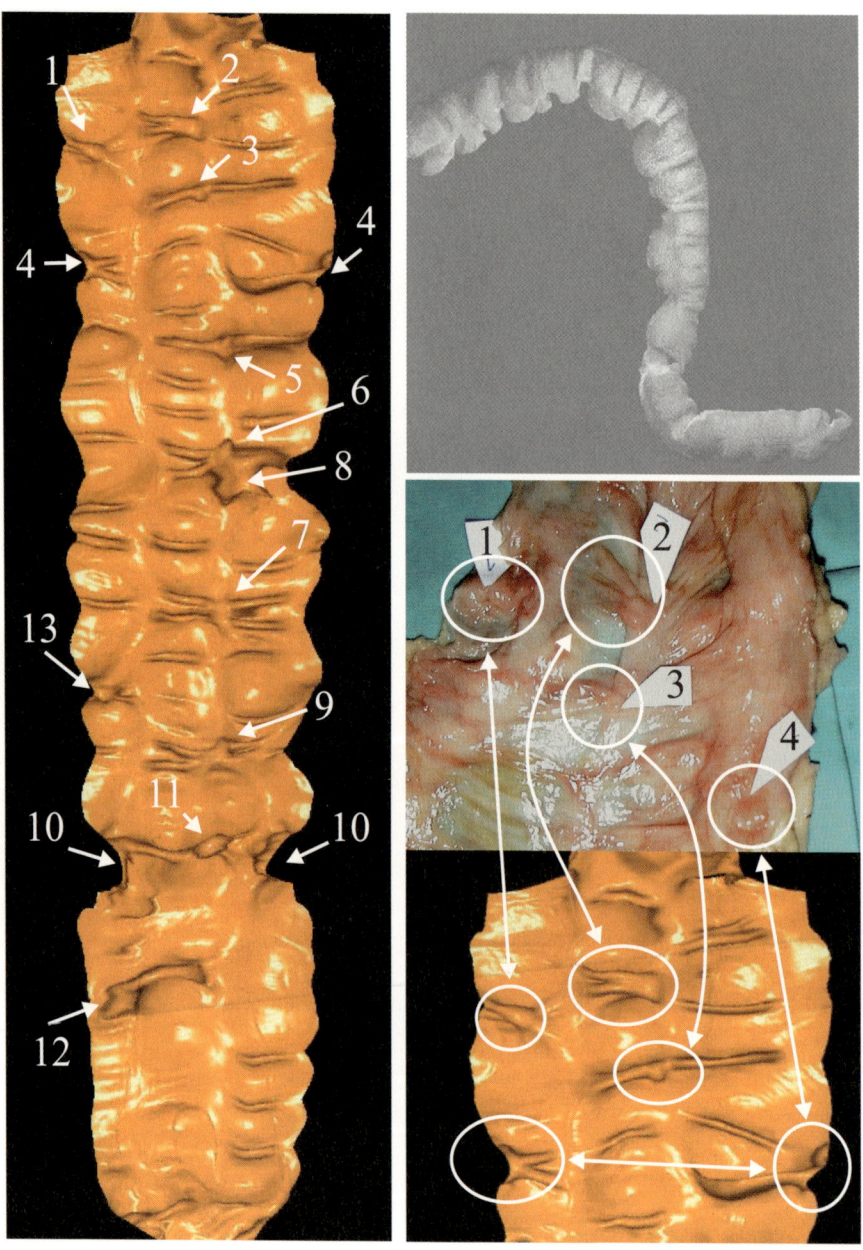

Fig. C.33. **Right top** Outside view of the segmented surface of an extracted colon CT data set with resolution 381x120x632. **Left** Virtually unfolded colon with the polyps numbered according to the real dissection. **Right middle and bottom** Qualitative comparison of the virtually unfolded colon with pictures taken from the real dissection. The orientation in which the pictures were taken does not correspond to the orientation of the virtually unfolded colon.
(VILANOVA AND GROLLER, PP. 453 469)